Advances in Molecular and Cellular Microbiology 19

Stress Response in Pathogenic Bacteria

Edited by

Stephen P. Kidd

University of Adelaide, Australia

www.cabi.org

 **Advances in Molecular and Cellular Microbiology**

Through the application of molecular and cellular microbiology we now recognize the diversity and dominance of microbial life forms on our planet, which exist in all environments. These microbes have many important planetary roles, but for we humans a major problem is their ability to colonize our tissues and cause disease. The same techniques of molecular and cellular microbiology have been applied to the problems of human and animal infection during the past two decades and have proved to be immensely powerful tools in elucidating how microorganisms cause human pathology. This series has the aim of providing information on the advances that have been made in the application of molecular and cellular microbiology to specific organisms and the diseases that they cause. The series is edited by researchers active in the application of molecular and cellular microbiology to human disease states. Each volume focuses on a particular aspect of infectious disease and will enable graduate students and researchers to keep up with the rapidly diversifying literature in current microbiological research.

Series Editors

Professor Brian Henderson
University College London

Professor Michael Wilson
University College London

Titles Available from CABI

17. *Helicobacter pylori* in the 21st Century
Edited by Philip Sutton and Hazel M. Mitchell

18. Antimicrobial Peptides: Discovery, Design and Novel Therapeutic Strategies
Edited by Guangshun Wang

19. Stress Response in Pathogenic Bacteria
Edited by Stephen P. Kidd

Titles Forthcoming from CABI

Lyme Disease: an Evidence-based Approach
Edited by John J. Halperin

Tuberculosis: Molecular Techniques in Diagnosis and Treatment
Edited by Timothy McHugh

Microbial Metabolomics
Edited by Silas Villas-Bôas and Katya Ruggiero

Earlier titles in the series are available from Cambridge University Press (www.cup.cam.ac.uk).

CABI is a trading name of CAB International

CABI Head Office
Nosworthy Way
Wallingford
Oxfordshire, OX10 8DE
UK

CABI North American Office
875 Massachusetts Avenue
7th Floor
Cambridge, MA 02139
USA

Tel: +44 (0)1491 832111
Fax: +44 (0)1491 833508
E-mail: cabi@cabi.org
Website: www.cabi.org

Tel: +1 617 395 4056
Fax: +1 617 354 6875
E-mail: cabi-nao@cabi.org

A catalogue record for this book is available from the British Library, London, UK.

Library of Congress Cataloging-in-Publication Data

Stress response in pathogenic bacteria / edited by Stephen P. Kidd.
    p. ; cm. -- (Advances in molecular and cellular microbiology ; 19)
  Includes bibliographical references and index.
  ISBN 978-1-84593-760-7 (alk. paper)
 1. Pathogenic bacteria. 2. Stress (Physiology) I. Kidd, Stephen P. II. Title.
III. Series: Advances in molecular and cellular microbiology ; 19.
  [DNLM: 1. Bacteria--pathogenicity. 2. Bacterial Physiological Phenomena.
3. Stress, Physiological--physiology. QW 52]

  QR46.S79 2011
  616.07'1--dc22

                                                                    2011003330

ISBN-13: 978 1 84593 760 7

Commissioning editor: Rachel Cutts
Production editors: Shankari Wilford and Simon Hill

Typeset by Columns Design XML Ltd, Reading, UK.
Printed and bound in the UK by CPI Antony Rowe, Chippenham.

Contents

Contributors

Brian J. Akerley, Department of Molecular Genetics and Microbiology, University of Massachusetts Medical School, 55 Lake Ave., N., S6-242, Worcester, MA, 01655, USA. E-mail: brian.akerley@umassmed.edu

Jennifer S. Cavet, Life Sciences, Michael Smith Building, University of Manchester, Oxford Road, Manchester M13 9PT, UK. E-mail: Jennifer.s.cavet@manchester.ac.uk

Peter T. Chivers, Department of Biochemistry and Molecular Biophysics, Washington University School of Medicine, Box 8231, 660 S. Euclid Ave., St Louis, MO 63110, USA. E-mail: chivers@wustl.edu

Selina R. Clayton, School of Biosciences, The University of Nottingham, Sutton Bonington Campus, Sutton Bonington, Loughborough LE12 5RD, UK. E-mail: selina.clayton@nottingham.ac.uk

Jeffery A. Cole, School of Biosciences, University of Birmingham, Edgbaston, Birmingham B15 2TT, UK. E-mail: j.a.cole@bham.ac.uk

Elisa Deriu, Department of Microbiology and Molecular Genetics and Institute for Immunology, University of California Irvine, Irvine, CA 92697-4025, USA. E-mail: ederiu@uci.edu

Matthew S. Francis, Department of Molecular Biology, Umeå University, SE-901 87 Umeå, Sweden. E-mail: matthew.francis@molbiol.umu.se

Hanan Gancz, Department of Microbiology and Immunology, Uniformed Services University of the Health Sciences, 4301 Jones Bridge Rd., Bethesda, Maryland, USA. E-mail: hgancz@usuhs.mil

Daniel Hassett, Department of Molecular Genetics, Biochemistry and Microbiology, University of Cincinnati, Cincinnati, OH 45267-0524, USA. E-mail: hassetdj@ucmail.uc.edu

Karin Heurlier, School of Biosciences, The University of Nottingham, Sutton Bonington Campus, Sutton Bonington, Loughborough LE12 5RD, UK. E-mail: Karin Heurlier@nottingham.ac.uk

Stuart A. Hill, Department of Biological Sciences, Northern Illinois University, DeKalb, IL 60115, USA. E-mail: sahill@niu.edu

Jon L. Hobman, School of Biosciences, The University of Nottingham, Sutton Bonington Campus, Sutton Bonington, Loughborough LE12 5RD, UK. E-mail: Jon.Hobman@nottingham.ac.uk

James A. Imlay, Department of Microbiology, University of Illinois at Urbana-Champaign, Urbana, IL 61801, USA. E-mail: jimlay@life.illinois.edu

Stephen P. Kidd, Research Centre for Infectious Diseases, School of Molecular and Biomedical Sciences, The University of Adelaide, Adelaide, South Australia 5005, Australia. E-mail: stephen.kidd@adelaide.edu.au

Janet Z. Liu, Department of Microbiology and Molecular Genetics and Institute for Immunology, University of California Irvine, Irvine, CA 92697-4025, USA. E-mail: jzliu@uci.edu

D. Scott Merrell, Uniformed Services University of Health Sciences F. Edward Hébert School of Medicine, Department of Microbiology and Immunology, 4301 Jones Bridge Rd., Bethesda, MD 20814, USA. E-mail: dmerrell@usuhs.mil

Taku Oshima, Department of Bioinformatics and Genomics, Graduate School of Information Science, Nara Advanced Institute of Science and Technology (NAIST), 8916-5 Takayama-cho, Ikoma, Nara 630-0192, Japan.

Adam Potter, Research Centre for Infectious Disease, The University of Adelaide, School of Molecular and Biomedical Science, Adelaide, South Australia 5005, Australia. E-mail: adam.potter@adelaide.edu.au

Manuela Raffatellu, Department of Microbiology and Molecular Genetics, Institute for Immunology, University of California Irvine, C135E Med Sci I, Irvine, CA 92697-4025, USA. E-mail: manuelar@uci.edu

Stephen Spiro, Department of Molecular and Cell Biology, The University of Texas at Dallas, 800 W Campbell Road, Richardson, Texas 75080, USA. E-mail: stephen.spiro@utdallas.edu

Glen Ulett, School of Medical Science, Centre for Medicine and Oral Health Campus, Gold Coast Campus, Griffith University, QLD 4222, Australia. E-mail: g.ulett@griffith.edu.au

Sandy M.S. Wong, Department of Molecular Genetics and Microbiology, University of Massachusetts Medical School, Worcester, MA 01655, USA. E-mail: Sandy.Wong@umassmed.edu

Preface

_____

Bacteria have an extraordinary ability to adapt to vastly diverse environmental conditions. These environments include ecological niches which experience extremes in physical and chemical conditions and these damage cell components and therefore create a stress the cell needs to overcome. Bacteria have evolved specific systems which allow them to respond to various stresses and, in so doing, nullify or detoxify the toxic agent, as well as systems which are designed for their metabolic adaptation to these harmful conditions, thereby allowing them to grow within these environments. In the context of pathogenic bacteria, these responses have implications for bacterial survival within their different environmental reservoirs, as well as their passage through food and preservation processes and then, importantly, within their human host. For bacteria which can survive in humans, either as harmless or even useful commensal bacteria or as disease-causing agents, central to this survival is their ability to use the conditions of the human body as a growth environment. The physical and chemical composition varies between tissues, organs and particular anatomical niches throughout the body. For some bacteria their site of entry to the body is vastly different to their end-site, and indeed different again from the environmental reservoir in which the bacteria have existed previously. At one level, this translates to a required variation in metabolic pathways and energy generation, due to a variation in carbon and energy source (this could be the presence of sugars, carbohydrates or amino acids, or the amount of oxygen present), as well as the availability of other macro- and micronutrients that the bacteria require to survive. With some of these nutrients, the host sequesters them tightly for its own use, creating an environment of nutrient starvation for the bacteria. In addition, there are conditions created by the host specifically to clear invading pathogens such as temperature and pH and then the presence of host-generated and damaging chemicals such as reactive oxygen and reactive nitrogen species. The interactions and combinations of these stresses generate a complex and harmful environment, which in many cases is specific to a particular part of the body, and thereby a bacterium which inhabits an anatomical niche has evolved specialized survival mechanisms tailored for that environment. While there are recent books which deal with many of these aspects, there is not one which covers the field of bacterial response and adaptation to the environmental stresses within the human host.

The ability of pathogenic bacteria to adapt to the various chemical, biochemical and physical conditions of the different anatomical niches within the human host, as well as an ability to respond to the direct stresses which are generated in these environments, is a central feature of infectious diseases and the outcome of bacterial infection. Our understanding of the mechanisms used by numerous bacteria for this survival, and indeed the nature of the stresses within the human host, has increased significantly over the past 5–10 years. These key aspects of this rapidly developing field will be the central theme of this book. Specifically, it will cover: the generation of stresses by the host immune system, the bacterial response to reactive chemicals as a result of the host response to bacterial infection (including reactive oxygen and reactive nitrogen species), bacterial metabolism as part of the stress response and the bacterial adaptation to environmental conditions of anatomical niches such as the gut, mouth and urogenital tract (these conditions include acid conditions, nutrient stress and low oxygen levels), and increasingly, the importance of different metal ions in the pathogenesis and survival of specific bacteria. This volume uses experts in the research of each of these areas to develop a book that provides a comprehensive outline of the current understanding of this field, the latest developments and where future research is likely to be directed.

Part 1
Oxidative and Nitrosative Stress

1 Oxidative and Nitrosative Stress Defence Systems in *Escherichia coli* and *Pseudomonas aeruginosa*: A Model Organism of Study Versus a Human Opportunistic Pathogen

James A. Imlay and Daniel J. Hassett[*]

1.1 Oxidative Stress in *Escherichia coli*

Escherichia coli has served as a model system for studies of oxidative stress, originally because of its genetic tractability. A more specific advantage is that if this facultative anaerobe is maintained in an anaerobic environment, workers can engineer mutations that sufficiently disrupt oxidative defences that are lethal during aerobic growth.

Most organisms cannot grow if they are exposed to oxygen levels that substantially exceed those of their native habitats. The basis of this phenomenon is not immediately obvious, since molecular oxygen does not damage amino acids, carbohydrates, lipids, or nucleic acids directly – in short, the basic molecules from which organisms are made. So, why is oxygen toxic? In 1954 Gershman *et al.* proposed that the problem was not oxygen per se, but rather the partially reduced forms into which it might be converted within cells (Gershman *et al.*, 1954). Indeed, organisms universally contain enzymes that are devoted to the scavenging of superoxide (O_2^-) and hydrogen peroxide

(H_2O_2), and these species are notably more reactive than is molecular oxygen. The truest test of this notion was the creation of mutants that lacked these scavenging enzymes. In 1986 Carlioz and Touati (1986) reported that mutants of *E. coli* that lacked superoxide dismutase (SOD) were unable to grow without specific amino acid supplements. Strikingly, the key supplements – branched-chain, aromatic and sulfurous amino acids – matched the supplements that were required by wild-type *E. coli* when they were exposed to hyperoxia. The implication was obvious: when oxygen levels were very high, the rate of O_2^- formation was also elevated so that even wild-type cells could not reduce O_2^- to a non-toxic concentration. More recently, mutants have been constructed that cannot scavenge H_2O_2 (Seaver and Imlay, 2001a). These strains, too, have characteristic growth defects.

1.1.1 Sources of O_2^- and H_2O_2

The properties of the various reduction products of oxygen and reactive oxygen

* Corresponding author.

intermediates are shown in Eqn 1.1 and Table 1.1. The univalent reduction of molecular oxygen forms O_2^-, and its divalent reduction forms H_2O_2 (Fig. 1.1):

$$O_2 \rightarrow O_2^- \rightarrow H_2O_2 \rightarrow HO^{\cdot} + H_2O \rightarrow 2H_2O \quad (1.1)$$

Inside cells, these species are likely to be formed when molecular oxygen collides with the reduced flavins of redox enzymes (Fig. 1.2).

NADH dehydrogenases, succinate dehydrogenase, pyruvate dehydrogenase and sulfite reductase are all among the many enzymes that display this behaviour *in vitro*, and presumably do so *in vivo* as well (Messner and Imlay, 1999, 2002). These reactions are not by design: these enzymes employ flavins to hold electrons momentarily in the process of transfer from one metabolite to another. However, if dissolved oxygen happens to collide with the solvent-exposed dihydro-flavin, then it can abstract either a single electron (forming O_2^-) or two consecutive electrons (forming H_2O_2). The endogenous fluxes of these species are therefore the sum of the contributions of all the inadvertent oxidations of these enzymes inside the cell.

During parts of its life cycle, *E. coli* inhabits air-saturated waters, such as ponds. Consequently, it has evolved to express sufficient scavenging enzymes were levels of even endogenous O_2^- and H_2O_2 are not debilitating. However, in special circumstances, these oxidants can rise to threatening levels. H_2O_2 accumulates in natural habitats through the gradual oxidation of sulfur and reduced metal compounds. It is

Table 1.1. Oxygen reduction products and other reactive oxygen intermediates.

Oxidant type	Properties
$O_2^{\cdot-}$, superoxide anion	One-electron (univalent) reduction product of O_2. Formed by enzymes of the electron transport chain such as fumarate reductase and succinate dehydrogenase (Messner and Imlay, 2002). Rather unreactive but can release Fe^{2+} from iron-sulfur proteins and ferritin. Undergoes dismutation to form H_2O_2 spontaneously at a rate of 10^5 M^{-1} s^{-1} or by enzymatic catalysis via SOD (10^9 M^{-1} s^{-1}) and is a precursor for metal-catalysed •OH formation. Poisons aconitase and fumarase of the TCA cycle (Gardner and Fridovich, 1991b; Liochev and Fridovich, 1992).
H_2O_2, hydrogen peroxide	Two-electron reduction product of molecular oxygen. Can be formed by spontaneous dismutation of $\bullet O_2^{\cdot-}$ at ~10^5 M^{-1} s^{-1} or by direct reduction of O_2. Also formed by fumarate reductase and aspartate oxidase (Messner and Imlay, 2002). Freely soluble in lipid bilayers and thus able to diffuse easily across membranes. Disrupts proton motive force.
HO·, hydroxyl radical	Vastly reactive radical species. Remarkably destructive, acts at near diffusion-limited rates and capable of damaging virtually every known biomolecule including protein, DNA/RNA, carbohydrate and lipid. Three-electron reduction product of O_2 and produced primarily via the Fenton reaction using either iron or copper as catalyst.
HOCl, hypochlorous acid	Formed by the oxidation of Cl^- *in vivo* by the action of MPO, H_2O_2 and NaCl. Lipid soluble and highly reactive. Will oxidize protein constituents readily, including thiol groups, amino groups and methionine.
ROOH, organic hydroperoxide	Formed by radical reactions with cellular components such as lipids and nucleobases.
$^1O_2$, singlet oxygen	The lowest excited state of the dioxygen.
ONOO-, peroxynitrite	Formed in a rapid reaction between $\bullet O_2^{\cdot-}$ and NO•. Lipid soluble and similar in reactivity to hypochlorous acid. Protonation forms peroxynitrous acid, which can undergo homolytic cleavage to form hydroxyl radical and nitrogen dioxide.

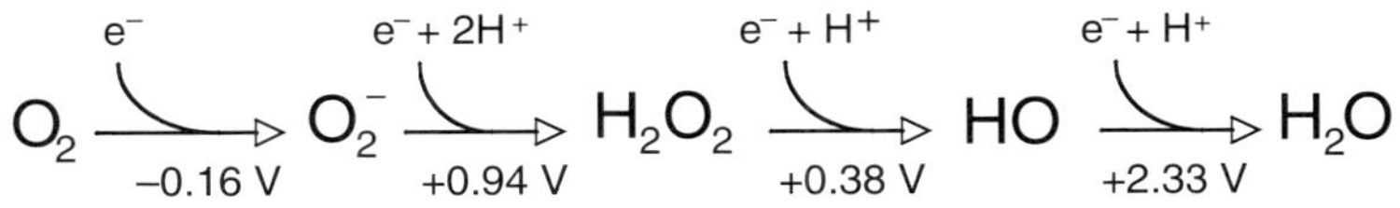

Fig. 1.1. The redox states of oxygen. Consecutive univalent reductions of oxygen generate superoxide, hydrogen peroxide, the hydroxyl radical and water.

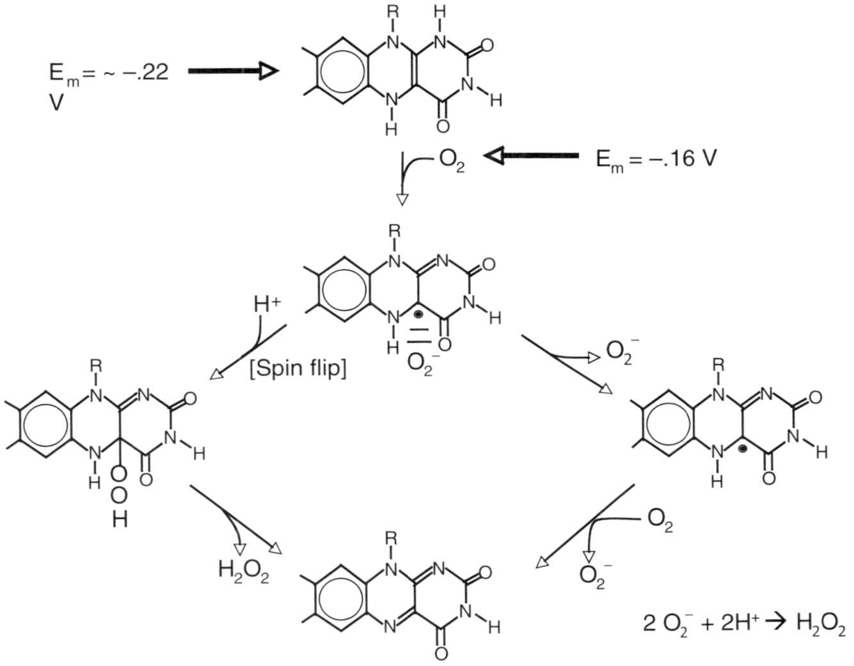

Fig. 1.2. Adventitious formation of O_2^- and H_2O_2 through flavin autoxidation. An initial electron transfer generates a caged radical pair of superoxide and flavosemiquinone, which can resolve via recombination and heterolytic cleavage (direct H_2O_2 formation, left pathway), or via superoxide escape and another univalent reaction (right pathway, which ultimately forms H_2O_2 within the cell by superoxide dismutation).

also generated by plant, amoebic and mammalian host responses to microbial pathogens (Fig. 1.3).

Because H_2O_2 is a small, uncharged molecule, it diffuses across membranes at a substantial rate (Seaver and Imlay, 2001b; Winterbourn *et al.*, 2006). *E. coli* can be toxified by as little as 5 μM extracellular H_2O_2. To defend itself, it engages the specialized OxyR stress response (see below).

O_2^- (pKa 4.8) is anionic at neutral pH and cannot traverse membranes effectively

(Korshunov and Imlay, 2002). Therefore, although it is also produced by hosts, its only direct action against pathogens must be limited to the cell surface or periplasm. Nevertheless, *E. coli* elaborates an antioxidant response that features the induction of cytoplasmic superoxide dismutase (SOD) (Gregory and Fridovich, 1973; Hassan and Fridovich, 1977). The reason for this is that both plants and other microbes attempt to poison their microbial competitors by excreting redox-cycling drugs. These

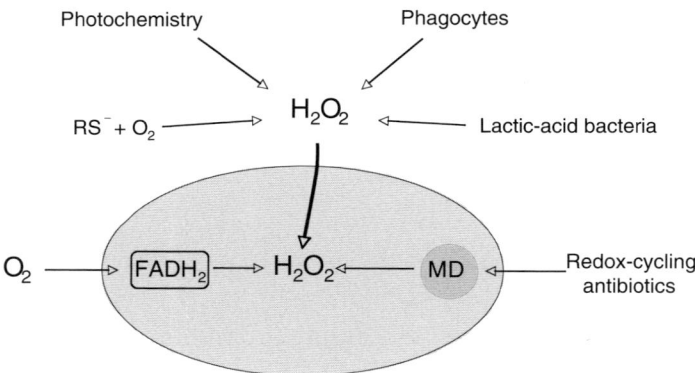

Fig. 1.3. Environmental sources of H_2O_2. Bacteria can be exposed to extracellular H_2O_2 by photochemical reduction of O_2, respiratory burst of phagocytes, lactic acid bacteria and reduction of O_2 by RS^-. Intracellularly, bacteria face H_2O_2 stress by reaction of O_2 with reduced flavins or by redox-cycling antibiotics such as menadione (MD).

compounds – including quinones and phenazines – diffuse across membranes and catalyse the oxidation of cytoplasmic redox enzymes, thereby generating toxic amounts of O_2^- (and H_2O_2). Activation of the SoxRS response is a key defensive measure.

1.1.2 Targets of O_2^- and H_2O_2

O_2^-

The *sodA sodB* mutants of *E. coli* lack cytoplasmic SOD activity and among their phenotypes are inabilities to synthesize branched-chain amino acids and to catabolize tricarboxylic acid (TCA)-cycle substrates (Carlioz and Touati, 1986). These traits led investigators to the discovery that O_2^- directly inactivates a family of dehydratases that includes two in the branched-chain biosynthetic pathway (dihydroxyacid dehydratase and isopropylmalate dehydratase) and three in the TCA cycle (aconitase B and fumarases A and B) (Kuo *et al.*, 1987; Gardner and Fridovich, 1991a,b; Liochev and Fridovich, 1992). These enzymes employ solvent-exposed [4Fe-4S] clusters that bind and abstract hydroxide anions from their substrates; however, when substrate is not bound, O_2^- is attracted to the cluster electrostatically. Once bound, it abstracts an

electron, which destabilizes the cluster. The cluster releases an iron atom and the resultant [3Fe-4S] cluster is catalytically inactive (Fig. 1.4) (Flint *et al.*, 1993b).

This accidental reaction has a very high rate constant, exceeding 10^6 M^{-1} s^{-1}. A consequence is that the cell must manufacture enough SOD to keep O_2^- levels as low as 10^{-10} M, or else the half-time of dehydratase inactivation will be less than an hour (Gort and Imlay, 1998). This is a notable result in that 10^{-10} M represents a time average of less than one molecule of O_2^- per *E. coli* cell. To achieve this the SOD titres must exceed 10 µM, which comprises a substantial dedication of cell resources.

Evolution has calibrated the basal rate of SOD synthesis to near the bare minimum: if the rate of O_2^- formation is increased even twofold, the consequent rise in O_2^- levels causes the activities of dehydratases to drop measurably. The upshot is that the basal SOD titres are inadequate in the face of any additional O_2^- stress – such as arises from redox-cycling antibiotics.

O_2^- also accelerates the rate of DNA damage (Farr *et al.*, 1986). This effect apparently arises from the iron that is released from dehydratase iron-sulfur clusters, since iron catalyses the formation of hydroxyl radicals ($HO^·$), the direct oxidants of DNA (Liochev and Fridovich, 1994; Keyer and

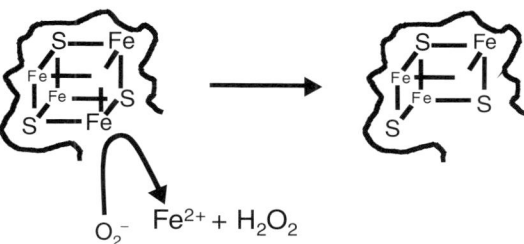

Fig. 1.4. Superoxide inactivates dehydratases by oxidizing their exposed catalytic cluster. The cluster is left in an inactive [3Fe-4S] form that lacks the substrate-binding iron atom. Such damaged enzymes are reactivated continuously in vivo by an undetermined process.

Imlay, 1996). Other consequences of O_2^- stress include problems in the synthesis of aromatic amino acids and in the reductive assimilation of sulfur atoms from sulfate (Imlay and Fridovich, 1992). The molecular bases of these last two phenotypes are not yet clear.

E. coli also synthesizes a periplasmic SOD cofactored by copper and zinc (Benov et al., 1995). Its role may include the scavenging of O_2^- that arises from the incidental oxidation of membrane-bound menaquinone (Korshunov and Imlay, 2006). The other half of the equation is unclear, however: no periplasmic molecule has been identified that is vulnerable to O_2^-. Dehydratases are not found outside of the cytoplasm. This is a point of active investigation, since data indicate that periplasmic SOD is critical to the ability of pathogens to withstand phagocytic O_2^- (below) (De Groote et al., 1997).

H_2O_2

E. coli synthesizes two catalases and at least one active NADH peroxidase, denoted Ahp (Loewen et al., 1985; Poole, 2005). Mutants that lack these enzymes are very poor at scavenging H_2O_2 and, like SOD mutants, display growth defects that have been traced back to molecular lesions. H_2O_2, like O_2^-, can bind, oxidize and inactivate dehydratases (Loewen et al., 1985). The rate constant ($c.10^4$ M^{-1} s^{-1}) implies that intracellular H_2O_2 levels need to be kept beneath 10 nM, and once again the titre of scavenging enzymes has been calibrated to achieve this mark.

Unlike O_2^-, which produces lesions that inhibit growth but do not diminish viability,

H_2O_2 has effects that are potentially lethal. Cells possess pools of ferrous iron that is available for incorporation into iron-containing enzymes; however, in the presence of H_2O_2, this iron can be oxidized through the Fenton reaction (Eqn 1.2), forming $HO^•$:

$$Fe^{2+} + H_2O_2 \rightarrow Fe^{3+} + OH^- + HO^• \qquad (1.2)$$

$HO^•$ is an extremely powerful univalent oxidant that oxidizes most organic molecules at nearly diffusion-limited rates. Thus, they predominantly oxidize biomolecules in the vicinity of ferrous iron – including nucleic acids, which bind iron efficiently. Consequently, H_2O_2-stressed cells accumulate DNA lesions that are mutagenic and, when abundant, lethal (Imlay and Linn, 1988). E. coli and other bacteria require several DNA repair pathways to survive H_2O_2 exposure (below).

The rate constant of the Fenton reaction depends on the ligands to the iron atom, but generally is in the order of 10^3 M^{-1} s^{-1} (Park et al., 2005). Profuse DNA damage occurs when the intracellular concentration of H_2O_2 rises to 1 µM. The rate of damage depends also on the concentration of unincorporated intracellular iron, and the cell employs a variety of mechanisms to keep the iron pool within limits (discussed below). Interestingly, at concentrations of H_2O_2 above 10 µM, the rate of DNA damage becomes independent of H_2O_2 concentration (Park et al., 2005), as the rate-limiting step in DNA damage is the reduction of ferric iron back to the ferrous form by intracellular reductants such as cysteine and reduced flavins (Woodmansee and Imlay, 2002; Park et al., 2005). Thus, workers should be careful not to imagine that if damage is limited at 1 mM H_2O_2 it will be insignificant at the micromolar concentrations that occur in natural habitats.

In addition to producing DNA lesions, H_2O_2 disrupts several processes that are required for robust growth. Like O_2^-, H_2O_2 can bind, oxidize and inactivate dehydratases of the [4Fe-4S] family, leading to dysfunction of both the branched-chain biosynthetic pathway and the TCA cycle. Recent data indicate that E. coli also employs ferrous iron as a non-redox mononuclear metal for other enzymes, including the ribulose-5-phosphate epimerase (Rpe) of the pentose phosphate pathway (Sobota and Imlay, 2011). The metal atom of Rpe binds its substrate in order to activate it for epimerization (Akana et al., 2006); as a consequence, the iron is solvent exposed and prone to oxidation by H_2O_2. The reaction converts the iron to an inactive ferric form; it also generates a hydroxyl radical that can damage active-site residues covalently. In either case, activity is lost. The number of such iron-using mononuclear enzymes in E. coli is almost certainly underestimated, since most workers typically have worked with aerobic buffers – in which iron is oxidized quickly to an insoluble ferric form – when testing the ability of enzymes to be activated by various metals.

The three damaging reactions that have been observed are all varieties of the Fenton reaction: the oxidation of DNA-bound iron, the inactivation of [4Fe-4S] dehydratases and the inactivation of mononuclear iron enzymes. To these must be added two other targets of H_2O_2, the Fur (ferric uptake regulator) and Isc systems. Fur is the master regulator of iron levels in E. coli and many other bacteria (Nielands, 1993). When iron levels are sufficient, Fur is metallated by ferrous iron and, in that form, binds DNA to repress transcription of iron import systems (e.g. siderophores). However, when H_2O_2 is present, the Fur:Fe^{2+} complex is oxidized, releasing iron (Varghese et al., 2007). This reaction inactivates the Fur repressor – potentially leading to unregulated iron import. Effectively, the presence of H_2O_2 is mistaken for the absence of iron. This mistake could be disastrous, as the combination of H_2O_2 and excessive iron import can lead to very high rates of DNA damage.

The Isc system builds the iron-sulfur clusters that are used in a wide variety of enzymes, including not only dehydratases but also electron-transfer enzymes (Barras et al., 2005). H_2O_2 disrupts this system, presumably by oxidizing the nascent iron-sulfur cluster either as it is assembled on the IscU scaffold protein or as it is transferred to the recipient client protein (Jang and Imlay, 2010). The OxyR response includes the induction of proteins whose role is to diminish the physiological impact of Fur and Isc inactivation (see below).

One might expect that new molecular H_2O_2 targets will be identified in the future. Thus far, there is no clear-cut evidence for lipid peroxidation. Standard models of lipid peroxidation require the participation of polyunsaturated fatty acids, which are thermodynamically most vulnerable to oxidation (Bielski et al., 1983). Unlike eukaryotes, bacteria lack these fatty acids, which might make bacterial membranes resistant to the effects of H_2O_2. It is more likely that H_2O_2 might inactivate some enzymes via the oxidation of key cysteine residues. This is unlikely to be a general effect, since free cysteine – and presumably typical cysteine residues – are much less reactive with H_2O_2 ($k \sim 2$–$20\,M^{-1}\,s^{-1}$) than are iron atoms (Winterbourn and Metodiewa, 1999). However, cysteinyl residues may be activated in the context of local protein structure. Depuydt et al. (2009) have shown that the catalytic cysteine residues of periplasmic transpeptidases can be oxidized by H_2O_2 in vitro. DsbG, a periplasmic isomerase, can reduce the resultant sulfenic acid. Notably, dsbG is part of the OxyR regulon – which implies that the oxidation of such proteins is a threat during environmental H_2O_2 stress. In contrast, no compelling evidence for the oxidation of cytoplasmic proteins has yet come to light. The elimination of cytosolic-reducing systems, including glutaredoxin 1 and thioredoxin 2, which are controlled similarly by OxyR, did not create any obvious growth defect during low-grade H_2O_2 stress (Martin and Imlay, unpublished results).

1.1.3 Cellular defences against oxidative stress

Defences against O_2^-

The most immediate way to defend a cell against oxygen species is to scavenge them. *E. coli* expresses three SODs. The MnSOD (encoded by *sodA*) and FeSOD (*sodB*) isozymes are both cytoplasmic, and their converse regulation in response to iron levels supports the conclusion that FeSOD is synthesized when iron is available and MnSOD provides the majority of SOD when iron is scarce (Compan and Touati, 1993; Masse and Gottesman, 2002). FeSOD is even synthesized under anaerobic conditions, apparently in preparation for entry into aerobic habitats (Masse and Gottesman, 2002). The choice of this isozyme for anaerobic synthesis is fitting because iron tends to be bioavailable in anaerobic environments. As mentioned, mutants that lack cytoplasmic SOD activity are poisoned by endogenous O_2^-.

A primary target of O_2^- is the iron-sulfur clusters of dehydratases, which it degrades to an inactive [3Fe-4S] form (Kuo *et al.*, 1987; Gardner and Fridovich, 1991a,b; Flint *et al.*, 1993a,b). The damaged clusters are repaired *in vivo*, with a half-time of several minutes. This process can be replicated *in vitro* by the provision of thiols and ferrous iron. Repair presumably proceeds in a similar fashion *in vivo*, although no protein machinery has yet been linked definitively to the process.

While the basal defence systems are adequate to defend *E. coli* against superoxide that is produced routinely by enzyme auto-oxidation, they are insufficient to fend off the much more extreme stress that can occur when *E. coli* is suffused with redox-cycling antibiotics. When added to laboratory cultures, these drugs can amplify intracellular superoxide formation by orders of magnitude (Hassan and Fridovich, 1979), although the likely concentrations of such drugs in real-world habitats is unknown. When challenged by these antibiotics, *E. coli* and other bacteria activate the SoxRS response. SoxR is a transcription factor (of the MerR family) that contains a [2Fe–2S] cluster that is oxidized during drug exposure; in its oxidized form, it binds specifically to the promoter region of *soxS*, activating its transcription (Ding and Demple, 1997; Gaudu *et al.*, 1997). SoxS protein is itself a transcription factor that activates expression of a broad regulon that has been estimated to include scores of genes (Pomposiello and Demple, 2000). When redox-cycling drugs are removed, SoxR reverts to its reduced form, and the rapid turnover of SoxS by proteases then ensures that the response ends (Griffith *et al.*, 2004; Shah and Wolf, 2006).

Because the SoxRS response was first associated with MnSOD induction, it seemed likely that O_2^- per se was the oxidant that activated SoxR. However, subsequent experiments have shown that the system is minimally induced in SOD mutants, despite the presence of toxic levels of O_2^-; further, overexpression of SOD does not diminish the induction of SoxS by redox-cycling drugs (Gort and Imlay, 1998). In fact, the SoxRS system can be activated under anaerobic conditions, when O_2^- cannot be formed (Gu and Imlay, 2011). The antibiotics themselves trigger the response by oxidizing SoxR directly.

The set of genes that SoxRS controls confirms that its normal trigger is the presence of redox-cycling antibiotics. The response includes several modifications that diminish the level of intracellular antibiotic. WaaYZ induction allows modification of the lipo-polysaccharide, which may impede the approach of antibiotics to the membrane surface (Lee *et al.*, 2009). Transcription of the porin gene encoding OmpF is downregulated through the induction of MicF, an antisense RNA (Chou *et al.*, 1993). AcrAB (acriflavine resistance) is a multi-drug efflux system that pumps antibiotics non-specifically out of the cell (White *et al.*, 1997). *E. coli* also synthesizes a quinone reductase that may suppress univalent redox cycling (Liu *et al.*, 2009). Collectively, these activities suppress the tendency of antibiotics to redox cycle.

The SoxRS system also elevates the expression of glucose-6-phosphate de-hydrogenase (Greenberg and Demple, 1989), the entry point into the pentose phosphate pathway, presumably in order to replenish NADPH that might be consumed through the antibiotic-mediated redox cycling of NADPH-reducible enzymes. This pathway does not seem particularly important for SOD mutants, and so it seems unlikely that this response addresses O_2^- toxicity itself.

In contrast, the induction of a cluster-free isozyme of fumarase is clearly a strategy that circumvents the O_2^- sensitivity of the housekeeping [4Fe–4S] fumarases (Liochev and Fridovich, 1992). And although the SoxS-induced aconitase A, like the standard aconitase B, has an iron-sulfur cluster, the aconitase A cluster appears to be relatively O_2^- resistant (Varghese et al., 2003). YggX somehow facilitates the repair of damaged dehydratase clusters (Gralnick and Downs, 2003). And finally, the induction of endonuclease IV, which helps to fix oxidative DNA lesions, ameliorates the impact that dehydratase-borne iron atoms have on the integrity of the genome (Cunningham et al., 1986; Demple et al., 1986).

At this point, then, it appears that the SoxRS response is designed to ward off the toxicity of redox-cycling antibiotics – and that a key element of that response is the induction of enzymes that defray the impact of the O_2^- that these antibiotics generate. Some studies have indicated that non-oxidative problems also arise from the antibiotics, which might include the depletion of NADPH, for example (Thompson, 1987). Thus, it will be interesting to learn whether some of the SoxRS-controlled genes whose activities are not yet identified might serve to minimize these non-oxidative stresses.

Defences against H_2O_2

E. coli has two catalases and several enzymes that have been regarded as peroxidases. This raises a question: What is the benefit of having so many scavengers of H_2O_2? The katG gene encodes a catalase (HPI) that is of the peroxidase class (Singh et al., 2008). Its peroxidase activity is a poor relative to its catalase activity. This is the predominant catalase during the exponential phase and it is induced strongly when the cell is exposed

to H_2O_2. The second catalase, HPII (encoded by katE), is induced by the stationary-phase sigma factor, RpoS (Loewen and Hengge-Aronis, 1994). The KatG catalase binds a molecule of NADH that might rescue it from a dead-end intermediate state, and one might imagine that KatG cannot work well in non-growing cells because NADH might be unavailable. However, this is mere speculation.

In log-phase cells, however, the primary scavenger of H_2O_2 is Ahp (alkyl hydroperoxide reductase), which is actually a true NADH peroxidase (Seaver and Imlay, 2001a; Poole, 2005). Mutants that lack this enzyme exhibit activation of the H_2O_2 stress response, indicating that in the absence of Ahp the endogenous H_2O_2 accumulates to a stressful level. Ahp function is lost when cells are exposed to high (>20 µM) H_2O_2, possibly because the catalytic cysteine residue is overoxidized to a sulfinic acid. This effect is not undesirable, however, since when H_2O_2 levels are very high, Ahp turnover would otherwise deplete NADH from the cell. Instead, when Ahp is saturated, catalase – that needs no reductant – becomes the primary scavenging activity. This complementarity rationalizes nicely why it is useful for cells to have both peroxidases and catalases.

E. coli mutants that lack Ahp, KatG and KatE excrete endogenous H_2O_2 and cannot scavenge exogenous H_2O_2 efficiently, which indicates that the cell has no other significant H_2O_2-scavenging enzymes, at least under laboratory growth conditions. This is somewhat puzzling, since the genome also encodes Bcp (a thiol-style peroxidase homologue, also known as bacterioferritin comigratory protein) (Jeong et al., 2000), a thiol peroxidase (Cha et al., 1996) and a cytochrome c peroxidase (Partridge et al., 2007). One possibility is that the natural substrates for these enzymes are organic hydroperoxides, rather than H_2O_2 itself.

Because Ahp is sufficient to drive endogenous H_2O_2 concentrations down to non-toxic levels, H_2O_2 stress arises primarily when E. coli enters environments in which external H_2O_2 is present. Under those conditions, the influx of H_2O_2 can elevate intracellular H_2O_2 above the concentration that triggers activation of the OxyR stress response – about 0.1 μM (Aslund et al., 1999). OxyR protein has a reactive cysteinyl residue that reacts rapidly with H_2O_2, being oxidized to a sulfenic acid form that condenses with another cysteine residue to form a disulfide bond (Zheng et al., 1998). This bridge locks the protein into a conformation that activates its capacity as a transcription factor. When H_2O_2 stress abates, the disulfide bond is reduced by glutaredoxins, and the response ends.

Gisela Storz and colleagues used microarrays and sequence analysis to identify genes under the control of OxyR (Zheng et al., 2001a,b). In principle, these genes should comprise a list of the strategies by which cells defend themselves against H_2O_2 – and, in fact, many of the genes on the E. coli list appear also on the lists of genes that either OxyR or PerR (another global regulator known to respond to peroxide stress) activates in other H_2O_2-stressed microbes (Imlay, 2008). Two of the induced proteins serve to suppress Fenton chemistry by limiting the amount of free iron in the cell: Dps, a ferritin-like protein that sequesters iron (Altuvia et al., 1994; Grant et al., 1998; Ilari et al., 2002), and Fur, the transcription factor that blocks synthesis of iron-import systems (Zheng et al., 1999). Induction of Fur compensates in part for its gradual inactivation by H_2O_2 (as discussed above). The absence of either dps or fur enables micromolar doses of intracellular

H_2O_2 to create overwhelming DNA damage (Park et al., 2005; Varghese et al., 2007).

The sufABCDSE operon encodes H_2O_2-resistant cluster-assembly machinery (Takahashi and Tokumoto, 2002; Nachin et al., 2003). It assists in the repair of damaged [4Fe-4S] clusters that have degraded beyond the [3Fe-4S] state, and it replaces the Isc machinery in catalysing the de novo assembly of iron-sulfur clusters in newly synthesized proteins (Jang and Imlay, 2010).

The induction of MntH, a manganese import system, is important for maintenance of the activity of mononuclear iron enzymes (Anjem et al., 2009; Sobota and Imlay, 2011). Manganese can substitute for iron in the active sites of these enzymes – which is valuable because manganese, unlike iron, does not participate in HO·-generating Fenton chemistry (Fig. 1.5). Mutants that lack mnt cannot tolerate even 0.5 μM intracellular H_2O_2.

The OxyR regulon also includes one of the two cellular thioredoxins, one of its three glutaredoxins and glutathione reductase. The apparent implication is that H_2O_2 stress creates unwanted disulfide bonds – but the deletion of these genes has not detectably diminished cell fitness during micromolar H_2O_2 stress (Imlay, 2008). It could be possible that the growth conditions that were employed happened not to require activity from disulfide-bonded enzymes. Alternatively, since OxyR uses a thiol to sense cysteine-directed stresses, perhaps it does 'double

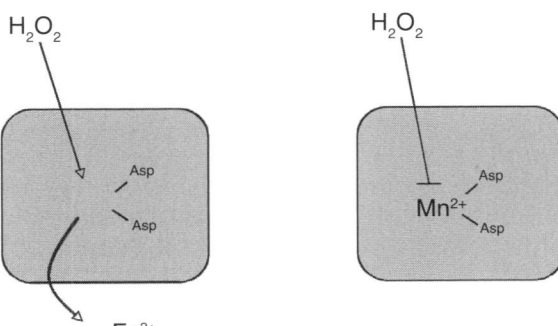

Fig. 1.5. Inactivation of iron enzymes. Mononuclear iron enzymes can be inactivated by H_2O_2, but the manganese-charged enzyme is active and resistant to Fenton chemistry-mediated oxidation.

duty' in responding to sulfhydryl stresses apart from H_2O_2. Clearly, more work must be done to resolve this issue.

Interestingly, DNA repair enzymes are essential for *E. coli* to tolerate H_2O_2 exposure, but they are not controlled by the OxyR system. Both recombinational repair, which depends on RecA and RecBCD, and excision repair, which depends on exonuclease III, are important (Ananthaswamy and Eisenstark, 1977; Demple *et al.*, 1983). Recombination is stimulated in H_2O_2-stressed cells via the activation of the SOS response. Perhaps the logic of the use of DNA damage, rather than H_2O_2 per se, as the signal to elevate repair capacity is that DNA damage can persist even after the H_2O_2 has been scavenged. Exonuclease III is controlled by RpoS, but there is no evidence that its titres respond to either H_2O_2 or to DNA lesions.

1.2 Nitrosative Stress

1.2.1 Sources of nitric oxide

Several reactive nitrogen species are harmful to bacteria. Research has focused primarily on nitric oxide (NO), nitrosothiols (RSNO) and peroxynitrite (HOONO). *E. coli* and other enterics reduce nitrite to ammonia; while this pathway does not involve NO as an obligatory intermediate, it appears to escape from the process in yields sufficient to stress bacteria (Corker and Poole, 2003). Thus, when nitrite is provided as a nitrogen source,

NO-scavenging enzymes (as discussed below) are induced and are necessary to preserve cell fitness. *E. coli* is exposed to exogenous NO when it cohabitates with denitrifying bacteria, since NO is an intermediate in the process by which they reduce nitrite to dinitrogen (Zumft, 1997). NO is also produced in toxic doses by the inducible NO synthase of eukaryotic phagocytes (Fang, 2004).

1.2.2 The hazards of nitrogen species

The threat that NO poses is several-fold (Fig. 1.6).

A radical itself, NO can react directly with radical enzymes such as ribonucleotide reductase (Kwon *et al.*, 1991; Lepoivre *et al.*, 1994). Because it binds tightly to metal centres, NO is a potent inhibitor of cytochrome oxidases, including both the cytochrome *o* and cytochrome *bd* oxidases of *E. coli* (Hori *et al.*, 1996; Butler *et al.*, 1997). A block in respiration is, therefore, one of its first and most consequential effects. NO also binds to the iron atoms of iron-sulfur clusters (Drapier, 1997; Gardner *et al.*, 1997; Ding and Demple, 2000; Cairo *et al.*, 2002; Rogers *et al.*, 2003; Duan *et al.*, 2009). Interestingly, this effect includes not only the solvent-exposed clusters of dehydratases (the targets for H_2O_2), but also oxidant-resistant clusters. In both these cases, binding leads to dissociation of bridging sulfur atoms and elimination of catalytic activity; thus, one would expect both catabolic and innumerable biosynthetic pathways to be

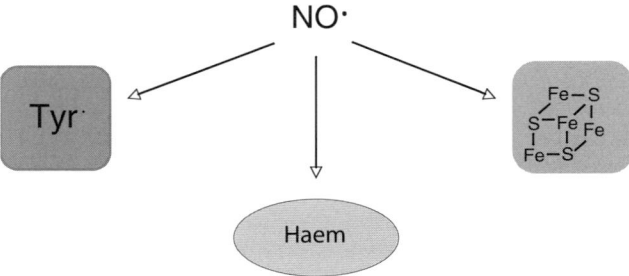

Fig. 1.6. Inactivation of enzymes by NO. Enzymes with radicals, haems and iron-sulfur clusters are vulnerable to inactivation by NO.

affected. Notably, these effects are bacteriostatic rather than lethal, and the inhibition of metal centres is reversed quickly when NO levels fall.

Much work has been done on the toxicity of nitrosothiol compounds; S-nitrosylated glutathione (GSNO) is a frequent choice as a stressor as it is an effector of the human innate immune system. Such a species is likely generated inside cells by the univalent oxidation of NO (to nitrosium, NO$^+$) by intracellular metals, either free or as essential elements of enzymes, and subsequent conjugation to glutathione (GSH) or other thiol species (Hogg, 2002). In any case, RSNO compounds transfer the NO$^+$ cation easily to other thiolates, potentially affecting the activities of enzymes with critical cysteinyl residues. Studies with GSNO are complicated by its gradual release of authentic NO, so that the toxic effects of the latter are overlaid on the direct effects of the GSNO itself.

NO can react rapidly with O_2^- to create HOONO (peroxynitrite), a strongly oxidizing species that can damage proteins, metal centres and even DNA (Ferrer-Sueta and Radi, 2009). HOONO is unlikely to be generated in quantity in bacteria that simply are stressed with NO, since the O_2^- level is so low. However, HOONO is a credible threat in phagocytes, where the NO synthase and NADPH oxidase might operate simultaneously.

1.2.3 Cellular defences against NO

E. coli expresses two enzymes that eliminate NO. Flavohaemoglobin (Hmp) is a haem/flavoenzyme that uses molecular oxygen to convert NO to nitrate (Gardner *et al.*, 1998). The enzyme is induced during exposure to NO (Poole *et al.*, 1996), and mutants that lack this enzyme scavenge NO poorly and are hypersensitive to growth inhibition. Under anaerobic conditions, the NADH-dependent NO-reducing enzyme, NorVW, is induced in response to NO, presumably because Hmp cannot function without oxygen (Gardner *et al.*, 2002; Gomes *et al.*, 2002).

The regulators of these responses have been the focus of close study. NsrR contains an apparent [2Fe-2S] cluster, and it seems likely that complexation and/or destruction of the cluster by NO is the event by which NO is sensed (Tucker *et al.*, 2008; Yukl *et al.*, 2008). NsrR represses Hmp synthesis, as well as that of several other genes (Bodenmiller and Spiro, 2006). Interestingly, *hmp* transcription is also repressed by MetR when bound to homocysteine – which indicates that homocysteine modification (presumably to RSNO status) is employed by cells as a proxy for the presence of NO (Membrillo-Hernandez *et al.*, 1998).

In contrast, NorVW is controlled by a mononuclear iron protein, NorR, which has the capacity to act as a positive transcription factor (Gardner *et al.*, 2003; Tucker *et al.*, 2005). In the resting state, the metal-binding domain blocks the transcriptional activity of the protein; however, complexation by NO disables this domain and unveils its transcriptional effect.

The NsrR regulon varies among bacteria; in *Vibrio fischeri* it includes an alternative, haem-independent oxidase that does not bind NO and therefore allows respiration to continue in NO-laden habitats (Dunn *et al.*, 2010). Recent microarray studies have suggested that in *E. coli*, NsrR also controls the synthesis of periplasmic nitrite and nitrate reductases, so that they are induced when NO levels rise (Filenko *et al.*, 2007). The implication may be that the cell wishes to eliminate the local pool of the nitrate/nitrite that denitrifying bacteria might convert to toxic NO.

Information is sparse about other defensive strategies. YtfE, another member of the NsrR regulon, is a di-iron protein that appears to assist the reconstruction of iron-sulfur clusters after they have been degraded by NO (Justino *et al.*, 2006, 2007). Strikingly, other metal-containing regulators are also affected by NO, including Fur, SoxR, Fnr and IscR (Spiro, 2007). The doses of NO that affect these proteins are much higher than the nanomolar doses that activate NsrR, and presumably NorR; however, by sensing stress to iron and iron-sulfur proteins, these transcription factors may elicit responses that are helpful. The same may be true of OxyR, which can be nitrosylated by RSNO and which then stimulates the synthesis of thioredoxin and glutaredoxin systems.

In general, our understanding of the physiology of nitrosative stress has not remotely approached that of reactive oxygen stress. The complicated chemistry of nitrogen species is part of the difficulty. However, it has become clear that adaptive responses to NO stress are more likely the rule than the exception among bacteria, and the pace of discovery continues to quicken.

1.3 *Pseudomonas aeruginosa*: Oxidative and Nitrosative Stress Systems in an Important Human Pathogen

1.3.1 Preface

Having set the stage with oxidative and nitrosative stress in *E. coli*, their impact on cellular constituents and viability, defence systems against them and the many regulators that control gene expression, we now move to parallel stress systems in the important human pathogen *Pseudomonas aeruginosa*. Although such systems are not nearly as broadly understood as those in *E. coli*, there has been a burgeoning literature on such systems since the first cloning of the *sodA* and *sodB* genes in *P. aeruginosa* in 1993 (Hassett *et al.*, 1993). This next section also brings to bear the importance of *P. aeruginosa* as a true opportunistic pathogen of humans, and we make some vital connections between oxidative and nitrosative stress with a disease that most frequently is a hallmark of *P. aeruginosa* infection, cystic fibrosis (CF).

P. aeruginosa: an important opportunistic pathogen

In contrast to *E. coli*, which is a natural inhabitant of human gut flora, *Pseudomonas* (Greek, 'false unit') *aeruginosa* (Latin, 'copper rust') (PA) is a Gram-negative bacillus that is capable of infecting a variety of hosts including humans, animals (mice, rats, mink), plants (*Arabidopsis thaliana*; Hendrickson *et al.*, 2001), nematodes (*Caenorhabditis elegans*;

Tan *et al.*, 1999; Tan and Ausubel, 2000), caterpillars (*Galleria mellonella*; Miyata *et al.*, 2003), insects (e.g. fruit flies, *Drosophila melanogaster*; Lau *et al.*, 2003) and fish and reptiles (Rahme *et al.*, 2000). Humans whose immune systems have been compromised, including those suffering from severe burns (Holder and Neely, 1989), cancer chemo-therapy (Vento *et al.*, 2008), neutropenic patients (Wang *et al.*, 2005), chronic alcoholics (Ormerod *et al.*, 1988) and HIV, and the elderly are particularly susceptible to PA infections. However, its greatest notoriety comes from often lifelong infections of the airways of CF patients.

When the original Bergey's manual of determinative bacteriology was published in 1923 by the then Society of American Bacteriologists (now called the American Society for Microbiology), the organism was initially called *Bacterium aeruginosum* by Schroter in 1872. However, with time, it has been renamed several times (refer to Table 1.2). Interestingly, and paradoxically pertinent to this chapter, is that PA was classified initially as an obligate aerobe. However, given its metabolic breadth and thousands of publications relating to its metabolic capacity, it is now classified as a facultative anaerobe.

Cystic fibrosis airway infection: a battle between P. aeruginosa and the human neutrophil

During airway infection of CF patients, a vast amount of neutrophils at titres reaching nearly 1500-fold higher than normal levels accumulate during the early aerobic phase of CF airway infection (Downey *et al.*, 2009). These innate immune cells generate copious quantities of O_2^-, as well as powerful biocides including antimicrobial hydrogen peroxide (H_2O_2) and (hypochlorous acid) HOCl during the antimicrobial respiratory burst (Hassett and Imlay, 2006). In fact, within the confines of the neutrophil phagolysosome, H_2O_2 concentrations have been estimated to be as high as 100 mM, a concentration that is known to kill easily both the free-swimming planktonic as well as the highly refractory biofilm form of PA, these are likely to be reduced to micromolar

Table 1.2. Former genus and species name of *P. aeruginosa*.

Bacterium aeruginosum	Schroeter, 1872
Bacterium aeruginosum	Cohn, 1872
Micrococcus pyocyaneus	Zopf, 1884a,b
Bacillus aeruginosus	(Schroeter, 1872) Trevisan, 1885
Bacillus pyocyaneus	(Zopf, 1884a) Flügge, 1886
Pseudomonas pyocyanea	(Zopf, 1884a) Migula, 1900
Bacterium pyocyaneum	(Zopf, 1884a) Lehmann and Neumann, 1896
Pseudomonas vendrelli	Various

levels (Brown *et al.*, 1995; Hassett *et al.*, 1999a,b; Ma *et al.*, 1999). In addition, the CF airway mucus is rife with myeloperoxidase and lacto-peroxidase, enzymes that oxidize halides such as Cl⁻, Br⁻ and I⁻ to highly antimicrobial hypohalous acids (Babior, 1978; Klebanoff, 1988). We will thus focus the second half of this review on this pathogen and its defences against oxidative and nitrosative stresses.

1.3.2 Cellular defences against oxidative stress by *P. aeruginosa*

Defences against O_2^-

SOD: FE-SOD AND MN-SOD. Like *E. coli*, PA is capable of synthesizing two dimeric SODs, one cofactored by iron, the other by manganese (Hassett *et al.*, 1992). Manganese can substitute for iron in the active site of Fe-SOD (Hassett *et al.*, 1992). In contrast to *E. coli*, there is no Cu,Zn-SOD in this organism. Fe-SOD, encoded by *sodB*, is expressed constitutively and has a very high specific activity in aerobic rich media or in a low phosphate succinate medium (65–170 IU mg⁻¹; Hassett *et al.*, 1992, 1993), yet it is also produced anaerobically. The latter event is likely an insurance measure were the organism to face ambient oxygen suddenly. Fe-SOD expression also is dependent on both the *las* and *rhl* quorum-sensing cascades (Hassett *et al.*, 1999b). In contrast, the *sodA* gene (encoding Mn-SOD) is induced primarily by iron deprivation and controlled in a repressive fashion by Fur (Hassett *et al.*, 1996). The *sodA* gene is a member of a four-gene operon including *fagA* (Fur-associated gene), *orfX*, *fumC* (encoding an oxygen-stable fumarase; Hassett *et al.*, 1997b) and *sodA* (Hassett *et al.*, 1997a,b). Unlike *E. coli*, the *sodA*

gene is activated only by iron deprivation, and not oxygen or redox-cycling antibiotics, and is dependent on the *las* quorum-sensing cascade (Hassett *et al.*, 1999b). In fact, *sodA* is one of the most dramatically induced genes upon iron deprivation, with transcript levels increasing nearly 45-fold relative to control bacteria (Ochsner *et al.*, 2002). Isogenic mutant organisms defective in *sodA* are only moderately sensitive to O_2^--generating agents. Yet, ~11% of clinical PA isolates produce a Mn-SOD variant that has a greater R_f value in non-denaturing gels (Britigan *et al.*, 2001). Furthermore, Mn-SOD activity is greatest in mucoid-type PA (Hassett *et al.*, 1993), an event linked to the fact that the extracellular alginate produced by these cells can act as a weak iron chelator (Kim *et al.*, 2003) and thereby activate the *fagA-fumC-sodA-orfX* operon. A *sodA* mutant of a mucoid PA CF isolate was shown to be more sensitive to paraquat (an O_2^--generating agent) and required longer growth periods to attain normal aerobic growth rates than its parental strain (Polack *et al.*, 1996). In contrast, *sodB*, and especially *sodA sodB*, mutants are susceptible to paraquat (Hassett *et al.*, 1995). Finally, *sodA sodB* mutants of PA, like their *E. coli* counterparts, are auxotrophic for branched-chain amino acids (Hassett *et al.*, 1995). However, these mutants are also incapable of producing the redox-active antibiotic called pyocyanin, which is produced when bacteria are starved for phosphate (Hassett *et al.*, 1992).

Defences against H_2O_2

CATALASE: KATA, KATB AND KATC. Enzymatic defences of PA against H_2O_2 are mediated by at least three catalases (KatA, KatB and KatC; Brown *et al.*, 1995; Ma *et al.*, 1999) and four alkyl hydroperoxide reductases, AHPs (AhpA, AhpB, AhpCF and Ohr), some of which have peroxidase and/or catalase activity.

KatA. KatA is phylogenetically a member of the Group III catalase family (Klotz *et al.*, 1997) and is the major constitutively expressed catalase in PA (Hassett *et al.*, 1992; Ma *et al.*, 1999). Still, *katA* transcription is regulated in part by quorum sensing (Hassett *et al.*, 1999b) and slightly inducible by aerobic growth in media containing high iron and exposure to H_2O_2 (Ma *et al.*, 1999). Although not significantly inducible as compared to *katB* transcription (see below), its optimal expression is also dependent on the LysR-type regulator, OxyR (Heo *et al.*, 2010).

The *iscR* gene product, encoding a homologue of the *E. coli* iron-sulfur assembly regulator, IscR, was also found to influence both catalase and SOD activity. In fact, both catalase and SOD activities were reduced significantly in a PA *iscR* mutant (Kim *et al.*, 2009). These results suggest that both IscR and OxyR are required for the optimal resistance to H_2O_2. These results were supported further in a transposon mutant study where KatA, IscR and OxyR genes were found to be essential for optimal H_2O_2 resistance (Choi *et al.*, 2007).

KatA was thought initially to be a heterotrimeric protein based on mass spectrometric analysis of proteolytic fragments of purified KatA (Ma *et al.*, 1999). At the time (in the late 1990s), this was a very novel finding and virtually unprecedented (Loewen *et al.*, 2000). Our unpublished studies indicate that KatA is a classical heterotrimer. At least in *E. coli*, the enzyme behaves as a classic tetramer, functioning with haem moieties embedded within each monomer (J. Wilson, S. Su, D.J. Hassett and R.A. Kovall, unpublished data). KatA is also highly metastable, a conclusion supported by two different research groups (Hassett *et al.*, 2000; Shin *et al.*, 2008) relative to the major catalases from the Gram-positive bacterium *Bacillus subtilis* and that its unique properties involving metastability and extracellular presence may contribute to the peroxide resistance of PA biofilm and

presumably to chronic infections. KatA has also been shown to be a moderately effective vaccine during acute respiratory infections by PA in rats (Thomasa *et al.*, 2000).

We have crystallized the KatA protein recently because it appears to have a protective function during anaerobic growth (data not shown). KatA has also been shown to be essential for full virulence in an intraperitoneal infection model in mice using the strain PA14 (Lee *et al.*, 2005). In contrast to KatA expression, KatC expression is a remarkable outlier, in that it is known to be inducible only by elevated temperature, the elevation of which mirrors the sensitivity of *katC* mutants to H_2O_2 (Mossialos *et al.*, 2006). However, the major response to H_2O_2 is governed by OxyR, a 34 kDa global transactivator that governs transcription of a myriad of genes that are independent of *katA* and *katC* expression, the most obvious of which are the antioxidants KatB (catalase), a 228 kDa tetramer (Brown *et al.*, 1995), AhpB (a periplasmic alkylhydroperoxide reductase) and a cytoplasmic AhpCF, respectively (Hassett *et al.*, 2000; Ochsner *et al.*, 2000). Given the vast repertoire of enzymes and regulatory systems governing the response of PA to H_2O_2, it is not surprising that the organism tolerates it at relatively high levels. However, the organism is far more resistant to H_2O_2 when cultured in biofilms. In fact, biofilm organisms are nearly 1000-fold more resistant to H_2O_2 than their planktonic counterparts (Panmanee *et al.*, 2008).

KatB. In contrast to KatA, KatB is generated only on extreme oxidative stress and is under the control of the global H_2O_2 responsive regulator, OxyR (Ochsner *et al.*, 2000). Isogenic mutants devoid of KatB have only a slight sensitivity to H_2O_2 and alkylhydroperoxides (Brown *et al.*, 1995; Ochsner *et al.*, 2000). However, because of the uncharacteristically inefficient KatA (H_2O_2 K_M, 44 mM), KatB is a very efficient cellular relief mechanism to KatA. In fact, we believe that KatB does not act alone in the massive efforts of cellular defence against copious quantities of H_2O_2 generated by human neutrophils and macrophages, or in cases of the environmental disinfection of problematic biofilms.

KatC. The only manuscript involving KatC describes a H_2O_2 sensitivity that appears to be temperature dependent. A British group revealed that inactivation of overexpression of a cyanide-insensitive oxidase caused a reduction or absence of KatC activity (Mossialos *et al.*, 2006). Furthermore, KatC was also absolutely dependent on DsbA, as a *dsbA* mutant produced no detectable KatC activity. This indicates that an intact disulfide is required for KatC activity.

AHPS IN *P. AERUGINOSA*: AHPB, AHPCF AND OHR. AHPs are highly efficient enzymes that possess the dual capacity for enzymatic defences against H_2O_2 and organic peroxides, and in PA these are mediated by at least four alkyl hydroperoxide reductase enzymes (AhpA, AhpB – Ochsner *et al.*, 2000; AhpCF – Ochsner *et al.*, 2000; Ohr – Ochsner *et al.*, 2001), some of which have peroxidase and/or catalase activity (Ochsner *et al.*, 2000). These enzymes act in an oxygen-dependent manner and are under tight regulation, controlled predominantly by the global regulator, OxyR. OxyR is known to regulate both *ahpB* and *ahpC* genes directly (Ochsner *et al.*, 2000), and this will be discussed in greater detail below.

OXYR. The *oxyR* gene of PA strain PAO1 was cloned nearly 10 years ago (Ochsner *et al.*, 2000). This LysR family regulator is responsible for the H_2O_2/AHP-mediated activation of the OxyR regulon, among which include the backup 224 kDa tetrameric catalase, KatB, the periplasmic AHP, AhpB, and the cytoplasmic AHP, AhpCF (Ochsner *et al.*, 2000). Despite possessing near wild-type catalase levels, an *oxyR* mutant is remarkably susceptible to H_2O_2. In fact, such organisms are unable to form isolated colonies on rich medium (e.g. Luria agar) under aerobic conditions (Fig. 1.7).

This is due to the fact that Luria agar contains auto-oxidizable components that generate ~1 microM H_2O_2 min^{-1} at room temperature. This level of H_2O_2 resembles that present in human blood, sufficient to kill *oxyR* mutant bacteria. We have engaged recently in an unpublished research endeavour to identify genes under the control of OxyR that would not be considered to be obvious (e.g. those *not* predicted to be involved in oxidative stress defence). The aforementioned members of the OxyR regulon are involved in the disposal of both H_2O_2 and alkylhydroperoxides. Using a bioinformatics approach, we have shown recently that many unexpected genes harbour signature OxyR-binding motifs within their promoter regions. Among these include the *bdlA* (biofilm dispersion locus) and *qscR* (quorum-sensing control regulator) genes that are involved in the active dispersion of bacteria housed in biofilms and negative regulator of the intercellular signalling

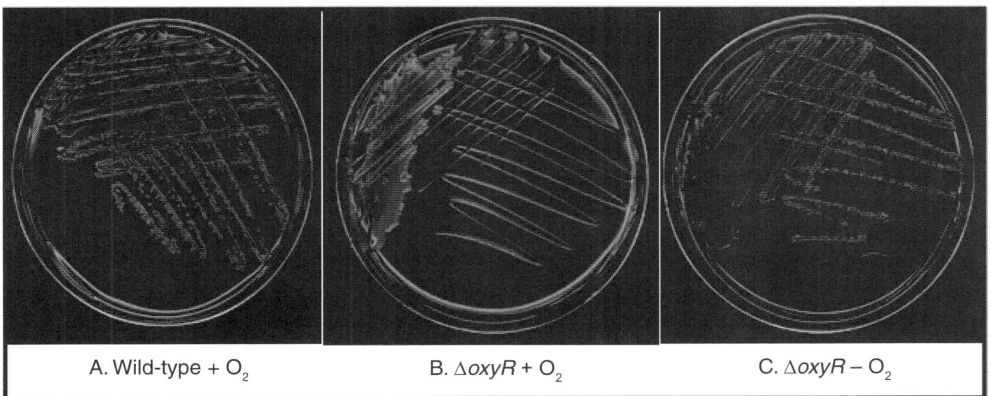

| A. Wild-type + O$_2$ | B. $\Delta oxyR$ + O$_2$ | C. $\Delta oxyR$ – O$_2$ |

Fig. 1.7. Colony morphology of wild-type and *oxyR* mutant bacteria on L-agar plates. Note in C that *oxyR* mutants can form isolated colonies during anaerobic but not aerobic (B) conditions.

process known as quorum sensing, respectively (Chugani *et al.*, 2001; Morgan *et al.*, 2006). This novel form of OxyR control sheds additional light on the global regulatory breadth of this regulator. The physiological importance is obvious; bacteria grown in a biofilm that are threatened with potentially lethal levels of H_2O_2 actively disperse from the biofilms while simultaneously reducing the energy required for quorum sensing regulated gene products. OxyR also appears to influence the production of the quorum sensing regulated gene pyocyanin and rhamnolipid, yet the mechanism underlying this phenomenon is unknown (Vinckx *et al.*, 2008). Finally, without OxyR, bacteria have a defect for growth in iron-limiting media. This is likely because of reduced KatA or an inability to obtain iron from multiple PA siderophores (Vinckx *et al.*, 2008).

1.3.3 Nitrosative stress in *P. aeruginosa*: how does this occur endogenously and exogenously?

NO production via anaerobic respiration

Anaerobic respiration in PA involves the reduction of nitrogen oxides to the final product, nitrogen gas (N_2). In the context of ATP generation, the greatest means to generate anaerobic energy is via nitrate (NO_3^-) respiration. The major factors involved in this process are the nitrate reductase (NAR), nitrite reductase (NIR), nitric oxide reductase (NOR) and nitrous oxide reductase (NOS). NO is an intermediate during this process. NO titres must be held in check; intracellular concentrations greater than 13 mM impair anaerobic growth severely. The major regulators as well as the critical enzymatic machinery involved in this process, the molecular mechanism and CF-relevant phenotypes of mutants lacking these enzymes are described in detail below.

Anaerobic regulatory machinery: follow the leader, ANR

The master regulatory switch for anaerobic NO_3^-/NO_2^- respiration as well as anaerobic arginine substrate level phosphorylation is the CRP/FNR-like global regulator known as ANR (<u>a</u>naerobic <u>n</u>itrate <u>r</u>egulator) that was discovered by Gallimand *et al.* in 1991 (Gallimand *et al.*, 1991). An *anr* mutant was incapable of anaerobic growth in the presence of NO_3^- or arginine. In 1995 Ye *et al.* also showed that an *anr* mutant harboured little or no NIR and NOR activity (Ye *et al.*, 1995). The initial nature of ANR-dependent gene regulation was targeted by Hiroki Arai and colleagues in Japan in 1994 (Arai *et al.*, 1994), and this productive group has continued to dissect unique features of anaerobic metabolism in PA. Subsequent to that, the same research group segregated genes regulated by ANR and also by downstream regulatory loci, such as *dnr* (Arai *et al.*, 1995). From 1995 to 1997 ANR was known to regulate only *nirS* (encoding the respiratory nitrite reductase) and *nirQ* (encoding a regulator immediately downstream of ANR and subsequently also regulated by DNR; Arai *et al.*, 1995). Paradoxically, we have shown that transcription of the *anr* gene varies little in microarray-based studies under aerobic versus anaerobic conditions (Platt *et al.*, 2008). This is a testament to the fact that ANR acts as a transcriptional activator and a repressor, as well as being regulated post-translationally. A list of genes under the control of ANR is provided by Hassett *et al.* (Hassett *et al.*, 2004).

Although its three-dimensional structure has not been solved, as was the *E. coli* FNR (<u>fu</u>marate <u>n</u>itrate <u>r</u>egulator), ANR is a dimeric protein, harbouring a $[4Fe-4S]^{2+}$ that is required for anaerobic transcriptional activation as well as repression of typically aerobically expressed genes (Yoon *et al.*, 2007; Kawakami *et al.*, 2010). Previously, we have used a bioinformatics approach to identify putative genes under the control of ANR. These were reproduced from a very recent work by Max Shobert and colleagues in Germany (Trunk *et al.*, 2010).

The ANR homologue in *E. coli*, FNR, typically can sense ~5–10 microM NO in bacterial cells (Cruz-Ramos *et al.*, 2002). This begs the question as to why a *norCB* mutant mysteriously survived during anaerobic growth when the intracellular concentration of NO was estimated at >16 µM (Yoon *et al.*, 2007), a concentration that has a significant negative impact on bacterial viability (Yoon *et al.*, 2006). This concentration of intra-cellular NO disrupts the [4Fe–4S]$^{2+}$ cluster of the PA ANR, eliciting a circuit breaker mechanism that allows for anaerobic survival. These results were supported by visible/ UV spectroscopy, Mossbauer spectroscopy, electrophoretic mobility shift assays using *narK1*, an ANR-specific promoter and *lacZ* reporter assays (Yoon *et al.*, 2007).

ANR and aerobic metabolism

ANR also has been found recently to act as a potential repressor of cyanide-insensitive oxidase genes (conveniently coined CIO) (Ugidos *et al.*, 2008; Kawakami *et al.*, 2010). Hassett *et al.* (1992) showed that this oxidase resisted even 1 mM KCN during oxygen respirometric experiments. Similarly, ANR has also been found to be involved in the repression of the *hmgA* and *hpd* genes that encode homogentisate-1,2-dioxygenase and 4-hydroxyphenylpyruvate dioxygenase,

respectively. Both HmgA and Hpd are part of the tyrosine degradation pathway that occurs predominantly under aerobic conditions and contribute to the protection of a *norCB* mutant that is unable to dispose of NO enzymatically (Fig. 1.8).

The aerobic oxygen-dependent enzyme, *hmgA*, encoding homogentisate-1,2-dio-xygenase, typically is repressed during anaerobic growth of wild-type bacteria (Yoon *et al.*, 2002). However, in an anaerobic *norCB* mutant, both HmgA and Hpd, part of the tyrosine degradation pathway, are dra-matically upregulated (Yoon *et al.*, 2007). Paradoxically, ANR has also been shown to act as a positive regulator of the aerotactic chemotaxis regulator, Aer (Hong *et al.*, 2004), a protein responsible for flagellum-mediated motility toward increasing oxygen levels.

DNR: the second tier of anaerobic regulation

ANR activates transcription of the *dnr* gene during anaerobic growth. The DNR protein, although considered to be a second tier regulator, can direct anaerobic growth when

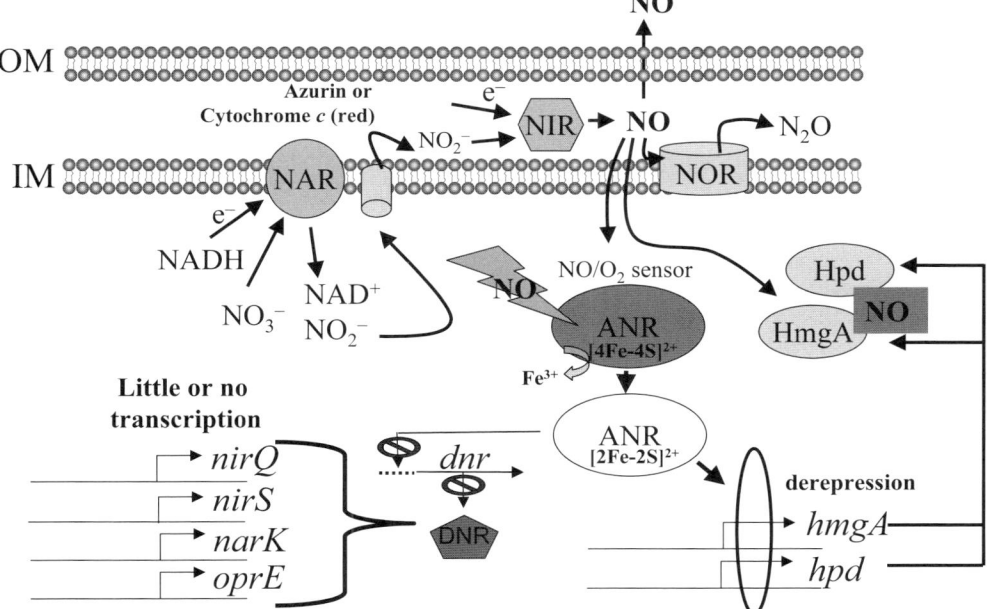

Fig. 1.8. The ANR and NO pathways in *P. aeruginosa*. ANR is involved in the transcriptional response to oxygen availability but this pathway is also important in NO protection.

activated constitutively in the absence of ANR (Arai *et al.*, 1997). Specifically, anaerobiosis, the presence of NO_3^-/NO_2^- and especially anaerobic NO, trigger transcription activation of the *dnr* gene (Hasegawa *et al.*, 1998; Giardina *et al.*, 2008). For activation of similar genes in *E. coli*, DNR was found to require not only NO, but also haem (Castiglione *et al.*, 2009). The DNR protein activates transcription of multiple genes including *nirS*, *nirQ*, *norCB*, *hemF* and *hemN* (Arai *et al.*, 1995; Rompf *et al.*, 1998). However, recent X-ray crystallographic studies elegantly solved by Francesca Cutruzollo and colleagues in Italy revealed that DNR could be post-translationally modified by NO (Giardina *et al.*, 2008). These studies used a C-terminal deletion that strategically lacked the DNA-binding domain. A *dnr* mutant has a similar anaerobic phenotype as an *anr* mutant, an inability to grow anaerobically with NO_3^- or arginine (Arai *et al.*, 1995). Because both ANR and DNR are dimers that sense NO, we suggest that if ANR is poisoned, a critical concentration of NO can still activate DNR to be an active transcription factor.

NarX-NarL: a two-component phospho-relay system and part of the third-tier regulatory network for anaerobic growth of P. aeruginosa

The impact of NO_3^- on PA is diverse, especially in the context of oxygen concentration and NO_3^-. Classical early reports by Rowe and colleagues have shed light on these two important factors (Hernandez and Rowe, 1987; Hernandez *et al.*, 1991). A recent report by Benkert *et al.* (2008) indicates that NarL functions to repress transcription of the *arcDABC* genes. While NarL typically functions as a transcriptional activator, PA inherently realizes that it can gain far more energy from anaerobic NO_3^- respiration than by arginine substrate level phosphorylation.

1.3.4 How does *P. aeruginosa* respond to NO?

NO plays a crucial role in the antimicrobial activity of host defence systems, and especially in the airway. Macrophage-generated NO is important for controlling several bacterial infections including *Salmonella typhimurium*, *Mycobacterium tuberculosis* and PA. Growth of a NO reductase-deficient (NOR) mutant of PA under a microaerobic condition was inhibited by the exogenous NO. Furthermore, the intracellular survival assay within the NO-producing RAW 264.7 macrophages revealed that the wild-type strain survived longer than the NO reductase-deficient mutant. These results suggest that the PA NO reductase (NOR) may contribute to the intracellular survival by acting as a counter component against the host's defence systems. Mutant organisms lacking NOR also have a significantly impaired anaerobic growth rate, yet they do not die (Yoon *et al.*, 2002). NO also appears to be a signalling agent in PA. For example, exposure to *S*-nitrosoglutathione (GSNO), an NO donor, activated the process of biofilms dispersal in PA (Barraud *et al.*, 2009). One of the major features of exposure to NO is the likelihood of *S*-nitrosylation, the conversion of thiol groups, including cysteine residues in proteins, to form *S*-nitrosothiols (RSNOs). *S*-nitrosylation is a mechanism for dynamic, post-translational regulation of most or all major classes of protein.

Sensitivity of ANR to endogenously generated metabolic NO

Like many regulators that have redox-active Fe-S centres such as ANR, inactivation of the regulatory capacity of such proteins can often be mediated via common metabolites within bacteria. These include both superoxide (O_2^-) and NO. Two TCA enzymes, aconitase and fumarase, are also inactivated by both of these oxygen and nitrogen reduction products (Gardner and Fridovich, 1991b; Liochev and Fridovich, 1993; Gardner *et al.*, 1997). In *E. coli*, SoxR and OxyR represent those that are under redox control. ANR also is involved in redox sensing. Yoon *et al.* (2007) have shown that ANR can be inactivated metabolically by physiological levels of molecular oxygen and NO. The molecular basis underlying the inactivation of ANR appears to involve the stripping of iron from both [2Fe–2S]$^{2+}$ clusters, causing dimer to monomer conversion (Yoon *et al.*, 2007). This is very much akin to how the related protein, FNR, functions in *E. coli*. FNR is also known to be sensitive to both oxygen and NO. For example, a mutant in the *fnr* gene that causes FNR to be a monomer can still activate anaerobic genes even under aerobic conditions (Lazazzera *et al.*, 1993). However, unlike other proteins, ANR can be reused. Enzymes known as cysteine desulfurases can reincorporate cysteine into the cluster, thereby causing reactivation of this class of proteins. These proteins typically do not enter the proteosome, somehow evading proteolysis within the cell. This is likely a function of the *anr* gene not having a secretion-mediated signal peptide.

1.4 Influence of Quorum Sensing on Unique Metabolic Properties of *P. aeruginosa*: Anaerobic Metabolism and Influence of NO on the Intra-bacterium Competition Between *P. aeruginosa* and *S. aureus*

We have reflected previously on the importance of quorum sensing (QS) vis-à-vis anaerobic metabolism of PA. The latest finding of interest to the overall biology of PA in the anaerobic mucus of the CF airways is a paper by Hoffman *et al.* demonstrating several unique properties of PA CF isolates that harbour *lasR* mutations (Hoffman *et al.*, 2010). The *lasR* mutants were found to have distinct advantages, especially via the ability to metabolize nitrogen-containing compounds, in particular NO_3^-, in the presence of reduced oxygen levels. The enhanced metabolism of NO_3^- relative to that of wild-type organisms favoured a strikingly important clinical phenotype, resistance to tobramycin and ciprofloxacin, two major front-line CF antibiotics. Another feature of these mutants is that they outcompete *S. aureus* during anaerobic culture by overproduction of respiratory NO. However, one Achilles' heel in the *lasR* mutant is its enhanced sensitivity to slightly acidified nitrite (A-NO_2^-) (Hoffman *et al.*, 2010). LasR controls transcription of the second tier of the *rhlR* gene in PA (Pesci *et al.*,

1997). Thus, *lasR* mutants are essentially *rhlR* mutants. This has been reflected by Yoon *et al.*, who have shown that anaerobic *rhlR* mutant bacteria grow much faster than wild-type bacteria, but commit a respiratory suicide by overproduction of anaerobic NO (Yoon *et al.*, 2006).

Anaerobic *rhlR* mutant bacteria in biofilms are killed rapidly and are protected by the NO-scavenger carboxy-PTIO, as well as genetically by construction of a double *rhlR nirS* mutant that is incapable of endogenous anaerobic NO generation. However, to date, the cause-and-effect relationship between NO and cell death has not been elucidated. Yet, there are several scenarios in hand. This includes the fact that RhlR-PAI-2 act as stators of the QS process, based on data showing that the stationary phase sigma factor, RpoS, is critical for carefully regulated *rhl* QS control (Schuster *et al.*, 2004).

Still, the most important question remains as to why the QS mutants die via anaerobic metabolic suicide, yet mutants lacking NOR survive? In the latter case, and as stated above, ANR is poisoned in the *norCB*. However, why does the same mechanism not hold true in the *rhlR* mutant? One feature of the *rhlR* mutant is that it appears to die anaerobically only in biofilms and not in planktonic culture. The reason behind this is likely the spatial arrangement of cells in biofilms where they are associated tightly with one another or in a tightly woven matrix. However, in planktonic cells there is nearly 100-fold more space for NO to diffuse, given that a stationary phase planktonic culture generates nearly 5×10^9 CFU ml^{-1}. The dimensions of PA have been estimated at 1×3 µm. Thus, 5×10^9 bacteria (notwithstanding the dead cells) in planktonic culture would represent only approximately <1:100 of the total volume of the planktonic cell culture (D.J. Hassett, unpublished results). However, in biofilms, these cells are packed tightly among one another, thereby influencing dramatically the rate of NO diffusion and the ensuing effects on adjacent organisms. Although clearly speculation at this juncture, we would argue that the likelihood of this event occurring is high.

1.5 Anaerobiosis, NO and Bacteriophage Activation

Platt *et al.*, using both a proteomic and microarray study, showed that some of the most highly upregulated genes during anaerobic growth in the presence of NO_3^- or NO_2^- were those encoding bacteriophage-related proteins (Platt *et al.*, 2008). The reason for this is unknown. However, we will take the liberty of offering a possible viable explanation for such a scenario in the following statements. Virtually all bacteriophages are activated by one form of cellular stress or another. This paradigm is supported by patterns of replication of known human viral pathogens including HIV, herpes, CMV and papillomavirus activation (Mollace *et al.*, 2002; Pereira *et al.*, 2003; Thompson *et al.*, 2009). Within the PA genome exist two primary operons which encode a variety of bacteriophage proteins that are commonly referred to as filamentous Pf1 (phage that bind to bacterial pili and flagella) bacteriophages. Two regions on the PA genome harbour Pf1 phage-encoding genes, PA0614-0648 and PA0712-0729, respectively.

Bacteriophage induction has been linked previously to oxidative or nitrosative stress (Barraud *et al.*, 2006, 2009). Recent work by Kjelleberg and colleagues showed that the Pf1 phage gene PA0712 was essential for the normal biofilm life cycle, as well as virulence in PA (Rice *et al.*, 2009). In fact, there were more phage particles detected in biofilm effluents from a *norCB* mutant (unable to remove NO) relative to wild-type bacteria and few if any phage particles from an *nirS* mutant (unable to generate endogenous NO). A Pf1 phage was involved in the killing of biofilms, and this killing was absent in an *rpoN* mutant that lacked the surface appendages, pili and flagella (Webb *et al.*, 2003; Barraud *et al.*, 2006). Thus, either NO donors under aerobic conditions or endogenously generated NO via anaerobic respiration trigger phage activation. This ensures a likely carefully timed seeding of fresh phage particles to infect or super-infect other organisms in the immediate or downstream settings. In many routine cases

in the laboratory, prolonged incubation of PA on solidified media triggers plaque formation on rich medium plates by an as yet unknown mechanism, but one that is universally thought to be mediated by stress.

1.6 Using a Natural Anaerobic Metabolite to Kill *P. aeruginosa*

Pathogenic bacteria are remarkably equipped to adapt to a variety of niches, both in the host and in the environment. A feature that is not considered in current or future treatment strategies is the use of common metabolites that are generated during normal metabolism. Arguably the most formidable foe in the context of treating problematic CF lung infections is mucoid PA. Mutations within the PA genome are the most common cause of mucoid conversion, an event that leads to overproduction of the viscous exopolysaccharide alginate. The most frequent mutations are those within the gene (mucoidy) encoding the membrane-spanning anti-sigma factor, MucA (Martin *et al.*, 1993). There are other mutations that occur *in vivo* including *mucB* (*algN*) (Goldberg *et al.*, 1993; Boucher *et al.*, 1996), *mucC* (Boucher *et al.*, 1996), *mucD* (Boucher *et al.*, 1996) and *algW* (Boucher *et al.*, 1996), respectively. During anaerobic growth, mucoidy is stabilized by an as yet unknown mechanism(s). In the absence of MucA, the σ factor AlgT(U) transcribes genes involved in alginate overproduction, regulatory elements and other stress-related genes. These organisms enjoy the apparent luxury of being inherently resistant to conventional antibiotics and phagocytosis by human neutrophils. Surprisingly, we found that mucoid, *mucA* mutant bacteria were extremely sensitive to acidified NO_2^-. We have taken the liberty of coining this defect the 'Achilles' heel' of mucoid PA. Yoon *et al.* first showed that the pH of the airway mucus was slightly acidic (6.4–6.5) (Yoon *et al.*, 2006). Amazingly, mucoid organisms could grow anaerobically at pH 6.5 using NO_3^- but not NO_2^-. In fact, NO_2^- killed mucoid, *mucA* mutant PA, in part because of low NIR activity. Without NIR,

NO_2^- is reduced chemically to HNO_2 and NO gas. NO gas kills mucoid, *mucA* mutant PA but not $MucA^+$ organisms. NO_2^- also kills mucoid bacteria in mixed culture with non-mucoid organisms, in anaerobic biofilms and in freshly excised CF airway mucus (Yoon *et al.*, 2006). Two mutations that occur routinely in the PA genome during CF airway disease, *rhlR* and *mucA*, offer some hope for alternative treatment strategies using $NaNO_2$, based on some recent developments in 2002 and 2006 (Yoon *et al.*, 2002, 2006).

Acknowledgements

This work is supported by National Institutes of Health Grant GM49640 to J.A.I. and grants from the Veteran's Administration Merit Award, Cystic Fibrosis Research, Inc, and Cure Finders, Inc, to D.J.H.

References

Akana, J., Fedorov, A.A., Fedorov, E., Novak, W.R., Babbitt, P.C., Almo, S.C. and Gerlt, J.A. (2006) D-ribulose 5-phosphate 3-epimerase: functional and structural relationships to members of the ribulose-phosphate binding (beta/alpha)8-barrel superfamily. *Biochemistry* 45, 2493–2503.

Altuvia, S., Almirón, M., Huisman, G., Kolter, R. and Storz, G. (1994) The *dps* promoter is activated by OxyR during growth and by IHF and σ^S in stationary phase. *Molecular Microbiology* 13, 265–272.

Ananthaswamy, H.N. and Eisenstark, A. (1977) Repair of hydrogen peroxide-induced single-strand breaks in *Escherichia coli* deoxyribonucleic acid. *Journal of Bacteriology* 130, 187–191.

Anjem, A., Varghese, S. and Imlay, J.A. (2009) Manganese import is a key element of the OxyR response to hydrogen peroxide in *Escherichia coli*. *Molecular Microbiology* 72, 844–858.

Arai, H., Igarashi, Y. and Kodama, T. (1994) Structure and ANR-dependent transcription of the *nir* genes for denitrification from *Pseudomonas aeruginosa*. *Bioscience, Biotechnology and Biochemistry* 58, 1286–1291.

Arai, H., Igarashi, Y. and Kodama, T. (1995) Expression of the *nir* and *nor* genes for denitrification of *Pseudomonas aeruginosa* requires a novel CRP/FNR-related transcriptional regulator, DNR, in addition to ANR. *FEBS Letters* 371, 73–76.

Arai, H., Kodama, T. and Igarashi, Y. (1997) Cascade regulation of the two CRP/FNR-related transcriptional regulators (ANR and DNR) and the denitrification enzymes in *Pseudomonas aeruginosa*. *Molecular Microbiology* 25, 1141–1148.

Aslund, F., Zheng, M., Beckwith, J. and Storz, G. (1999) Regulation of the OxyR transcription factor by hydrogen peroxide and the cellular thiol-disulfide status. *Proceedings of the National Academy of Sciences of the United States of America* 96, 6161–6165.

Babior, B. (1978) Oxygen-dependent microbial killing by phagocytes. *New England Journal of Medicine* 298, 659–668 and 721–725.

Barras, F., Loiseau, L. and Py, B. (2005) How *Escherichia coli* and *Saccharomyces cerevisiae* build Fe/S proteins. *Advances in Microbial Physiology* 50, 41–101.

Barraud, N., Hassett, D.J., Hwang, S.H., Rice, S.A., Kjelleberg, S. and Webb, J.S. (2006) Involvement of nitric oxide in biofilm dispersal of *Pseudomonas aeruginosa*. *Journal of Bacteriology* 188, 7344–7353.

Barraud, N., Schleheck, D., Klebensberger, J., Webb, J.S., Hassett, D.J., Rice, S.A. and Kjelleberg, S. (2009) Nitric oxide signaling in *Pseudomonas aeruginosa* biofilms mediates phosphodiesterase activity, decreased cyclic di-GMP levels, and enhanced dispersal. *Journal of Bacteriology* 191, 7333–7342.

Benkert, B., Quack, N., Schreiber, K., Jaensch, L., Jahn, D. and Schobert, M. (2008) Nitrate-responsive NarX-NarL represses arginine-mediated induction of the *Pseudomonas aeruginosa* arginine fermentation *arcDABC* operon. *Microbiology* 154, 3053–3060.

Benov, L., Chang, L.Y., Day, B. and Fridovich, I. (1995) Copper, zinc superoxide dismutase in *Escherichia coli* : periplasmic localization. *Archives in Biochemistry and Biophysics* 319, 508–511.

Bielski, B.H., Arudi, R.L. and Sutherland, M.W. (1983) A study of the reactivity of HO_2/O_2^- with unsaturated fatty acids. *Journal of Biological Chemistry* 258, 4759–4761.

Bodenmiller, D.M. and Spiro, S. (2006) The *yjeB* (*nsrR*) gene of *Escherichia coli* encodes a nitric oxide-sensitive transcriptional regulator. *Journal of Bacteriology* 188, 874–881.

Boucher, J.C., Martinez-Salazar, J., Schurr, M.J., Mudd, M.H., Yu, H. and Deretic, V. (1996) Two distinct loci affecting conversion to mucoidy in

Pseudomonas aeruginosa in cystic fibrosis encode homologs of the serine protease HtrA. *Journal of Bacteriology* 178, 511–523.

Britigan, B.E., Miller, R.A., Hassett, D.J., Pfaller, M.A., McCormick, M.L. and Rasmussen, G.T. (2001) Antioxidant enzyme expression in clinical isolates of *Pseudomonas aeruginosa*: identification of an atypical form of manganese superoxide dismutase. *Infection and Immunity* 69, 7396–7401.

Brown, S.M., Howell, M.L., Vasil, M.L., Anderson, A.J. and Hassett, D.J. (1995) Cloning and characterization of the *katB* gene of *Pseudomonas aeruginosa* encoding a hydrogen peroxide-inducible catalase: purification of KatB, cellular localization, and demonstration that it is essential for optimal resistance to hydrogen peroxide. *Journal of Bacteriology* 177, 6536–6544.

Butler, C.S., Seward, H.E., Greenwood, C. and Thomson, A.J. (1997) Fast cytochrome *bo* from *Escherichia coli* binds two molecules of nitric oxide at CuB. *Biochemistry* 36, 16259–16266.

Cairo, G., Recalcati, S., Pietrangelo, A. and Minotti, G. (2002) The iron regulatory proteins: targets and modulators of free radical reactions and oxidative damage. *Free Radical Biology in Medicine* 32, 1237–1243.

Carlioz, A. and Touati, D. (1986) Isolation of superoxide dismutase mutants in *Escherichia coli*: is superoxide dismutase necessary for aerobic life? *EMBO Journal* 5, 623–630.

Castiglione, N., Rinaldo, S., Giardina, G. and Cutruzzola, F. (2009) The transcription factor DNR from *Pseudomonas aeruginosa* specifically requires nitric oxide and haem for the activation of a target promoter in *Escherichia coli*. *Microbiology* 155, 2838–2844.

Cha, M.K., Kim, H.K. and Kim, I.H. (1996) Mutation and mutagenesis of thiol peroxidase of *Escherichia coli* and a new type of thiol peroxidase family. *Journal of Bacteriology* 178, 5610–5614.

Choi, Y.S., Shin, D.H., Chung, I.Y., Kim, S.H., Heo, Y.J. and Cho, Y.H. (2007) Identification of *Pseudomonas aeruginosa* genes crucial for hydrogen peroxide resistance. *Journal of Microbiology and Biotechnology* 17, 1344–1352.

Chou, J.H., Greenberg, J.T. and Demple, B. (1993) Post-transcriptional repression of *Escherichia coli* OmpF protein in response to redox stress: positive control of the *micF* antisense RNA by the *soxRS* locus. *Journal of Bacteriology* 175, 1026–1031.

Chugani, S.A., Whiteley, M., Lee, K.M., D'Argenio, D., Manoil, C. and Greenberg, E.P. (2001) QscR,

a modulator of quorum-sensing signal synthesis and virulence in *Pseudomonas aeruginosa*. *Proceedings of the National Academy of Sciences of the United States of America* 98, 2752–2757.

Cohn, F. (1872) Untersuchungen fiber bacterien. *Beitrage zur Biologie der Pflan-zen 1, Heft* 2, 127–224.

Compan, I. and Touati, D. (1993) Interaction of six global transcription regulators in expression of manganese superoxide dismutase in *Escherichia coli* K-12. *Journal of Bacteriology* 175, 1687–1696.

Corker, H. and Poole, R.K. (2003) Nitric oxide formation by *Escherichia coli*. Dependence on nitrite reductase, the NO-sensing regulator Fnr, and flavohemoglobin Hmp. *Journal of Biological Chemistry* 278, 31584–31592.

Cruz-Ramos, H., Crack, J., Wu, G., Hughes, M.N., Scott, C., Thomson, A.J., Green, J. and Poole, R.K. (2002) NO sensing by FNR: regulation of the *Escherichia coli* NO-detoxifying flavo-haemoglobin, Hmp. *EMBO Journal* 21, 3235–3244.

Cunningham, R.P., Saporito, S.M., Spitzer, S.G. and Weiss, B.M. (1986) Endonuclease IV (*nfo*) mutant of *Escherichia coli*. *Journal of Bacteriology* 168, 1120–1127.

De Groote, M.A., Ochsner, U.A., Shiloh, M.U., Nathan, C., McCord, J.M., Dinauer, M.C., Libby, S.J., Vazquez-Torres, A., Xu, Y. and Fang, F.C. (1997) Periplasmic superoxide dismutase protects *Salmonella* from products of phagocyte NADPH-oxidase and nitric oxide synthase. *Proceedings of the National Academy of Sciences of the United States of America* 94, 13997–14001.

Demple, B., Halbrook, J. and Linn, S. (1983) *Escherichia coli xth* mutants are hypersensitive to hydrogen peroxide. *Journal of Bacteriology* 153, 1079–1082.

Demple, B., Johnson, A. and Fung, D. (1986) Exonuclease III and endonuclease IV remove 3' blocks from DNA synthesis primers in H_2O_2-damaged *Escherichia coli*. *Proceedings of the National Academy of Sciences of the United States of America* 83, 7731–7735.

Depuydt, M., Leonard, S.E., Vertommen, D., Denoncin, K., Morsomme, P., Wahni, K., Messens, J., Carroll, K.S. and Collet, J.F. (2009) A periplasmic reducing system protects single cysteine residues from oxidation. *Science* 326, 1109–1111.

Ding, H. and Demple, B. (1997) *In vivo* kinetics of a redox-regulated transcriptional switch. *Proceedings of the National Academy of Sciences* of the United States of America 94, 8445–8449.

Ding, H. and Demple, B. (2000) Direct nitric oxide signal transduction via nitrosylation of iron-sulfur centers in the SoxR transcription activator. *Proceedings of the National Academy of Sciences of the United States of America* 97, 5146–5150.

Downey, D.G., Bell, S.C. and Elborn, J.S. (2009) Neutrophils in cystic fibrosis. *Thorax* 64, 81–88.

Drapier, J.C. (1997) Interplay between NO and [Fe-S] clusters: relevance to biological systems. *Methods* 11, 319–329.

Duan, X., Yang, J., Ren, B., Tan, G. and Ding, H. (2009) Reactivity of nitric oxide with the [4Fe-4S] cluster of dihydroxyacid dehydratase from *Escherichia coli*. *Biochemical Journal* 417, 783–789.

Dunn, A.K., Karr, E.A., Wang, Y., Batton, A.R., Ruby, E.G. and Stabb, E.V. (2010) The alternative oxidase (AOX) gene in *Vibrio fischeri* is controlled by NsrR and upregulated in response to nitric oxide. *Molecular Microbiology* 77(1), 44–55.

Fang, F.C. (2004) Antimicrobial reactive oxygen and nitrogen species: concepts and controversies. *Nature Reviews Microbiology* 2, 820–832.

Farr, S.B., D'Ari, R. and Touati, D. (1986) Oxygen-dependent mutagenesis in *Escherichia coli* lacking superoxide dismutase. *Proceedings of the National Academy of Sciences of the United States of America* 83, 8268–8272.

Ferrer-Sueta, G. and Radi, R. (2009) Chemical biology of peroxynitrite: kinetics, diffusion, and radicals. *ACS Chemical Biology* 4, 161–177.

Filenko, N., Spiro, S., Browning, D.F., Squire, D., Overton, T.W., Cole, J. and Constantinidou, C. (2007) The NsrR regulon of *Escherichia coli* K-12 includes genes encoding the hybrid cluster protein and the periplasmic, respiratory nitrite reductase. *Journal of Bacteriology* 189, 4410–4417.

Flint, D.H., Emptage, M.H., Finnegan, M.G., Fu, W. and Johnson, M.K. (1993a) The role and properties of the iron-sulfur cluster in *Escherichia coli* dihydroxy-acid dehydratase. *Journal of Biological Chemistry* 268, 14732–14742.

Flint, D.H., Tuminello, J.F. and Emptage, M.H. (1993b) The inactivation of Fe-S cluster containing hydro-lases by superoxide. *Journal of Biological Chemistry* 268, 22369–22376.

Flugge, C. (1886) *Die mikroorganismen*. 2 Auflage. F.C.W. Vogel, Leipzig, Germany, pp. 1–692.

Gallimand, M., Gamper, M., Zimmermann, A. and Hass, D. (1991) Positive FNR-like control of anaerobic arginine degradation and nitrate

respiration in *Pseudomonas aeruginosa*. *Journal of Bacteriology* 173, 1598–1606.

Gardner, A.M., Helmick, R.A. and Gardner, P.R. (2002) Flavorubredoxin, an inducible catalyst for nitric oxide reduction and detoxification in *Escherichia coli*. *Journal of Biological Chemistry* 277, 8172–8177.

Gardner, A.M., Gessner, C.R. and Gardner, P.R. (2003) Regulation of the nitric oxide reduction operon (*norRVW*) in *Escherichia coli*. Role of NorR and sigma54 in the nitric oxide stress response. *Journal of Biological Chemistry* 278, 10081–10086.

Gardner, P.R. and Fridovich, I. (1991a) Superoxide sensitivity of the *Escherichia coli* 6-phosphogluconate dehydratase. *Journal of Biological Chemistry* 266, 1478–1483.

Gardner, P.R. and Fridovich, I. (1991b) Superoxide sensitivity of the *Escherichia coli* aconitase. *Journal of Biological Chemistry* 266, 19328–19333.

Gardner, P.R., Costantino, G., Szabo, C. and Salzman, A.L. (1997) Nitric oxide sensitivity of the aconitases. *Journal of Biological Chemistry* 272, 25071–25076.

Gardner, P.R., Gardner, A.M., Martin, L.A. and Salzman, A.L. (1998) Nitric oxide dioxygenase: an enzymic function for flavohemoglobin. *Proceedings of the National Academy of Sciences of the United States of America* 95, 10378–10383.

Gaudu, P., Moon, N. and Weiss, B. (1997) Regulation of the *soxRS* oxidative stress regulon. Reversible oxidation of the Fe-S centers of SoxR *in vivo*. *Journal of Biological Chemistry* 272, 5082–5086.

Gershman, R., Gilber, D.L., Nye, S.W., Dwyer, P. and Fenn, W.O. (1954) Oxygen poisoning and X-irradiation: a mechanism in common. *Science* 119, 623–626.

Giardina, G., Rinaldo, S., Johnson, K.A., Di Matteo, A., Brunori, M. and Cutruzzola, F. (2008) NO sensing in *Pseudomonas aeruginosa*: structure of the transcriptional regulator DNR. *Journal of Molecular Biology* 378, 1002–1015.

Goldberg, J.B., Gorman, W.L., Flynn, J.L. and Ohman, D.E. (1993) A mutation in *algN* permits trans activation of alginate production by *algT* in *Pseudomonas aeruginosa*. *Journal of Bacteriology* 175, 1303–1308.

Gomes, C.M., Giuffre, A., Forte, E., Vicente, J.B., Saraiva, L.M., Brunori, M. and Teixeira, M. (2002) A novel type of nitric-oxide reductase. *Escherichia coli* flavorubredoxin. *Journal of Biological Chemistry* 277, 25273–25276.

Gort, A.S. and Imlay, J.A. (1998) Balance between endogenous superoxide stress and antioxidant defenses. *Journal of Bacteriology* 180, 1402–1410.

Gralnick, J.A. and Downs, D.M. (2003) The YggX protein of *Salmonella enterica* is involved in Fe(II) trafficking and minimizes the DNA damage caused by hydroxyl radicals: residue CYS-7 is essential for YggX function. *Journal of Biological Chemistry* 278, 20708–20715.

Grant, R.A., Filman, D.J., Finkel, S.E., Kolter, R. and Hogle, J.M. (1998) The crystal structure of Dps, a ferritin homolog that binds and protects DNA. *Nature Structural Biology* 5, 294–303.

Greenberg, J.T. and Demple, B. (1989) A global response induced in *Escherichia coli* by redox-cycling agents overlaps with that induced by peroxide stress. *Journal of Bacteriology* 171, 3933–3939.

Gregory, E.M. and Fridovich, I. (1973) Induction of superoxide dismutase by molecular oxygen. *Journal of Bacteriology* 114, 543–548.

Griffith, K.L., Shah, I.M. and Wolf, R.E. Jr (2004) Proteolytic degradation of *Escherichia coli* transcription activators SoxS and MarA as the mechanism for reversing the induction of the superoxide (SoxRS) and multiple antibiotic resistance (Mar) regulons. *Molecular Microbiology* 51, 1801–1816.

Hasegawa, N., Arai, H. and Igarashi, Y. (1998) Activation of a consensus FNR-dependent promoter by DNR of *Pseudomonas aeruginosa* in response to nitrite. *FEMS Microbiology Letters* 166, 213–217.

Hassan, H.M. and Fridovich, I. (1977) Regulation of the synthesis of superoxide dismutase in *Escherichia coli*: induction by methyl viologen. *Journal of Biological Chemistry* 252, 7667–7672.

Hassan, H.M. and Fridovich, I. (1979) Intracellular production of superoxide radical and of hydrogen peroxide by redox active compounds. *Archives in Biochemistry and Biophysics* 196, 385–395.

Hassett, D.J. and Imlay, J.A. (2006) Oxidative stress systems in bacteria: four model systems. In: Nickerson, C. and Schurr, M.J. (eds) *Molecular Paradigm of Infectious Disease: A Bacterial Perspective*, Chapter 14. Kluwer Academic-Plenum Publishers, New York–Boston–Dordrecht–London–Moscow, pp. 544–573.

Hassett, D.J., Charniga, L., Bean, K.A., Ohman, D.E. and Cohen, M.S. (1992) Antioxidant defense mechanisms in *Pseudomonas aeruginosa*: resistance to the redox-active antibiotic pyocyanin and demonstration of a manganese-cofactored superoxide dismutase. *Infection and Immunity* 60, 328–336.

Hassett, D.J., Woodruff, W.A., Wozniak, D.J., Vasil,

M.L., Cohen, M.S. and Ohman, D.E. (1993) Cloning of the *sodA* and *sodB* genes encoding manganese and iron superoxide dismutase in *Pseudomonas aeruginosa*: demonstration of increased manganese superoxide dismutase activity in alginate-producing bacteria. *Journal of Bacteriology* 175, 7658–7665.

Hassett, D.J., Schweizer, H.P. and Ohman, D.E. (1995) *Pseudomonas aeruginosa sodA* and *sodB* mutants defective in manganese- and iron-cofactored superoxide dismutase activity demonstrate the importance of the iron-cofactored form in aerobic metabolism. *Journal of Bacteriology* 177, 6330–6337.

Hassett, D.J., Sokol, P., Howell, M.L., Ma, J.-F., Schweizer, H.P., Ochsner, U. and Vasil, M.L. (1996) Ferric uptake regulator (Fur) mutants of *Pseudomonas aeruginosa* demonstrate defective siderophore-mediated iron uptake and altered aerobic metabolism. *Journal of Bacteriology* 178, 3996–4003.

Hassett, D.J., Howell, M.L., Ochsner, U., Johnson, Z., Vasil, M. and Dean, G.E. (1997a) An operon containing *fumC* and *sodA* encoding fumarase C and manganese superoxide dismutase is controlled by the ferric uptake regulator (Fur) in *Pseudomonas aeruginosa*: *fur* mutants produce elevated alginate levels. *Journal of Bacteriology* 179, 1452–1459.

Hassett, D.J., Howell, M.L., Sokol, P.A., Vasil, M. and Dean, G.E. (1997b) Fumarase C activity is elevated in response to iron deprivation and in mucoid, alginate-producing *Pseudomonas aeruginosa*: cloning and characterization of *fumC* and purification of native FumC. *Journal of Bacteriology* 179, 1442–1451.

Hassett, D.J., Elkins, J.G., Ma, J.-F. and McDermott, T.R. (1999a) *Pseudomonas aeruginosa* biofilm sensitivity to biocides: use of hydrogen peroxide as model antimicrobial agent for examining resistance mechanisms. *Methods in Enzymology* 310, 599–608.

Hassett, D.J., Ma, J.-F., Elkins, J.G., McDermott, T.R., Ochsner, U.A., West, S.E.H., Huang, C.-T., Fredericks, J., Burnett, S., Stewart, P.S., McPheters, G., Passador, L. and Iglewski, B.H. (1999b) Quorum sensing in *Pseudomonas aeruginosa* controls expression of catalase and superoxide dismutase genes and mediates biofilm susceptibility to hydrogen peroxide. *Molecular Microbiology* 34, 1082–1093.

Hassett, D.J., Alsabbagh, E., Parvatiyar, K., Howell, M.L., Wilmott, R.W. and Ochsner, U.A. (2000) A protease-resistant catalase, KatA, that is released upon cell lysis during stationary phase, is essential for aerobic survival of a *Pseudomonas aeruginosa oxyR* mutant at low cell densities. *Journal of Bacteriology* 182, 4557–4563.

Hassett, D.J., Lymar, S.V., Rowe, J.J., Schurr, M.J., Passador, L., Herr, A.B., Winsor, G.L., Brinkman, F.S.L., Lau, G.W., Yoon, S.S. and Hwang, S.H. (2004) Anaerobic metabolism by *Pseudomonas aeruginosa* in cystic fibrosis airway biofilms: role of nitric oxide, quorum sensing and alginate production. In: Nakano, M.M. and Zuber, P. (eds) *Strict and Facultative Anaerobes: Medical and Environmental Aspects*. Horizon Bioscience, UK, pp. 87–108.

Hendrickson, E.L., Plotnikova, J., Mahajan-Miklos, S., Rahme, L.G. and Ausubel, F.M. (2001) Differential roles of the *Pseudomonas aeruginosa* PA14 *rpoN* gene in pathogenicity in plants, nematodes, insects, and mice. *Journal of Bacteriology* 183, 7126–7134.

Heo, Y.J., Chung, I.Y., Cho, W.J., Lee, B.Y., Kim, J.H., Choi, K.H., Lee, J.W., Hassett, D.J. and Cho, Y.H. (2010) The major catalase gene (*katA*) of *Pseudomonas aeruginosa* PA14 is under both positive and negative control of the global transactivator OxyR in response to hydrogen peroxide. *Journal of Bacteriology* 192, 381–390.

Hernandez, D, and Rowe, J.J. (1987) Oxygen regulation of nitrate uptake in denitrifying *Pseudomonas aeruginosa*. *Applied and Environmental Microbiology* 53(4), 745–750.

Hernandez, D., Dias, F.M. and Rowe, J.J. (1991) Nitrate transport and its regulation by O2 in *Pseudomonas aeruginosa*. *Archives of Biochemistry and Biophysics* 286(1), 159–163.

Hoffman, L.R., Richardson, A.R., Houston, L.S., Kulasekara, H.D., Martens-Habbena, W., Klausen, M., Burns, J.L., Stahl, D.A., Hassett, D.J., Fang, F.C. and Miller, S.I. (2010) Nutrient availability as a mechanism for selection of antibiotic tolerant *Pseudomonas aeruginosa* within the CF airway. *PLoS Pathogens* 6, e1000712.

Hogg, N. (2002) The biochemistry and physiology of *S*-nitrosothiols. *Annual Reviews in Pharmacology and Toxicology* 42, 585–600.

Holder, I.A. and Neely, A.N. (1989) Combined host and specific anti-*Pseudomonas*-directed therapy for *Pseudomonas aeruginosa* infections in burned mice: experimental results and theoretic considerations. *Journal of Burns Care and Rehabilitation* 10, 131–136.

Hong, C.S., Kuroda, A., Ikeda, T., Takiguchi, N., Ohtake, H. and Kato, J. (2004) The aerotaxis transducer gene *aer*, but not *aer-2*, is transcriptionally regulated by the anaerobic regulator ANR in *Pseudomonas aeruginosa*.

Journal of Bioscience and Bioengineering 97, 184–190.

Hori, H., Tsubaki, M., Mogi, T. and Anraku, Y. (1996) EPR study of NO complex of *bd*-type ubiquinol oxidase from *Escherichia coli*. *Journal of Biological Chemistry* 271, 9254–9258.

Ilari, A., Ceci, P., Ferrari, D., Rossi, G.L. and Chiancone, E. (2002) Iron incorporation into *Escherichia coli* Dps gives rise to a ferritin-like microcrystalline core. *Journal of Biological Chemistry* 277, 37619–37623.

Imlay, J.A. (2008) Cellular defenses against superoxide and hydrogen peroxide. *Annual Reviews of Biochemistry* 77, 755–776.

Imlay, J.A. and Fridovich, I. (1992) Suppression of oxidative envelope damage by pseudoreversion of a superoxide dismutase-deficient mutant of *Escherichia coli*. *Journal of Bacteriology* 174, 953–961.

Imlay, J.A. and Linn, S. (1988) DNA damage and oxygen radical toxicity. *Science* 240, 1302–1309.

Jang, S. and Imlay, J.A. (2010) Hydrogen peroxide inactivates the *Escherichia coli* Isc iron-sulphur assembly system, and OxyR induces the Suf system to compensate. *Molecular Microbiology* 78, 1448–1467.

Jeong, W., Cha, M.K. and Kim, I.H. (2000) Thioredoxin-dependent hydroperoxide peroxidase activity of bacterioferritin comigratory protein (BCP) as a new member of the thiol-specific antioxidant protein (TSA)/Alkyl hydroperoxide peroxidase C (AhpC) family. *Journal of Biological Chemistry* 275, 2924–2930.

Justino, M.C., Almeida, C.C., Goncalves, V.L., Teixeira, M. and Saraiva, L.M. (2006) *Escherichia coli* YtfE is a di-iron protein with an important function in assembly of iron-sulphur clusters. *FEMS Microbiology Letters* 257, 278–284.

Justino, M.C., Almeida, C.C., Teixeira, M. and Saraiva, L.M. (2007) *Escherichia coli* di-iron YtfE protein is necessary for the repair of stress-damaged iron-sulfur clusters. *Journal of Biological Chemistry* 282, 10352–10359.

Kawakami, T., Kuroki, M., Ishii, M., Igarashi, Y. and Arai, H. (2010) Differential expression of multiple terminal oxidases for aerobic respiration in *Pseudomonas aeruginosa*. *Environmental Microbiology* 12(6), 1399–1412.

Keyer, K. and Imlay, J.A. (1996) Superoxide accelerates DNA damage by elevating free-iron levels. *Proceedings of the National Academy of Sciences of the United States of America* 93, 13635–13640.

Kim, E.J., Sabra, W. and Zeng, A.P. (2003) Iron deficiency leads to inhibition of oxygen transfer and enhanced formation of virulence factors in cultures of *Pseudomonas aeruginosa* PAO1. *Microbiology* 149, 2627–2634.

Kim, S.H., Lee, B.Y., Lau, G.W. and Cho, Y.H. (2009) IscR modulates catalase A (KatA) activity, peroxide resistance and full virulence of *Pseudomonas aeruginosa* PA14. *Journal of Microbiology and Biotechnology* 19, 1520–1526.

Klebanoff, S.J. (1988) Phagocytic cells: products of oxygen metabolism. In: Gallin, J.I., Goldstein, I.M. and Snyderman, R. (eds) *Inflammation: Basic Principles and Clinical Correlates*. Raven, New York, pp. 391–444.

Klotz, M.G., Klassen, G.R. and Loewen, P.C. (1997) Phylogenetic relationships among prokaryotic and eukaryotic catalases. *Molecular Biology and Evolution* 14, 951–958.

Korshunov, S.S. and Imlay, J.A. (2002) A potential role for periplasmic superoxide dismutase in blocking the penetration of external superoxide into the cytosol of Gram-negative bacteria. *Molecular Microbiology* 43, 95–106.

Korshunov, S. and Imlay, J.A. (2006) Detection and quantification of superoxide formed within the periplasm of *Escherichia coli*. *Journal of Bacteriology* 188, 6326–6334.

Kuo, C.-F., Mashino, T. and Fridovich, I. (1987) α,β-dihydroxyisovalerate dehydratase: a superoxide sensitive enzyme. *Journal of Biological Chemistry* 262, 4724–4727.

Kwon, N.S., Stuehr, D.J. and Nathan, C.F. (1991) Inhibition of tumor cell ribonucleotide reductase by macrophage-derived nitric oxide. *Journal of Experimental Medicine* 174, 761–767.

Lau, G.W., Goumnerov, B.C., Walendziewicz, C.L., Hewitson, J., Xiao, W., Mahajan-Miklos, S., Tompkins, R.G., Perkins, L.A. and Rahme, L.G. (2003) The *Drosophila melanogaster* toll pathway participates in resistance to infection by the Gram-negative human pathogen *Pseudomonas aeruginosa*. *Infection and Immunity* 71, 4059–4066.

Lazazzera, B.A., Bates, D.M. and Kiley, P.J. (1993) The activity of the *Escherichia coli* transcription factor FNR is regulated by a change in oligomeric state. *Genes and Development* 7, 1993–2005.

Lee, J.H., Lee, K.L., Yeo, W.S., Park, S.J. and Roe, J.H. (2009) SoxRS-mediated lipopolysaccharide modification enhances resistance against multiple drugs in *Escherichia coli*. *Journal of Bacteriology* 191, 4441–4450.

Lee, J.S., Heo, Y.J., Lee, J.K. and Cho, Y.H. (2005) KatA, the major catalase, is critical for osmoprotection and virulence in *Pseudomonas aeruginosa* PA14. *Infection and Immunity* 73, 4399–4403.

Lehmann, K.B. and Neumann, R. (1896) *Atlas und grund-riss der bakteriotogie und lehrbuch der specietlen bakteriologischen diagnostik*. 1 Auflage, Teil 11. J.F. Lehmann, Munchen, pp. 1–448.

Lepoivre, M., Flaman, J.M., Bobe, P., Lemaire, G. and Henry, Y. (1994) Quenching of the tyrosyl free radical of ribonucleotide reductase by nitric oxide. Relationship to cytostasis induced in tumor cells by cytotoxic macrophages. *Journal of Biological Chemistry* 269, 21891–21897.

Liochev, S.I. and Fridovich, I. (1992) Fumarase C, the stable fumarase of *Escherichia coli*, is controlled by the *soxRS* regulon. *Proceedings of the National Academy of Sciences of the United States of America* 89, 5892–5896.

Liochev, S.I. and Fridovich, I. (1993) Modulation of the fumarases of *Escherichia coli* in response to oxidative stress. *Archives in Biochemistry and Biophysics* 301, 379–384.

Liochev, S.I. and Fridovich, I. (1994) The role of $O2^{\cdot-}$ in the production of HO$\cdot$: *in vitro* and *in vivo*. *Free Radical Biology in Medicine* 16, 29–33.

Liu, G., Zhou, J., Fu, Q.S. and Wang, J. (2009) The *Escherichia coli* azoreductase AzoR is involved in resistance to thiol-specific stress caused by electrophilic quinones. *Journal of Bacteriology* 191, 6394–6400.

Loewen, P.C. and Hengge-Aronis, R. (1994) The role of the sigma factor sigma S (KatF) in bacterial global regulation. *Annual Reviews in Microbiology* 48, 53–80.

Loewen, P.C., Klotz, M.G. and Hassett, D.J. (2000) Catalase – an 'old' enzyme that continues to surprise us. *American Society of Microbiology News* 66, 76–82.

Loewen, T.C., Switala, J. and Triggs-Raine, B.L. (1985) Catalases HPI and HPII in *Escherichia coli* are induced independently. *Archives in Biochemistry and Biophysics* 243, 144–149.

Ma, J.-F., Ochsner, U.A., Klotz, M.G., Nanayakkara, V.K., Howell, M.L., Johnson, Z., Posey, J., Vasil, M.L., Monaco, J.J. and Hassett, D.J. (1999) Bacterioferritin A modulates catalase A (KatA) activity and resistance to hydrogen peroxide in *Pseudomonas aeruginosa*. *Journal of Bacteriology* 181, 3730–3742.

Martin, D.W., Schurr, M.J., Mudd, M.H., Govan, J.R., Holloway, B.W. and Deretic, V. (1993) Mechanism of conversion to mucoidy in *Pseudomonas aeruginosa* infecting cystic fibrosis patients. *Proceedings of the National Academy of Sciences of the United States of America* 90, 8377–8381.

Masse, E. and Gottesman, S. (2002) A small RNA regulates the expression of genes involved in iron metabolism in *Escherichia coli*. *Proceedings* of the National Academy of Sciences of the United States of America 99, 4620–4625.

Membrillo-Hernandez, J., Coopamah, M.D., Channa, A., Hughes, M.N. and Poole, R.K. (1998) A novel mechanism for upregulation of the Escherichia coli K-12 hmp (flavo-haemoglobin) gene by the 'NO releaser', S-nitrosoglutathione: nitrosation of homocysteine and modulation of MetR binding to the glyA-hmp intergenic region. Molecular Microbiology 29, 1101–1112.

Messner, K.R. and Imlay, J.A. (1999) The identification of primary sites of superoxide and hydrogen peroxide formation in the aerobic respiratory chain and sulfite reductase complex of Escherichia coli. Journal of Biological Chemistry 274, 10119–10128.

Messner, K.R. and Imlay, J.A. (2002) Mechanism of superoxide and hydrogen peroxide formation by fumarate reductase, succinate dehydrogenase, and aspartate oxidase. Journal of Biological Chemistry 277, 42563–42571.

Migula, W. (1900) System der Backterien, Vol 2. Gustav Fisher, Jena.

Miyata, S., Casey, M., Frank, D.W., Ausubel, F.M. and Drenkard, E. (2003) Use of the Galleria mellonella caterpillar as a model host to study the role of the type III secretion system in Pseudomonas aeruginosa pathogenesis. Infection and Immunity 71, 2404–2413.

Mollace, V., Salvemini, D., Riley, D.P., Muscoli, C., Iannone, M., Granato, T., Masuelli, L., Modesti, A., Rotiroti, D., Nistico, R., Bertoli, A., Perno, C.F. and Aquaro, S. (2002) The contribution of oxidative stress in apoptosis of human-cultured astroglial cells induced by supernatants of HIV-1-infected macrophages. Journal of Leukocyte Biology 71, 65–72.

Morgan, R., Kohn, S., Hwang, S.H., Hassett, D.J. and Sauer, K. (2006) BdlA, a chemotaxis regulator essential for biofilm dispersion in Pseudomonas aeruginosa. Journal of Bacteriology 188, 7335–7343.

Mossialos, D., Tavankar, G.R., Zlosnik, J.E. and Williams, H.D. (2006) Defects in a quinol oxidase lead to loss of KatC catalase activity in Pseudomonas aeruginosa: KatC activity is temperature dependent and it requires an intact disulphide bond formation system. Biochemical and Biophysical Research Communications 341, 697–702.

Nachin, L., Loiseau, L., Expert, D. and Barras, F. (2003) SufC: an unorthodox cytoplasmic ABC/ATPase required for [Fe-S] biogenesis under oxidative stress. Embo Journal 22, 427–437.

Nielands, J.B. (1993) Siderophores. Archives in Biochemistry and Biophysics 302, 1–3.

Ochsner, U.A., Vasil, M.L., Alsabbagh, E., Parvatiyar, K. and Hassett, D.J. (2000) Role of the Pseudomonas aeruginosa oxyR-recG operon in oxidative stress defense and DNA repair: OxyR-dependent regulation of katB, ahpB, and ahpCF. Journal of Bacteriology 182, 4533–4544.

Ochsner, U.A., Hassett, D.J. and Vasil, M.L. (2001) Genetic and physiological characterization of ohr, encoding a protein involved in organic hydroperoxide resistance in Pseudomonas aeruginosa. Journal of Bacteriology 183, 773–778.

Ochsner, U.A., Wilderman, P.J., Vasil, A.I. and Vasil, M.L. (2002) GeneChip expression analysis of the iron starvation response in Pseudomonas aeruginosa: identification of novel pyoverdine biosynthesis genes. Molecular Microbiology 45, 1277–1287.

Ormerod, L.D., Gomez, D.S., Schanzlin, D.J. and Smith, R.E. (1988) Chronic alcoholism and microbial keratitis. Brazilian Journal of Ophthalmology 72, 155–159.

Panmanee, W., Gomez, F., Witte, D., Vijay, P., Britigan, B.E. and Hassett, D.J. (2008) The peptidoglycan-associated lipoprotein, OprL, helps protect a Pseudomonas aeruginosa mutant devoid of the transactivator, OxyR, from hydrogen peroxide-mediated killing during planktonic and biofilm culture. Journal of Bacteriology 190, 3658–3669.

Park, S., You, X. and Imlay, J.A. (2005) Substantial DNA damage from submicromolar intracellular hydrogen peroxide detected in Hpx-mutants of Escherichia coli. Proceedings of the National Academy of Sciences of the United States of America 102, 9317–9322.

Partridge, J.D., Poole, R.K. and Green, J. (2007) The Escherichia coli yhjA gene, encoding a predicted cytochrome c peroxidase, is regulated by FNR and OxyR. Microbiology 153, 1499–1507.

Pereira, D.B., Antoni, M.H., Danielson, A., Simon, T., Efantis-Potter, J., Carver, C.S., Duran, R.E., Ironson, G., Klimas, N. and O'Sullivan, M.J. (2003) Life stress and cervical squamous intraepithelial lesions in women with human papillomavirus and human immunodeficiency virus. Psychosomology Medicine 65, 427–434.

Pesci, E.C., Pearson, J.P., Seed, P.C. and Iglewski, B.H. (1997) Regulation of las and rhl quorum sensing in Pseudomonas aeruginosa. Journal of Bacteriology 179, 3127–3132.

Platt, M.D., Schurr, M.J., Sauer, K., Vazquez, G., Kukavica-Ibrulj, I., Potvin, E., Levesque, R.C., Fedynak, A., Brinkman, F.S., Schurr, J., Hwang, S.H., Lau, G.W., Limbach, P.A., Rowe, J.J.,

Lieberman, M.A., Barraud, N., Webb, J., Kjelleberg, S., Hunt, D.F. and Hassett, D.J. (2008) Proteomic, microarray, and signature-tagged mutagenesis analyses of anaerobic *Pseudomonas aeruginosa* at pH 6.5, likely representing chronic, late-stage cystic fibrosis airway conditions. *Journal of Bacteriology* 190, 2739–2758.

Polack, B., Dacheux, D., Delic-Attree, I., Toussaint, B. and Vignais, P.M. (1996) Role of manganese superoxide dismutase in a mucoid isolate of *Pseudomonas aeruginosa*: adaptation to oxidative stress. *Infection and Immunity* 64, 2216–2219.

Pomposiello, P.J. and Demple, B. (2000) Identification of SoxS-regulated genes in *Salmonella enterica* serovar Typhimurium. *Journal of Bacteriology* 182, 23–29.

Poole, L.B. (2005) Bacterial defenses against oxidants: mechanistic features of cysteine-based peroxidases and their flavoprotein reductases. *Archives in Biochemistry and Biophysics* 433, 240–254.

Poole, R.K., Anjum, M.F., Membrillo-Hernandez, J., Kim, S.O., Hughes, M.N. and Stewart, V. (1996) Nitric oxide, nitrite, and Fnr regulation of hmp (flavohemoglobin) gene expression in *Escherichia coli* K-12. *Journal of Bacteriology* 178, 5487–5492.

Rahme, L.G., Ausubel, F.M., Cao, H., Drenkard, E., Goumnerov, B.C., Lau, G.W., Mahajan-Miklos, S., Plotnikova, J., Tan, M.W., Tsongalis, J., Walendziewicz, C.L. and Tompkins, R.G. (2000) Plants and animals share functionally common bacterial virulence factors. *Proceedings of the National Academy of Sciences of the United States of America* 97, 8815–8821.

Rice, S.A., Tan, C.H., Mikkelsen, P.J., Kung, V., Woo, J., Tay, M., Hauser, A., McDougald, D., Webb, J.S. and Kjelleberg, S. (2009) The biofilm life cycle and virulence of *Pseudomonas aeruginosa* are dependent on a filamentous prophage. *ISME Journal* 3, 271–282.

Rogers, P.A., Eide, L., Klungland, A. and Ding, H. (2003) Reversible inactivation of *E. coli* endonuclease III via modification of its [4Fe-4S] cluster by nitric oxide. *DNA Repair (Amst)* 2, 809–817.

Rompf, A., Hungerer, C., Hoffmann, T., Lindenmeyer, M., Romling, U., Gross, U., Doss, M.O., Arai, H., Igarashi, Y. and Jahn, D. (1998) Regulation of *Pseudomonas aeruginosa hemF* and *hemN* by the dual action of the redox response regulators Anr and Dnr. *Molecular Microbiology* 29, 985–997.

Schroeter, J. (1872) Ueber einige durch bacterien gebildete Pigmente, In: Cohn, F. (ed.) *Beitrage zur Biologie der Pflanzen. 1, Heft* 2, pp. 109–126.

Schuster, M., Hawkins, A.C., Harwood, C.S. and Greenberg, E.P. (2004) The *Pseudomonas aeruginosa* RpoS regulon and its relationship to quorum sensing. *Molecular Microbiology* 51, 973–985.

Seaver, L.C. and Imlay, J.A. (2001a) Alkyl hydroperoxide reductase is the primary scavenger of endogenous hydrogen peroxide in *Escherichia coli. Journal of Bacteriology* 183, 7173–7181.

Seaver, L.C. and Imlay, J.A. (2001b) Hydrogen peroxide fluxes and compartmentalization inside growing *Escherichia coli. Journal of Bacteriology* 183, 7182–7189.

Shah, I.M. and Wolf, R.E. Jr (2006) Inhibition of Lon-dependent degradation of the *Escherichia coli* transcription activator SoxS by interaction with 'soxbox' DNA or RNA polymerase. *Molecular Microbiology* 60, 199–208.

Shin, D.H., Choi, Y.S. and Cho, Y.H. (2008) Unusual properties of catalase A (KatA) of *Pseudomonas aeruginosa* PA14 are associated with its biofilm peroxide resistance. *Journal of Bacteriology* 190, 2663–2670.

Singh, R., Wiseman, B., Deemagarn, T., Jha, V., Switala, J. and Loewen, P.C. (2008) Comparative study of catalase-peroxidases (KatGs). *Archives in Biochemistry and Biophysics* 471, 207–214.

Sobota, J.M. and Imlay, J.A. (2011) Iron enzyme ribulose-5-phosphate 3-epimerase in *Escherichia coli* is rapidly damaged by hyrdogen peroxide but can be protected by mangenese. *Proceedings of the National Academy of Science of the United States of America* 108, 5402–5407.

Spiro, S. (2007) Regulators of bacterial responses to nitric oxide. *FEMS Microbiology Reviews* 31, 193–211.

Takahashi, Y. and Tokumoto, U. (2002) A third bacterial system for the assembly of iron-sulfur clusters with homologs in archaea and plastids. *Journal of Biological Chemistry* 277, 28380–28383.

Tan, M.W. and Ausubel, F.M. (2000) *Caenorhabditis elegans*: a model genetic host to study *Pseudomonas aeruginosa* pathogenesis. *Current Opinion in Microbiology* 3, 29–34.

Tan, M.W., Mahajan-Miklos, S. and Ausubel, F.M. (1999) Killing of *Caenorhabditis elegans* by *Pseudomonas aeruginosa* used to model mammalian bacterial pathogenesis. *Proceedings of the National Academy of Sciences of the United States of America* 96, 715–720.

Thomasa, L.D., Dunkleyb, M.L., Moorea, R., Reynoldsa, S., Bastina, D.A., Kyda, J.M. and

Crippsa, A.W. (2000) Catalase immunization from *Pseudomonas aeruginosa* enhances bacterial clearance in the rat lung. *Vaccine* 19, 348–357.

Thompson, R.H. (1987) *Naturally Occurring Quinones III. Recent Advances.* Chapman and Hall, London.

Thompson, R.L., Preston, C.M. and Sawtell, N.M. (2009) *De novo* synthesis of VP16 coordinates the exit from HSV latency *in vivo. PLoS Pathogens* 5, e1000352.

Trevisan, V. (1885) Caratteri di alcuni nuovi generi dibatteriacee. *Atti della Accademia Fisio-Medico-Statistica in Milano, Serie 4* 2, 92–106.

Trunk, K., Enkert, B., Quäck, N., Münch, R., Scheer, M., Garbe, J., Jänsch, L., Trost, M., Wehland, J., Buer, J., Jahn, M., Schobert, M. and Jahn, D. (2010) Anaerobic adaptation in *Pseudomonas aeruginosa*: definition of the Anr and Dnr regulons. *Environmental Microbiology* 12, 1719–1733.

Tucker, N., D'Autreaux, B., Spiro, S. and Dixon, R. (2005) DNA binding properties of the *Escherichia coli* nitric oxide sensor NorR: towards an understanding of the regulation of flavorubredoxin expression. *Biochemical Society Transactions* 33, 181–183.

Tucker, N.P., Hicks, M.G., Clarke, T.A., Crack, J.C., Chandra, G., Le Brun, N.E., Dixon, R. and Hutchings, M.I. (2008) The transcriptional repressor protein NsrR senses nitric oxide directly via a [2Fe-2S] cluster. *PLoS One* 3, e3623.

Ugidos, A., Morales, G., Rial, E., Williams, H.D. and Rojo, F. (2008) The coordinate regulation of multiple terminal oxidases by the *Pseudomonas putida* ANR global regulator. *Environmental Microbiology* 10, 1690–1702.

Varghese, S., Tang, Y. and Imlay, J.A. (2003) Contrasting sensitivities of *Escherichia coli* aconitases A and B to oxidation and iron depletion. *Journal of Bacteriology* 185, 221–230.

Varghese, S., Wu, A., Park, S., Imlay, K.R. and Imlay, J.A. (2007) Submicromolar hydrogen peroxide disrupts the ability of Fur protein to control free-iron levels in *Escherichia coli. Molecular Microbiology* 64, 822–830.

Vento, S., Cainelli, F. and Temesgen, Z. (2008) Lung infections after cancer chemotherapy. *Lancet Oncology* 9, 982–992.

Vinckx, T., Matthijs, S. and Cornelis, P. (2008) Loss of the oxidative stress regulator OxyR in *Pseudomonas aeruginosa* PAO1 impairs growth under iron-limited conditions. *FEMS Microbiology Letters* 288, 258–265.

Wang, F.D., Lin, M.L. and Liu, C.Y. (2005)

Bacteremia in patients with hematological malignancies. *Chemotherapy* 51, 147–153.

Webb, J.S., Thompson, L.S., James, S., Charlton, T., Tolker-Nielsen, T., Koch, B., Givskov, M. and Kjelleberg, S. (2003) Cell death in *Pseudomonas aeruginosa* biofilm development. *Journal of Bacteriology* 185, 4585–4592.

White, D.G., Goldman, J.D., Demple, B. and Levy, S.B. (1997) Role of the *acrAB* locus in organic solvent tolerance mediated by expression of *marA, soxS,* or *robA* in *Escherichia coli. Journal of Bacteriology* 179, 6122–6126.

Winterbourn, C.C. and Metodiewa, D. (1999) Reactivity of biologically important thiol compounds with superoxide and hydrogen peroxide. *Free Radical Biology in Medicine* 27, 322–328.

Winterbourn, C.C., Hampton, M.B., Livesey, J.H. and Kettle, A.J. (2006) Modeling the reactions of superoxide and myeloperoxidase in the neutrophil phagosome: implications for microbial killing. *Journal of Biological Chemistry* 281, 39860–39869.

Woodmansee, A.N. and Imlay, J.A. (2002) Reduced flavins promote oxidative DNA damage in non-respiring *Escherichia coli* by delivering electrons to intracellular free iron. *Journal of Biological Chemistry* 277, 34055–34066.

Ye, R.W., Haas, D., Ka, J.-O., Krishnalillai, V., Zimerman, A., Baird, C. and Tiedje, J.M. (1995) Anaerobic activation of the entire denitrification pathway in *Pseudomonas aeruginosa* requires Anr, an analog of Fnr. *Journal of Bacteriology* 177, 3606–3609.

Yoon, S.S., Hennigan, R.F., Hilliard, G.M., Ochsner, U.A., Parvatiyar, K., Kamani, M.C., Allen, H.L., DeKievit, T.R., Gardner, P.R., Schwab, U., Rowe, J.J., Iglewski, B.H., McDermott, T.R., Mason, R.P., Wozniak, D.J., Hancock, R.E., Parsek, M.R., Noah, T.L., Boucher, R.C. and Hassett, D.J. (2002) *Pseudomonas aeruginosa* anaerobic respiration in biofilms: relationships to cystic fibrosis pathogenesis. *Development Cell* 3, 593–603.

Yoon, S.S., Coakley, R., Lau, G.W., Lymar, S.V., Gaston, B., Karabulut, A.C., Hennigan, R.F., Hwang, S.H., Buettner, G., Schurr, M.J., Mortensen, J.E., Burns, J.L., Speert, D., Boucher, R.C. and Hassett, D.J. (2006) Anaerobic killing of mucoid *Pseudomonas aeruginosa* by acidified nitrite derivatives under cystic fibrosis airway conditions. *Journal Clinical Investigation* 116, 436–446.

Yoon, S.S., Karabulut, A.C., Lipscomb, J.D., Hennigan, R.F., Lymar, S.V., Groce, S.L., Herr, A.B., Howell, M.L., Kiley, P.J., Schurr, M.J.,

Gaston, B., Choi, K.H., Schweizer, H.P. and Hassett, D.J. (2007) Two-pronged survival strategy for the major cystic fibrosis pathogen, *Pseudomonas aeruginosa*, lacking the capacity to degrade nitric oxide during anaerobic respiration. *EMBO Journal* 26, 3662–3672.

Yukl, E.T., Elbaz, M.A., Nakano, M.M. and Moenne-Loccoz, P. (2008) Transcription factor NsrR from *Bacillus subtilis* senses nitric oxide with a 4Fe-4S cluster. *Biochemistry* 47(49), 13084–13092.

Zheng, M., Åslund, F. and Storz, G. (1998) Activation of the OxyR transcription factor by reversible disulfide bond formation. *Science* 279, 1718–1721.

Zheng, M., Doan, B., Schneider, T.D. and Storz, G. (1999) OxyR and SoxRS regulation of *fur*. *Journal of Bacteriology* 181, 4639–4643.

Zheng, M., Wang, X., Doan, B., Lewis, K.A., Schneider, T.D. and Storz, G. (2001a) Computation-directed identification of OxyR DNA binding sites in *Escherichia coli*. *Journal of Bacteriology* 183, 4571–4579.

Zheng, M., Wang, X., Templeton, L.J., Smulski, D.R., LaRossa, R.A. and Storz, G. (2001b) DNA microarray-mediated transcriptional profiling of the *Escherichia coli* response to hydrogen peroxide. *Journal of Bacteriology* 183, 4562–4570.

Zopf, W. (1884a) *Die spaltpilze*. 2 Auflage. E. Trewendt, Breslau, Poland, pp. 1–98.

Zopf, W. (1884b) *Die Spaltpilze*. 2 Auflage. E. Trewendt, Breslau, Poland, 98 pp.

Zumft, W.G. (1997) Cell biology and molecular basis of denitrification. *Microbiology Molecular Biology Reviews* 61, 533–616.

2 Coordinated Regulation of Stress and Virulence Adaptations in Stages of *Haemophilus* Pathogenesis

Sandy M.S. Wong and Brian J. Akerley*

2.1 Summary

The ability to sense and adapt to environmental conditions is an important survival strategy for bacteria. *Haemophilus influenzae* is a respiratory pathogen of humans and transits between distinct niches in the host, including the airway surface, middle ear and bloodstream; environments that are known to expose bacteria to diverse physiological conditions and stresses.

Recent evidence supports a model in which coordinated regulation of gene expression in response to changes in oxygen levels allows *H. influenzae* to control appropriately genes needed to evade and resist host defences. The *H. influenzae* ArcAB (anoxic redox control) and FNR (formate-dependent nitrite reductase regulator) systems control oxidative and nitrosative stress defence mechanisms in response to reduction or oxidation (redox) signals and are most active under anaerobic environmental conditions. The ArcAB two-component signal transduction system activates a ferritin-like gene required for pre-emptive defence against transition from low to high oxygen conditions and the associated oxidative stress, a transition that *H. influenzae* encounters on exposure to phagocytic cells. Such stress defences are likely to be important in sites of colonization such as the bloodstream, in which ArcA has been implicated in virulence in animal models. FNR activates a di-iron protein that defends against exogenous nitric oxide (NO) donors and NO generated by macrophages activated by interferon-gamma under anaerobiosis. This nitrosative stress defence strategy could protect *H. influenzae* after cytokine production in inflammation where it is likely to encounter activated macrophages. In the lung, *H. influenzae* employs alternative oxidative stress defence mechanisms appropriate to a highly aerobic environment. A high-throughput methodology that utilized whole-genome transposon mutagenesis in conjunction with massively parallel deep sequencing identified the *H. influenzae* genes required for survival in the mouse lung. Products of many of these genes protect against oxidative stress, including *oxyR*, global regulator of antioxidant defences, an *oxyR*-regulated glutathione-dependent peroxidase and DNA recombination and repair genes. Genes mediating carbohydrate extension of the lipopolysaccharide (LPS) are also required in the lung model, but not to the extent that they are needed for bacteraemia, consistent with the observation that *H. influenzae* elaborates lower levels of

* Corresponding author.

certain LPS structures under high oxygen conditions than in microaerobiosis. Thus, *H. influenzae*'s defences against stressors encountered during infection and its production of surface structures involved in immune evasion are modulated in response to redox conditions. Moreover, the requirement for these bacterial adaptations in distinct sites of pathogenesis is consistent with their patterns of redox-responsive regulation. Together, this information supports the view that *H. influenzae* senses varied oxygen levels in diverse sites of infection to express the repertoire of genes needed for optimal growth and survival during pathogenesis.

2.2 Introduction

H. influenzae is a Gram-negative bacterium that has no identified natural niche outside of the human host. It colonizes primarily the nasopharyngeal mucosa, but can also disseminate to other sites to cause otitis media (OM), upper and lower respiratory tract infections, septicaemia and meningitis (Klein, 1997; Moxon and Murphy, 2000). During infection, *H. influenzae* likely encounters environmental niches of varying redox levels as it transits between the relatively high oxygen levels of the airway surface to sites lower in oxygen such as an interstitial location during traversal of the mucosal epithelium, spread to the middle ear, or entry into the bloodstream. Therefore, modulation of gene expression in response to varying oxygen levels in diverse environments represents a potential mechanism by which *H. influenzae* could coordinate gene expression profiles needed for efficient colonization and pathogenesis.

Consistent with this hypothesis, *H. influenzae* utilizes redox-responsive regulatory systems to sense low oxygen environments, and recent studies indicate that this response influences its pathogenic potential. *H. influenzae* contains two global transcriptional regulatory systems, ArcAB and FNR, which are similar to their respective homologues in *Escherichia coli* (Shaw *et al.*, 1983; Lynch and Lin, 1996b) and which modulate the expression of the genes needed for adaptation

to changes in oxygen availability. In this chapter, we discuss evidence for a role of redox sensing in *H. influenzae* pathogenesis. The ArcAB system directs a strategy to protect the cell from oxidative stress in a pre-emptive manner during transition from low to high oxygen conditions, a role previously not described for ArcAB in bacteria possessing this system (Wong *et al.*, 2007). FNR is needed for defence of *H. influenzae* against nitrosative stress under anaerobic conditions (Harrington *et al.*, 2009). The regulatory effects of ArcA could protect cells growing under relatively low oxygen levels, a condition that *H. influenzae* likely encounters in venous blood, against sudden oxidative stresses such as exposure to the oxidative defences of phagocytic cells. Resistance to reactive nitrogen species (RNS) is also likely to be an important adaptation for *H. influenzae* in the nasopharynx, the primary colonization site for this organism. NO-producing cells such as macrophages and epithelial cells are abundant in human sinuses and, in fact, NO levels are normally very high in nasal sinuses, implicating it in local host defence (Lundberg *et al.*, 1995; Lundberg, 2008).

In parallel with stress adaptations, cell surface structures involved in pathogenesis show a similar mode of regulation. *H. influenzae* can respond to different levels of oxygen availability to modulate the phosphorylcholine epitope display on its lipooligosaccharide (LOS), an important virulence factor of this bacterium (Moxon and Maskell, 1992). Similar to LPS of other Gram-negative bacteria, except that LOS lacks extended O antigen structures, *H. influenzae* LOS is a glycolipid consisting of lipid A, an inner core comprised of several sugars including a single 3-deoxy-D-*manno*-octulosonic acid linked to three conserved heptose residues, and a variable outer core containing a heteropolymer of glucose and galactose generally not exceeding six residues and modified with sialic acid, *N*-acetylgalactosamine and phosphorylcholine (*P*Cho) (Fig. 2.1; Hood *et al.*, 1999; Risberg *et al.*, 1999).

Global expression profiling and gene-specific RNA analysis led to the model by

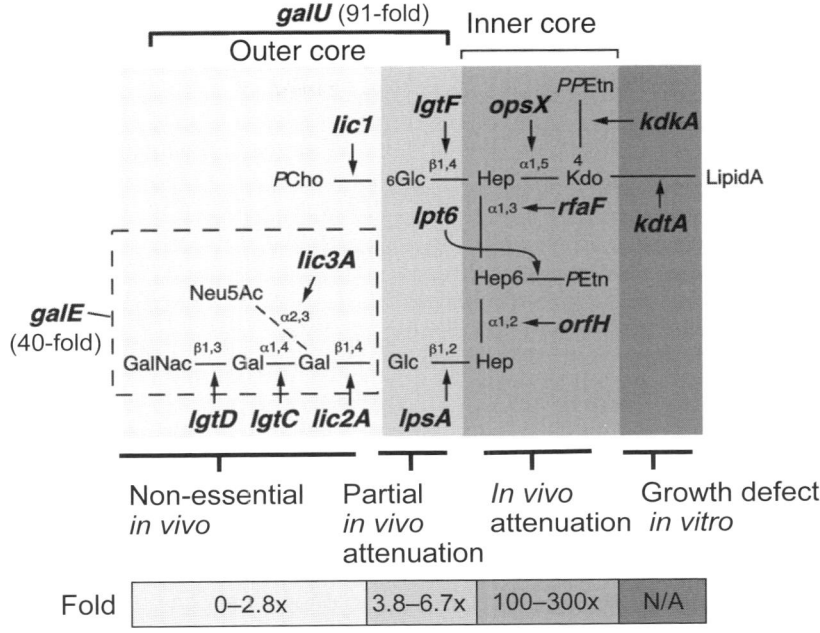

Fig. 2.1. *H. influenzae* Rd LOS biosynthesis genes required in the lung. Diagram illustrates fold-attenuation of LOS mutants in the mouse lung relative to the degree of truncation of the outer versus inner core residues (Gawronski *et al.*, 2009). Fold-attenuation was calculated from HITS data as the ratio of *mariner* transposon insertions detected in each gene before *in vivo* selection divided by the number of insertions detected in the same gene within the mutant bank recovered from infected mouse lungs. Information regarding the structure of the *H. influenzae* LOS and the genes responsible for adding the specific linkages (in bold) has been described previously (Risberg *et al.*, 1999; Hood *et al.*, 2001). Gal, galactose; GalNac, *N*-acetylgalactosamine; Glc, glucose; Hep, heptose; Kdo, 2-keto-3-deoxyoctulosonic acid; Neu5Ac, sialic acid; *P*Cho, phosphorylcholine; *P*Etn, phosphoethanolamine; *PP*Etn, pyrophosphoethanolamine.

which sugar precursor abundance for hexose extensions of the LOS was decreased under high oxygen conditions and increased under microaerobic or lower oxygen conditions (Wong and Akerley, 2005). Elaboration of the LOS outer core can protect *H. influenzae* against bactericidal mechanisms such as the complement pathway, but these structures are also targets of host immunity. For example, *P*Cho can be recognized by C-reactive protein and natural antibodies (Leon and Young, 1971; Weiser *et al.*, 1998; Shaw *et al.*, 2000). Survival of the bacterium during infection is likely to involve appropriate expression of LOS structures in the context of immune effectors present in different environments, and this chapter will discuss recent evidence

to suggest niche-specific requirements for LOS modifications in relation to redox responses.

2.3 ArcAB Signal Transduction System

In bacteria, two-component signal transduction systems are employed to respond to changes in the environment. The genome of the *H. influenzae* Rd strain possesses few two-component signal transduction systems (there are four that are known) as compared to bacteria with larger genomes such as *E. coli* (where there are greater than 20) (KEGG Pathway Maps; http://www.genome.jp/kegg/

pathway.html). However, *H. influenzae* possesses an ArcAB system that has biochemical and regulatory functions similar to those of *E. coli* ArcAB in modulating gene expression in response to redox conditions of growth (Manukhov *et al.*, 2000; Georgellis *et al.*, 2001b).

In *E. coli*, the well-studied ArcAB (anoxic redox control) two-component signal transduction system responds to redox conditions of growth to modulate expression of genes involved in respiratory or fermentative metabolism (Lynch and Lin, 1996b). Under anoxic or microaerobic growth conditions, the membrane-bound ArcB sensor kinase autophosphorylates and transfers a phosphoryl group to ArcA, a DNA-binding protein that can act as either repressor or activator depending on the configuration of the target promoter (Lynch and Lin, 1996a,b). ArcB senses the availability of oxygen indirectly by monitoring the redox status of the ubiquinone:ubiquinol pool, central electron carriers of respiration (Georgellis *et al.*, 2001a; Malpica *et al.*, 2004). The activity of ArcB is dependent on the formation of intermolecular disulfide bonds via two cysteine residues (Cys-180 and Cys-241) of ArcB monomers that are oxidized by the quinone pool. Under high levels of aerobiosis, ArcB is least active, as the presence of oxygen leads to the oxidation of the quinone pool, which induces the formation of the inter-molecular disulfide bonds of the cysteine residues in ArcB. ArcB is most active under anaerobiosis as the quinone pool shifts to the reduced state and the two cysteine residues in the ArcB monomers are reduced. Recently, it has been found that the response of the *E. coli* ArcAB system to oxygen availability and to the quinone pool is dependent on both ubiquinone and menaquinone (Bekker *et al.*, 2010). This finding is relevant to *H. influenzae* biology because its genome does not contain the set of genes encoding enzymes of ubiquinone biosynthesis, but does contain genes for menaquinone biosynthesis (Fleischmann *et al.*, 1995), suggesting that the endogenous signals available to *H. influenzae* ArcB are likely to differ from those present in *E. coli*. Ubiquinone in *E. coli* functions at higher levels of oxygen supply than mena-

quinone and a model for redox responses by *E. coli* ArcB under both high (80% aerobiosis) and low oxygen (<20% aerobiosis) has been proposed (Bekker *et al.*, 2010). In *H. influenzae*, which possesses only menaquinone, ArcB should respond primarily to redox transitions occurring at low oxygen supply levels (i.e. <20% aerobiosis) consistent with available data that have detected ArcB activity in *H. influenzae* under microaerobic conditions but not in highly aerobic conditions (Manukhov *et al.*, 2000; Georgellis *et al.*, 2001b; Wong *et al.*, 2007). Along with the absence of a number of genes of the aerobic respiratory pathways that are found in *E. coli*, the metabolic profile of *H. influenzae* suggests that the organism favours a low oxygen environment (Tatusov *et al.*, 1996; Raghunathan *et al.*, 2004).

2.3.1 ArcAB in *H. influenzae* pathogenesis and oxidative stress defence

A role in pathogenesis has been implicated for *H. influenzae* ArcA in mouse models of bacteraemia as the *arcA* mutant exhibited a higher median lethal dose compared with wild-type (De Souza-Hart *et al.*, 2003) and a persistence defect in the mouse bloodstream (Wong *et al.*, 2007). ArcA mutants of other pathogens such as *Vibrio cholerae*, which causes diarrhoea in humans, and *Actino-bacillus pleuropneumoniae*, causative agent of porcine pleuropneumonias, have also shown reduced virulence in animal models (Sengupta *et al.*, 2003; Buettner *et al.*, 2008) .

The way in which ArcA participates in pathogenesis is not well understood; however, insight into potential mechanisms was obtained by genomic expression profiling under an anaerobic condition of a *H. influenzae* arcA mutant in comparison to an isogenic parental strain and a complemented strain in which the *arcA* gene was restored *in trans* (Wong *et al.*, 2007). In addition to the anticipated role of ArcA in negative control of aerobic respiration and genes of the TCA (tricarboxylic acid) cycle predicted from studies in *E. coli* (Lynch and Lin, 1996b; Patschkowski *et al.*, 2000; Oshima *et al.*, 2002; Liu and De Wulf, 2004; Salmon *et al.*, 2005), the microarray data revealed an unexpected finding of positive control of oxidative stress resistance genes (Wong *et al.*, 2007). All four genes, HI0592, *speF*, *potE* and a *dps*-like gene (HI1349) whose expression was down-regulated in the *arcA* mutant anaerobically, potentially were involved in resistance to oxidative stress. The *speF* and *potE* genes encode predicted proteins of polyamine putrescine biosynthesis and transport pathways, respectively. HI0592 is a gene of unknown function located in the same operon as *speF* and *potE*, suggesting a related function. Polyamines (e.g. putrescine, spermidine) have been implicated in a wide variety of biological processes and their optimal cellular concentrations are maintained by polyamine biosynthesis, degradation and transport (Igarashi and Kashiwagi, 1999). Polyamines have also been implicated in resistance to oxidative stress in *E. coli* and other organisms (Tkachenko *et al.*, 2001; Chattopadhyay *et al.*, 2003; Jung and Kim, 2003). HI1349 contains a conserved iron-binding domain (Ilari *et al.*, 2000; Pulliainen *et al.*, 2005) found in members of the ferritin-like Dps protein family. Dps was first discovered in *E. coli* and functions to protect DNA from hydrogen peroxide-induced oxidative damage (Almiron *et al.*, 1992). Dps proteins bind iron and oxidize Fe^{2+} with hydrogen peroxide to form a stable ferric oxide mineral core within the cavity of dodecameric shell assemblies composed of this protein, thereby avoiding generation of toxic hydroxyl radicals mediated by Fenton chemistry (Zhao *et al.*, 2002). The downregulation of *speF*, *potE* and *dps* in the *arcA* mutant was unexpected and novel in that these genes had not been documented in the literature as being part of the ArcA regulon in any other bacterial species.

The gene expression profile of the *H. influenzae arcA* mutant suggests that ArcA functions to suppress the generation of reactive oxygen intermediates (ROI). Because ArcA represses genes of aerobic respiration implicated in generation of ROI, a prediction is that the *arcA* mutant could generate more endogenous ROI compared to wild-type when shifted from low to high oxygen conditions due to increased expression of aerobic respiration genes, thereby causing increased oxidative stress during initial exposure to oxygen. Normally on transition from low to high oxygen, genes of aerobic respiration are induced in conjunction with genes involved in antioxidant defence, such that ROI toxicity is minimized. For example, in *E. coli* adaptation to exposure to air after anaerobic growth occurs several minutes following the transition and requires the expression of peroxide stress defence genes (Partridge *et al.*, 2006). Potential compounding effects of increased ROI generation in the *arcA* mutant, along with decreased expression of *speF*, *potE* and *dps*, could lead to decreased levels of resistance to oxidative stress. These observations support a model in which ArcA of *H. influenzae* participates in a mechanism to protect the bacterium during rapid transition from a low oxygen niche to an oxidative stress condition, transitions they likely experience within the mammalian host.

Support for this model was provided by the observation that the *H. influenzae arcA* mutant had a ~15-fold sensitivity to hydrogen peroxide (H_2O_2)-mediated oxidative stress compared to the parent strain when shifted to this stress condition after anaerobic growth (Wong *et al.*, 2007). As expected, no appreciable difference in sensitivity was observed between the mutant and wild-type when bacteria were grown under aerobic conditions that led to decreased ArcAB activity. Sensitivity of the *arcA* mutant towards H_2O_2 challenge could result from the combined effects of increased ROI generation by respiration, decreased cellular polyamine levels and reduced expression of the putative Dps protein. A mutant containing triple deletions in HI0592/*speF*/*potE* had only a modest effect on H_2O_2 sensitivity. This result could be attributed to the presence in *H. influenzae* of alternative genes/pathways for polyamine production such as HI0949 and HI0946 for 1,3-diaminopropane biosynthesis (Ikai and Yamamoto, 1998) and *potABCD* for potential spermidine and putrescine uptake as reported functions for the probable *E. coli* homologues (Furuchi *et al.*, 1991). In contrast, a mutation in the putative *dps* homologue (HI1349) conferred a ~9-fold increase in H_2O_2 sensitivity compared to wild-type when cells were cultured anaerobically but not aerobically, similar to that seen in the *arcA* mutant.

Consistent with the requirement for *arcA* and *dps* in facilitating resistance to oxidative stress during transition to aerobiosis from anaerobiosis, ArcA-mediated regulation of Dps in *H. influenzae* was found to be important for this phenotype. Dps was expressed constitutively in the *arcA* mutant under the transcriptional control of a promoter from the *hel* (*HI0693*) locus, which was expressed similarly in ArcA$^+$ versus ArcA$^-$ cells. Constitutive expression complemented the H$_2$O$_2$ resistance phenotype to near parental levels (Wong and Akerley, unpublished data), indicating that *dps* played a major role in ArcA regulated adaptation during transition from anaerobiosis to oxidative stress conditions.

Although the role of ArcA in anaerobic activation of oxidative stress defence genes may at first seem paradoxical, this strategy by *H. influenzae* to induce resistance mechanisms under conditions when its survival is not yet threatened could represent an important means of overcoming future unfavourable conditions. The regulatory effects of ArcA could protect cells growing with relatively low oxygen levels, a condition they likely encounter in venous blood or submucosal tissues, against sudden oxidative stresses such as exposure to the oxidative defences of phagocytic cells. Further investigation of signalling properties of the *H. influenzae* ArcAB system and its ability to control bacterial countermeasures against host defence is required to understand fully its physiological role and contribution to *H. influenzae* pathogenesis.

2.4 Role of FNR in Response to Nitrosative Stress in *H. influenzae*

In addition to the ArcAB system, *H. influenzae* also possesses the global oxygen-responsive regulator, FNR (Harrington *et al.*, 2009). In *E. coli* and other species, FNR, a cyclic AMP receptor protein/catabolite activator protein-like (CRP) transcriptional regulatory protein

participates in the transition between aerobic and anaerobic growth environments by activating and repressing the expression of multiple genes under anaerobic conditions (Shaw *et al.*, 1983). Whereas the ArcAB system senses oxygen levels indirectly, FNR senses oxygen directly via its iron-sulfur centre [4Fe-4S] which, under anaerobiosis, promotes dimerization and DNA binding (Lazazzera *et al.*, 1996). On oxidation, the iron-sulfur cluster becomes unstable, leading to the formation of inactive FNR monomers and inhibition of DNA binding (Crack *et al.*, 2004; Dibden and Green, 2005).

The role of FNR in pathogenesis may be related to a class of FNR-regulated genes of nitrosative defence identified in several organisms, including *Campylobacter jejuni* (Elvers *et al.*, 2005), *Salmonella typhimurium* (Fink *et al.*, 2007; Gilberthorpe and Poole, 2008) and *E. coli* (Poole *et al.*, 1996; Constantinidou *et al.*, 2006; Pullan *et al.*, 2007). In fact, the *E. coli* FNR can respond directly to NO (Pullan *et al.*, 2007). Moreover, *H. influenzae* FNR was shown to participate in regulation of defence against reactive nitrogen species (Harrington *et al.*, 2009). The *H. influenzae fnr* mutant is sensitive to two donors of NO, S-nitrosoglutathione (GNSO) (Singh *et al.*, 1996) and acidified nitrite (ASN) (Samouilov *et al.*, 1998) under anaerobic conditions compared to parent and complemented strains. No measurable growth inhibition differences were observed for wild-type, *fnr* mutant and complemented strains when the GNSO and ASN sensitivity assays were conducted under aerobic conditions, consistent with activity of FNR under low but not high oxygen conditions.

The mechanism of FNR-mediated resistance to NO was investigated by identifying FNR-regulated genes. A bioinformatics approach using an FNR position weighted matrix (PWM) with characterized *E. coli* FNR-binding site sequences (FNR box) was used to predict the potential FNR-regulated genes in *H. influenzae* (Tan *et al.*, 2001). One potential FNR regulon member of *H. influenzae*, HI1677 (annotated as a hypothetical gene), was of interest as it encoded a predicted protein having 57% amino acid identity to the *ytfE* gene of *E. coli*, which encoded a di-iron protein required to repair iron-sulfur centres damaged by oxidative and nitrosative stress (Justino *et al.*, 2005, 2006, 2007; Overton *et al.*, 2008). HI1677 contains a potential FNR box in its promoter region matching 13 of 14 bp to the *E. coli* FNR box (Eiglmeier *et al.*, 1989; Melville and Gunsalus, 1996) and is regulated positively by FNR in *H. influenzae* at the transcriptional level (Harrington *et al.*, 2009). A deletion mutant lacking this gene in *H. influenzae* is markedly sensitive to the two NO donors, GSNO and ASN (Harrington *et al.*, 2009). These results indicate that HI1677 (*ytfE*) functions in NO defence like the YtfE of other species and implicate the activation of this *ytfE* in *H. influenzae* as a mechanism for FNR-mediated resistance to reactive nitrogen species under anaerobic conditions. Unlike other species such as *E. coli* in which FNR regulates *ytfE* negatively (Justino *et al.*, 2006), *H. influenzae* FNR mediates positive control of *ytfE* expression. The difference in regulation of nitrosative stress may reflect that these bacteria have very different life cycles, the intestine versus the human respiratory tract, and are likely to encounter NO in different contexts.

Release of NO by activated macrophages provides a defence against invading pathogens and is thought to be particularly important under anaerobic conditions in which NO is stable and reactive oxygen species are less effective (MacMicking *et al.*, 1997; Bogdan *et al.*, 2000). For example, NO was shown to be bacteriostatic against *E. coli* under fully anaerobic conditions by damaging iron-sulfur centres of branched-chain amino acid synthesis enzymes (Ren *et al.*, 2008). Phagocytic cells are critical to host defence against *H. influenzae* infections (Toews *et al.*, 1985; Bakaletz *et al.*, 1987; Noel *et al.*, 1992; Sun *et al.*, 2000; Winter and Barenkamp, 2006) and recent evidence shows that FNR and YtfE are required by *H. influenzae* in nitrosative defence against macrophage-generated NO (Harrington *et al.*, 2009). Specifically, FNR and YtfE were required for the ability of *H. influenzae* to resist antibacterial effects of bone marrow-derived murine macrophages

stimulated with interferon-gamma (IFN-γ) or LPS, which trigger inducible nitric oxide synthase (iNOS) expression and thus NO production (Lorsbach and Russell, 1992). In addition, treatment of activated macrophages with N_ω-nitro-L-arginine methyl ester HCL (L-NAME), a competitive inhibitor of NO synthases and thus inhibition of NO production, eliminated the survival defect of the *ytfE* mutant, which was now able to resist bacteriostatic effects comparable to wild-type levels (Harrington *et al.*, 2009).

These studies indicate that *H. influenzae* FNR and the FNR-regulated *ytfE* gene promote resistance to NO under anaerobic conditions. The *fnr* and *ytfE* genes are likely required to resist the antimicrobial effects of reactive nitrogen species generated by host immune cells during infection. This represents a strategy available to *H. influenzae* for defence against nitrosative stress under low oxygen conditions. A distinct pathway was implicated in nitrosative stress defence under aerobic conditions. NmlR of *H. influenzae*, a member of the MerR-like family of transcriptional regulators that respond to environmental signals including oxidative stress (Brown *et al.*, 2003) and an *nmlR*-regulated gene, *adhC*, encoding a glutathione-dependent alcohol dehydrogenase were shown to be required for nitrosative stress defence against GSNO under aerobic conditions (Kidd *et al.*, 2007). The FNR system expands the capability of *H. influenzae* to negotiate its environment as it may experience low oxygen levels during invasion into submucosal sites or in the bloodstream. FNR appears to be important for pathogenesis by diverse bacteria as *fnr* mutants of *S. enterica* serovar Typhimurium (Fink *et al.*, 2007), *Neisseria meningitides* (Bartolini *et al.*, 2006), *A. pleuropneumoniae* (Baltes *et al.*, 2005) and *Bordetella pertussis* (Wood *et al.*, 1998) are attenuated in animal models of infection. Regulatory targets and their patterns of regulation by FNR differ in *H. influenzae* compared to other bacteria, and it will be of interest to investigate the extent of the involvement of the *H. influenzae* FNR regulon in virulence gene regulation required for establishing infection and evading host defences.

2.5 Global Analysis of *H. influenzae* Genes Required in a Mouse Lung Environment

A better understanding of adaptive responses by bacteria to their environments *in vivo* would benefit from more complete knowledge of genes needed in different stages of infection. Information applicable to this gap in knowledge was obtained recently by application of a methodology termed HITS (High-throughput Insertion Tracking by Deep Sequencing) to identify the genes required for lung colonization/persistence in *H. influenzae* (Gawronski *et al.*, 2009). This approach exploits a negative selection strategy to identify on a genomic scale the genes essential for growth or survival under a condition of interest *in vitro* or during infection in a model host. Identification of these genes is based on the relative decrease in abundance of mutants deficient in specific genes in a large mutant library subjected to selection in the host in comparison to pre-selective growth conditions (i.e. *in vitro*). The power of the HITS technology derives from its ability to query the entire genome rapidly and comprehensively to determine the relative importance of each gene in bacterial survival in a host environment. The profile of genes implicated in bacterial survival in a particular host site provides knowledge about the selection conditions that the bacterium must overcome within that site.

The model that emerges from analysis of *H. influenzae* genes needed in the lung is that this presumably aerobic location exposes *H. influenzae* to a variety of stressors, including nutrient limitation and ROI (Gawronski *et al.*, 2009). Under aerobic conditions bacteria need to cope with oxygen toxicity that results from the formation of reactive intermediates such as hydrogen peroxide, superoxide and hydroxyl radicals, all by-products of aerobic metabolism which can damage proteins, nucleic acids and cell membranes (Storz and Imlay, 1999; Imlay, 2003). Pathogenic bacteria face additional challenges from exogenous oxidants produced by phagocytes as part of innate immune responses. DNA recombination and repair genes, specifically *recC*, *ruvA* and *ruvC*, which have been shown to be protective against oxidative stress caused by hydrogen peroxide in *N. gonorrhoeae* (Stohl and Seifert, 2006), were implicated for survival adaptation in this lung model. The *pgdX* gene encoding a glutathione-dependent peroxidase (Pauwels *et al.*, 2003) was also required in the lung, implicating metabolism of reactive oxygen species (Gawronski *et al.*, 2009).

In addition to genes involved directly in protection from oxidative stress, the oxidative stress responsive regulator, OxyR, was required, indicating that the ability of *H. influenzae* to sense and respond to these conditions was important in the lung model (Gawronski *et al.*, 2009). OxyR, a global transcriptional regulator of antioxidant mechanisms, is found in many bacterial species (Pomposiello and Demple, 2001). Under normal, favourable growth conditions, the two redox-active cysteine residues of the regulatory domain of OxyR are maintained in the reduced state by the dithiol protein glutaredoxin in the presence of glutathione. However, on exposure to oxidative stress such as exposure to hydrogen peroxide, these two cysteines are oxidized, forming a disulfide bond. The active oxidized form of OxyR binds to promoter regions of target genes and activates transcription. The OxyR regulon in *H. influenzae* includes a class of genes that are upregulated in response to hydrogen peroxide exposure (Harrison *et al.*, 2007) and have known or probable roles in detoxifying reactive oxygen species including *hktE* encoding catalase (Bishai *et al.*, 1994b), *pgdX* encoding glutathione-dependent peroxidase (Pauwels *et al.*, 2003), *dps* encoding an iron-binding protein (Almiron *et al.*, 1992) and *pnt* encoding NADP transhydrogenase (Brumaghim *et al.*, 2003; Sauer *et al.*, 2004). Mutational inactivation of the most highly regulated gene of the OxyR regulon, *hktE* (Harrison *et al.*, 2007), which was very sensitive to exogenous H_2O_2 exposure (Bishai *et al.*, 1994a), did not cause attenuation in colonization in the mouse lung (Gawronski *et al.*, 2009). These results are consistent with an earlier report of the lack of a virulence defect of a *hktE* mutant in the nasal passage of infant rats (Bishai *et al.*, 1994a). Of the genes known to be within the *H. influenzae* OxyR regulon, only *pdgX* was required in the lung. Perhaps in the lung there is some redundancy in defence systems that are used in resistance against oxidative stress. Consistent with the model that *H. influenzae* lung infection primarily involves aerobic adaptations at 24 h, the anaerobic regulator ArcA is not required at this site, thus reinforcing the concept that oxidative stress defences are highly context dependent during infection.

In contrast to the aerobic environment implicated in the lung model, results from models of bacteraemia support the view that bloodstream colonization by *H. influenzae* requires microaerobic adaptations. As discussed above, the anaerobic regulator ArcA was required in models of bacteraemia (Wong *et al.*, 2007), but not in the lung (Gawronski *et al.*, 2009). Moreover, the OxyR-regulated *pgdX* gene in *H. influenzae* was not required in an infant rat model of bacteraemia (Vergauwen *et al.*, 2006). Similarly, the superoxide dismutase (*sodA*) mutant in *H. influenzae* participates in oxidative stress defence *in vitro* and colonization in the nasopharynx, but not in the bloodstream in infant rats (D'Mello *et al.*, 1997). Collectively, these observations indicate that survival of *H. influenzae* in the lung requires a greater degree of oxidative stress adaptations compared to the bloodstream and that different stress defence strategies are used by *H. influenzae* within these two environments.

Results with HITS in the lung in conjunction with previous studies provide support for the concept that *H. influenzae* responds to aeration conditions not only to resist sources of oxidative stress, but also to control genes involved in immune evasion during infection. Components of the bacterial cell surface are frequently the most direct participants in host–pathogen interactions. Our laboratory previously obtained evidence that redox conditions influenced levels of carbohydrate structural modification of cell surface LOS. Specifically, high oxygen conditions led to a decrease in the levels of LOS carbohydrate extensions in comparison to microaerobic conditions (Wong and Akerley, 2005). LOS is essential in models of *H. influenzae* pathogenesis in the middle-ear and bloodstream and contributes to numerous aspects of non-typeable *H. influenzae* (NTHi) infection including evasion of complement and antimicrobial peptides (Lysenko *et al.*, 2000; Figueira *et al.*, 2007; Ho *et al.*, 2007). Nevertheless, LOS structures are also targets for components of innate and acquired immunity, and downregulation of these structures may be beneficial in sites of infection in which they are not essential for bacterial survival. HITS data provided an indication of the relative importance of different LOS biosynthesis genes in the lung, expressed as a fold-attenuation (f.a.) representing the ratio of mutants corresponding to a particular gene that were detected before infection to those detected after selection in the lung (e.g. input/output; as shown in Fig. 2.1). High f.a. values indicate a greater defect in this model. Genes needed for extension of the LOS inner core structures (*opsX*, *rfaF* and *orfH*; f.a. 100–300 times) were most important *in vivo* in the lung, as were genes required for precursor production for LOS carbohydrate outer core hexose extensions (*galU* and *galE*) (f.a. 91 times and 40 times, respectively). The *galU* encoded enzyme converts glucose-1-phosphate to uridine diphosphate glucose (UDP-glucose), the substrate for glucose residue addition to the LOS. A *galU* mutant of *H. influenzae* does not contain hexose in the outer core (Wong and Akerley, unpublished data). The enzyme encoded by *galE* converts UDP-glucose to UDP-galactose for galactose addition to the LOS. A *galE* mutant synthesizes low levels of galactose containing epitopes in the outer core (Maskell *et al.*, 1992). Therefore, the results indicate that truncation of the inner core or inability to extend hexose residues from the inner core results in enhanced clearance of *H. influenzae* from the lung.

Despite the importance of LOS for *H. influenzae* in the lung, genes in which mutations resulted in less dramatic truncations of the LOS, including genes required for hexose extensions from the first heptose, *lgtF* (f.a. = 6.6), or from the terminal heptose of the inner core (Hood *et al.*, 2001, 2004), *lpsA* (f.a. = 3.8), were partially required, and a trend of moderate attenuation (~1.5-fold) was also observed in single strain infections with a defined *lpsA* mutant

(Gawronski *et al.*, 2009). Distal modifications of the LOS outer core structure mediated by genes such as *lic3A* which added sialic acid or the *lic1* locus responsible for addition of phosphorylcholine appeared to be non-essential *in vivo* in these experiments. In contrast to the minimal attenuation in the lung model of mutants producing partial outer core structures, these structures are critical in bloodstream models (Herbert *et al.*, 2002). For example, disruption of the *lpsA* gene results in a 36- and 90-fold attenuation in a bacteraemia model at 24 and 48 h post-inoculation, respectively, relative to the same parental strain used for HITS (Wong *et al.*, 2007). Therefore, the expression profile of LOS structures in response to aeration conditions is consistent with the *in vivo* importance of the genes responsible for generating those LOS structures during infection in niches predicted to differ in oxygen levels. It will therefore be of interest to determine the precise role of this regulatory profile in immune evasion and pathogenesis.

2.6 Conclusion

H. influenzae can sense its environment in response to varying oxygen levels to express genes needed for optimal growth and survival. Coordinated regulation of oxidative stress resistance pathways represents a probable defence against exposure to phagocytes whereby ArcA promotes survival of *H. influenzae* during bloodstream colonization. In addition, the discovery of FNR mediated activation of *ytfE* expression in response to nitrosative stress under anaerobiosis provides information concerning mechanisms by which *H. influenzae* may defend itself against activated host immune cells such as macrophages in relatively low oxygen environments. The profile of genes required for survival in the mouse lung indicates that this environment demands oxidative stress adaptation by *H. influenzae*. Genes that have known functions in combating oxidative stress (i.e. *oxyR*, *pgdX*) were seen to be required in the lungs. Conversely, ArcA, which was most active under lower oxygen environments, was

implicated in adaptation to conditions in the bloodstream, a niche that might reflect conditions present in submucosal sites of nasopharyngeal colonization with respect to oxygen levels. In parallel with physiological adaptations, redox responses by *H. influenzae* are implicated in modulating LOS, a classical virulence factor.

The regulatory profile of LOS structures, together with the relative fitness of LOS deficient mutants in different sites in the host, supports the view that *H. influenzae* coordinates expression of genes needed for pathogenesis, at least in part, in response to oxygen levels the bacterium encounters in each site of infection. A more detailed understanding of this process will provide insight into the factors influencing the transition of a respiratory pathogen between asymptomatic modes of colonization and disease states.

Acknowledgements

This work was supported in part by a grant from the NIH (AI49437) to B.J.A.

References

Almiron, M., Link, A.J., Furlong, D. and Kolter, R. (1992) A novel DNA-binding protein with regulatory and protective roles in starved *Escherichia coli*. *Genes and Development* 6, 2646–2654.

Bakaletz, L.O., Demaria, T.F. and Lim, D.J. (1987) Phagocytosis and killing of bacteria by middle ear macrophages. *Archives of Otolaryngology – Head and Neck Surgery* 113, 138–144.

Baltes, N., N'diaye, M., Jacobsen, I.D., Maas, A., Buettner, F.F. and Gerlach, G.F. (2005) Deletion of the anaerobic regulator HlyX causes reduced colonization and persistence of *Actinobacillus pleuropneumoniae* in the porcine respiratory tract. *Infection and Immunity* 73, 4614–4619.

Bartolini, E., Frigimelica, E., Giovinazzi, S., Galli, G., Shaik, Y., Genco, C., Welsch, J.A., Granoff, D.M., Grandi, G. and Grifantini, R. (2006) Role of FNR and FNR-regulated, sugar fermentation genes in *Neisseria meningitidis* infection. *Molecular Microbiology* 60, 963–972.

Bekker, M., Alexeeva, S., Laan, W., Sawers, G., Teixeira De Mattos, J. and Hellingwerf, K. (2010) The ArcBA two-component system of

Escherichia coli is regulated by the redox state of both the ubiquinone and the menaquinone pool. *Journal of Bacteriology* 192, 746–754.

Bishai, W.R., Howard, N.S., Winkelstein, J.A. and Smith, H.O. (1994a) Characterization and virulence analysis of catalase mutants of *Haemophilus influenzae*. *Infection and Immunity* 62, 4855–4860.

Bishai, W.R., Smith, H.O. and Barcak, G.J. (1994b) A peroxide/ascorbate-inducible catalase from *Haemophilus influenzae* is homologous to the *Escherichia coli katE* gene product. *Journal of Bacteriology* 176, 2914–2921.

Bogdan, C., Rollinghoff, M. and Diefenbach, A. (2000) The role of nitric oxide in innate immunity. *Immunological Reviews* 173, 17–26.

Brown, N.L., Stoyanov, J.V., Kidd, S.P. and Hobman, J.L. (2003) The MerR family of transcriptional regulators. *FEMS Microbiology Reviews* 27, 145–163.

Brumaghim, J.L., Li, Y., Henle, E. and Linn, S. (2003) Effects of hydrogen peroxide upon nicotinamide nucleotide metabolism in *Escherichia coli*: changes in enzyme levels and nicotinamide nucleotide pools and studies of the oxidation of NAD(P)H by Fe(III). *Journal of Biological Chemistry* 278, 42495–42504.

Buettner, F.F., Maas, A. and Gerlach, G.F. (2008) An *Actinobacillus pleuropneumoniae arcA* deletion mutant is attenuated and deficient in biofilm formation. *Veterinary Microbiology* 127, 106–115.

Chattopadhyay, M.K., Tabor, C.W. and Tabor, H. (2003) Polyamines protect *Escherichia coli* cells from the toxic effect of oxygen. *Proceedings of the National Academy of Sciences of the United States of America* 100, 2261–2265.

Constantinidou, C., Hobman, J.L., Griffiths, L., Patel, M.D., Penn, C.W., Cole, J.A. and Overton, T.W. (2006) A reassessment of the FNR regulon and transcriptomic analysis of the effects of nitrate, nitrite, NarXL, and NarQP as *Escherichia coli* K12 adapts from aerobic to anaerobic growth. *Journal of Biological Chemistry* 281, 4802–4815.

Crack, J., Green, J. and Thomson, A.J. (2004) Mechanism of oxygen sensing by the bacterial transcription factor fumarate-nitrate reduction (FNR). *Journal of Biological Chemistry* 279, 9278–9286.

D'Mello, R.A., Langford, P.R. and Kroll, J.S. (1997) Role of bacterial Mn-cofactored superoxide dismutase in oxidative stress responses, nasopharyngeal colonization, and sustained bacteremia caused by *Haemophilus influenzae* type b. *Infection and Immunity* 65, 2700–2706.

De Souza-Hart, J.A., Blackstock, W., Di Modugno, V., Holland, I.B. and Kok, M. (2003) Two-component systems in *Haemophilus influenzae*: a regulatory role for ArcA in serum resistance. *Infection and Immunity* 71, 163–172.

Dibden, D.P. and Green, J. (2005) *In vivo* cycling of the *Escherichia coli* transcription factor FNR between active and inactive states. *Microbiology* 151, 4063–4070.

Eiglmeier, K., Honore, N., Iuchi, S., Lin, E.C. and Cole, S.T. (1989) Molecular genetic analysis of FNR-dependent promoters. *Molecular Microbiology* 3, 869–878.

Elvers, K.T., Turner, S.M., Wainwright, L.M., Marsden, G., Hinds, J., Cole, J.A., Poole, R.K., Penn, C.W. and Park, S.F. (2005) NssR, a member of the Crp-Fnr superfamily from *Campylobacter jejuni*, regulates a nitrosative stress-responsive regulon that includes both a single-domain and a truncated haemoglobin. *Molecular Microbiology* 57, 735–750.

Figueira, M.A., Ram, S., Goldstein, R., Hood, D.W., Moxon, E.R. and Pelton, S.I. (2007) Role of complement in defense of the middle ear revealed by restoring the virulence of nontypeable *Haemophilus influenzae siaB* mutants. *Infection and Immunity* 75, 325–333.

Fink, R.C., Evans, M.R., Porwollik, S., Vazquez-Torres, A., Jones-Carson, J., Troxell, B., Libby, S.J., Mcclelland, M. and Hassan, H.M. (2007) FNR is a global regulator of virulence and anaerobic metabolism in *Salmonella enterica* serovar *Typhimurium* (ATCC 14028s). *Journal of Bacteriology* 189, 2262–2273.

Fleischmann, R.D., Adams, M.D., White, O., Clayton, R.A., Kirkness, E.F., Kerlavage, A.R., Bult, C.J., Tomb, J.F., Dougherty, B.A., Merrick, J.M., McKenney, K., Suton, G., FitzHugh, W., Fields, C., Gocayne, J.D., Scot, J., Shirley, R., Liu, L., Glodek, A., Keley, J.M., Weidman, J.F., Philips, C.A., Sprigs, T., Hedblom, E., Cotton, M.D., Utterback, T.R., Hanna, M.C., Nguyen, D.T., Saudek, D.M., Brandon, R.C., Fine, L.D., Fritchman, J.L., Fuhrmann, J.L., Geoghagen, N.S.M., Gnehm, C.L., McDonald, L.A., Small, K.V., Fraser, C.M., Smith, H.O. and J.C. Venter. (1995) Whole-genome random sequencing and assembly of *Haemophilus influenzae* Rd. *Science* 269, 496–512.

Furuchi, T., Kashiwagi, K., Kobayashi, H. and Igarashi, K. (1991) Characteristics of the gene for a spermidine and putrescine transport system that maps at 15 min on the *Escherichia coli* chromosome. *Journal of Biological Chemistry* 266, 20928–20933.

Gawronski, J.D., Wong, S.M., Giannoukos, G., Ward, D.V. and Akerley, B.J. (2009) Tracking insertion mutants within libraries by deep sequencing and a genome-wide screen for *Haemophilus* genes required in the lung. *Proceedings of the National Academy of Sciences of the United States of America* 106, 16422–16427.

Georgellis, D., Kwon, O. and Lin, E.C. (2001a) Quinones as the redox signal for the Arc two-component system of bacteria. *Science* 292, 2314–2316.

Georgellis, D., Kwon, O., Lin, E.C., Wong, S.M. and Akerley, B.J. (2001b) Redox signal transduction by the ArcB sensor kinase of *Haemophilus influenzae* lacking the PAS domain. *Journal of Bacteriology* 183, 7206–7212.

Gilberthorpe, N.J. and Poole, R.K. (2008) Nitric oxide homeostasis in *Salmonella typhimurium*: roles of respiratory nitrate reductase and flavohemoglobin. *Journal of Biological Chemistry* 283, 11146–11154.

Harrington, J.C., Wong, S.M., Rosadini, C.V., Garifulin, O., Boyartchuk, V. and Akerley, B.J. (2009) Resistance of *Haemophilus influenzae* to reactive nitrogen donors and gamma interferon-stimulated macrophages requires the formate-dependent nitrite reductase regulator-activated *ytfE* gene. *Infection and Immunity* 77, 1945–1958.

Harrison, A., Ray, W.C., Baker, B.D., Armbruster, D.W., Bakaletz, L.O. and Munson, R.S. Jr (2007) The OxyR regulon in nontypeable *Haemophilus influenzae*. *Journal of Bacteriology* 189, 1004–1012.

Herbert, M.A., Hayes, S., Deadman, M.E., Tang, C.M., Hood, D.W. and Moxon, E.R. (2002) Signature tagged mutagenesis of *Haemophilus influenzae* identifies genes required for *in vivo* survival. *Microbial Pathogenesis* 33, 211–223.

Ho, D.K., Ram, S., Nelson, K.L., Bonthuis, P.J. and Smith, A.L. (2007) *lgtC* expression modulates resistance to C4b deposition on an invasive nontypeable *Haemophilus influenzae*. *Journal of Immunology* 178, 1002–1012.

Hood, D.W., Makepeace, K., Deadman, M.E., Rest, R.F., Thibault, P., Martin, A., Richards, J.C. and Moxon, E.R. (1999) Sialic acid in the lipopolysaccharide of *Haemophilus influenzae*: strain distribution, influence on serum resistance and structural characterization. *Molecular Microbiology* 33, 679–692.

Hood, D.W., Cox, A.D., Wakarchuk, W.W., Schur, M., Schweda, E.K., Walsh, S.L., Deadman, M.E., Martin, A., Moxon, E.R. and Richards, J.C. (2001) Genetic basis for expression of the major globotetraose-containing lipopolysaccharide from *H. influenzae* strain Rd (RM118). *Glycobiology* 11, 957–967.

Hood, D.W., Deadman, M.E., Cox, A.D., Makepeace, K., Martin, A., Richards, J.C. and Moxon, E.R. (2004) Three genes, *lgtF*, *lic2C* and *lpsA*, have a primary role in determining the pattern of oligosaccharide extension from the inner core of *Haemophilus influenzae* LPS. *Microbiology* 150, 2089–2097.

Igarashi, K. and Kashiwagi, K. (1999) Polyamine transport in bacteria and yeast. *Biochemical Journal* 344 Pt 3, 633–642.

Ikai, H. and Yamamoto, S. (1998) Two genes involved in the 1,3-diaminopropane production pathway in *Haemophilus influenzae*. *Biological and Pharmaceutical Bulletin* 21, 170–173.

Ilari, A., Stefanini, S., Chiancone, E. and Tsernoglou, D. (2000) The dodecameric ferritin from *Listeria innocua* contains a novel intersubunit iron-binding site. *Nature Structural Biology* 7, 38–43.

Imlay, J.A. (2003) Pathways of oxidative damage. *Annual Review of Microbiology* 57, 395–418.

Jung, I.L. and Kim, I.G. (2003) Transcription of *ahpC*, *katG*, and *katE* genes in *Escherichia coli* is regulated by polyamines: polyamine-deficient mutant sensitive to H_2O_2-induced oxidative damage. *Biochemical and Biophysical Research Communications* 301, 915–922.

Justino, M.C., Vicente, J.B., Teixeira, M. and Saraiva, L.M. (2005) New genes implicated in the protection of anaerobically grown *Escherichia coli* against nitric oxide. *Journal of Biological Chemistry* 280, 2636–2643.

Justino, M.C., Almeida, C.C., Goncalves, V.L., Teixeira, M. and Saraiva, L.M. (2006) *Escherichia coli* YtfE is a di-iron protein with an important function in assembly of iron-sulphur clusters. *FEMS Microbiology Letters* 257, 278–284.

Justino, M.C., Almeida, C.C., Teixeira, M. and Saraiva, L.M. (2007) *Escherichia coli* di-iron YtfE protein is necessary for the repair of stress-damaged iron-sulfur clusters. *Journal of Biological Chemistry* 282, 10352–10359.

Kidd, S.P., Jiang, D., Jennings, M.P. and Mcewan, A.G. (2007) Glutathione-dependent alcohol dehydrogenase AdhC is required for defense against nitrosative stress in *Haemophilus influenzae*. *Infection and Immunity* 75, 4506–4513.

Klein, J.O. (1997) Role of nontypeable *Haemophilus influenzae* in pediatric respiratory tract infections. *Pediatric Infectious Disease Journal* 16, S5–8.

Lazazzera, B.A., Beinert, H., Khoroshilova, N., Kennedy, M.C. and Kiley, P.J. (1996) DNA binding and dimerization of the Fe-S-containing FNR protein from *Escherichia coli* are regulated by oxygen. *Journal of Biological Chemistry* 271, 2762–2768.

Leon, M.A. and Young, N.M. (1971) Specificity for

phosphorylcholine of six murine myeloma proteins reactive with Pneumococcus C polysaccharide and beta-lipoprotein. *Biochemistry* 10, 1424–1429.

Liu, X. and De Wulf, P. (2004) Probing the ArcA-P modulon of *Escherichia coli* by whole genome transcriptional analysis and sequence recognition profiling. *Journal of Biological Chemistry* 279, 12588–12597.

Lorsbach, R.B. and Russell, S.W. (1992) A specific sequence of stimulation is required to induce synthesis of the antimicrobial molecule nitric oxide by mouse macrophages. *Infection and Immunity* 60, 2133–2135.

Lundberg, J.O. (2008) Nitric oxide and the paranasal sinuses. *Anatomical Record (Hoboken)* 291, 1479–1484.

Lundberg, J.O., Farkas-Szallasi, T., Weitzberg, E., Rinder, J., Lidholm, J., Anggaard, A., Hokfelt, T., Lundberg, J.M. and Alving, K. (1995) High nitric oxide production in human paranasal sinuses. *Nature Medicine* 1, 370–373.

Lynch, A.S. and Lin, E.C. (1996a) Transcriptional control mediated by the ArcA two-component response regulator protein of *Escherichia coli*: characterization of DNA binding at target promoters. *Journal of Bacteriology* 178, 6238–6249.

Lynch, A.S. and Lin, E.C.C. (1996b) Responses to molecular oxygen. In: Neidhardt, F.C., Curtiss, R. III, Ingraham, J.L., Lin, E.C.C., Low, K.B., Magasanik, B., Reznikoff, W.S., Riley, M., Schaechter, M. and Umbarger, H.E. (eds) *Escherichia coli and Salmonella: Cellular and Molecular Biology*. ASM Press, Washington, DC, pp. 1526–1538.

Lysenko, E.S., Gould, J., Bals, R., Wilson, J.M. and Weiser, J.N. (2000) Bacterial phosphorylcholine decreases susceptibility to the antimicrobial peptide LL-37/hCAP18 expressed in the upper respiratory tract. *Infection and Immunity* 68, 1664–1671.

MacMicking, J., Xie, Q.W. and Nathan, C. (1997) Nitric oxide and macrophage function. *Annual Review of Immunology* 15, 323–350.

Malpica, R., Franco, B., Rodriguez, C., Kwon, O. and Georgellis, D. (2004) Identification of a quinone-sensitive redox switch in the ArcB sensor kinase. *Proceedings of the National Academy of Sciences of the United States of America* 101, 13318–13323.

Manukhov, I.V., Bertsova, Y.V., Trofimov, D.Y., Bogachev, A.V. and Skulachev, V.P. (2000) Analysis of HI0220 protein from *Haemophilus influenzae*, a novel structural and functional analog of ArcB protein from *Escherichia coli*. *Biochemistry* 65, 1321–1326.

Maskell, D.J., Szabo, M.J., Deadman, M.E. and Moxon, E.R. (1992) The *gal* locus from *Haemophilus influenzae*: cloning, sequencing and the use of *gal* mutants to study lipopolysaccharide. *Molecular Microbiology* 6, 3051–3063.

Melville, S.B. and Gunsalus, R.P. (1996) Isolation of an oxygen-sensitive FNR protein of *Escherichia coli*: interaction at activator and repressor sites of FNR-controlled genes. *Proceedings of the National Academy of Sciences of the United States of America* 93, 1226–1231.

Moxon, E.R. and Maskell, D. (1992) *Haemophilus influenzae* lipopolysaccharide: the biochemistry and biology of a virulence factor. In: Hormaeche, C.E., Penn, C.W. and Smyths, C.J. (eds) *Molecular Biology of Bacterial Infection: Current Status and Future Perspectives*. Cambridge University Press, Cambridge, UK, pp. 75–96.

Moxon, E.R. and Murphy, T.F. (2000) *Haemophilus influenzae*. In: Mandell, G.L., Bennett, J.R. and Dolin, R. (eds) *Mandell, Douglas, and Bennett's Principles and Practices of Infectious Diseases*. Churchill Livingstone, New York, pp. 2369–2378.

Noel, G.J., Hoiseth, S.K. and Edelson, P.J. (1992) Type b capsule inhibits ingestion of *Haemophilus influenzae* by murine macrophages: studies with isogenic encapsulated and unencapsulated strains. *Journal of Infectious Diseases* 166, 178–182.

Oshima, T., Aiba, H., Masuda, Y., Kanaya, S., Sugiura, M., Wanner, B.L., Mori, H. and Mizuno, T. (2002) Transcriptome analysis of all two-component regulatory system mutants of *Escherichia coli* K-12. *Molecular Microbiology* 46, 281–291.

Overton, T.W., Justino, M.C., Li, Y., Baptista, J.M., Melo, A.M., Cole, J.A. and Saraiva, L.M. (2008) Widespread distribution in pathogenic bacteria of di-iron proteins that repair oxidative and nitrosative damage to iron-sulfur centers. *Journal of Bacteriology* 190, 2004–2013.

Partridge, J.D., Scott, C., Tang, Y., Poole, R.K. and Green, J. (2006) *Escherichia coli* transcriptome dynamics during the transition from anaerobic to aerobic conditions. *Journal of Biological Chemistry* 281, 27806–27815.

Patschkowski, T., Bates, D.M. and Kiley, P.J. (2000) Mechanisms for sensing and responding to oxygen deprivation. In: Storz, G. and Hengge-Aronis, R. (eds) *Bacterial Stress Responses*. ASM Press, Washington, DC, pp. 61–78.

Pauwels, F., Vergauwen, B., Vanrobaeys, F., Devreese, B. and Van Beeumen, J.J. (2003) Purification and characterization of a chimeric enzyme from *Haemophilus influenzae* Rd that exhibits glutathione-dependent peroxidase activity. *Journal of Biological Chemistry* 278, 16658–16666.

Pomposiello, P.J. and Demple, B. (2001) Redox-operated genetic switches: the SoxR and OxyR transcription factors. *Trends in Biotechnology* 19, 109–114.

Poole, R.K., Anjum, M.F., Membrillo-Hernandez, J., Kim, S.O., Hughes, M.N. and Stewart, V. (1996) Nitric oxide, nitrite, and Fnr regulation of *hmp* (flavohemoglobin) gene expression in *Escherichia coli* K-12. *Journal of Bacteriology* 178, 5487–5492.

Pullan, S.T., Gidley, M.D., Jones, R.A., Barrett, J., Stevanin, T.M., Read, R.C., Green, J. and Poole, R.K. (2007) Nitric oxide in chemostat-cultured *Escherichia coli* is sensed by Fnr and other global regulators: unaltered methionine biosynthesis indicates lack of S nitrosation. *Journal of Bacteriology* 189, 1845–1855.

Pulliainen, A.T., Kauko, A., Haataja, S., Papageorgiou, A.C. and Finne, J. (2005) Dps/Dpr ferritin-like protein: insights into the mechanism of iron incorporation and evidence for a central role in cellular iron homeostasis in *Streptococcus suis*. *Molecular Microbiology* 57, 1086–1100.

Raghunathan, A., Price, N.D., Galperin, M.Y., Makarova, K.S., Purvine, S., Picone, A.F., Cherny, T., Xie, T., Reilly, T.J., Munson, R. Jr, Tyler, R.E., Akerley, B.J., Smith, A.L., Palsson, B.O. and Kolker, E. (2004) *In silico* metabolic model and protein expression of *Haemophilus influenzae* strain Rd KW20 in rich medium. *Omics* 8, 25–41.

Ren, B., Zhang, N., Yang, J. and Ding, H. (2008) Nitric oxide-induced bacteriostasis and modification of iron-sulphur proteins in *Escherichia coli*. *Molecular Microbiology* 70, 953–964.

Risberg, A., Masoud, H., Martin, A., Richards, J.C., Moxon, E.R. and Schweda, E.K. (1999) Structural analysis of the lipopolysaccharide oligosaccharide epitopes expressed by a capsule-deficient strain of *Haemophilus influenzae* Rd. *European Journal of Biochemistry* 261, 171–180.

Salmon, K.A., Hung, S.P., Steffen, N.R., Krupp, R., Baldi, P., Hatfield, G.W. and Gunsalus, R.P. (2005) Global gene expression profiling in *Escherichia coli* K12: effects of oxygen availability and ArcA. *Journal of Biological Chemistry* 280, 15084–15096.

Samouilov, A., Kuppusamy, P. and Zweier, J.L. (1998) Evaluation of the magnitude and rate of nitric oxide production from nitrite in biological systems. *Archives of Biochemistry and Biophysics* 357, 1–7.

Sauer, U., Canonaco, F., Heri, S., Perrenoud, A. and Fischer, E. (2004) The soluble and membrane-bound transhydrogenases UdhA and PntAB have divergent functions in NADPH metabolism of *Escherichia coli*. *Journal of Biological Chemistry* 279, 6613–6619.

Sengupta, N., Paul, K. and Chowdhury, R. (2003) The global regulator ArcA modulates expression of virulence factors in *Vibrio cholerae*. *Infection and Immunity* 71, 5583–5589.

Shaw, D.J., Rice, D.W. and Guest, J.R. (1983) Homology between CAP and Fnr, a regulator of anaerobic respiration in *Escherichia coli*. *Journal of Molecular Biology* 166, 241–247.

Shaw, P.X., Horkko, S., Chang, M.K., Curtiss, L.K., Palinski, W., Silverman, G.J. and Witztum, J.L. (2000) Natural antibodies with the T15 idiotype may act in atherosclerosis, apoptotic clearance, and protective immunity. *Journal of Clinical Investigation* 105, 1731–1740.

Singh, R.J., Hogg, N., Joseph, J. and Kalyanaraman, B. (1996) Mechanism of nitric oxide release from S-nitrosothiols. *Journal of Biological Chemistry* 271, 18596–18603.

Stohl, E.A. and Seifert, H.S. (2006) *Neisseria gonorrhoeae* DNA recombination and repair enzymes protect against oxidative damage caused by hydrogen peroxide. *Journal of Bacteriology* 188, 7645–7651.

Storz, G. and Imlay, J.A. (1999) Oxidative stress. *Current Opinion in Microbiology* 2, 188–194.

Sun, J., Chen, J., Cheng, Z., Robbins, J.B., Battey, J.F. and Gu, X.X. (2000) Biological activities of antibodies elicited by lipooligosaccharide based-conjugate vaccines of nontypeable *Haemophilus influenzae* in an otitis media model. *Vaccine* 18, 1264–1272.

Tan, K., Moreno-Hagelsieb, G., Collado-Vides, J. and Stormo, G.D. (2001) A comparative genomics approach to prediction of new members of regulons. *Genome Research* 11, 566–584.

Tatusov, R.L., Mushegian, A.R., Bork, P., Brown, N.P., Hayes, W.S., Borodovsky, M., Rudd, K.E. and Koonin, E.V. (1996) Metabolism and evolution of *Haemophilus influenzae* deduced from a whole-genome comparison with *Escherichia coli*. *Current Biology* 6, 279–291.

Tkachenko, A., Nesterova, L. and Pshenichnov, M. (2001) The role of the natural polyamine putrescine in defense against oxidative stress in *Escherichia coli*. *Archives of Microbiology* 176, 155–157.

Toews, G.B., Vial, W.C. and Hansen, E.J. (1985) Role of C5 and recruited neutrophils in early clearance of nontypable *Haemophilus influenzae* from murine lungs. *Infection and Immunity* 50, 207–212.

Vergauwen, B., Herbert, M. and Van Beeumen, J.J. (2006) Hydrogen peroxide scavenging is not a virulence determinant in the pathogenesis of *Haemophilus influenzae* type b strain Eagan. *BMC Microbiology* 6, 3.

Weiser, J.N., Pan, N., Mcgowan, K.L., Musher, D., Martin, A. and Richards, J. (1998) Phosphorylcholine on the lipopolysaccharide of *Haemophilus influenzae* contributes to persistence in the respiratory tract and sensitivity to serum killing mediated by C-reactive protein. *Journal of Experimental Medicine* 187, 631–640.

Winter, L.E. and Barenkamp, S.J. (2006) Antibodies specific for the high-molecular-weight adhesion proteins of nontypeable *Haemophilus influenzae* are opsonophagocytic for both homologous and heterologous strains. *Clinical and Vaccine Immunology* 13, 1333–1342.

Wong, S.M. and Akerley, B.J. (2005) Environmental and genetic regulation of the phosphorylcholine epitope of *Haemophilus influenzae* lipooligosaccharide. *Molecular Microbiology* 55, 724–738.

Wong, S.M., Alugupalli, K.R., Ram, S. and Akerley, B.J. (2007) The ArcA regulon and oxidative stress resistance in *Haemophilus influenzae*. *Molecular Microbiology* 64, 1375–1390.

Wood, G.E., Khelef, N., Guiso, N. and Friedman, R.L. (1998) Identification of Btr-regulated genes using a titration assay. Search for a role for this transcriptional regulator in the growth and virulence of *Bordetella pertussis*. *Gene* 209, 51–58.

Zhao, G., Ceci, P., Ilari, A., Giangiacomo, L., Laue, T.M., Chiancone, E. and Chasteen, N.D. (2002) Iron and hydrogen peroxide detoxification properties of DNA-binding protein from starved cells. A ferritin-like DNA-binding protein of *Escherichia coli*. *Journal of Biological Chemistry* 277, 27689–27696.

3 Nitric Oxide Stress in *Escherichia coli* and *Salmonella*

Stephen Spiro

3.1 Introduction

Pathogenic bacteria trapped inside the phagosome of host professional phagocytes are exposed to a variety of 'stressful' conditions that are sufficiently noxious to kill many species efficiently. In the battlefield of the phagosome the host deploys a range of weapons in its fight against the invading pathogen, including acidification of the phagosome lumen, synthesis of oxygen and nitrogen radicals, depletion of essential nutrients and synthesis of antimicrobial proteins and peptides (Flannagan *et al.*, 2009). Some pathogens mount a successful counter-offensive and so are able to survive phagocyte killing. The strategies pathogens employ to do this include the inhibition of phagocytosis and phagosome maturation, escape from the phagosome, development of resistance to antimicrobial activities of the phagosome, inhibition of the synthesis of oxygen and nitrogen radicals and repair of the damage to macromolecules caused by reagents synthesized in the phagosome (Flannagan *et al.*, 2009). This chapter is devoted to a description of one particular weapon deployed by phagocytes, nitric oxide (NO), and the mechanisms used by enteric pathogens to avoid and counteract the effects of NO.

The activated phagocyte assembles a multi-subunit phagocyte oxidase (Phox) in the phagosome membrane. This enzyme couples the oxidation of NADPH with the reduction of oxygen to superoxide, which is released in the phagocyte lumen. There, the superoxide is dismutated to hydrogen peroxide, which subsequently may be converted to hydroxyl radicals and singlet oxygen. H_2O_2 may also be converted to hypochlorous acid by myeloperoxidase. Several lines of evidence point to the importance of these reactive oxygen species (ROS) in the killing of invading pathogens. *Salmonella enterica* serovar Typhimurium (*S.* Typhimurium) encodes proteins in the *Salmonella* pathogenicity island 2 (SPI2) that interfere with localization of Phox to the phagosome membrane, and expresses periplasmic and cytosolic superoxide dismutases that have important roles in survival in macrophages (Vazquez-Torres *et al.*, 2000; Craig and Slauch, 2009). Transgenic mice with a defective Phox are hyper-susceptible to *Salmonella* infection, and humans with mutations affecting Phox activity suffer from chronic granulomatous disease, which is characterized by recurrent bacterial and fungal infections (Vazquez-Torres and Fang, 2001).

Besides superoxide, the activated phagocyte also synthesizes NO as a product of the activity of the inducible NO synthase (iNOS, also known as NOS2). Initial difficulties replicating findings from murine macrophages in human cell lines have led to suggestions that NO is not synthesized in human macrophages (Schneemann and Schoeden, 2007). However, the growing consensus in the literature is that human macrophages do indeed express an active iNOS and make NO in response to stimulation (Fang and Vazquez-Torres, 2002; Fang and Nathan, 2007). The dimeric NOS enzymes catalyse the oxygen-dependent oxidation of the guanidino nitrogen of arginine, producing NO and citrulline. The reaction proceeds in two mono-oxygenation steps, with the electrons required for oxygen activation originating from NADPH (Mowat *et al.*, 2010). The iNOS is expressed in phagocytes in response to microbial stimulation; regulation is predominantly, though not exclusively, at the level of transcription (Fang, 2004). iNOS activity might also be limited by the supply of arginine and cofactors, which may have some physiological significance (see below). Either spontaneously, or by enzyme-catalysed reactions, NO can be converted to a range of other nitrogen radicals, which are referred to collectively as reactive nitrogen species (RNS) or NO congeners. Of particular relevance is the reaction of NO with superoxide (the product of Phox) to generate peroxynitrite (ONOO⁻), which can undergo subsequent conversion to nitrogen dioxide (NO_2) and hydroxyl radicals. Since the products of the reaction of NO with superoxide are potently toxic *in vitro*, it is often assumed that the NO and Phox pathways contribute synergistically to killing. In the case of *Salmonella* infections, the production of superoxide and NO is separated temporally, such that Phox-mediated killing occurs prior to iNOS-mediated killing. This suggests that the two pathways are not synergistic in this system (Fang, 2004). On the other hand, the periplasmic superoxide dismutase, SodC1, protects *S.* Typhimurium against host-derived NO, most likely by diverting superoxide away from peroxynitrite formation (De Groote *et al.*, 1997), an observation which

suggests synergistic actions of NO and superoxide. In any case, in what follows it is important to consider microbial reactions to the RNS produced from NO, in addition to NO per se.

3.2 Toxicity of NO and RNS

NO has a particular avidity for metal centres in proteins, and the most sensitive targets for inhibition by NO seem to be the respiratory oxidases and the dehydratases that contain labile [4Fe-4S] clusters of the type found in aconitase (Hyduke *et al.*, 2007). Other types of damage (for example, to protein thiols and tyrosines, lipids and DNA) sometimes ascribed in the literature to NO are more likely to result from the actions of other RNS. High-throughput studies are beginning to shed light on the specific molecular targets for peroxynitrite (McClean *et al.*, 2010a) and other RNS derived from NO (Brandes *et al.*, 2007). It is well established that peroxynitrite causes nitration of tyrosine residues (Abello *et al.*, 2009), and SPI2 mutants of *Salmonella* (which are less able than wild type to interfere with iNOS localization) suffer from protein tyrosine nitration in infected macrophages (Chakravortty *et al.*, 2002). Peroxynitrite and other RNS, though not NO itself, can damage DNA. *Salmonella* mutants defective in base excision repair are attenuated for virulence in mice, and attenuation requires iNOS-derived NO (Richardson *et al.*, 2009). Nitrosative stress imposed by *S*-nitrosoglutathione (GSNO) inhibits replication, which may be ascribed to the loss of zinc from zinc-containing DNA-binding proteins (Schapiro *et al.*, 2003). The toxic effects of superoxide and NO can, at least in part, be avoided by expressing enzymes that detoxify these radicals by converting them to less toxic products. In *E. coli* at least three enzymes have been shown to detoxify NO by oxidizing it (to nitrate) or reducing it (to ammonia or N_2O). These enzymes are the flavo-haemoglobin, Hmp, the flavorubredoxin, FlRd, and the respiratory nitrite reductase, Nrf, each of which will be considered briefly in turn in the following sections.

Paradoxically, exposure to NO may also help some pathogens to survive the phagosome environment. In *Bacillus anthracis* NO exposure protects against oxidative stress and so helps the pathogen to survive macrophage-mediated killing (Shatalin *et al.*, 2008). NO achieves this effect by activating a catalase directly and by suppressing the Fenton reaction by which DNA-damaging hydroxyl radicals are derived from hydrogen peroxide (Gusarov and Nudler, 2005). In *S.* Typhimurium NO also protects against oxidative stress but, in this case, does so by inhibiting respiration and so elevating levels of NADH, which provides the reducing power for antioxidative processes (Husain *et al.*, 2008).

3.3 Detoxification of NO and RNS

3.3.1 Flavohaemoglobin

The flavohaemoglobin (Hmp) is the best characterized of the NO detoxification activities in enteric bacteria. The finding that *hmp* expression in *E. coli* is upregulated by sources of NO suggested a role in NO metabolism (Poole *et al.*, 1996). Subsequently, it was found that *E. coli* mutants selected for resistance to NO overexpressed Hmp and that Hmp was associated with an NO-inducible oxygen-dependent activity that converted NO to nitrate (Gardner *et al.*, 1998). Hmp proteins consist of a C-terminal FAD and NAD(P)H binding domain, fused to an N-terminal globin-like domain. Electrons released by the oxidation of NAD(P)H are transferred to ligands bound to the haem in the globin domain. The enzymatic mechanism of Hmp has proved to be controversial. In the NO dioxygenase mechanism, oxy-Hmp reacts with NO to form nitrate (Gardner *et al.*, 1998; Gardner, 2005). On the other hand, it has been argued that at physiological oxygen and NO concentrations, NO binds to Hmp preferentially. The haem-bound nitroxyl (NO⁻) equivalent is then oxidized to nitrate. This mechanism has been called the O_2 nitroxylase or denitrosylase (Hausladen *et al.*, 2001). Under anaerobic conditions the reaction product is N_2O (Kim *et al.*, 1999),

though this reaction is much slower and its physiological significance has been a matter of debate. A detailed description of Hmp biochemistry is beyond the scope of this article and interested readers may consult a number of reviews (Poole and Hughes, 2000; Wu *et al.*, 2003; Gardner, 2005; Poole, 2005; Mowat *et al.*, 2010).

S. Typhimurium *hmp* mutants are sensitized to sources of NO *in vitro* and Hmp protects respiration against inhibition by NO (Stevanin *et al.*, 2002). The *S.* Typhimurium *hmp* gene is upregulated after 8 h growth in macrophages, a time point which coincides with the maximum of NO synthesis (Eriksson *et al.*, 2003), pointing to a possible role for Hmp in the detoxification of iNOS-derived NO. An *S.* Typhimurium *hmp* mutant is two- to threefold more sensitive to killing by human macrophages than a wild-type strain. This difference between strains disappeared in the presence of an iNOS inhibitor, suggesting that iNOS-derived NO was responsible for sensitivity of the mutant (Stevanin *et al.*, 2002). Subsequently, it was shown that Hmp was required for virulence in mice and that virulence could be restored to a *hmp* mutant by treatment of mice with an iNOS inhibitor (Bang *et al.*, 2006). This observation confirms that the *hmp* mutant is attenuated by host-synthesized NO. Another observed role for the *S.* Typhimurium Hmp is to reduce the partially iNOS-dependent formation of low molecular weight *S*-nitrosothiols (SNO) in mouse macrophages. This inhibition of SNO formation may represent a new mechanism by which *S.* Typhimurium damages host cells (Laver *et al.*, 2010).

The role played by Hmp in virulence in the case of pathogenic *E. coli* strains has been less well studied, although Hmp is required for the ability of K12 strains to resist killing by human macrophages (Stevanin *et al.*, 2007). The *hmp* gene was upregulated in the transcriptome of a uropathogenic *E. coli* (UPEC) isolated from mouse urine (Snyder *et al.*, 2004). Similarly, the *hmp* gene was upregulated in both asymptomatic and UPEC strains isolated from the urinary tracts of humans (Roos and Klemm, 2006; Svensson *et al.*, 2010). These results may indicate that the

UPEC and asymptomatic strains are exposed to NO made by host cells, which is a reasonable supposition given that NO-generating neutrophils infiltrate the urinary tracts of infected animals, and that iNOS is expressed in urinary tract epithelial cells (Sivick and Mobley, 2010). On the other hand, at least for the asymptomatic strains, genes encoding the enzymes of nitrate respiration are also upregulated *in vivo* (Roos and Klemm, 2006), which leaves open the possibility that the induction of *hmp* is mediated by NO generated endogenously in the bacteria as a by-product of nitrate and nitrite respiration.

The UPEC strain UTI89 is partially resistant to the nitrosative stress as exerted by 3 mM nitrite at pH 5 (acidified nitrite, ASN), a treatment which prevents growth of a non-pathogenic K12 strain. Isolation of mutants sensitive to ASN suggests that the synthesis of polyamines is one factor contributing to ASN resistance, though the mechanistic basis for protection is not known (Bower and Mulvey, 2006). The *hmp* gene is upregulated in UTI89 cells exposed to ASN for a prolonged period, though the role that Hmp plays in ASN resistance has not been examined (Bower *et al.*, 2009). UPEC 'conditioned' by pre-exposure to ASN was better able to establish infection in a mouse model (Bower *et al.*, 2009). A *hmp* mutant of UPEC strain J96 was out-competed significantly by its wild-type parent in a mouse model of urinary tract infection (Svensson *et al.*, 2010). There is a report that K12 strains are killed by NO more effectively than UPEC strains (Svensson *et al.*, 2006) and that *hmp* mutants of UPEC strains are more sensitive to killing by NO than their wild-type parent (Svensson *et al.*, 2010). NO is not normally cytotoxic and these strains were exposed to an NO donor for a prolonged period (24 or 5 h, respectively), suggesting that the killing agent was an RNS derived from NO oxidation, rather than NO per se. In experiments carried out in the author's laboratory, the UPEC strain CFT073 was found to be no more sensitive to an NO donor than a K12 strain (Fig. 3.1). NO resistance of CFT073 could be ascribed at least partially to Hmp since a *hmp* mutant was defective in the

recovery from NO exposure and an *nsrR* mutant was resistant to NO, presumably because of constitutive expression of *hmp* (Fig. 3.1). Current evidence therefore suggests that UPEC strains are more resistant to nitrosative stress than laboratory strains (Bower and Mulvey, 2006; Bower *et al.*, 2009), though not necessarily to NO per se (Fig. 3.1).

3.3.2 Flavorubredoxin

The flavorubredoxin (FlRd) of *E. coli* (the product of the *norV* gene) consists of a flavo di-iron protein fused to a C-terminal rubredoxin-like domain (Wasserfallen *et al.*, 1998). This protein has sequence similarity to the flavo di-iron and rubredoxin subunits of a soluble oxygen-reducing complex from *Desulfovibrio gigas* (Gomes *et al.*, 1997). The *norV* gene of *E. coli* is co-transcribed with *norW*, which encodes a rubredoxin reductase that mediates electron transfer between NADH and FlRd (Gomes *et al.*, 2000). The FlRd/FlRd-reductase complex couples NADH oxidation to oxygen reduction *in vitro*, leading to the suggestion that this enzyme might be a physiological oxidase responsible for scavenging low concentrations of oxygen, as is the case for the homologous *D. gigas* enzyme (Gomes *et al.*, 2000). Later experiments demonstrated that the FlRd/FlRd-reductase complex acted as an NO reductase *in vitro*, with a turnover number comparable to those of the respiratory NO reductases from denitrifying bacteria (Gomes *et al.*, 2002). Mutation of *norV* or *norW* eliminated or impaired, respectively, an NO inducible NO detoxification activity and growth of the mutants was more sensitive to exogenous NO (Gardner *et al.*, 2002). Thus, the *norVW* gene products are a physiological NO reductase, which operates under anoxic or micro-oxic conditions, perhaps because the activity is oxygen sensitive (Gardner *et al.*, 2002). The *D. gigas* flavo di-iron protein/rubredoxin complex also has NO reductase activity and complements the NO-sensitive phenotype of an *E. coli norV* mutant (Rodrigues *et al.*, 2006). Expression of *norVW* in *E. coli* is activated in the presence of NO by the product of the *norR*

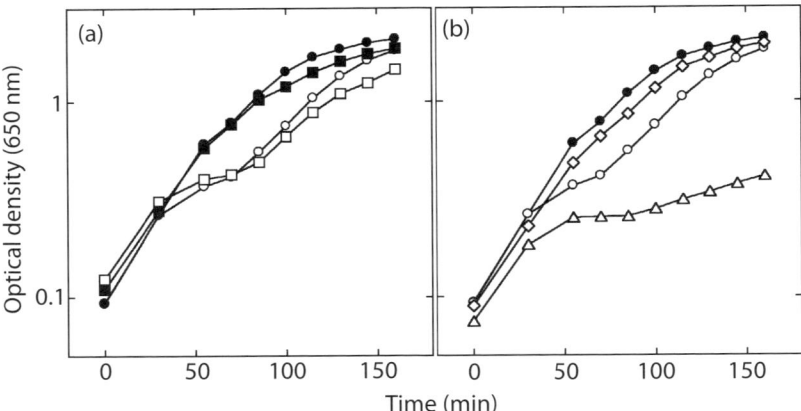

Fig. 3.1. Sensitivity of E. coli UPEC and K12 strains to NO. Cultures were grown aerobically in L broth and were exposed to NO at the 30 min time point. The NO source was 0.5 mM diethylenetriamine NONOate, which liberates two equivalents of NO with a half-life of 20 h at 37°C (www.caymanchem. com). The use of a relatively high concentration of a compound that decomposes very slowly ensures that the rate of NO release is approximately constant over the duration of the experiment. (a) The strains were *E. coli* K12 strain MG1655 (square symbols) and UPEC strain CFT073 (circles), and the cultures were exposed to NO (open symbols) or left untreated (filled symbols). (b) The strains were wild-type CFT073 (circles) and mutant derivatives deleted for *nsrR* (diamonds) or *hmp* (triangles). Cultures were exposed to NO (open symbols) or left untreated (filled symbols).

regulatory gene, which is transcribed divergently from *norVW* (Hutchings *et al.*, 2002; Gardner *et al.*, 2002, 2003). The properties of the NorR protein are described below.

The available evidence suggests that FlRd of pathogens does not play a major role in the detoxification of host-derived NO. The S. Typhimurium FlRd contributes to NO resistance in cultures grown under anaerobic conditions (Mills *et al.*, 2008), but the *norVW* genes are not virulence determinants in a mouse model of infection (Bang *et al.*, 2006). On the other hand, the *norVW* genes are upregulated in S. Typhimurium grown in murine macrophages (Eriksson *et al.*, 2003). In pathogenic strains of *E. coli* it is quite common for *norV* to be a pseudogene that is inactivated by one or more mutations. In the uropathogenic strain CFT073, *norV* is inactivated by a frameshift mutation, which along with several missense mutations creates a novel open reading frame on the opposite strand (Welch *et al.*, 2002). In the closely related uropathogenic UTI89, *norV* is intact. In 100 isolates of *E. coli* O157:H7, 58% had a

204 bp in-frame deletion in the *norV* gene that was presumed (though not proven) to inactivate FlRd. Occurrence of the *norV* deletion correlated strongly with the presence of the Shiga toxin *stx₁* gene, leading to speculation that there was a selective pressure for *norV* inactivation in strains that had *stx₁* (Kulasekara *et al.*, 2009). In this context it is interesting that NO inhibits expression of the *stx₂* gene in *E. coli* O157:H7 strain EDL933, an isolate which has the 204 bp deletion in *norV* (Vareille *et al.*, 2007; Kulasekara *et al.*, 2009).

3.3.3 Nitrite reductase

The NrfA catalytic subunit of the respiratory nitrite reductase is a homo-dimer of two penta-haem cytochromes that is located in the periplasm and interacts with membrane-associated subunits of the Nrf complex. The six electron reduction of nitrite to ammonia is proposed to proceed through enzyme-bound intermediates, NO and hydroxyla-mine (Einsle *et al.*, 2002). This mechanism is consistent with the observations that purified NrfA can reduce both NO and hydroxyla-mine to ammonia (Costa *et al.*, 1990; Rudolf *et al.*, 2002; van Wonderen *et al.*, 2008). The Nrf complex functions as a physiological NO reductase *in vivo* in *E. coli* if cultures are grown under anaerobic conditions that favour the expression of this enzyme, and Nrf protects against exogenous sources of NO (Poock *et al.*, 2002). Transcription of the *nrf* operon that encodes the components of the Nrf complex is regulated by the NO-sensitive repressor, NsrR (Filenko *et al.*, 2007; Browning *et al.*, 2010). However, mechanisms that regulate *nrf* expression according to the function of Nrf as a nitrite reductase seem to dominate, so that although Nrf is an NO reductase, its expression is not strongly upregulated by exposure to NO. In *S.* Typhimurium, Nrf also protects against NO in anaerobic cultures and there is evidence that Nrf expression is upregulated by exposure to NO (Mills *et al.*, 2008). Thus, in this case, Nrf may be a *bona fide* NO-inducible NO detoxification system, though the mechanism of *nrf* operon regulation by NO has not been examined. Recent evidence suggests that the *S.* Typhimurium *nrf* promoter is not regulated by NsrR (Browning *et al.*, 2010). The *nrf* genes are not required for *S.* Typhimurium virulence in mouse (Bang *et al.*, 2006) and there is no evidence that Nrf has a role to play in detoxifying host-derived NO.

3.3.4 Peroxynitrite metabolism

S. Typhimurium and *E. coli* express an alkyl hydroperoxide reductase which reduces organic peroxides to the corresponding alcohol and H_2O_2 to water and oxygen. The enzyme is an $\alpha_2\beta_2$ tetramer, comprising two copies of an NAD(P)H-linked flavoprotein (AhpF) and two copies of a peroxiredoxin (AhpC). AhpC protects *S.* Typhimurium against RNIs (derived from GSNO or ASN) in an AhpF-independent fashion (Chen *et al.*, 1998). Purified AhpC functions as a peroxynitrite reductase (the reaction product being nitrite), with electron transfer from NADH mediated either by AhpF or by

thioredoxin and thioredoxin reductase (Bryk *et al.*, 2000). This redundancy may explain the earlier observation of the AhpF independence of RNI resistance mediated by AhpC. The *ahpCF* genes were upregulated by growth of *S.* Typhimurium in macrophages (Francis *et al.*, 1997), but were not required for virulence in a mouse model, with the caveat that different results were obtained with two different mutant strains (Taylor *et al.*, 1998). The lack of attenuation of the *ahpC* knockout mutant may be accounted for by the fact that *S.* Typhimurium has a second alkyl hydro-peroxidase designated TsaA; a mutant lacking both enzymes is defective for growth in macrophages but is not attenuated for virulence in mouse (Hébrard *et al.*, 2009). In any case, AhpC and TsaA probably contribute to H_2O_2 scavenging in the host (Hébrard *et al.*, 2009), making their contribution to per-oxynitrite removal difficult to evaluate.

The hydroperoxidase I (HPI or KatG) of *S.* Typhimurium is a catalase-peroxidase, which recently has been shown also to have a peroxynitritase activity. The reaction product was not identified; the authors argued against it being nitrite and suggested isomerization to nitrate as an alternative possibility (McLean *et al.*, 2010b). The *katG* gene is not required for growth in phagocytes or for virulence in a mouse model (Papp-Szabò *et al.*, 1994; Buchmeier *et al.*, 1995). Perhaps the existence of redundant peroxynitrite detoxification activities accounts for the lack of attenuation of single mutants lacking one of these activities.

A recent study has shown that expression of *ahpCF* and *katG* is upregulated by exposure of *E. coli* to peroxynitrite, with regulation being mediated by the OxyR protein (McLean *et al.*, 2010a). Thus, both AhpC and KatG appear to be peroxynitrite inducible defence mechanisms.

3.3.5 Superoxide dismutase

The prophage-associated *sodC1* gene of *S.* Typhimurium encodes a periplasmic copper-zinc superoxide dismutase (Ammendola *et al.*, 2008). A *sodC1* mutant is sensitive to sources of superoxide and NO *in vitro* and these agents kill the *sodC1* mutant synergistically. The mutant is attenuated for virulence in a mouse model and virulence is restored by administration of an iNOS inhibitor (De Groote *et al.*, 1997). Collectively, these observations suggest that SodC1 protects against the toxic effects of peroxynitrite by detoxifying superoxide and so preventing peroxynitrite formation (De Groote *et al.*, 1997).

3.4 Repair of the Damage Caused by Reactive Nitrogen Species

3.4.1 Repair of iron-sulfur proteins

As has been mentioned, [Fe-S] clusters are sensitive targets for NO and RNIs and the ability to return damaged or modified clusters to their native state may be an important component of the recovery from NO exposure. The [4Fe-4S] clusters of dehydratases (such as aconitase) are sensitive to inactivation by NO (Gardner *et al.*, 1997; Woodmansee and Imlay, 2003; Justino *et al.*, 2007; Ren *et al.*, 2008) and peroxynitrite (Keyer and Imlay, 1997). NO reacts *in vitro* with purified [Fe-S] proteins to form protein-bound dinitrosyl iron complexes (DNIC) with a characteristic signal in electron paramagnetic resonance (EPR) spectra. Enzymes inactivated by DNIC formation can regain activity *in vivo*, in the absence of new protein synthesis, with the repair being accelerated considerably or observed only in the presence of oxygen (Gardner *et al.*, 1997; Ren *et al.*, 2008).

Two rather different (though not necessarily mutually exclusive) views have emerged in the recent literature concerning the mechanism of repair of nitrosylated clusters. On the one hand, the di-iron rubrethryin-like protein YtfE has been suggested (based on evidence from *in vivo* and *in vitro* experiments) to be required for the repair of clusters damaged by H_2O_2 or NO (Justino *et al.*, 2007). Expression of the *E. coli ytfE* gene is regulated by NsrR (Bodenmiller and Spiro, 2006), which is consistent with the role of YtfE in the repair of clusters damaged by NO, though the biochemical details of the suggested repair

process have not been described. Like the NsrR-regulated *hmp* gene, the *S.* Typhimurium *ytfE* gene is upregulated at the 8 h time point in macrophages (Eriksson *et al.*, 2003) and *ytfE* is upregulated in UPEC and asymptomatic strains of *E. coli* in the urinary tract (Snyder *et al.*, 2004; Roos and Klemm, 2006). Thus, *ytfE* may have a role during infection in the response to host-derived NO. The [Fe-S] cluster repair activity of YtfE activity may be required to return NsrR to its resting state in cells recovering from NO exposure (Overton *et al.*, 2008). In other studies, the repair of nitrosylated clusters has been reported to occur in aerobically growing cells in the presence of a protein synthesis inhibitor, with repair requiring only L-cysteine and no protein factors (Ren *et al.*, 2008). Iron-sulfur cluster biogenesis is a virulence determinant in the plant pathogen *Erwinia chrysanthemi* (Rincon-Enriquez *et al.*, 2008); whether the same is true for enteric pathogens with animal hosts has not been addressed. Given that both oxygen radicals and RNIs attack [Fe-S] clusters, the intracellular environments of phagocytes is likely to be inimical for clusters and cluster repair/biogenesis processes are likely to be important.

3.4.2 Methionine sulfoxide repair processes

Both oxygen radicals and RNIs can oxidize protein methionine residues to methionine sulfoxide, Met(O), which exists in two enantiomers referred to as Met-R(O) and Met-S(O). MsrA and MsrB are methionine sulfoxide reductases that are unrelated in sequence and are specific for Met-S(O) and Met-R(O), respectively (Weissbach *et al.*, 2002; Boschi-Muller *et al.*, 2008). An *msrA* mutant of *E. coli* is hypersensitive to the nitrosative stress imposed by GSNO or acidified nitrite, though only in the presence of oxygen, suggesting that the oxidant involved is peroxynitrite (John *et al.*, 2001). The methionine sulfoxide reductases are virulence determinants in several plant and animal pathogens, though the basis for the attenuation of *msrA* and *msrB* mutants is not well understood (Ezraty *et al.*, 2005) and the role that these genes play in *Salmonella* and pathogenic *E. coli* strains has not been addressed.

3.4.3 DNA repair and NO

NO itself is not known to damage DNA, but other RNIs, including peroxynitrite, can modify DNA directly and NO acts synergistically with hydrogen peroxide to damage DNA (Wink *et al.*, 1991; Woodmansee and Imlay, 2003; Niles *et al.*, 2006). The mutagenic effects of NO require oxygen, confirming that NO congeners rather than NO per se are the mutagenic agent (Weiss, 2006). Assay of the NO sensitivity of mutant strains suggested that processing of abasic sites, the SOS response, recombinational repair and double-strand end repair all played roles in the *E. coli* response to RNIs (Spek *et al.*, 2001). The *E. coli* SoxRS regulon protects against macrophage-derived NO and includes genes with roles in DNA repair (Nunoshiba *et al.*, 1993). In *S.* Typhimurium the base excision repair pathway is required for growth in macrophages and for virulence in a mouse model, and protects against the genotoxic effects of host-derived RNIs (Richardson *et al.*, 2009; Suvarnapunya *et al.*, 2003). Thus, there is a substantial body of evidence to indicate that DNA repair processes play an important role in enteric pathogens exposed to NO and its congeners in host cells.

3.4.4 Nitrotyrosine formation and repair

The peroxynitrite that results from the reaction of NO and superoxide can nitrate tyrosine residues in proteins, a modification that inactivates some enzymes. The extent to which tyrosine nitration contributes to the killing activity of phagocytes is not known. *Salmonella*-induced septic shock causes nitration of specific host proteins in a mouse model (Ghesquière *et al.*, 2009), but the extent to which pathogen proteins are nitrated is not known. One study confirmed the production of peroxynitrite in *S.* Typhimurium-infected macrophages, but also showed that nitro-

tyrosine did not co-localize with bacteria (Vazquez-Torres *et al.*, 2000). Another study drew a similar conclusion for wild-type *S.* Typhimurium, but also showed that after infection with a SPI2-deficient mutant (which failed to interfere with the localization of iNOS), nitrotyrosine staining could be observed in the vicinity of bacteria (Chakravortty *et al.*, 2002). If tyrosine nitration is a significant component of the antimicrobial activity of phagocytes, then it might also be the case that pathogens such as *S.* Typhimurium have a mechanism to repair nitrated proteins. In general, very little is known about the fate of nitrated proteins or of free nitrotyrosine in any experimental system. One study found no evidence of an ability to repair nitrated proteins in *E. coli* (Lightfoot *et al.*, 2000). Nitration of the ArgR repressor *in vitro* causes its degradation (McClean *et al.*, 2010a). In eukaryotes, there is evidence of a denitrase activity that converts protein nitrotyrosine to tyrosine (Kamisaki *et al.*, 1998; Smallwood *et al.*, 2007) and of a metabolic pathway that catabolizes free nitrotyrosine (Blanchard-Fillion *et al.*, 2006). A similar pathway may exist in *E. coli*, though the enzymes involved are not widely distributed in strains of enteric pathogens (Rankin *et al.*, 2008). The contribution (if any) that nitrotyrosine repair and/or metabolism make to bacterial resistance to phagocyte killing remains unknown.

3.5 Other Ways to Avoid Killing by iNOS

3.5.1 Inhibition of iNOS activity and expression

Unlike Phox, iNOS is not associated with the phagosome membrane and is thought generally to have a cytosolic location. Nevertheless, there is evidence that iNOS might be trafficked to the phagosomes of *S.* Typhimurium-infected macrophages and that (unidentified) products of the SPI2 pathogenicity island have a role to play in inhibiting this iNOS trafficking (Chakravortty *et al.*, 2002). Thus, iNOS co-localizes with SPI2-deficient mutants, but has a more diffuse distribution in cells infected with wild-type *S.* Typhimurium. This mechanism may be analogous to the SPI2-mediated inhibition of the trafficking of Phox (Vazquez-Torres *et al.*, 2000).

Besides inhibition of trafficking, enteric pathogens may also be able to influence the expression and activity of iNOS. Interferon-γ (IFN-γ) regulates expression of iNOS, along with genes involved in the biosynthesis of arginine and tetrahydrobiopterin (iNOS substrate and cofactor, respectively). The SPI2 product, SpiC, is a secreted virulence factor that inhibits IFN-γ signalling indirectly (Uchiya and Nikai, 2005); thus, SpiC acts to inhibit the upregulation of iNOS and so reduce the production of NO in IFN-γ activated macrophages. Conversely, iNOS-derived NO inhibits transcription of *spiC* and other SPI2 genes (McCollister *et al.*, 2005). The mechanism by which NO inhibits *spiC* expression is not known, though it may involve the SPI2-encoded two-component regulatory system, SsrA-SsrB. The data indicate that NO-mediated inhibition of SPI2 activity is required for the killing of *S.* Typhimurium by IFN-γ-stimulated macrophages (McCollister *et al.*, 2005). A recent study indicates that *S.* Typhimurium uses the NirC transport protein to import nitrite from the *Salmonella*-containing vacuole (a modified phagosome) and then reduces it to ammonia. Since NO can be oxidized to nitrite, NirC acts to lower the concentration of NO and is therefore proposed to counteract the NO-mediated inhibition of *spiC* expression (Das *et al.*, 2009). Pathogenic strains of *E. coli* and *Citrobacter rodentium* are also able to interfere with the IFN-γ-mediated stimulation of iNOS expression in intestinal epithelial cells (Vallance *et al.*, 2002; Maresca *et al.*, 2005; Vareille *et al.*, 2008).

The enteropathogenic and enterohaemorrhagic strains of *E. coli* (EPEC and EHEC, respectively) and the mouse pathogen, *C. rodentium*, are non-invasive; rather, they attach to the surface of intestinal epithelial cells. These pathogens form attaching and effacing (A/E) lesions, which are characterized by the formation of pedestal-like structures beneath the attached bacteria. A major strategy employed by the A/E pathogens to avoid the effects of NO seems to be the inhibition of iNOS upregulation in epithelial cells (Vallance *et al.*, 2002; Maresca *et al.*, 2005; Vareille *et al.*, 2008). The role, if any, that Hmp plays in this group of pathogens has not been examined.

3.5.2 Arginase

The production of NO by iNOS requires arginine, raising the possibility that arginine catabolism might be a defence strategy for bacterial pathogens. This strategy has been best characterized in *Helicobacter pylori*, which uses its own arginase to decrease arginine availability in host cells (Gobert *et al.*, 2001; Chaturvedi *et al.*, 2007). *H. pylori* also induces upregulation of the host arginase II and ornithine decarboxylase (ODC) in macrophages (Gobert *et al.*, 2002). Polyamines generated by the arginase/ODC pathway inhibit the translation of the iNOS mRNA (Bussiere *et al.*, 2005) and arginine depletion lowers the efficiency of translation of the iNOS mRNA by a mechanism that has not been defined (Lewis *et al.*, 2010). *S.* Typhimurium also stimulates expression of arginase II in macrophages; in this case, there was no apparent effect on the abundance of iNOS, suggesting that competition for arginine might be a significant contributor to the reduction of NO levels by arginase II (Lahiri *et al.*, 2008).

3.6 Regulatory Mechanisms

At least some of the NO defence systems described above are induced by exposure of cells to a source of NO. This effect can be seen in very simple experiments. For example, a culture exposed to NO and allowed to recover from the resulting transient growth inhibition becomes resistant to a second dose of NO (Hyduke *et al.*, 2007). Similarly, oxygen uptake by washed cell suspensions becomes largely resistant to NO if cultures are pretreated with a source of NO prior to harvest (Stevanin *et al.*, 2000). These and other observations point to the existence of regulatory proteins that, in the presence of NO, switch on the expression of genes encoding enzymes that detoxify NO. In the following sections, the various regulatory proteins that are known to mediate transcriptional responses to NO and nitrosative stress are described.

3.6.1 FNR

FNR from *E. coli*, a founding member of the FNR/CRP family, is a sensor of molecular oxygen that contains an oxygen labile [4Fe-4S] cluster (Kiley and Beinert, 2003). Besides O_2, FNR can also be inactivated by NO by the formation of dinitrosyl iron complexes that result from reaction of NO with the [Fe-S] cluster (Cruz-Ramos *et al.*, 2002). Since FNR is a repressor of the *hmp* gene (Poole *et al.*, 1996), inactivation of FNR by NO provides a mechanism for NO-mediated derepression of a defence activity (Fig. 3.2). Inactivation of FNR by NO may also lead to upregulation of the *cydAB* genes encoding the cytochrome *bd* oxidase, which endows a degree of NO resistance on respiration (Pullan *et al.*, 2007;

Mason *et al.*, 2009). An alternative possibility is that *cydAB* upregulation results from an increased activity of the ArcBA sensor–regulator pair when respiration is inhibited by NO (Hyduke *et al.*, 2007). Mutation of *arcA* in *S. enterica* serovar Enteritidis causes an increased sensitivity to RNS derived from GSNO, though the basis for this sensitivity is not known (Lu *et al.*, 2002).

3.6.2 NorR

NorR was first identified as an activator of the respiratory NO reductase gene in the denitrifying organism, *Ralstonia eutropha* (Pohlmann *et al.*, 2000). NorR activity required the endogenous production of NO from nitrite, or exogenous provision of sodium nitroprusside, suggesting that NorR was an NO sensor (Pohlmann *et al.*, 2000). The *norR* gene of *E. coli* is transcribed divergently from the *norVW* genes encoding FlRd and its redox partner, and NorR activates transcription of *norVW* in response to sources of NO or

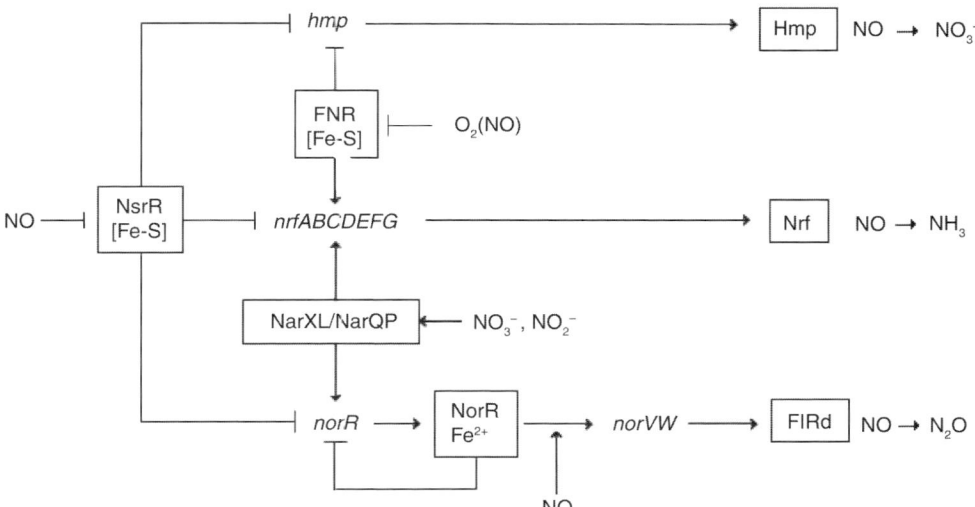

Fig. 3.2. Regulatory networks controlling expression of the three NO detoxification systems in *E. coli*. The NO-sensitive repressor, NsrR, regulates expression of the *hmp* gene (Bodenmiller and Spiro, 2006) and probably serves to fine-tune expression of the *nrf* operon encoding the respirarory nitrite reductase (Filenko *et al.*, 2007). Expression of the flavorubredoxin gene, *norV*, is activated in response to NO by NorR, and the *norR* gene is subject to negative autoregulation in an NO-independent fashion. Unpublished data indicate that *norR* expression is also modulated in response to NO by NsrR, and is regulated by the nitrate and nitrite sensing systems, NarXL and NarQP.

nitrosative stress (Gardner *et al.*, 2002, 2003; Hutchings *et al.*, 2002; Mukhopadhyay *et al.*, 2004). There are three NorR-binding sites upstream of *norVW* and the *norAB* genes of *R. eutropha* (Büsch *et al.*, 2004; Tucker *et al.*, 2004). All three binding sites are required for transcription activation (Büsch *et al.*, 2004; Tucker *et al.*, 2009).

NorR is a σ^{54}-dependent activator with a three-domain structure typical for that family of proteins: a C-terminal DNA-binding domain, a central domain from the AAA⁺ family that interacts with RNA polymerase to drive the ATP-dependent transition from closed to open complex and an N-terminal signalling domain. In NorR, the N-terminal GAF domain contains a mononuclear non-haem iron that is the binding site for a single molecule of NO. NO binding relieves the intramolecular inhibition exerted by the GAF domain on the central domain, which then becomes active for ATP hydrolysis and makes a productive interaction with RNA polymerase (D'Autréaux *et al.*, 2005). Site-directed mutagenesis, assays of iron binding, spectroscopy and structural modelling of the NorR proteins from *E. coli* and *R. eutropha* have suggested that the iron is coordinated by the side chains of three aspartate residues, an arginine and a cysteine (Klink *et al.*, 2007; Tucker *et al.*, 2008a).

3.6.3 NsrR

NsrR is a transcriptional repressor from the Rrf2 family; it was discovered in the nitrifying organism *Nitrosomonas europaea*, where it behaved as a nitrite-sensitive repressor of the gene encoding a copper nitrite reductase that reduced nitrite to NO (Beaumont *et al.*, 2004). NsrR was later shown to be an NO-sensitive repressor of the *E. coli hmp* gene, as was also predicted by comparative genomics (Rodionov *et al.*, 2005; Bodenmiller and Spiro, 2006). The balance of evidence suggests that the effector for NsrR is NO; in the case of *N. europaea*, it is probably the NO generated by nitrite reduction, rather than nitrite itself (Spiro, 2007). NsrR has since been shown also to regulate the expression of *hmp* in *S.* Typhimurium (Bang *et al.*, 2006; Gilberthorpe

et al., 2007) and genes involved in NO metabolism in a variety of other species (Rodionov *et al.*, 2005; Spiro, 2007; Tucker *et al.*, 2010). As far as is known, NorR activates only a single target in enteric bacteria, while NsrR controls a regulon of multiple genes in cases that have been examined experimentally and computationally (Rodionov *et al.*, 2005). Confirmed additional targets for NsrR regulation in *E. coli* and *S.* Typhimurium include the *hcp-hcr* genes encoding the hybrid cluster protein and its redox partner, the *ytfE* gene implicated in [Fe-S] cluster repair (see above) and a gene of unknown function, *ygbA* (Bodenmiller and Spiro, 2006; Filenko *et al.*, 2007; Gilberthorpe *et al.*, 2007). In *E. coli*, chromatin immunoprecipitation and microarray analysis (ChIP-chip) has identified ~60 NsrR-binding sites in intergenic regions in the chromosome, implicating the adjacent genes as possible NsrR targets. Confirmed NsrR targets from this data set include genes involved in the degradation of aromatic amines and genes involved in motility (Rankin *et al.*, 2008; Partridge *et al.*, 2009). The ChIP-chip data also revealed the presence of an NsrR-binding site in the *norR-norVW* intergenic region (Partridge *et al.*, 2009). Unpublished data from the author's laboratory confirms that NsrR acts as a repressor of *norR*, though the effect is quite small (twofold elevated *norR* expression in an *nsrR* mutant). Thus, the two major NO-responsive regulatory networks of *E. coli* are not completely independent of one another, but apparently are arranged in a hierarchical fashion (Fig. 3.2). Additional unpublished data indicate that the two-component nitrate and nitrite sensing systems, NarXL and NarQP, also serve to fine-tune the expression of *norR* (Fig. 3.2).

NsrR proteins from three organisms have been expressed and purified from *E. coli*; in all cases, the purified protein contains an [Fe-S] cluster. The proteins from *Neisseria gonorrhoeae* and *Streptomyces coelicolor* purified in the presence of oxygen contained a [2Fe-2S] cluster (Tucker *et al.*, 2008b; Isabella *et al.*, 2009). The *B. subtilis* protein, modified with a hexa-histidine tag and purified in the absence of oxygen, contained a [4Fe-4S] cluster (Yukl *et al.*, 2008). The proteins from *S. coelicolor* and *B. subtilis* react with NO to form dinitrosyl iron complexes, and reaction of the S. *coelicolor* and N. *gonorrhoeae* proteins with NO lowers their affinity for DNA (Tucker *et al.*, 2008b; Yukl *et al.*, 2008; Isabella *et al.*, 2009). Exposure of the B. *subtilis* NsrR to oxygen led to conversion of the [4Fe-4S] cluster to a [2Fe-2S] cluster, which was itself unstable to oxygen (Yukl *et al.*, 2008). In contrast, the [2Fe-2S] proteins from S. *coelicolor* and N. *gonorrhoeae* appear to be air stable (Tucker *et al.*, 2008b; Isabella *et al.*, 2009). These observations suggest either that there is heterogeneity in NsrR proteins from different organisms in the nature of the [Fe-S] cluster, or that the [2Fe-2S] proteins are degradation products that result from isolating the protein in the presence of oxygen. In the latter case, it is noteworthy that the [2Fe-2S] proteins retain the ability to bind to DNA in an NO-sensitive fashion (Tucker *et al.*, 2008b; Isabella *et al.*, 2009).

3.6.4 SoxR, IscR, Fur, OxyR and others

NO is a reactive molecule, which interacts readily with (for example) metal centres in proteins, especially haem, non-haem iron, [Fe-S] clusters and zinc. A consequence is that regulatory proteins containing these metals may have activities that are sensitive to NO, even if the proteins do not necessarily mediate a physiological response to NO exposure. Deciding whether a particular protein falls into this category or is a true NO sensor is not necessarily straightforward. A case in point is SoxR, which is activated by nitrosylation of its [2Fe-2S] cluster (Ding and Demple, 2000). Genes activated by SoxR (*via* SoxS) were reported to be required for resistance to macrophage-derived NO (Nunoshiba *et al.*, 1993). However, SoxRS-regulated genes are not upregulated consistently by NO in microarray data (Spiro, 2007), casting some doubt on the physiological relevance of the reaction of SoxR with NO.

IscR (a homologue of NsrR) is another [2Fe-2S] protein that reacts with NO (at least as judged by *in vivo* data), a reaction that might be physiologically significant. IscR-regulated genes are de-repressed by exposure of cultures to a source of NO (Hyduke *et al.*, 2007; Pullan *et al.*, 2007). Since the IscR regulon includes genes involved in the biogenesis of [Fe-S] clusters (Schwartz *et al.*, 2001), a possible physiological rationale for this response is that IscR-regulated genes play a role in the repair of clusters damaged or modified by NO (Hyduke *et al.*, 2007). An *iscR* mutant was more resistant than wild type to NO, which was suggested to be due to constitutive expression of the [Fe-S] cluster biogenesis genes (Hyduke *et al.*, 2007).

The ferric uptake regulator, Fur, is a transcriptional repressor that regulates genes involved in iron homeostasis in enteric bacteria (Andrews *et al.*, 2003). Fur contains a structural zinc site and a ferrous iron-binding site that, when occupied, stimulates the DNA-binding activity of the protein. The repressor activity of Fur is sensitive to NO *in vivo* and *in vitro*, and NO reacts with Fur to form iron–nitrosyl complexes (D'Autreaux *et al.*, 2002, 2004). Effects of NO or nitrosative stress on members of the Fur regulon are apparent in some transcriptomics experiments (Mukhopadhyay *et al.*, 2004; Justino *et al.*, 2005; Pullan *et al.*, 2007), but not others (Flatley *et al.*, 2005; Hyduke *et al.*, 2007; Jarboe *et al.*, 2008), leaving open the question of whether Fur mediates a physiological response to NO and RNS. Fur is a target for NO in S. Typhimurium, as is the PhoPQ two-component regulatory system (Bourret *et al.*, 2008, 2009).

Many investigators interested in responses to NO have used compounds such as GSNO as an NO mimic or source. In fact, while GSNO does release NO slowly by homolytic decomposition, it may also nitrosate cysteine thiols directly, an activity not normally associated with NO per se. Thus, GSNO and related compounds (such as S-nitrosocysteine) impose a 'nitrosative stress', this being defined originally as a stress caused by a treatment that results in nitrosation of protein thiols (Hausladen *et al.*, 1996). Therefore, NO itself does not cause nitrosative stress, *sensu stricto* (Hausladen *et al.*, 1996). Indeed, comparison of array data for cultures treated with NO (Hyduke *et al.*, 2007; Pullan *et al.*, 2007) and GSNO (Flatley *et al.*, 2005; Jarboe *et al.*, 2008) reveals overlapping but non-identical effects. Importantly, some regulatory proteins may be sensitive to nitrosative stress (because they contain reactive thiols) but not to NO per se. One such example may be OxyR (Hausladen *et al.*, 1996; Kim *et al.*, 2002), a case that has been discussed extensively in the literature (Poole *et al.*, 2004). OxyR is required for the transcriptional response to peroxynitrite and regulates expression of two peroxynitrite detoxification activities (McLean *et al.*, 2010a). Treatment with GSNO probably perturbs the cellular pools of cysteine and homocysteine, which has downstream effects on regulatory proteins such as MetR, MetJ and CysB (Flatley *et al.*, 2005; Jarboe *et al.*, 2008). GSNO also inhibits the ArcBA sensor regulator system (Jarboe *et al.*, 2008), perhaps because the activity of the sensor kinase ArcB depends on two redox active cysteine residues (Malpica *et al.*, 2004). Note that the direct inactivation of ArcB by GSNO contrasts with the indirect activation by NO mentioned earlier (Jarboe *et al.*, 2008). Treatment with GSNO also inactivates zinc proteins (Schapiro *et al.*, 2003).

3.7 Conclusions and Prospects

At this point in time there seems little doubt that NO and the RNS derived from NO are an important component of the antimicrobial activity of eukaryotic phagocytes. Moreover, it is well established that the ability of some pathogens to resist phagocyte killing involves interfering with NO production by iNOS, detoxifying NO and/or repairing the damage caused by NO and RNS. However, many interesting questions remain unanswered. For example, if NO and RNS actually kill intraphagosomal bacteria, what are the killing mechanisms and what are the principal molecular targets involved? Are the Phox and iNOS pathways synergistic (Fang, 2004) and, therefore, do RNS such as peroxynitrite contribute to killing? Is nitration of protein tyrosine residues significant and do pathogens have mechanisms to repair nitrated proteins and/or to dispose of nitrotyrosine? Are other repair activities (for example, for damaged iron-sulfur clusters) important? Why are there multiple regulatory proteins dedicated to sensing NO (such as NsrR and NorR) and multiple detoxification activities? What contribution does NO make to the killing of extracellular pathogens, such as those colonizing the urinary tract? What contribution does NO make in non-phagocytic cells, for example epithelial cells of the urinary tract that may be invaded by UPEC? The role that NO plays in host interactions with other pathogens is also, of course, of considerable interest. While the focus of this chapter has been on *E. coli* and *Salmonella* species, other chapters in this book illustrate that the importance of NO extends to other pathogens, including *Pseudomonas aeruginosa* and *Neisseria* species, and Gram-positive pathogens. Indeed, responses to NO and nitrosative stress are important in fungal pathogens (Missall *et al.*, 2006); doubtless, there is much that remains to be learned about the key role of NO in the host–pathogen interaction.

Acknowledgements

I am grateful to Jarrod Payne, Hou-Pu Liu and Jonathan Partridge for contributing unpublished data used in Figs 3.1 and 3.2, and to Alain Gobert for comments on the manuscript. Work in my laboratory is funded by award MCB-0702858 from the National Science Foundation. We recently re-sequenced the *norV* gene of UPEC strain CFT073 and found that the frameshift mutation reported in sequence databases is an error. Our data indicate that the *norV* gene is intact in this strain. We are able to detect FIRd activity in CFT073 extracts, confirming that the *norVW* genes express functional products.

Recently, it was shown that DNA binding by the *S. typhimurium* SP12-encoded response regulator SsrB is inactivated by modification of a cysteine residue by nitrogen radicals. Substitution of the cysteine residue by serine reduced the fitness of *S. typhimurium* in a mouse model (Husain *et al.*, 2010).

References

Abello, N., Kerstjens, H.A.M., Postma, D.S. and Bischoff, R. (2009) Protein tyrosine nitration: selectivity, physicochemical and biological consequences, denitration, and proteomics methods for the identification of tyrosine-nitrated proteins. *Journal of Proteome Research* 8, 3222–3238.

Ammendola, S., Pasquali, P., Pacello, F., Rotilio, G., Castor, M., Libby, S.J., Figueroa-Bossi, N., Bossi, L., Fang, F.C. and Battistoni, A. (2008) Regulatory and structural differences in the Cu,Zn-superoxide dismutases of *Salmonella enterica* and their significance for virulence. *Journal of Biological Chemistry* 283, 13688–13699.

Andrews, S., Robinson, A. and Rodriguez-Quinones, F. (2003) Bacterial iron homeostasis. *FEMS Microbiology Reviews* 27, 215–237.

Bang, I., Liu, L., Vazquez-Torres, A., Crouch, M., Stamler, J.S. and Fang, F.C. (2006) Maintenance of nitric oxide and redox homeostasis by the *Salmonella* flavohemoglobin *hmp*. *Journal of Biological Chemistry* 281, 28039–28047.

Beaumont, H.J., Lens, S.I., Reijnders, W.N., Westerhoff, H.V. and van Spanning, R.J. (2004) Expression of nitrite reductase in *Nitrosomonas europaea* involves NsrR, a novel nitrite-sensitive transcription repressor. *Molecular Microbiology* 54, 148–158.

Blanchard-Fillion, B., Prou, D., Polydoro, M., Spielberg, D., Tsika, E., Wang, Z., Hazen, S.L., Koval, M., Przedborski, S. and Ischiropoulos, H. (2006) Metabolism of 3-nitrotyrosine induces apoptotic death in dopaminergic cells. *Journal of Neuroscience* 26, 6124–6130.

Bodenmiller, D.M. and Spiro, S. (2006) The *yjeB* (*nsrR*) gene of *Escherichia coli* encodes a nitric oxide sensitive transcriptional regulator. *Journal of Bacteriology* 188, 874–881.

Boschi-Muller, S., Gand, A. and Branlant, G. (2008) The methionine sulfoxide reductases: catalysis and substrate specificities. *Archives of Biochemistry and Biophysics* 474, 266–273.

Bourrett, T.J., Porwollik, S., McClelland, M., Zhao, R., Greco, T., Ischiropoulos, H. and Vazquez-Torres, A. (2008) Nitric oxide antagonizes the acid tolerance response that protects *Salmonella* against innate gastric defenses. *PLoS ONE* 3, e1833.

Bourrett, T.J., Song, M. and Vazquez-Torres, A. (2009) Codependent and independent effects of nitric oxide-mediated suppression of PhoPQ and *Salmonella* pathogenicity island 2 on intracellular *Salmonella enterica* serovar Typhimurium survival. *Infection and Immunity* 77, 5107–5115.

Bower, J.M. and Mulvey, M.A. (2006) Polyamine-mediated resistance of uropathogenic *Escherichia coli* to nitrosative stress. *Journal of Bacteriology* 188, 928–933.

Bower, J.M., Gordon-Raagas, H.B. and Mulvey,

M.A. (2009) Conditioning of uropathogenic *Escherichia coli* for enhanced colonization of host. *Infection and Immunity* 77, 2104–2112.

Brandes, N., Rinck, A., Leichert, L.I. and Jakob, U. (2007) Nitrosative stress treatment of *E. coli* targets distinct set of thiol-containing proteins. *Molecular Microbiology* 66, 901–914.

Browning, D.F., Lee, D.J., Spiro, S. and Busby, S.J. (2010) Down-regulation of the *Escherichia coli* K-12 *nrf* promoter by binding of the NsrR nitric oxide-sensing transcription repressor to an upstream site. *Journal of Bacteriology* 192, 3824–3828.

Bryk, R., Griffin, P. and Nathan, C. (2000) Peroxynitrite reductase activity of bacterial peroxiredoxins. *Nature* 407, 211–215.

Buchmeier, N.A., Libby, S.J., Xu, Y., Loewen, P.C., Switala, J., Guiney, D.G. and Fang, F.C. (1995) DNA repair is more important than catalase for *Salmonella* virulence in mice. *Journal of Clinical Investigation* 95, 1047–1053.

Büsch, A., Pohlmann, A., Friedrich, B. and Cramm, R. (2004) A DNA region recognized by the nitric oxide-responsive transcriptional activator NorR is conserved in ß- and γ-proteobacteria. *Journal of Bacteriology* 186, 7980–7987.

Bussiere, F.I., Chaturvedi, R., Cheng, Y., Gobert, A.P., Asim, M., Blumberg, D.R., Xu, H., Hacker, A., Casero, R.A. Jr and Wilson, K.T. (2005) Spermine causes loss of innate immune response to *Helicobacter pylori* by inhibition of inducible nitric oxide synthase translation. *Journal of Biological Chemistry* 280, 2409–2412.

Chakravortty, D., Hansen-Wester, I. and Hensel, M. (2002) *Salmonella* Pathogenicity Island 2 mediates protection of intracellular *Salmonella* from reactive nitrogen intermediates. *Journal of Experimental Medicine* 195, 1155–1166.

Chaturvedi, R., Asim, M., Lewis, N.D., Algood, H.M., Cover, T.L., Kim, P.Y. and Wilson, K.T. (2007) L-arginine availability regulates inducible nitric oxide synthase-dependent host defense against *Helicobacter pylori*. *Infection and Immunity* 75, 4305–4315.

Chen, L., Xie, Q.W. and Nathan, C. (1998) Alkyl hydroperoxide reductase subunit C (AhpC) protects bacterial and human cells against reactive nitrogen intermediates. *Molecular Cell* 1, 795–805.

Costa, C., Moura, J.J., Moura, I., Liu, M.Y., Peck, H.D., LeGall, J., Wang, Y.N. and Huynh, B.H. (1990) Hexaheme nitrite reductase from *Desulfovibrio desulfuricans*. Mössbauer and EPR characterization of the heme groups. *Journal of Biological Chemistry* 265, 14382–14388.

Craig, M. and Slauch, J.M. (2009) Phagocytic superoxide specifically damages an extracytoplasmic target to inhibit or kill *Salmonella*. *PLoS ONE* 4, e4975.

Cruz-Ramos, H., Crack, J., Wu, G., Hughes, M.N., Scott, C., Thomson, A.J., Green, J. and Poole, R.K. (2002) NO sensing by FNR: regulation of the *Escherichia coli* NO-detoxifying flavo-haemoglobin, Hmp. *EMBO Journal* 21, 3235–3244.

D'Autréaux, B., Touati, D., Bersch, B., Latour, J.M. and Michaud-Soret, I. (2002) Direct inhibition by nitric oxide of the transcriptional ferric uptake regulation protein via nitrosylation of the iron. *Proceedings of the National Academy of Sciences of the United States of America* 99, 16619–16624.

D'Autréaux, B., Horner, O., Oddou, J., Jeandey, C., Gambarelli, S., Berthomieu, C., Latour, J.M. and Michaud-Soret, I. (2004) Spectroscopic description of the two nitrosyl-iron complexes responsible for Fur inhibition by nitric oxide. *Journal of the American Chemical Society* 126, 6005–6016.

D'Autréaux, B., Tucker, N.P., Dixon, R. and Spiro, S. (2005) A non-haem iron centre in the transcription factor NorR senses nitric oxide. *Nature* 437, 769–772.

Das, P., Lahiri, A., Lahiri, A. and Chakravortty, D. (2009) Novel role of the nitrite transporter NirC in *Salmonella* pathogenesis: SPI2-dependent suppression of inducible nitric oxide synthase in activated macrophages. *Microbiology* 155, 2476–2489.

De Groote, M.A., Ochsner, U.A., Shiloh, M.U., Nathan, C., McCord, J.M., Dinauer, M.C., Libby, S.J., Vazquez-Torres, A., Xu, Y. and Fang, F.C. (1997) Periplasmic superoxide dismutase protects *Salmonella* from products of phagocyte NADPH-oxidase and nitric oxide synthase. *Proceedings of the National Academy of Sciences of the United States of America* 94, 13997–14001.

Ding, H. and Demple, B. (2000) Direct nitric oxide signal transduction via nitrosylation of iron-sulfur centers in the SoxR transcription activator. *Proceedings of the National Academy of Sciences of the United States of America* 97, 5146–5150.

Einsle, O., Messerschmidt, A., Huber, R., Kroneck, P.M.H. and Neese, F. (2002) Mechanism of the six-electron reduction of nitrite to ammonia by cytochrome *c* nitrite reductase. *Journal of the American Chemical Society* 124, 11737–11745.

Eriksson, S., Lucchini, S., Thompson, A., Rhen, M. and Hinton, J.C. (2003) Unravelling the biology of macrophage infection by gene expression profiling of intracellular *Salmonella enterica*. *Molecular Microbiology* 47, 103–118.

Ezraty, B., Aussel, L. and Barras, F. (2005) Methionine sulfoxide reductases in prokaryotes. *Biochimica et Biophysica Acta* 1703, 221–229.

Fang, F.C. (2004) Antimicrobial reactive oxygen and nitrogen species: concepts and controversies. *Nature Reviews Microbiology* 2, 820–832.

Fang, F.C. and Nathan, C.F. (2007) Man is not a mouse: reply. *Journal of Leukocyte Biology* 81, 580.

Fang, F.C. and Vazquez-Torres, A. (2002) Nitric oxide production by human macrophages: there's NO doubt about it. *American Journal of Physiology – Lung Cellular and Molecular Physiology* 282, L941–943.

Filenko, N., Spiro, S., Browning, D.F., Squire, D., Overton, T.W., Cole, J. and Constantinidou, C. (2007) The NsrR regulon of *Escherichia coli* K-12 includes genes encoding the hybrid cluster protein and the periplasmic, respiratory nitrite reductase. *Journal of Bacteriology* 189, 4410–4417.

Flannagan, R.S., Cosío, G. and Grinstein, S. (2009) Antimicrobial mechanisms of phagocytes and bacterial evasion strategies. *Nature Reviews Microbiology* 7, 355–366.

Flatley, J., Barrett, J., Pullan, S., Hughes, M.N., Green, J. and Poole, R.K. (2005) Transcriptional responses of *Escherichia coli* to *S*-nitrosoglutathione under defined chemostat conditions reveal major changes in methionine biosynthesis. *Journal of Biological Chemical* 280, 10065–10072.

Francis, K.P., Taylor, P.D., Inchley, C.J. and Gallagher, M.P. (1997) Identification of the *ahp* operon of *Salmonella typhimurium* as a macrophage-induced locus. *Journal of Bacteriology* 179, 4046–4048.

Gardner, A.M., Helmick, R.A. and Gardner, P.R. (2002) Flavorubredoxin, an inducible catalyst for nitric oxide reduction and detoxification in *Escherichia coli*. *Journal of Biological Chemistry* 277, 8172–8177.

Gardner, A.M., Gessner, C.R. and Gardner, P.R. (2003) Regulation of the nitric oxide reduction operon (*norRVW*) in *Escherichia coli*. Role of NorR and σ^{54} in the nitric oxide stress response. *Journal of Biological Chemistry* 278, 10081–10086.

Gardner, P.R. (2005) Nitric oxide dioxygenase function and mechanism of flavohemoglobin, hemoglobin, myoglobin and their associated reductases. *Journal of Inorganic Biochemistry* 99, 247–266.

Gardner, P.R., Costantino, G., Szabo, C. and

Salzman, A.L. (1997) Nitric oxide sensitivity of the aconitases. *Journal of Biological Chemistry* 272, 25071–25076.

Gardner, P.R., Gardner, A.M., Martin, L.A. and Salzman, A.L. (1998) Nitric oxide dioxygenase: an enzymic function for flavohemoglobin. *Proceedings of the National Academy of Sciences of the United States of America* 95, 10378–10383.

Ghesquière, B., Colaert, N., Helsens, K., Dejager, L., Vanhaute, C., Verleysen, K., Kas, K., Timmerman, E., Goethals, M., Libert, C., Vandekerckhove, J. and Gevaert, K. (2009) *In vitro* and *in vivo* protein-bound tyrosine nitration characterized by diagonal chromatography. *Molecular and Cellular Proteomics* 8, 2642–2652.

Gilberthorpe, N.J., Lee, M.E., Stevanin, T.M., Read, R.C. and Poole, R.K. (2007) NsrR: a key regulator circumventing *Salmonella enterica* serovar Typhimurium oxidative and nitrosative stress *in vitro* and in IFN-γ-stimulated J774.2 macrophages. *Microbiology* 153, 1756–1771.

Gobert, A.P., McGee, D.J., Akhtar, M., Mendz, G.L., Newton, J.C., Cheng, Y., Mobley, H.L. and Wilson, K.T. (2001) *Helicobacter pylori* arginase inhibits nitric oxide production by eukaryotic cells: a strategy for bacterial survival. *Proceedings of the National Academy of Sciences of the United States of America* 98, 13844–13849.

Gobert, A.P., Cheng, Y., Wang, J.Y., Boucher, J.L., Iyer, R.K., Cederbaum, S.D., Casero, R.A. Jr, Newton, J.C. and Wilson, K.T. (2002) *Helicobacter pylori* induces macrophage apoptosis by activation of arginase II. *Journal of Immunology* 168, 4692–4700.

Gomes, C.M., Silva, G., Oliveira, S., LeGall, J., Liu, M.Y., Xavier, A.V., Rodrigues-Pousada, C. and Teixeira, M. (1997) Studies on the redox centers of the terminal oxidase from *Desulfovibrio gigas* and evidence for its interaction with rubredoxin. *Journal of Biological Chemistry* 272, 22502–22508.

Gomes, C.M., Vicente, J.B., Wasserfallen, A. and Teixeira, M. (2000) Spectroscopic studies and characterization of a novel electron-transfer chain from *Escherichia coli* involving a flavorubredoxin and its flavoprotein reductase partner. *Biochemistry* 39, 16230–16237.

Gomes, C.M., Giuffre, A., Forte, E., Vicente, J.B., Saraiva, L.M., Brunori, M. and Teixeira, M. (2002) A novel type of nitric-oxide reductase. *Escherichia coli* flavorubredoxin. *Journal of Biological Chemistry* 277, 25273–25276.

Gusarov, I. and Nudler, E. (2005) NO-mediated cytoprotection: instant adaptation to oxidative stress in bacteria. *Proceedings of the National Academy of Sciences of the United States of America* 102, 13855–13860.

Hausladen, A., Privalle, C.T., Keng, T., DeAngelo, J. and Stamler, J.S. (1996) Nitrosative stress: activation of the transcription factor OxyR. *Cell* 86, 719–729.

Hausladen, A., Gow, A.J. and Stamler, J.S. (2001) Flavohemoglobin denitrosylase catalyzes the reaction of a nitroxyl equivalent with molecular oxygen. *Proceedings of the National Academy of Sciences of the United States of America* 98, 10108–10112.

Hébrard, M., Viala, J.P.M., Méresse, S., Barras, F. and Aussel, L. (2009) Redundant hydrogen peroxide scavengers contribute to *Salmonella* virulence and oxidative stress resistance. *Journal of Bacteriology* 191, 4605–4614.

Husain, M., Bourrett, T.J., McCollister, B.D., Jones-Carson, J., Laughlin, J. and Vazquez-Torres, A. (2008) Nitric oxide evokes an adaptive response to oxidative stress by arresting respiration. *Journal of Biological Chemistry* 283, 7682–7689.

Hutchings, M.I., Mandhana, N. and Spiro, S. (2002) The NorR protein of *Escherichia coli* activates expression of the flavorubredoxin gene *norV* in response to reactive nitrogen species. *Journal of Bacteriology* 184, 4640–4643.

Hyduke, D.R., Jarboe, L.R., Tran, L.M., Chou, K.J.Y. and Liao, J.C. (2007) Integrated network analysis identifies nitric oxide response networks and dihydroxyacid dehydratase as a crucial target in *Escherichia coli*. *Proceedings of the National Academy of Sciences of the United States of America* 104, 8484–8489.

Isabella, V.M., Lapek, J.D. Jr, Kennedy, E.M. and Clark, V.L. (2009) Functional analysis of NsrR, a nitric oxide-sensing Rrf2 repressor in *Neisseria gonorrhoeae*. *Molecular Microbiology* 71, 227–239.

Jarboe, L.R., Hyduke, D.R., Tran, L.M., Chou, K.J.Y. and Liao, J.C. (2008) Determination of the *Escherichia coli* S-nitrosoglutathione response network using integrated biochemical and systems analysis. *Journal of Biological Chemistry* 283, 5148–5157.

John, G.S., Brot, N., Ruan, J., Erdjument-Bromage, H., Tempst, P., Weissbach, H. and Nathan, C. (2001) Peptide methionine sulfoxide reductase from *Escherichia coli* and *Mycobacterium tuberculosis* protects bacteria against oxidative damage from reactive nitrogen intermediates. *Proceedings of the National Academy of Sciences of the United States of America* 98, 9901–9906.

Justino, M.C., Vicente, J.B., Teixeira, M. and

Saraiva, L.M. (2005) New genes implicated in the protection of anaerobically grown *Escherichia coli* against nitric oxide. *Journal of Biological Chemistry* 280, 2636–2643.

Justino, M.C., Almeida, C.C., Teixeira, M. and Saraiva, L.M. (2007) *Escherichia coli* di-iron YtfE protein is necessary for the repair of stress-damaged iron-sulfur clusters. *Journal of Biological Chemistry* 282, 10352–10359.

Kamisaki, Y., Wada, K., Bian, K., Balabanli, B., Davis, K., Martin, E., Behbod, F., Lee, Y.-C. and Murad, F. (1998) An activity in rat tissues that modifies nitrotyrosine-containing proteins. *Proceedings of the National Academy of Sciences of the United States of America* 95, 11584–11589.

Keyer, K. and Imlay, J.A. (1997) Inactivation of dehydratase [4Fe-4S] clusters and disruption of iron homeostasis upon cell exposure to peroxynitrite. *Journal of Biological Chemistry* 272, 27652–27659.

Kiley, P.J. and Beinert, H. (2003) The role of Fe-S proteins in sensing and regulation in bacteria. *Current Opinion in Microbiology* 6, 181–185.

Kim, S.O., Orii, Y., Lloyd, D., Hughes, M.N. and Poole, R.K. (1999) Anoxic function for the *Escherichia coli* flavohaemoglobin (Hmp): reversible binding of nitric oxide and reduction to nitrous oxide. *FEBS Letters* 445, 389–394.

Kim, S.O., Merchant, K., Nudelman, R., Beyer, W.F., Keng, T., DeAngelo, J., Hausladen, A. and Stamler, J.S. (2002) OxyR: a molecular code for redox-related signaling. *Cell* 109, 383–396.

Klink, A., Elsner, B., Strube, K. and Cramm, R. (2007) Characterization of the signaling domain of the NO-responsive regulator NorR from *Ralstonia eutropha* H16 by site-directed mutagenesis. *Journal of Bacteriology* 189, 2743–2749.

Kulasekara, B.R., Jacobs, M., Zhou, Y., Wu, Z., Sims, E., Saenphimmachak, C., Rohmer, L., Ritchie, J.M., Radey, M., McKevitt, M., Freeman, T.L., Hayden, H., Haugen, E., Gillett, W., Fong, C., Chang, J., Beskhlebnaya, V., Waldor, M.K., Samadpour, M., Whittam, T.S., Kaul, R., Brittnacher, M. and Miller S.I. (2009) Analysis of the genome of the *Escherichia coli* O157:H7 2006 spinach-associated outbreak isolate indicates candidate genes that may enhance virulence. *Infection and Immunity* 77, 3713–3721.

Lahiri, A., Das, P. and Chakravortty, D. (2008) Arginase modulates *Salmonella* induced nitric oxide production in RAW264.7 macrophages and is required for *Salmonella* pathogenesis in mice model of infection. *Microbes and Infection* 10, 1166–1174.

Laver, J.R., Stevanin, T.M., Messenger, S.L., Lunn, A.D., Lee, M.E., Moir, J.W.B., Poole, R.K. and Read, R.C. (2010) Bacterial nitric oxide detoxification prevents host cell S-nitrosothiol formation: a novel mechanism of bacterial pathogenesis. *FASEB Journal* 24, 286–295.

Lewis, N.D., Asim, M., Barry, D.P., Singh, K., de Sablet, T., Boucher, J., Gobert, A.P., Chaturvedi, R. and Wilson, K.T. (2010) Arginase II restricts host defense to *Helicobacter pylori* by attenuating inducible nitric oxide synthase translation in macrophages. *Journal of Immunology* 184, 2572–2582.

Lightfoot, R.T., Shuman, D. and Ischiropoulos, H. (2000) Oxygen-insensitive nitroreductases of *Escherichia coli* do not reduce 3-nitrotyrosine. *Free Radical Biology and Medicine* 28, 1132–1136.

Lu, S., Killoran, P.B., Fang, F.C. and Riley, L.W. (2002) The global regulator ArcA controls resistance to reactive nitrogen and oxygen intermediates in *Salmonella enterica* Serovar Enteritidis. *Infection and Immunity* 70, 451–461.

McCollister, B.D., Bourret, T.J., Gill, R., Jones-Carson, J. and Vázquez-Torres, A. (2005) Repression of SPI2 transcription by nitric oxide-producing, IFNgamma-activated macrophages promotes maturation of *Salmonella* phago-somes. *Journal of Experimental Medicine* 202, 625–635.

McLean, S., Bowman, L.A.H., Sanguinetti, G., Read, R.C. and Poole, R.K. (2010a) Peroxynitrite toxicity in *Escherichia coli* K-12 elicits expression of oxidative stress responses, and protein nitration and nitrosylation. *Journal of Biological Chemistry* 285, 20724–20731.

McLean, S., Bowman, L.A.H. and Poole, R.K. (2010b) KatG from *Salmonella* Typhimurium is a peroxynitritase. *FEBS Letters* 584, 1628–1632.

Malpica, R., Franco, B., Rodriguez, C., Kwon, O. and Georgellis, D. (2004) Identification of a quinone-sensitive redox switch in the ArcB sensor kinase. *Proceedings of the National Academy of Sciences of the United States of America* 101, 13318–13323.

Maresca, M., Miller, D., Quitard, S., Dean, P. and Kenny, B. (2005) Enteropathogenic *Escherichia coli* (EPEC) effector-mediated suppression of antimicrobial nitric oxide production in a small intestinal epithelial model system. *Cellular Microbiology* 7, 1749–1762.

Mason, M.G., Shepherd, M., Nicholls, P., Dobbin, P.S., Dodsworth, K.S., Poole, R.K. and Cooper, C.E. (2009) Cytochrome *bd* confers nitric oxide resistance to *Escherichia coli*. *Nature Chemical Biology* 5, 94–96.

Mills, P.C., Rowley, G., Spiro, S., Hinton, J.C.D. and

Richardson, D.J. (2008) A combination of cytochrome *c* nitrite reductase (NrfA) and flavorubredoxin (NorV) protects *Salmonella enterica* serovar Typhimurium against killing by NO in anoxic environments. *Microbiology* 154, 1218–1228.

Missall, T.A., Pusateri, M.E., Donlin, M.J., Chambers, K.T., Corbett, J.A. and Lodge, J.K. (2006) Post-translational, translational, and transcriptional responses to nitric oxide stress in *Cryptococcus neoformans*: implications for virulence. *Eukaryotic Cell* 5, 518–529.

Mowat, C.G., Gazur, B., Campbell, L.P. and Chapman, S.K. (2010) Flavin-containing heme enzymes. *Archives of Biochemistry and Biophysics* 493, 37–52.

Mukhopadhyay, P., Zheng, M., Bedzyk, L.A., LaRossa, R.A. and Storz, G. (2004) Prominent roles of the NorR and Fur regulators in the *Escherichia coli* transcriptional response to reactive nitrogen species. *Proceedings of the National Academy of Sciences of the United States of America* 101, 745–750.

Niles, J.C., Wishnok, J.S. and Tannenbaum, S.R. (2006) Peroxynitrite-induced oxidation and nitration products of guanine and 8-oxoguanine: structures and mechanisms of product formation. *Nitric Oxide* 14, 109–121.

Nunoshiba, T., DeRojas-Walker, T., Wishnok, J.S., Tannenbaum, S.R. and Demple, B. (1993) Activation by nitric oxide of an oxidative-stress response that defends *Escherichia coli* against activated macrophages. *Proceedings of the National Academy of Sciences of the United States of America* 90, 9993–9997.

Overton, T.W., Justino, M.C., Li, Y., Baptista, J.M., Melo, A.M.P., Cole, J.A. and Saraiva, L.M. (2008) Widespread distribution in pathogenic bacteria of di-iron proteins that repair oxidative and nitrosative damage to iron-sulfur centers. *Journal of Bacteriology* 190, 2004–2013.

Papp-Szabò, E., Firtel, M. and Josephy, P.D. (1994) Comparison of the sensitivities of *Salmonella typhimurium oxyR* and *katG* mutants to killing by human neutrophils. *Infection and Immunity* 62, 2662–2668.

Partridge, J.D., Bodenmiller, D.M., Humphrys, M.S. and Spiro, S. (2009) NsrR targets in the *Escherichia coli* genome: new insights into DNA sequence requirements for binding and a role for NsrR in the regulation of motility. *Molecular Microbiology* 73, 680–694.

Pohlmann, A., Cramm, R., Schmelz, K. and Friedrich, B. (2000) A novel NO-responding regulator controls the reduction of nitric oxide in *Ralstonia eutropha*. *Molecular Microbiology* 38, 626–638.

Poock, S., Leach, E.R., Moir, J.W., Cole, J.A. and Richardson, D.J. (2002) Respiratory detoxification of nitric oxide by the cytochrome *c* nitrite reductase of *Escherichia coli*. *Journal of Biological Chemistry* 277, 23664–23669.

Poole, L.B., Karplus, P.A. and Claiborne, A. (2004) Protein sulfenic acids in redox signaling. *Annual Review of Pharmacology and Toxicology* 44, 325–347.

Poole, R.K. (2005) Nitric oxide and nitrosative stress tolerance in bacteria. *Biochemical Society Transactions* 33, 176–180.

Poole, R.K. and Hughes, M.N. (2000) New functions for the ancient globin family: bacterial responses to nitric oxide and nitrosative stress. *Molecular Microbiology* 36, 775–783.

Poole, R.K., Anjum, M.F., Membrillo-Hernandez, J., Kim, S.O., Hughes, M.N. and Stewart, V. (1996) Nitric oxide, nitrite, and Fnr regulation of *hmp* (flavohemoglobin) gene expression in *Escherichia coli* K-12. *Journal of Bacteriology* 178, 5487–5492.

Pullan, S.T., Gidley, M.D., Jones, R.A., Barrett, J., Stevanin, T.M., Read, R.C., Green, J. and Poole, R.K. (2007) Nitric oxide in chemostat-cultured *Escherichia coli* is sensed by Fnr and other global regulators: unaltered methionine biosynthesis indicates lack of *S* nitrosation. *Journal of Bacteriology* 189, 1845–1855.

Rankin, L.D., Bodenmiller, D.M., Partridge, J.D., Nishino, S.F., Spain, J.C. and Spiro, S. (2008) *Escherichia coli* NsrR regulates a pathway for the oxidation of 3-nitrotyramine to 4-hydroxy-3-nitrophenylacetate. *Journal of Bacteriology* 190, 6170–6177.

Ren, B., Zhang, N., Yang, J. and Ding, H. (2008) Nitric oxide-induced bacteriostasis and modification of iron-sulphur proteins in *Escherichia coli*. *Molecular Microbiology* 70, 953–964.

Richardson, A.R., Soliven, K.C., Castor, M.E., Barnes, P.D., Libby, S.J. and Fang, F.C. (2009) The base excision repair system of *Salmonella enterica* serovar Typhimurium counteracts DNA damage by host nitric oxide. *PLoS Pathogens* 5, e1000451.

Rincon-Enriquez, G., Crété, P., Barras, F. and Py, B. (2008) Biogenesis of Fe/S proteins and pathogenicity: IscR plays a key role in allowing *Erwinia chrysanthemi* to adapt to hostile conditions. *Molecular Microbiology* 67, 1257–1273.

Rodionov, D.A., Dubchak, I.L., Arkin, A.P., Alm, E.J. and Gelfand, M.S. (2005) Dissimilatory metabolism of nitrogen oxides in bacteria: comparative reconstruction of transcriptional networks. *PLoS Computational Biology* 1, e55.

Rodrigues, R., Vicente, J.B., Félix, R., Oliveira, S., Teixeira, M. and Rodrigues-Pousada, C. (2006) *Desulfovibrio gigas* flavodiiron protein affords protection against nitrosative stress *in vivo*. *Journal of Bacteriology* 188, 2745–2751.

Roos, V. and Klemm, P. (2006) Global gene expression profiling of the asymptomatic bacteriuria *Escherichia coli* strain 83972 in the human urinary tract. *Infection and Immunity* 74, 3565–3575.

Rudolf, M., Einsle, O., Neese, F. and Kroneck, P.M.H. (2002) Pentahaem cytochrome *c* nitrite reductase: reaction with hydroxylamine, a potential reaction intermediate and substrate. *Biochemical Society Transactions* 30, 649–653.

Schapiro, J.M., Libby, S.J. and Fang, F.C. (2003) Inhibition of bacterial DNA replication by zinc mobilization during nitrosative stress. *Proceedings of the National Academy of Sciences of the United States of America* 100, 8496–8501.

Schneemann, M. and Schoeden, G. (2007) Macrophage biology and immunology: man is not a mouse. *Journal of Leukocyte Biology* 81, 579.

Schwartz, C.J., Giel, J.L., Patschkowski, T., Luther, C., Ruzicka, F.J., Beinert, H. and Kiley, P.J. (2001) IscR, an Fe-S cluster-containing transcription factor, represses expression of *Escherichia coli* genes encoding Fe-S cluster assembly proteins. *Proceedings of the National Academy of Sciences of the United States of America* 98, 14895–14900.

Shatalin, K., Gusarov, I., Avetissova, E., Shatalina, Y., McQuade, L.E., Lippard, S.J. and Nudler, E. (2008) *Bacillus anthracis*-derived nitric oxide is essential for pathogen virulence and survival in macrophages. *Proceedings of the National Academy of Sciences of the United States of America* 105, 1009–1013.

Sivick, K.E. and Mobley, H.L.T. (2010) Waging war against uropathogenic *Escherichia coli*: winning back the urinary tract. *Infection and Immunity* 78, 568–585.

Smallwood, H.S., Lourette, N.M., Boschek, C.B., Bigelow, D.J., Smith, R.D., Pasa-Toli, L. and Squier, T.C. (2007) Identification of a denitrase activity against calmodulin in activated macrophages using high-field liquid chromatography-FTICR mass spectrometry. *Biochemistry* 46, 10498–10505.

Snyder, J.A., Haugen, B.J., Buckles, E.L., Lockatell, C.V., Johnson, D.E., Donnenberg, M.S., Welch, R.A. and Mobley, H.L.T. (2004) Transcriptome of uropathogenic *Escherichia coli* during urinary tract infection. *Infection and Immunity* 72, 6373–6381.

Spek, E.J., Wright, T.L., Stitt, M.S., Taghizadeh, N.R., Tannenbaum, S.R., Marinus, M.G. and Engelward, B.P. (2001) Recombinational repair is critical for survival of *Escherichia coli* exposed to nitric oxide. *Journal of Bacteriology* 183, 131–138.

Spiro, S. (2007) Regulators of bacterial responses to nitric oxide. *FEMS Microbiology Reviews* 31, 193–211.

Stevanin, T.M., Ioannidis, N., Mills, C.E., Kim, S.O., Hughes, M.N. and Poole, R.K. (2000) Flavohemoglobin Hmp affords inducible protection for *Escherichia coli* respiration, catalyzed by cytochromes *bo'* or *bd*, from nitric oxide. *Journal of Biological Chemistry* 275, 35868–35875.

Stevanin, T.M., Poole, R.K., Demoncheaux, E.A. and Read, R.C. (2002) Flavohemoglobin Hmp protects *Salmonella enterica* serovar Typhimurium from nitric oxide-related killing by human macrophages. *Infection and Immunity* 70, 4399–4405.

Stevanin, T.M., Read, R.C. and Poole, R.K. (2007) The *hmp* gene encoding the NO-inducible flavohaemoglobin in *Escherichia coli* confers a protective advantage in resisting killing within macrophages, but not *in vitro*: links with swarming motility. *Gene* 398, 62–68.

Suvarnapunya, A.E., Lagassé, H.A.D. and Stein, M.A. (2003) The role of DNA base excision repair in the pathogenesis of *Salmonella enterica* serovar Typhimurium. *Molecular Microbiology* 48, 549–559.

Svensson, L., Marklund, B.-I., Poljakovic, M. and Persson, K. (2006) Uropathogenic *Escherichia coli* and tolerance to nitric oxide: the role of flavohemoglobin. *Journal of Urology* 175, 749–753.

Svensson, L., Poljakovic, M., Säve, S., Gilberthorpe, N., Schön, T., Strid, S., Corker, H., Poole, R.K. and Persson, K. (2010) Role of flavohemoglobin in combating nitrosative stress in uropathogenic *Escherichia coli* – implications for urinary tract infection. *Microbial Pathogenesis* In press

Taylor, P.D., Inchley, C.J. and Gallagher, M.P. (1998) The *Salmonella typhimurium* AhpC polypeptide is not essential for virulence in BALB/c mice but is recognized as an antigen during infection. *Infection and Immunity* 66, 3208–3217.

Tucker, N.P., D'Autréaux, B., Studholme, D.J., Spiro, S. and Dixon, R. (2004) DNA binding activity of the *Escherichia coli* nitric oxide sensor NorR suggests a conserved target sequence in diverse proteobacteria. *Journal of Bacteriology* 186, 6656–6660.

Tucker, N.P., D'Autreaux, B., Yousafzai, F.K., Fairhurst, S.A., Spiro, S. and Dixon, R. (2008a)

Analysis of the nitric oxide-sensing non-heme iron center in the NorR regulatory protein. *Journal of Biological Chemistry* 283, 908–918.

Tucker, N.P., Hicks, M.G., Clarke, T.A., Crack, J.C., Chandra, G., Le Brun, N.E., Dixon, R. and Hutchings, M.I. (2008b) The transcriptional repressor protein NsrR senses nitric oxide directly via a [2Fe-2S] cluster. *PLoS One* 3, e3623.

Tucker, N.P., Ghosh, T., Bush, M., Zhang, X. and Dixon, R. (2009) Essential roles of three enhancer sites in σ^{54}-dependent transcription by the nitric oxide sensing regulatory protein NorR. *Nucleic Acids Research* 38, 1182–1194.

Tucker, N.P., Le Brun, N.E., Dixon, R. and Hutchings, M.I. (2010) There's NO stopping NsrR, a global regulator of the bacterial NO stress response. *Trends in Microbiology* 18, 149–156.

Uchiya, K. and Nikai, T. (2005) Salmonella pathogenicity island 2-dependent expression of suppressor of cytokine signaling 3 in macrophages. *Infection and Immunity* 73, 5587–5594.

Vallance, B.A., Deng, W., De Grado, M., Chan, C., Jacobson, K. and Finlay, B.B. (2002) Modulation of inducible nitric oxide synthase expression by the attaching and effacing bacterial pathogen *Citrobacter rodentium* in infected mice. *Infection and Immunity* 70, 6424–6435.

Vareille, M., de Sablet, T., Hindré, T., Martin, C. and Gobert, A.P. (2007) Nitric oxide inhibits Shiga-toxin synthesis by enterohemorrhagic *Escherichia coli*. *Proceedings of the National Academy of Sciences of the United States of America* 104, 10199–10204.

Vareille, M., Rannou, F., Thélier, N., Glasser, A.-L., de Sablet, T., Martin, C. and Gobert, A.P. (2008) Heme oxygenase-1 is a critical regulator of nitric oxide production in enterohemorrhagic *Escherichia coli*-infected human enterocytes. *Journal of Immunology* 180, 5720–5726.

Vazquez-Torres, A. and Fang, F.C. (2001) Oxygen-dependent anti-*Salmonella* activity of macrophages. *Trends in Microbiology* 9, 29–33.

Vazquez-Torres, A., Xu, Y., Jones-Carson, J., Holden, D.W., Lucia, S.M., Dinauer, M.C., Mastroeni, P. and Fang, F.C. (2000a) *Salmonella* pathogenicity island 2-dependent evasion of the phagocyte NADPH oxidase. *Science* 287, 1655–1658.

Vazquez-Torres, A., Jones-Carson, J., Mastroeni, P., Ischiropoulos, H. and Fang, F.C. (2000b)

Antimicrobial actions of the NADPH phagocyte oxidase and inducible nitric oxide synthase in experimental Salmonellosis. I. Effects on microbial killing by activated peritoneal macrophages *in vitro*. *Journal of Experimental Medicine* 192, 227–236.

Wasserfallen, A., Ragettli, S., Jouanneau, Y. and Leisinger, T. (1998) A family of flavoproteins in the domains Archaea and Bacteria. *European Journal of Biochemistry* 254, 325–332.

Weiss, B. (2006) Evidence for mutagenesis by nitric oxide during nitrate metabolism in *Escherichia coli*. *Journal of Bacteriology* 188, 829–833.

Weissbach, H., Etienne, F., Hoshi, T., Heinemann, S.H., Lowther, W.T., Matthews, B., St John, G., Nathan, C. and Brot, N. (2002) Peptide methionine sulfoxide reductase: structure, mechanism of action, and biological function. *Archives of Biochemistry and Biophysics* 397, 172–178.

Welch, R.A., Burland, V., Plunkett, G., Redford, P., Roesch, P., Rasko, D., Buckles, E.L., Liou, S., Boutin, A., Hackett, J., Stroud, D., Mayhew, G.F., Rose, D.J., Zhou, S., Schwartz, D.C., Perna, N.T., Mobley, H.L.T., Donnenberg, M.S. and Blattner, F.R. (2002) Extensive mosaic structure revealed by the complete genome sequence of uropathogenic *Escherichia coli*. *Proceedings of the National Academy of Sciences of the United States of America* 99, 17020–17024.

Wink, D.A., Kasprzak, K.S., Maragos, C.M., Elespuru, R.K., Misra, M., Dunams, T.M., Cebula, T.A., Koch, W.H., Andrews, A.W. and Allen, J.S. (1991) DNA deaminating ability and genotoxicity of nitric oxide and its progenitors. *Science* 254, 1001–1003.

van Wonderen, J.H., Burlat, B., Richardson, D.J., Cheesman, M.R. and Butt, J.N. (2008) The nitric oxide reductase activity of cytochrome *c* nitrite reductase from *Escherichia coli*. *Journal of Biological Chemistry* 283, 9587–9594.

Woodmansee, A.N. and Imlay, J.A. (2003) A mechanism by which nitric oxide accelerates the rate of oxidative DNA damage in *Escherichia coli*. *Molecular Microbiology* 49, 11–22.

Wu, G., Wainwright, L.M. and Poole, R.K. (2003) Microbial globins. *Advances in Microbial Physiology* 47, 255–310.

Yukl, E.T., Elbaz, M.A., Nakano, M.M. and Moënne-Loccoz, P. (2008) Transcription factor NsrR from *Bacillus subtilis* senses nitric oxide with a 4Fe-4S cluster. *Biochemistry* 47, 13084–13092.

4 Nitric Oxide and Gram-positive Pathogens: Host Triggers and Bacterial Defence Mechanisms

Glen Ulett[*] and Adam Potter

4.1 Nitric Oxide in the Host Response to Infection

4.1.1 Nitric oxide in innate immunity to Gram-positive pathogens

Nitric oxide (NO) and other reactive nitrogen species (RNS) derived from arginine (Nathan, 1992; Moncada and Higgs, 1993) have been a focus of bacterial pathogenesis studies since the gene encoding for the synthesis of NO in response to environmental stimuli, inducible nitric oxide synthase (iNOS) (Nathan and Xie, 1994a,b) formally entered the immunology scene in the host response to infection between 1985 and 1990 (Nathan, 1992; MacMicking et al., 1997; Bogdan, 2001). Derivatives of arginine metabolism including NO are key elements of innate immune responses to bacteria functioning not only in direct effector capacity but also as immunoregulatory molecules (Lyons, 1995). The antimicrobial actions of NO are now regarded as among the most broad-spectrum mechanisms of host resistance operating in innate immunity to pathogenic bacteria (Fang, 2004). NO displays detrimental effects towards viruses, intracellular pathogens, protozoa, fungi (Liew et al., 1990; Liew and Cox, 1991; Croen, 1993;

MacMicking et al., 1995; MacLean et al., 1998) and Gram-positive pathogens including Staphylococci (Malawista et al., 1992), Bacillus (Weaver et al., 2007), Listeria (Boockvar et al., 1994) and Streptococci (Hoehn et al., 1998). While the precise antimicrobial effector mechanisms of NO are not defined fully at the molecular level (Fang, 1997) (see Spiro, Chapter 3, this volume), NO diffuses readily across most biological membranes (Subczynski and Wisniewska, 2000) and displays diverse cytotoxic actions towards several intracellular targets. For example, NO destabilizes several types of proteins, modifies tyrosine residues, lipids and DNA (Wink et al., 1991; Butler and Megson, 2002; Schopfer et al., 2003), modulates gene transcription and signal transduction (Bingisser et al., 1998; Denninger and Marletta, 1999) and can inhibit transcription factors (Kroncke and Carlberg, 2000; Marshall and Stamler, 2001). The mechanisms of NO-induced protein in-stability, certainly within a host–pathogen environment, are largely unknown (Hucke et al., 2004). Other specific actions and targets of NO include the inhibition of mitochondrial electron transfer via effects on iron-containing enzymes (e.g. aconitase) and retardation of microbial metabolism via effects on reductase enzymes (e.g. ribonucleotide reductase)

* Corresponding author.

(Hibbs *et al.*, 1990; Lancaster and Hibbs, 1990) and acetyl-CoA generation (Richardson *et al.*, 2008). Reactions of NO with haem moieties can cause activation or inhibition of the affected enzymes (Cooper, 1999). Many of the antimicrobial actions of NO are thought to involve interactions of NO with oxygen that can result in the generation of reactive oxygen intermediates (ROI) with potent antimicrobial properties. In addition, NO has key immunoregulatory functions in the host [as reviewed in (Bogdan, 2001)], including recently described regulation of IgA production (Tezuka *et al.*, 2007), maintenance of immune tolerance (Oates and Gilkeson, 2006) and is used to define a functionally distinct dendritic cell subset (Serbina *et al.*, 2003).

Considering the broad range of bio-activities of NO, it is important to recognize that these depend on the molecular target (Moncada and Higgs, 1991; Moncada *et al.*, 1991; Kerwin *et al.*, 1995), and the net effect of NO towards a cell can range from cytotoxicity to cytoprotection, depending on specific conditions (Demple, 1999; Thomas *et al.*, 2002; Schopfer *et al.*, 2003; Radi, 2004). Host cells are not immune to high concentrations of NO. The existence of a detoxification mechanism based on an evolutionarily conserved *S*-nitrosoglutathione (GSNO) reductase pathway does provide some tolerance against nitrosative damage in many cell types (Liu *et al.*, 2001). Nevertheless, this means that among the RNS, NO is a prime example of a molecule that can exert either beneficial or detrimental effects when it is synthesized as part of the host inflammatory response. On the one hand, NO promotes bactericidal capacity of phagocytes including neutrophils (Malawista *et al.*, 1992, 1996; Moffat *et al.*, 1996) and macrophages (Hibbs *et al.*, 1988; Liew *et al.*, 1990; Liew and Cox, 1991; Nathan and Hibbs, 1991); however, NO also suppresses infection-induced lymphocyte proliferation (Isobe and Nakashima, 1992; Liew, 1995) and cytokine production by T helper cell clones (Taylor-Robinson *et al.*, 1994). While these activities suggest a role in moderating immune responses that are potentially damaging to the host (McInnes *et al.*, 1998), in states of immune-complex disease

NO directly causes tissue injury (Mulligan *et al.*, 1991), illustrating clear potential for immunopathology.

NO contributes to the pathogenesis of some tumours and chronic degenerative diseases (Lyons, 1995; Burgner *et al.*, 1999; Singh *et al.*, 2000; Xu *et al.*, 2002) and also mediates impaired neutrophil migration during Gram-positive bacterial sepsis, which can be reversed by iNOS inhibitors (Crosara-Alberto *et al.*, 2002). Such iNOS inhibitors have received considerable attention as possible therapeutics for the treatment of NO-mediated pathologies, such as occurs during endotoxaemia (Draisma *et al.*, 2008), acute renal failure (Heemskerk *et al.*, 2009) and chronic obstructive pulmonary disease (COPD) (Brindicci *et al.*, 2009); however, complex effects of NO hampered therapeutic uses, as will be discussed later. NO also inhibits gamma interferon-induced indoleamine 2,3-dioxygenase-mediated bacteriostatic effects of normal human cells that provide a mechanism for growth arrest of tryptophan-dependent micro-organisms (Daubener *et al.*, 1999; Hucke *et al.*, 2004), underscoring the beneficial/detrimental potential of NO.

High-level NO-producing cells such as macrophages secrete NO after encountering antigens such as lipopolysaccharide (LPS), lipoteichoic acid (LTA) (Kengatharan *et al.*, 1998), peptidoglycan and *Staphylococcal* enterotoxin B (SEB) (Moncada and Higgs, 1993; Nathan and Xie, 1994a,b; McInnes *et al.*, 1996). Several cell types in addition to macrophages also produce and respond to NO (Bogdan, 2001; Poon *et al.*, 2003; Ghaffari *et al.*, 2006). Secretion of NO is potentiated by pro-inflammatory cytokines including tumour necrosis factor alpha, interleukin-1, gamma interferon (Hibbs *et al.*, 1988; Stuehr *et al.*, 1991; Gazzinelli *et al.*, 1993; Green *et al.*, 1993; Cunha *et al.*, 1994; Paul-Clark *et al.*, 2006), granulocyte macrophage colony-stimulating factor (Carryn *et al.*, 2004) and interleukin-18 (Neighbors *et al.*, 2001), and some of these cytokines directly induce transcription of iNOS (Liew, 1994). With respect to NO induction by LTA of Gram-positive bacteria, some studies have identified LPS contamination in commercial LTA preparations, obscuring interpretation of LTA NO-triggering activity (Gao *et al.*, 2001). More recent studies using purification methods such as butanol extraction, which preserve LTA structure, enable LPS-free preparations and have facilitated a better understanding of NO responses in LTA-based assays (Morath *et al.*, 2002; Paul-Clark *et al.*, 2006). The structure of LTA appears to be key in determining the induction of iNOS in response to Gram-positive bacteria. This is in contrast to peptidoglycan, which amplifies the iNOS response due to NAG-NAM-L-ala-D-isoglutamine moieties (Kengatharan *et al.*, 1998). In other words, the origin of LTA but not peptidoglycan determines whether the cell wall components from a Gram-positive organism will synergize to induce iNOS (Kengatharan *et al.*, 1998). Several other distinct virulence factors and unique bacterial structures from Gram-positive pathogens have also been identified as NO triggers, including SEB (Florquin *et al.*, 1994), listeriolysin O (Rose *et al.*, 2001) and *Streptococcal* beta-haemolysin/cytolysin (Ring *et al.*, 2002).

4.1.2 Pathogenesis studies in animal models: role of nitric oxide

Animal model studies using Gram-positive pathogens have provided key knowledge on RNS and the role of NO in associated disease processes; most have focused on *Staphylococci*, *Bacillus*, *Listeria* and *Streptococci*. These have highlighted the complex beneficial and detrimental potential of NO during infection. In *Staphylococcal* septic arthritis for example, high-output NO synthesis is protective against disease (McInnes *et al.*, 1998); in contrast, *Bacillus* sp. models have demonstrated that NO protects invading bacteria and promotes virulence (Shatalin *et al.*, 2008; Gusarov *et al.*, 2009). Such diverse roles of NO in disease due to Gram-positive pathogens are discussed here against the background of pathogenesis studies mainly in mice over the past 15 years.

The *Staphylococcus aureus* cell wall component LTA induces iNOS (Auguet *et al.*, 1992; De Kimpe *et al.*, 1995a,b; Kengatharan

et al., 1996b; Hattor *et al.*, 1997) and synergizes with peptidoglycan to trigger high-level NO release (De Kimpe *et al.*, 1995b; Kengatharan *et al.*, 1996a). NO contributes to protection against SEB-induced shock (Florquin *et al.*, 1994) and injection of the *Staphylococcal* toxic shock syndrome toxin-1, which, like SEB, leads to T lymphocyte-mediated shock, is exacerbated by exogenous iNOS inhibitors. These observations support a net protective role of NO in *S. aureus* toxin-triggered shock syndromes. Similarly, intravenous inoculation with *S. aureus* in iNOS-deficient mice induces significantly increased frequency and severity of arthritis and septicaemia (McInnes *et al.*, 1998), underscoring broad-ranging protective roles for NO in the host response to *S. aureus*. These findings may reflect a direct role of NO in *Staphylococcal* killing or anti-*Staphylococcal* effects through the generation of peroxynitrite and superoxide (McInnes *et al.*, 1998). The inhibition of iNOS activity in mice leading to increased mortality following systemic infection with *S. aureus* appears to be unrelated to bacterial load in blood and other organs, indicating indirect effects of NO on *Staphylococcal* persistence (Sasaki *et al.*, 1998).

In a skin infection model, iNOS activity is also important for control of *S. aureus*-related disease and, in this case, relies on the quorum-sensing abilities of *S. aureus*; quorum sensing-deficient *S. aureus* are unaffected by NO, highlighting a unique link between NO-mediated host protection and a target that is a distinct bacterial virulence factor (Rothfork *et al.*, 2004). Collectively, these observations show the potential benefits of NO in antimicrobial defence against bacteria such as *S. aureus*, as highlighted elsewhere (Nathan and Hibbs, 1991). Importantly, however, NO also exhibits the potential for deleterious effects on the host during *S. aureus* infection. In multiple organ failure associated with host responses to LTA and peptidoglycan from *S. aureus*, for example, lethality occurs via an NO-dependent mechanism (De Kimpe *et al.*, 1995b). In addition, metabolic adaptation to nitrosative stress by *S. aureus* through an NO-inducible bacterial lactate dehydrogenase enables the organism to evade innate immunity, which promotes disease (Richardson *et al.*, 2008). This may represent a mechanism whereby NO induces increased pathogen virulence and compares to an inducible defence mechanism against NO in *Candida albicans* (Ullmann *et al.*, 2004). A separate study found that NO aggravated *S. aureus*-induced septic arthritis (Sakiniene *et al.*, 1997). Finally, T cell activation in response to *S. aureus* SEB is suppressed by macrophages through an NO-dependent mechanism; while suppression of lymphocyte activation by NO may help to moderate immune responses (Isobe and Nakashima, 1992), the prior examples illustrate the clear potential for NO as both a beneficial and detrimental component of the host response to *S. aureus* infection.

Models of *B. anthracis* infection in mice have shown that NO is produced as part of the inflammatory response to infection, and this contributes to killing of the pathogen (Kang *et al.*, 2008). These antibacterial actions probably occur via an NO-dependent bactericidal programme in infected macrophages, but the precise mechanism is unknown (Weaver *et al.*, 2007). The exosporium of *B. anthracis* acts to suppress NO synthesis in infected macrophages and this protects the bacteria from macrophage-mediated killing (Weaver *et al.*, 2007). Compared to the quorum sensing by *S. aureus*, this is a contrasting example of a direct link between an NO-based antimicrobial host response and subversion of this response by expression of a distinct bacterial virulence factor, namely *B. anthracis* exosporium. Intriguingly, *B. anthracis* bacilli and its endospores also exhibit arginase activity, which may drive competition for the iNOS product substrate, L-arginine, with the host (Raines *et al.*, 2006).

It is well known that many Gram-positive bacteria contain their own NO synthases, although their role in physiology and/or pathogenesis has been largely unknown. Recent studies of *B. anthracis* have revealed the surprising finding that this organism utilizes its own NO-synthase in a proactive manner to promote the organism's ability to survive within human phagocytes (Shatalin *et al.*, 2008). Could this be due to outcompeting the host for L-arginine, thereby dampening

host NO antimicrobial responses? Probably not: avirulence of anthrax spores deficient in NO production and related to the lack of bacterial NO is associated with *B. anthracis* NO-mediating activation of the antioxidant enzyme catalase. Activation of catalase suppresses the Fenton reaction, which would provide a defence mechanism against the highly toxic host-driven antibacterial oxidative burst in phagocytic cells infected with *B. anthracis*. NO is also synthesized by *B. subtilis* (Santolini *et al.*, 2006) and provides protection against sudden oxidative damage (Gusarov and Nudler, 2005). In a broader regard, endogenous NO produced by the Gram-positive *Deinococcus radiodurans* acts as a transcriptional regulator of genes involved in stress responses and growth to facilitate recovery of the bacteria after radiation damage (Patel *et al.*, 2009). Bacterial NO synthases in Gram-positive species also appear to increase antibiotic resistance through the alleviation of oxidative stress, which probably provides a fitness advantage in polymicrobial environments (Gusarov *et al.*, 2009).

Studies of listeriosis have provided conflicting data on the role of NO in the pathogenesis of disease. On the one hand, there are multiple observations that iNOS contributes little to the elimination of the causal organism, *Listeria monocytogenes*, and that NO triggered in response to the bacteria is not involved in protection against disease (Tanaka *et al.*, 1995; Fehr *et al.*, 1997) nor killing of the bacteria in either human macrophage precursors (Leenen *et al.*, 1994) or hepatocytes (Szalay *et al.*, 1995). In fact, NO expression induced by *L. monocytogenes* in macrophages and endothelial cells, which is attributable mainly to the organism's listeriolysin O (Rose *et al.*, 2001), appears to cause death of the infected host cell under some circumstances (Zwaferink *et al.*, 2008). On the other hand, several seemingly contradictory studies have shown that NO plays an important role in host control of *L. monocytogenes* infection in mouse models (Boockvar *et al.*, 1994; MacMicking *et al.*, 1995; Shiloh *et al.*, 1999; Jin *et al.*, 2001) and control of encephalitis in rats (Remer *et al.*, 2001). Findings of protection using *in vivo* models correlate closely with *in vitro* assays that have shown a protective role for NO. For example, using metabolic inhibitors to prevent NO production caused increased susceptibility to listeriosis after a few days in studies on gamma interferon-primed macrophage responses to the organism (Beckerman *et al.*, 1993). Similar *in vitro* models show cellular killing of *Listeria* via nitrite formation and nitrosation of 2,3-diaminonaphthalene (Ogawa *et al.*, 2001). Using gamma interferon-primed human monocytes, Ouadrhiri *et al.* found that inhibition of intracellular growth of *L. monocytogenes* depended only on NO and hydrogen peroxide (Ouadrhiri *et al.*, 1999). Macrophages from infected mice produce NO when cultured and exhibit listericidal effects mediated by NO (Pollock *et al.*, 1995). Macrophages appear to utilize CD14 in cooperation with TLR2 to mount an effective NO-based antibacterial response to *L. monocytogenes* (Janot *et al.*, 2008). Another study found that TLR signalling was required for macrophage NO production in response to *Listeria*, but killing of *Listeria* by activated macrophages occurred independently of TLR2 (Edelson and Unanue, 2002). Localized RNS also inhibit escape of *L. monocytogenes* from vacuoles in activated macrophages and this provides evidence of antibacterial function of NO against the bacteria (Myers *et al.*, 2003). Similar results of NO-mediated protection *in vivo* were obtained in mice lacking the gene for iNOS (MacMicking *et al.*, 1995, 1997). Killing of *Listeria* by RNS is enhanced by reactive oxygen produced by phagocyte oxidase (Shiloh *et al.*, 1999; Janot *et al.*, 2008). NO is also critical to CpG-mediated protection against the bacteria (Ito *et al.*, 2005) and impairs the intracellular growth of *L. monocytogenes* in human monocytes via a mechanism involving granulocyte macrophage colony-stimulating factor and tumour necrosis factor alpha (TNF-α) (Carryn *et al.*, 2004). Notable studies that have indicated the existence of alternative NO-independent mechanisms of killing of *L. monocytogenes* are discussed elsewhere (Fehr *et al.*, 1997; Boyartchuk *et al.*, 2004) and point to both NO⁻ dependent and NO⁻ independent pathways for the control of *L. monocytogenes* (Ogawa *et*

al., 2001). Finally, it is recognized that the stage of disease has implications for NO in antibacterial activity and pathogenesis in listeriosis (Rogers *et al.*, 1996; Dinauer *et al.*, 1997) and these effects need to be considered against the background of NO as a beneficial, detrimental or bystander molecule in listeriosis.

The role of NO in disease due to *Streptococci* has received considerable attention across several species including Lancefield Groups A (*Streptococcus pyogenes*) and B *Streptococcus* (*S. agalactiae*), as well as *S. pneumoniae*. Here, only the latter two species will be discussed. Group B *Streptococci* (GBS) cell wall components in concert with the organism's beta-haemolysin/cytolysin stimulate NO production in macrophages (Ring *et al.*, 2002), which contributes to mechanisms of apoptosis in the infected host cell (Ulett *et al.*, 2005). Phagocytosis-induced TNF-α is responsible for GBS-induced NO production in interferon-primed macrophages (Goodrum *et al.*, 1995), which is mediated through complement receptor type 3 (Goodrum *et al.*, 1994). GBS cell wall also induces toxicity in neurons by stimulating the production of NO in astrocytes and microglia (Kim and Tauber, 1996), which is dependent on interaction of TLR2 (Lehnardt *et al.*, 2006, 2007). In GBS meningitis, the bacteria induce iNOS expression in cerebral microvascular endothelial cells in pigs (Glibetic *et al.*, 2001) and NO is beneficial in a model of experimental GBS meningitis in rats by reducing cerebral ischaemia (Leib *et al.*, 1998). In mice infected intranasally with GBS, dramatic increases in iNOS expression in alveolar macrophages (Fong *et al.*, 2008) stimulate prostaglandin synthesis via cyclo-oxygenase (Maloney *et al.*, 2000; Natarajan *et al.*, 2007). Human lung epithelial cells also secrete NO in response to GBS (Doran *et al.*, 2002; Goodrum and Poulson-Dunlap, 2002), and indeed, NO inhalation therapy has been suggested as a potentially beneficial therapeutic for the clinical management of pulmonary hypertension secondary to GBS sepsis in neonates for control of GBS lung infection (Dabrowska *et al.*, 2005). GBS appears to be susceptible to gaseous NO and reportedly has a less active protection mechanism against gaseous NO cytotoxicity compared to other pathogens (Ghaffari *et al.*, 2006). However, NO also contributes to overproduction of pro-inflammatory mediators including interleukin-6 (IL-6) during GBS-induced lung inflammation and this may initiate lung injury (Raykova *et al.*, 2003). Inhibition of NO production in a mouse GBS sepsis model resulted in higher mortality and more frequent and severe arthritis (Puliti *et al.*, 2004), which compared to similar findings in iNOS-deficient mice suffering arthritis and septicaemia due to *S. aureus* (McInnes *et al.*, 1998). In this model, NO appears to function both as a direct effector and as an immunoregulatory molecule in response to GBS sepsis (Puliti *et al.*, 2004). In contrast, inhibition of iNOS during GBS sepsis in neonatal piglets was beneficial and suggestive of possible clinical benefits for neonatal sepsis (Gibson *et al.*, 1994; Rudinsky *et al.*, 1994). A subsequent study cast doubt on the inhibition of iNOS as a therapeutic tool in neonatal septic shock due to exacerbation of many of the adverse haemodynamic consequences of systemic GBS (Meadow *et al.*, 1995). Interestingly, clinical findings in neonates with GBS sepsis that showed a decrease in concentrations of NO between birth and days 5–7 of life have indicated that GBS may actually attenuate endogenous NO synthesis, which may affect disease progression (Endo *et al.*, 1999). *In vitro*, NO contributes to control of GBS growth under some conditions (Hoehn *et al.*, 1998) but not others (Ulett and Adderson, 2005), suggesting GBS strains may possess unique sets of NO-detoxifying genes, or an effect of dissimilar experimental conditions between studies. Phagocytosis of *S. pneumoniae* also triggers NO and this response contributes to killing of internalized pneumococci (Kerr *et al.*, 2004; Marriott *et al.*, 2004). Production of NO in response to *S. pneumoniae* pneumolysin occurs via a mechanism dependent on gamma interferon (Baba *et al.*, 2002). The acyl chains of pneumococci are a critical determinant of NO production in response to *S. pneumoniae* LTA (Kim *et al.*, 2005). Similar dependence on

the opacity variant type of pneumococci for NO production in an otitis media model has also been described (Long *et al.*, 2003).

4.1.3 Infection-induced NO in human cells

The seminal study that identified RNS as a potential factor in microbial disease was that of Stuehr and Marletta, which showed cytokine-inducible NO production in murine macrophages in response to *Escherichia coli* LPS (Stuehr and Marletta, 1985). Early studies on certain infections in humans revealed increased levels of NO, which provided some suggestion of an antimicrobial role of NO in human disease (Fang, 1997). Many subsequent studies investigated iNOS expression and the role of host NO synthesis in bacterial pathogenesis by focusing on the innate immune response to infection in mice which readily mount an iNOS response to a wide range of pathogens and other environmental stimuli (reviewed in Bogdan *et al.*, 2000; Bogdan, 2001). However, a collection of analogous studies based largely on primary human monocyte-derived macrophages from either healthy subjects or infected patients and cultured *ex vivo* failed to detect appreciable NO synthesis, suggesting that iNOS might not be expressed in human phagocytes (Cameron *et al.*, 1990; James *et al.*, 1990; Murray and Teitelbaum, 1992; Schneemann and Schoedon, 2002). Negative findings of NO responses in human mononuclear cells continued, despite broad use of antigenic and non-antigenic stimuli (Bermudez, 1993; Schneemann *et al.*, 1993; Ulett and Adderson, 2005). One detailed characterization of human mononuclear phagocytes from normal donors found that these cells could generate only very low levels of NO (Weinberg *et al.*, 1995), which supported the view that iNOS induction and generation of NO was not part of the antimicrobial armamentarium in human mononuclear cells. Considerable controversy over the issue of NO production by human macrophages involved recognition of the effects of differences in methodologies and cell types (MacMicking *et al.*, 1997; Fang and

Vazquez-Torres, 2002) and debate over differences between species (Schneemann and Schoedon, 2002). Since that time, however, several hundred studies have demonstrated iNOS mRNA, associated protein activity, or NO itself has a biological function in human monocytic cells/macrophages when appropriate conditions are met (Denis, 1991; Munoz-Fernandez *et al.*, 1992; Dumarey *et al.*, 1994; Reiling *et al.*, 1994; Zembala *et al.*, 1994; Weinberg, 1998; Shay *et al.*, 2003; Carryn *et al.*, 2004). Nevertheless, human mononuclear cells appear to express lower levels of iNOS than rodent cells, and some controversy remains (Schneemann and Schoedon, 2002). In contrast, studies with human hepatocytes have demonstrated NO production equal to that in rodent hepatocytes (Nussler *et al.*, 1991, 1992). This is particularly relevant considering the pathogenesis of liver conditions such as cirrhosis and hepatic damage in which iNOS has a distinct role (Billiar *et al.*, 1990; McNaughton *et al.*, 2002).

Specific detection of NO has now been demonstrated in patient tissues or primary human cells infected with *S. aureus* (Malawista *et al.*, 1992; Choi *et al.*, 1998), *S. pneumoniae* (Marriott *et al.*, 2004), *Mycobacterium tuberculosis* (Nicholson *et al.*, 1996; Sharma *et al.*, 2004) and *M. avium* (Denis, 1991). Bacterial infection also induces NO synthase in human neutrophils (Wheeler *et al.*, 1997) and this has been implicated in antimicrobial function towards *S. aureus* (Malawista *et al.*, 1992, 1996). These examples do not represent an exhaustive list but are representative of a body of current literature that illustrates human monocytes and macrophages can express functional iNOS and produce NO in response to bacterial infection (MacMicking *et al.*, 1997; Fang and Vazquez-Torres, 2002). The relevance of observations in rodent models will continue to benefit clinical studies that will prove essential to provide a more complete understanding of iNOS signalling events and the role of NO in infection of human cells. NO from iNOS has also been demonstrated in patients with non-infectious diseases such as rheumatoid and osteoarthritis (McInnes *et al.*, 1996; St Clair *et al.*, 1996; Kroncke *et al.*, 2000) and is thought to play a role in these conditions.

4.1.4 Clinical applications of nitric oxide in infection

The role of NO in bacterial disease pathogenesis and infection has led to investigation of therapeutic strategies that have taken advantage of the antimicrobial properties of NO, as well as its beneficial effects towards vasodilation, angiogenesis and wound healing (Moncada and Higgs, 1995). The two divergent approaches to therapy are increasing NO bioactivity where it is desirable to increase or mimic the actions of NO and decreasing or neutralizing NO bioactivity where there is overproduction of NO that causes excessive nitrosative stress to host cells (reviewed in Moncada and Higgs, 1995). The former includes topical formulations, transmembrane NO-generation systems (Hardwick *et al.*, 2001) and the more widely studied inhalation therapy utilizing NO gas. NO-releasing sol-gels have also proven effective as antibacterial coatings for orthopaedic implants (Nablo *et al.*, 2005). Studies on the potential of topical NO therapy in enhancing wound healing and reducing bacterial burden have used NO donors such as polyethyleneimine cellulose NONOate polymer, *S*-nitroso-*N*-acetylpenicillamine (SNAP), sodium nitroprusside (SNP) and molsidomine (nethoxycarbomyl-3-morpholinyl-sidnonimine). These agents have been administered topically with some success for various clinical conditions. For example, topical application of an acidified nitrite was used as an NO-based therapy to aid the resolution of several distinct cutaneous infections due to common bacterial and fungal pathogens (Weller *et al.*, 2001). Separate from topical applications of NO donors such as these as sources of exogenous therapeutic NO, gaseous application of air dilutions of medical-grade pure NO gas has been studied and used clinically. For example, inhaled NO has been part of the therapy for persistent pulmonary hypertension/hypoxic respiratory failure in neonates for years (Neonatal Inhaled Nitric Oxide Study Group, 1997; Cook and Stewart, 2005). This is effective in enhancing the killing of *S. aureus* and *P. aeruginosa* (Ghaffari *et al.*, 2006) and is beneficial for therapy of neonatal listeriosis where it involves severe respiratory failure (Ichiba *et al.*, 2000). In a model of GBS infection in piglets (Berger *et al.*, 1993) NO reversed pulmonary hypertension, which was worsened if endogenous NO production was inhibited (Barrington *et al.*, 2000). Pure exogenous NO gas potentially can overcome some of the limitations encountered in an NO donor system (as used in topical formulations), such as prolonged and consistent delivery of NO, dose control and toxicity of topically applied carrier molecules (Ghaffari *et al.*, 2006). In addition, for topical applications, it is difficult to predict the cytotoxicity of the end products at the infection site after the release of NO. The chemistry of NO and its derivatives is complex, so, finally, the release of NO by donors such as SNAP and SNP is dependent on temperature, pH and the nature of the carrier molecule (Maragos *et al.*, 1991), and these variables are not well understood.

The contrasting approach in targeting NO for clinical therapeutic benefits is the inhibition of NO bioactivity in diseases in which it contributes to pathology and tissue damage. This approach has been considered unsuitable for diseases such as pneumonia, in which NO inhibition would be undesirable because of its role in regulating pulmonary vascular resistance (Natarajan *et al.*, 2007). In theory, however, reduction of excess NO levels by iNOS inhibitors (Thiemermann, 1998) may benefit diseases such as septic shock by, for example, increasing systemic vascular resistance (Petros *et al.*, 1994; Brun-Buisson *et al.*, 1996). In reality, many iNOS inhibitors have had mixed success in animal models; one such agent increased mortality in a clinical trial (Richardson *et al.*, 2008) and another suggested that treatments involving suppression of iNOS induction might be unsuitable in cases of severe *S. aureus* infection as well as *Staphylococcal* superantigen-induced shock (Sasaki *et al.*, 1998). Data from a recent study of *S. aureus*-induced septic arthritis also cautioned against the clinical use of selective iNOS inhibitors in the management of septic arthritis (McInnes *et al.*, 1998). Suppression of iNOS induction and NO production also appears to be unsuitable in the management of GBS infection (Puliti *et al.*, 2004). Thus, despite inhibitors of iNOS having received

much attention for the treatment of septic shock, these seem unlikely to provide benefit in septic arthritis. An alternative to iNOS inhibitors is NO scavengers (Fischer *et al.*, 1998; Harbrecht, 2006). These are predicted to leave sufficient free NO for normal physiological functions and have provided promising data in a non-rodent *Drosophila melanogaster* model (Broderick *et al.*, 2006). Overall, however, a better understanding of NO is required to realize the significant therapeutic benefits of increasing or neutralizing NO bioactivity during infection. Therapeutic strategies need to minimize the potentially deleterious effects of NO such as inhibitory actions on neutrophil migration, while enhancing the contributions of NO to antimicrobial activity (Crosara-Alberto *et al.*, 2002). Current gaps in knowledge about NO represent significant hurdles for therapeutic application. These, when amplified in conditions such as severe polymicrobial infections (Weinstein *et al.*, 1983), reveal an even more complex physiologic role of NO; one that will require a more intricate and detailed understanding of NO to enable practical targeted therapy.

4.2 Bacterial Stress Responses to Nitric Oxide

Damage to bacterial cells as a result of exposure to RNS can occur via a number of mechanisms including nitrosation of protein thiols, nitrosylation of metal centres (particularly iron) (Wink and Mitchell, 1998), nitration of protein tyrosine residues (Schopfer *et al.*, 2003), peroxidation of lipids (Radi *et al.*, 1991) and deamination of DNA bases (Wink *et al.*, 1991). Studies on the nitrosative stress response of bacteria have focused largely on Gram-negative organisms (see earlier chapters in this volume: Imlay and Hassett, Chapter 1 and Spiro, Chapter 3), with strategies such as maintenance and replenishment of cytosolic thiol pools, altered metal homeostasis, activation of specific DNA repair processes and increased enzymatic NO detoxification well described in organisms such as *E. coli* and *Salmonella* sp. (Mukhopadhyay *et al.*, 2004; Flatley *et al.*, 2005; Justino *et al.*, 2005).

Mechanisms employed by Gram-positive bacteria to defend against RNS are less well characterized and the majority of studies have been undertaken using *S. aureus* as the model organism. Strategies characterized thus far can be classed broadly into four categories: (i) inhibition of NO production; (ii) enzymatic detoxification of NO and other RNS; (iii) altered central metabolism in response to NO stress; and (iv) repair of damage caused by RNS.

4.2.1 Inhibition of NO production

Inhibition of host NO production as a mechanism of bacterial survival during infection was first demonstrated in the Gram-negative pathogens *Salmonella typhimurium* and *Helicobacter pylori*. The type III secretion system encoded by Salmonella pathogenicity island 2 (SPI-2) was shown to interfere with the co-localization of iNOS with *S. typhimurium* in macrophages (Chakravortty *et al.*, 2002). *H. pylori* uses a different approach by expressing an enzyme (arginase) that degrades the iNOS substrate L-arginine to L-ornithine and urea, which results in the inhibition of NO production by activated macrophages (Gobert *et al.*, 2001). Proactive use of arginase by pathogenic bacteria to dampen the production of NO by the host has also been described in the Gram-positive *B. anthracis*. Endospores produced by *B. anthracis* are phagocytized by resident macrophages in the lung following inhalation and these must overcome phagocyte-mediated host defences in order to germinate, replicate and disseminate as vegetative bacilli. The endospore of *B. anthracis* contains a number of proteins that are tightly associated with the exosporium (Redmond *et al.*, 2004). Some of these proteins are known to have a role in spore germination, while a family of super-oxide dismutases (SODs) has also been identified, suggesting the exosporium is important for preventing oxidative damage (Passalacqua *et al.*, 2006). Arginase has also been identified in the exosporium of *B. anthracis* (Weaver *et al.*, 2007), suggesting a role for the exosporium in preventing damage by RNS.

4.2.2 Enzymatic detoxification of NO

A common mechanism used by microbes to counteract the deleterious effects of NO is through the induction of NO-metabolizing enzymes (see Spiro, Chapter 3, this volume). Flavo-di-iron NO reductases and flavohaemoglobins (Hmp) are able to detoxify NO directly. Flavohaemoglobins exhibit a two-domain architecture composed of an N-terminal haemoglobin-like domain and a C-terminal redox-active domain containing binding sites for NAD(P)H and FAD (Frey and Kallio, 2003). Hmp can detoxify NO aerobically or anaerobically using distinct reaction mechanisms. Under aerobic conditions the haem group of Hmp acts as a dioxygenase, whereby an O_2 molecule is incorporated into the substrate (NO) to produce nitrate. In this pathway ferrous haem is regenerated using electrons from NADH (Hausladen *et al.*, 1998). Under anaerobic conditions, NO binds to Hmp to form a nitrosyl adduct and subsequently is reduced to N_2O (Kim *et al.*, 1999). As a model for this in Gram-positive pathogens, in contrast to the Gram-negative systems (discussed in Spiro, Chapter 3, this volume), Hmp has been characterized in *S. aureus* (Nobre *et al.*, 2007) and this does appear to be involved in NO detoxification only under low oxygen conditions where expression of *hmp* is enhanced. Hmp is a key determinant of the ability of *S. aureus* to replicate in the presence of high levels of NO *in vitro* and this also translates to being an essential virulence factor during murine infection (Richardson *et al.*, 2006). Indeed, the role of Hmp in combating host-derived NO directly is evidenced clearly by the fact that virulence of an *S. aureus hmp*-deficient mutant strain is restored in iNOS-deficient mice (Richardson *et al.*, 2006). Although NO-dependent induction of *hmp* in *S. aureus* has been demonstrated (Richardson *et al.*, 2006), the mechanism behind this regulation is unknown (although it is worth noting that it seems to be co-regulated with its divergent gene, the NO-resistant and inducible lactate dehydrogenase). NO-dependent regulation of *hmp* in *B. subtilis* involves the two-component regulator, ResDE (Nakano, 2002), as well as the Rrf2 family transcription factor, NsrR (Rogstam *et al.*, 2007). The ResDE homologue in *S. aureus* (SrrAB) has only a minor effect on expression of *hmp* (Richardson *et al.*, 2006). While common NsrR target genes have been identified in a broad range of Gram-negative pathogens including *E. coli* (Tucker *et al.*, 2010), there is currently no data on the role of the *Staphylococcal* NsrR homologue in the response to RNS.

A second NO-metabolizing enzymatic system has been well characterized in Gram-negative enteric pathogens, namely the cytoplasmic enzyme flavorubredoxin (NorV) and its NADH-dependent flavorubredoxin reductase partner (NorW). These detoxify NO under anaerobic conditions (Gardner *et al.*, 2003). Reduction of NO is proposed to occur through binding of two NO molecules to the di-ferrous centre of NorV, with each subsequently reduced to form nitroxyl anions (NO⁻) and a di-ferric centre. The nitroxyl molecules can then combine to form N_2O and water, while NorW could then supply two electrons to the di-ferric centre via the rubredoxin domain and proximal flavin mononucleotide (FMN) in flavorubredoxin (Gardner *et al.*, 2002). There is some evidence that flavorubredoxin-mediated NO reduction may also occur in Gram-positive pathogens. A set of genes has been identified in the genome of *Corynebacterium diptheriae* that encode functional homologues of NorV and NorW, as well as the NO-sensing transcriptional regulator, NorR (Gupta *et al.*, 2007). However, further work is needed to define the role of these genes in the response of *C. diptheriae* to NO stress.

A novel family of nitroreductases has been characterized recently, some members of which have a dual function in protecting bacteria from transnitrosylation mediated by S-nitrosogluathione (GSNO) in addition to promoting nitrofuran activation (Tavares *et al.*, 2009). Nitrofuran derivatives are used in the treatment of *Staphylococcal* infections and the cytotoxicity of these compounds is dependent on their reduction by nitro-reductases expressed by the bacterial cell. The expression of the nitroreductase encoding *ntrA* in *S. aureus* is enhanced upon challenge with GSNO, while deletion of *ntrA* renders *S.*

aureus susceptible to killing with GSNO (Tavares *et al.*, 2009). Purification of NtrA revealed that this enzyme exhibited NAD(P) H-dependent GSNO reductase activity in addition to nitroreductase activity (Tavares *et al.*, 2009). Thus, NtrA appears to be important for metabolizing GSNO formed endogenously under conditions of nitrosative stress.

Another family of enzymes that possesses GSNO reductase activity is class III alcohol dehydrogenases. These enzymes catalyse the NADH-dependent reduction of GSNO in both bacterial and eukaryotic cells (Liu *et al.*, 2001). AdhC of *S. pneumoniae* possesses GSNO reductase activity and is required for resistance of pneumococci to challenge with GSNO *in vitro*, as well as for the ability of pneumococci to cause systemic disease in mice (Stroeher *et al.*, 2007). Expression of *adhC* is regulated by an MerR/NmlR family transcription factor (NmlR$_{sp}$) (see Kidd, Chapter 5, this volume), first described in *Neisseria gonorrhoeae* (Kidd *et al.*, 2005). NmlR of *N. gonorrhoeae* regulates the expression of thioredoxin reductase (*trxB*) and esterase D (*estD*), both of which have been shown to be important for resistance of gonococci to nitrosative stress (Potter *et al.*, 2009a,b). NmlR-like proteins have also been shown to regulate the expression of class III alcohol dehydrogenases in *Haemophilus influenzae* (Kidd *et al.*, 2007) and *N. meningitidis* (Potter *et al.*, 2007). The mechanism used by this regulator to sense RNS has not been defined; however, the conservation of certain cysteine residues among the NmlR family suggests a thiol-based switch may be involved (Kidd *et al.*, 2005).

4.2.3 Altered central metabolism in response to NO stress

Microarray and proteomic technology has shed light on the global cellular response of *S. aureus* to nitrosative stress (Richardson *et al.*, 2006; Hochgrafe *et al.*, 2008). Many aspects of the *Staphylococcal* response to NO differ significantly from the response of its non-pathogenic Gram-positive counterpart, *B. subtilis*. For example, nitrosative stress induces the σ^B-controlled general stress regulon in *B. subtilis* (Moore *et al.*, 2004; Rogstam *et al.*, 2007); this is not seen in *Staphylococci*. Of the 66 genes that are induced in *S. aureus* in response to NO, almost half are likely to be involved in iron homeostasis, which is consistent with the global transcriptional response of other bacterial species exposed to NO (Moore *et al.*, 2004; Mukhopadhyay *et al.*, 2004). This is thought to occur largely through the action of the ferric uptake regulator (Fur). NO reacts with Fur-bound ferrous iron, which results in the formation of dinitrosyl iron complexes (DNICs) and de-repression of the Fur regulon (D'Autreaux *et al.*, 2002). However, mutation of *fur* enhances significantly the susceptibility of *Staphylococci* to NO (Richardson *et al.*, 2006), suggesting that the effect of NO on the Fur regulon may be a detrimental consequence of nitrosative stress as opposed to an adaptive response. Genes involved in hypoxic/ fermentative metabolism in *S. aureus* also display altered expression under nitrosative stress (Richardson *et al.*, 2006; Hochgrafe *et al.*, 2008). The Staphylococcal respiratory regulator (SrrAB) is important for the expression of many genes involved in anaerobic metabolism (Yarwood *et al.*, 2001). Mutation of *srrAB* enhances susceptibility to NO significantly and results in decreased virulence in mice. However, virulence of *srrAB*-deficient *S. aureus* cannot be restored in iNOS-deficient mice, indicating it may have additional roles *in vivo*.

A shift to fermentative metabolism most likely reflects the fact that NO inhibits respiration by binding competitively to cytochrome haems of terminal oxidases. A recent study has demonstrated that specific induction of homolactic fermentation is an important metabolic adaptation that occurs in *S. aureus* during nitrosative stress (Richardson *et al.*, 2008). *S. aureus* expresses two lactate dehydrogenases, one of which is the NO-inducible L-lactate dehydrogenase (Ldh1). Analysis of the metabolic end products excreted by either respiring or fermenting *S. aureus* revealed almost exclusive production of L-lactate in the presence of NO. Nitrosative stress appears to restrict the glucose catabolic pathways available to *S. aureus* by inhibiting the activity of both

pyruvate dehydrogenase and pyruvate formate lyase. This would result in limited acetyl Co-A production and, subsequently, an inability to maintain redox homeostasis through NAD+ regeneration. Both lactate dehydrogenases expressed by *S. aureus* remain active in the presence of NO; thus, this pathway is essential for cells to maintain redox balance during nitrosative stress. *ldh1* is transcribed divergently from the gene encoding the NO-metabolizing enzyme, Hmp. However, the mechanism of induction of *ldh1* in the presence of NO has not been characterized, although it appears to be independent of SrrAB. Strains lacking the NO-inducible Ldh1 are attenuated in virulence, while an *ldh1/2*-deficient double mutant is essentially avirulent. Virulence of the *ldh1*-deficient mutant is restored in iNOS-deficient mice, indicating its importance in responding to host-derived NO. Interestingly, the expression of an NO-inducible lactate dehydrogenase is specific to the pathogenic *S. aureus* and is not exhibited by the closely related commensal species, *S. epidermidis* or *S. saprophyticus*.

4.2.4 Repair of damage caused by nitric oxide stress

Iron-sulfur (Fe-S) clusters are among the most ubiquitous and functionally diverse prosthetic groups in nature (Johnson *et al.*, 2005) and are particularly vulnerable to inactivation by RNS. Fe-S clusters react with NO to form a stable dinitrosyl-iron complex (DNIC) that remains bound to the protein (Ren *et al.*, 2008). The degree of oxidative/nitrosative damage of Fe-S clusters is greatly dependent on the surrounding polypeptide matrix, with those that are more solvent exposed such as enzymes of the TCA cycle (aconitase, fumarase) or the branched-chain amino acid biosynthesis pathway (dihydroxyacid dehydratase) most susceptible (Justino *et al.*, 2009). Microarray analysis revealed that a putative iron-containing protein (ScdA) was induced by NO in *S. aureus* (Richardson *et al.*, 2006). Subsequent work characterizing an *S. aureus scdA*-deficient mutant strain found that the activity of the [4Fe-4S]-containing aconitase was much lower in this strain compared to wild-type cells (Overton *et al.*, 2008). In addition, inhibition of enzyme activity in the presence of NO was greater in this strain and ScdA was found to be essential for recovery of aconitase activity following nitrosative damage (Overton *et al.*, 2008). ScdA subsequently was purified and shown to be a dimer containing two iron atoms per monomer, which was highly similar to the YtfE protein of *E. coli* (Todorovic *et al.*, 2008). Indeed, ScdA is able to repair NO-damaged fumarase in *E. coli ytfE* mutant cell lysates (Overton *et al.*, 2008). ScdA and YtfE are members of a large family (designated Ric, repair of iron centres) of proteins present in a number of bacterial species, including the Gram-positive pathogens *B. anthracis* and Clostridia (Overton *et al.*, 2008). The exact chemical reactions catalysed by this family of proteins during the repair of NO-damaged Fe-S clusters are yet to be determined. It is also unclear if the activity is restricted to repairing Fe-S centres, or if other iron-binding proteins may also act as substrates.

References

Auguet, M., Lonchampt, M.O., Delaflotte, S., Goulin-Schulz, J., Chabrier, P.E. and Braquet, P. (1992) Induction of nitric oxide synthase by lipoteichoic acid from *Staphylococcus aureus* in vascular smooth muscle cells. *FEBS Letters* 297, 183–185.

Baba, H., Kawamura, I., Kohda, C., Nomura, T., Ito, Y., Kimoto, T., Watanabe, I., Ichiyama, S. and Mitsuyama, M. (2002) Induction of gamma interferon and nitric oxide by truncated pneumolysin that lacks pore-forming activity. *Infection and Immunity* 70, 107–113.

Barrington, K.J., Etches, P.C., Schulz, R., Talbot, J.A., Graham, A.J., Pearson, R.J. and Cheung, P.Y. (2000) The hemodynamic effects of inhaled nitric oxide and endogenous nitric oxide synthesis blockade in newborn piglets during infusion of heat-killed group B streptococci. *Critical Care in Medicine* 28, 800–808.

Beckerman, K.P., Rogers, H.W., Corbett, J.A., Schreiber, R.D., McDaniel, M.L. and Unanue, E.R. (1993) Release of nitric oxide during the T cell-independent pathway of macrophage activation. Its role in resistance to *Listeria monocytogenes*. *Journal of Immunology* 150, 888–895.

Berger, J.I., Gibson, R.L., Redding, G.J., Standaert, T.A., Clarke, W.R. and Truog, W.E. (1993) Effect of inhaled nitric oxide during group B streptococcal sepsis in piglets. *American Reviews in Respiratory Diseases* 147, 1080–1086.

Bermudez, L.E. (1993) Differential mechanisms of intracellular killing of *Mycobacterium avium* and *Listeria monocytogenes* by activated human and murine macrophages. The role of nitric oxide. *Clinical and Experimental Immunology* 91, 277–281.

Billiar, T.R., Curran, R.D., Harbrecht, B.G., Stuehr, D.J., Demetris, A.J. and Simmons, R.L. (1990) Modulation of nitrogen oxide synthesis in vivo: NG-monomethyl-L-arginine inhibits endotoxin-induced nitrate/nitrate biosynthesis while promoting hepatic damage. *Journal of Leukocyte Biology* 48, 565–569.

Bingisser, R.M., Tilbrook, P.A., Holt, P.G. and Kees, U.R. (1998) Macrophage-derived nitric oxide regulates T cell activation via reversible disruption of the Jak3/STAT5 signaling pathway. *Journal of Immunology* 160, 5729–5734.

Bogdan, C. (2001) Nitric oxide and the immune response. *Nature Immunology* 2, 907–916.

Bogdan, C., Rollinghoff, M. and Diefenbach, A. (2000) Reactive oxygen and reactive nitrogen intermediates in innate and specific immunity. *Current Opinion in Immunology* 12, 64–76.

Boockvar, K.S., Granger, D.L., Poston, R.M., Maybodi, M., Washington, M.K., Hibbs, J.B. Jr and Kurlander, R.L. (1994) Nitric oxide produced during murine listeriosis is protective. *Infection and Immunity* 62, 1089–1100.

Boyartchuk, V., Rojas, M., Yan, B.S., Jobe, O., Hurt, N., Dorfman, D.M., Higgins, D.E., Dietrich, W.F. and Kramnik, I. (2004) The host resistance locus *sst1* controls innate immunity to *Listeria monocytogenes* infection in immunodeficient mice. *Journal of Immunology* 173, 5112–5120.

Brindicci, C., Ito, K., Torre, O., Barnes, P.J. and Kharitonov, S.A. (2009) Effects of aminoguanidine, an inhibitor of inducible nitric oxide synthase, on nitric oxide production and its metabolites in healthy control subjects, healthy smokers, and COPD patients. *Chest* 135, 353–367.

Broderick, K.E., Feala, J., McCulloch, A., Paternostro, G., Sharma, V.S., Pilz, R.B. and Boss, G.R. (2006) The nitric oxide scavenger cobinamide profoundly improves survival in a *Drosophila melanogaster* model of bacterial sepsis. *FASEB Journal* 20, 1865–1873.

Brun-Buisson, C., Doyon, F. and Carlet, J. (1996)

Bacteremia and severe sepsis in adults: a multicenter prospective survey in ICUs and wards of 24 hospitals. French Bacteremia-Sepsis Study Group. *American Journal of Respiratory Critical Care in Medicine* 154, 617–624.

Burgner, D., Rockett, K. and Kwiatkowski, D. (1999) Nitric oxide and infectious diseases. *Archives of Diseases in Children* 81, 185–188.

Butler, A.R. and Megson, I.L. (2002) Non-heme iron nitrosyls in biology. *Chemical Reviews* 102, 1155–1166.

Cameron, M.L., Granger, D.L., Weinberg, J.B., Kozumbo, W.J. and Koren, H.S. (1990) Human alveolar and peritoneal macrophages mediate fungistasis independently of L-arginine oxidation to nitrite or nitrate. *American Reviews in Respiratory Diseases* 142, 1313–1319.

Carryn, S., Van de Velde, S., Van Bambeke, F., Mingeot-Leclercq, M.P. and Tulkens, P.M. (2004) Impairment of growth of *Listeria monocytogenes* in THP-1 macrophages by granulocyte macrophage colony-stimulating factor: release of tumor necrosis factor-alpha and nitric oxide. *Journal of Infectious Diseases* 189, 2101–2109.

Chakravortty, D., Hansen-Wester, I. and Hensel, M. (2002) Salmonella pathogenicity island 2 mediates protection of intracellular Salmonella from reactive nitrogen intermediates. *Journal of Experimental Medicine* 195, 1155–1166.

Choi, K.C., Jeong, T.K., Lee, S.C., Kim, S.W., Kim, N.H. and Lee, K.Y. (1998) Nitric oxide is a marker of peritonitis in patients on continuous ambulatory peritoneal dialysis. *Advances in Peritological Dialysis* 14, 173–179.

Cook, L.N. and Stewart, D.L. (2005) Inhaled nitric oxide in the treatment of persistent pulmonary hypertension/ hypoxic respiratory failure in neonates: an update. *Journal of the Kentucky Medical Association* 103, 138–147.

Cooper, C.E. (1999) Nitric oxide and iron proteins. *Biochimica et Biophysica Acta* 1411, 290–309.

Croen, K.D. (1993) Evidence for antiviral effect of nitric oxide. Inhibition of herpes simplex virus type 1 replication. *Journal of Clinical Investigations* 91, 2446–2452.

Crosara-Alberto, D.P., Darini, A.L., Inoue, R.Y., Silva, J.S., Ferreira, S.H. and Cunha, F.Q. (2002) Involvement of NO in the failure of neutrophil migration in sepsis induced by *Staphylococcus aureus*. *Brazilian Journal of Pharmacology* 136, 645–658.

Cunha, F.Q., Assreuy, J., Moss, D.W., Rees, D., Leal, L.M., Moncada, S., Carrier, M., O'Donnell, C.A. and Liew, F.Y. (1994) Differential induction of nitric oxide synthase in various organs of the mouse during endotoxaemia: role of TNF-alpha and IL-1-beta. *Immunology* 81, 211–215.

D'Autreaux, B., Touati, D., Bersch, B., Latour, J.M. and Michaud-Soret, I. (2002) Direct inhibition by nitric oxide of the transcriptional ferric uptake regulation protein via nitrosylation of the iron. *Proceedings of the National Academy of Sciences of the United States of America* 99, 16619–16624.

Dabrowska, K., Hehre, D., Young, K.C., Navarette, C., Ladino, J.F., Bancalari, E. and Suguihara, C. (2005) Effects of a nebulized NONOate, DPTA/ NO, on group B streptococcus-induced pulmonary hypertension in newborn piglets. *Pediatric Research* 57, 378–383.

Daubener, W., Hucke, C., Seidel, K., Hadding, U. and MacKenzie, C.R. (1999) Interleukin-1 inhibits gamma interferon-induced bacteriostasis in human uroepithelial cells. *Infection and Immunity* 67, 5615–5620.

De Kimpe, S.J., Hunter, M.L., Bryant, C.E., Thiemermann, C. and Vane, J.R. (1995a) Delayed circulatory failure due to the induction of nitric oxide synthase by lipoteichoic acid from *Staphylococcus aureus* in anaesthetized rats. *Brazilian Journal of Pharmacology* 114, 1317–1323.

De Kimpe, S.J., Kengatharan, M., Thiemermann, C. and Vane, J.R. (1995b) The cell wall components peptidoglycan and lipoteichoic acid from *Staphylococcus aureus* act in synergy to cause shock and multiple organ failure. *Proceedings of the National Academy of Sciences of the United States of America* 92, 10359–10363.

Demple, B. (1999) Genetic responses against nitric oxide toxicity. *Brazilian Journal of Medical Biological Research* 32, 1417–1427.

Denis, M. (1991) Tumor necrosis factor and granulocyte macrophage-colony stimulating factor stimulate human macrophages to restrict growth of virulent *Mycobacterium avium* and to kill avirulent *M. avium*: killing effector mechanism depends on the generation of reactive nitrogen intermediates. *Journal of Leukocyte Biology* 49, 380–387.

Denninger, J.W. and Marletta, M.A. (1999) Guanylate cyclase and the NO/cGMP signaling pathway. *Biochimica et Biophysica Acta* 1411, 334–350.

Dinauer, M.C., Deck, M.B. and Unanue, E.R. (1997) Mice lacking reduced nicotinamide adenine dinucleotide phosphate oxidase activity show increased susceptibility to early infection with *Listeria monocytogenes*. *Journal of Immunology* 158, 5581–5583.

Doran, K.S., Chang, J.C., Benoit, V.M., Eckmann, L. and Nizet, V. (2002) Group B streptococcal beta-hemolysin/cytolysin promotes invasion of human lung epithelial cells and the release of interleukin-8. *Journal of Infectious Diseases* 185, 196–203.

Draisma, A., Dorresteijn, M., Pickkers, P. and van der Hoeven, H. (2008) The effect of systemic iNOS inhibition during human endotoxemia on the development of tolerance to different TLR-stimuli. *Innate Immunity* 14, 153–159.

Dumarey, C.H., Labrousse, V., Rastogi, N., Vargaftig, B.B. and Bachelet, M. (1994) Selective *Mycobacterium avium*-induced production of nitric oxide by human monocyte-derived macrophages. *Journal of Leukocyte Biology* 56, 36–40.

Edelson, B.T. and Unanue, E.R. (2002) MyD88-dependent but Toll-like receptor 2-independent innate immunity to Listeria: no role for either in macrophage listericidal activity. *Journal of Immunology* 169, 3869–3875.

Endo, A., Masunaga, K., Ayusawa, M., Minato, M., Takada, M., Takahashi, S. and Harada, K. (1999) Does Group B streptococcal sepsis attenuate endogenous nitric oxide synthesis in neonates? *Pediatrics* 103, 1313.

Fang, F.C. (1997) Perspectives series: host/pathogen interactions. Mechanisms of nitric oxide-related antimicrobial activity. *Journal of Clinical Investigations* 99, 2818–2825.

Fang, F.C. (2004) Antimicrobial reactive oxygen and nitrogen species: concepts and controversies. *Nature Reviews Microbiology* 2, 820–832.

Fang, F.C. and Vazquez-Torres, A. (2002) Nitric oxide production by human macrophages: there's NO doubt about it. *American Journal of Physiology of the Lung Cell and Molecular Physiology* 282, L941–943.

Fehr, T., Schoedon, G., Odermatt, B., Holtschke, T., Schneemann, M., Bachmann, M.F., Mak, T.W., Horak, I. and Zinkernagel, R.M. (1997) Crucial role of interferon consensus sequence binding protein, but neither of interferon regulatory factor 1 nor of nitric oxide synthesis for protection against murine listeriosis. *Journal of Experimental Medicine* 185, 921–931.

Fischer, S.R., Bone, H.G., Harada, M., Jourdain, M. and Traber, D.L. (1998) Nitric oxide scavengers in sepsis. *Sepsis* 1, 135–143.

Flatley, J., Barrett, J., Pullan, S.T., Hughes, M.N., Green, J. and Poole, R.K. (2005) Transcriptional responses of *Escherichia coli* to S-nitrosoglutathione under defined chemostat conditions reveal major changes in methionine biosynthesis. *Journal of Biological Chemistry* 280, 10065–10072.

Florquin, S., Amraoui, Z., Dubois, C., Decuyper, J. and Goldman, M. (1994) The protective role of

endogenously synthesized nitric oxide in staphylococcal enterotoxin B-induced shock in mice. *Journal of Experimental Medicine* 180, 1153–1158.

Fong, C.H., Bebien, M., Didierlaurent, A., Nebauer, R., Hussell, T., Broide, D., Karin, M. and Lawrence, T. (2008) An antiinflammatory role for IKKbeta through the inhibition of 'classical' macrophage activation. *Journal of Experimental Medicine* 205, 1269–1276.

Frey, A.D. and Kallio, P.T. (2003) Bacterial hemoglobins and flavohemoglobins: versatile proteins and their impact on microbiology and biotechnology. *FEMS Microbiology Reviews* 27, 525–545.

Gao, J.J., Xue, Q., Zuvanich, E.G., Haghi, K.R. and Morrison, D.C. (2001) Commercial preparations of lipoteichoic acid contain endotoxin that contributes to activation of mouse macrophages in vitro. *Infection and Immunity* 69, 751–757.

Gardner, A.M., Helmick, R.A. and Gardner, P.R. (2002) Flavorubredoxin, an inducible catalyst for nitric oxide reduction and detoxification in *Escherichia coli. Journal of Biological Chemistry* 277, 8172–8177.

Gardner, A.M., Gessner, C.R. and Gardner, P.R. (2003) Regulation of the nitric oxide reduction operon (*norRVW*) in *Escherichia coli* – role of *norR* and sigma(54) in the nitric oxide stress response. *Journal of Biological Chemistry* 278, 10081–10086.

Gazzinelli, R.T., Eltoum, I., Wynn, T.A. and Sher, A. (1993) Acute cerebral toxoplasmosis is induced by *in vivo* neutralization of TNF-alpha and correlates with the down-regulated expression of inducible nitric oxide synthase and other markers of macrophage activation. *Journal of Immunology* 151, 3672–3681.

Ghaffari, A., Miller, C.C., McMullin, B. and Ghahary, A. (2006) Potential application of gaseous nitric oxide as a topical antimicrobial agent. *Nitric Oxide* 14, 21–29.

Gibson, R.L., Berger, J.I., Redding, G.J., Standaert, T.A., Mayock, D.E. and Truog, W.E. (1994) Effect of nitric oxide synthase inhibition during group B streptococcal sepsis in neonatal piglets. *Pediatric Research* 36, 776–783.

Glibetic, M., Samlalsingh-Parker, J., Raykova, V., Ofenstein, J. and Aranda, J.V. (2001) Group B Streptococci and inducible nitric oxide synthase: modulation by nuclear factor kappa B and ibuprofen. *Seminars in Perinatology* 25, 65–69.

Gobert, A.P., McGee, D.J., Akhtar, M., Mendz, G.L., Newton, J.C., Cheng, Y.L., Mobley, H.L.T. and Wilson, K.T. (2001) *Helicobacter pylori* arginase inhibits nitric oxide production by eukaryotic cells: a strategy for bacterial survival.

Proceedings of the National Academy of Sciences of the United States of America 98, 13844–13849.

Goodrum, K.J. and Poulson-Dunlap, J. (2002) Cytokine responses to group B streptococci induce nitric oxide production in respiratory epithelial cells. Infection and Immunity 70, 49–54.

Goodrum, K.J., McCormick, L.L. and Schneider, B. (1994) Group B streptococcus-induced nitric oxide production in murine macrophages is CR3 (CD11b/CD18) dependent. Infection and Immunity 62, 3102–3107.

Goodrum, K.J., Dierksheide, J. and Yoder, B.J. (1995) Tumor necrosis factor alpha acts as an autocrine second signal with gamma interferon to induce nitric oxide in group B streptococcus-treated macrophages. Infection and Immunity 63, 3715–3717.

Green, S.J., Nacy, C.A., Schreiber, R.D., Granger, D.L., Crawford, R.M., Meltzer, M.S. and Fortier, A.H. (1993) Neutralization of gamma interferon and tumor necrosis factor alpha blocks in vivo synthesis of nitrogen oxides from L-arginine and protection against Francisella tularensis infection in Mycobacterium bovis BCG-treated mice. Infection and Immunity 61, 689–698.

Gupta, S., Bansal, S., Deb, J.K. and Kundu, B. (2007) Interplay between DtxR and nitric oxide reductase activities: a functional genomics approach indicating involvement of homologous protein domains in bacterial pathogenesis. International Journal of Experimental Pathology 88, 377–385.

Gusarov, I. and Nudler, E. (2005) NO-mediated cytoprotection: instant adaptation to oxidative stress in bacteria. Proceedings of the National Academy of Sciences of the United States of America 102, 13855–13860.

Gusarov, I., Shatalin, K., Starodubtseva, M. and Nudler, E. (2009) Endogenous nitric oxide protects bacteria against a wide spectrum of antibiotics. Science 325, 1380–1384.

Harbrecht, B.G. (2006) Therapeutic use of nitric oxide scavengers in shock and sepsis. Current Pharmacology Dessertations 12, 3543–3549.

Hardwick, J.B., Tucker, A.T., Wilks, M., Johnston, A. and Benjamin, N. (2001) A novel method for the delivery of nitric oxide therapy to the skin of human subjects using a semi-permeable membrane. Clinical Sciences (London) 100, 395–400.

Hattor, Y., Kasai, K., Akimoto, K. and Thiemermann, C. (1997) Induction of NO synthesis by lipoteichoic acid from Staphylococcus aureus in J774 macrophages: involvement of a CD14-dependent pathway. Biochemical and Biophysical Research Communications 233, 375–379.

Hausladen, A., Gow, A.J. and Stamler, J.S. (1998) Nitrosative stress: metabolic pathway involving the flavohemoglobin. Proceedings of the National Academy of Sciences of the United States of America 95, 14100–14105.

Heemskerk, S., Masereeuw, R., Russel, F.G. and Pickkers, P. (2009) Selective iNOS inhibition for the treatment of sepsis-induced acute kidney injury. Nature Reviews Nephrology 5, 629–640.

Hibbs, J.B. Jr, Taintor, R.R., Vavrin, Z. and Rachlin, E.M. (1988) Nitric oxide: a cytotoxic activated macrophage effector molecule. Biochemical and Biophysical Research Communications 157, 87–94.

Hibbs, J.B. Jr, Taintor, R.R., Vavrin, Z., Granger, D.L., Drapier, J.C., Amber, I.J. and Lancaster, J.R.J. (1990) Synthesis of nitric oxide from a terminal guanidino nitrogen atom of L-arginine: a molecular mechanism regulating cellular proliferation that targets intracellular iron. In: Moncada, S. and Higgs, E.A. (eds) Nitric Oxide from L-arginine: A Bioregulatory System. Elsevier, New York, pp. 189–223.

Hochgrafe, F., Wolf, C., Fuchs, S., Liebeke, M., Lalk, M., Engelmann, S. and Hecker, M. (2008) Nitric oxide stress induces different responses but mediates comparable protein thiol protection in Bacillus subtilis and Staphylococcus aureus. Journal of Bacteriology 190, 4997–5008.

Hoehn, T., Huebner, J., Paboura, E., Krause, M. and Leititis, J.U. (1998) Effect of therapeutic concentrations of nitric oxide on bacterial growth in vitro. Critical Care in Medicine 26, 1857–1862.

Hucke, C., MacKenzie, C.R., Adjogble, K.D., Takikawa, O. and Daubener, W. (2004) Nitric oxide-mediated regulation of gamma interferon-induced bacteriostasis: inhibition and degradation of human indoleamine 2,3-dioxygenase. Infection and Immunity 72, 2723–2730.

Ichiba, H., Fujioka, H., Saitoh, M. and Shintaku, H. (2000) Neonatal listeriosis with severe respiratory failure responding to nitric oxide inhalation. Pediatrics International 42, 696–698.

Isobe, K. and Nakashima, I. (1992) Feedback suppression of staphylococcal enterotoxin-stimulated T-lymphocyte proliferation by macrophages through inductive nitric oxide synthesis. Infection and Immunity 60, 4832–4837.

Ito, S., Ishii, K.J., Ihata, A. and Klinman, D.M. (2005) Contribution of nitric oxide to CpG-mediated protection against Listeria monocytogenes. Infection and Immunity 73, 3803–3805.

James, S.L., Cook, K.W. and Lazdins, J.K. (1990) Activation of human monocyte-derived macrophages to kill schistosomula of *Schistosoma mansoni in vitro*. *Journal of Immunology* 145, 2686–2690.

Janot, L., Secher, T., Torres, D., Maillet, I., Pfeilschifter, J., Quesniaux, V.F., Landmann, R., Ryffel, B. and Erard, F. (2008) CD14 works with toll-like receptor 2 to contribute to recognition and control of *Listeria monocytogenes* infection. *Journal of Infectious Diseases* 198, 115–124.

Jin, Y., Dons, L., Kristensson, K. and Rottenberg, M.E. (2001) Neural route of cerebral *Listeria monocytogenes* murine infection: role of immune response mechanisms in controlling bacterial neuroinvasion. *Infection and Immunity* 69, 1093–1100.

Johnson, D.C., Dean, D.R., Smith, A.D. and Johnson, M.K. (2005) Structure, function, and formation of biological iron-sulfur clusters. *Annual Review of Biochemistry* 74, 247–281.

Justino, M.C., Vicente, J.B., Teixeira, M. and Saraiva, L.M. (2005) New genes implicated in the protection of anaerobically grown *Escherichia coli* against nitric oxide. *Journal of Biological Chemistry* 280, 2636–2643.

Justino, M., Baptista, J. and Saraiva, L. (2009) Di-iron proteins of the Ric family are involved in iron-sulfur cluster repair. *Biometals* 22, 99–108.

Kang, T.J., Basu, S., Zhang, L., Thomas, K.E., Vogel, S.N., Baillie, L. and Cross, A.S. (2008) *Bacillus anthracis* spores and lethal toxin induce IL-1beta via functionally distinct signaling pathways. *European Journal of Immunology* 38, 1574–1584.

Kengatharan, K.M., De Kimpe, S.J. and Thiemermann, C. (1996a) Role of nitric oxide in the circulatory failure and organ injury in a rodent model of Gram-positive shock. *Brazilian Journal of Pharmacology* 119, 1411–1421.

Kengatharan, M., De Kimpe, S.J. and Thiemermann, C. (1996b) Analysis of the signal transduction in the induction of nitric oxide synthase by lipoteichoic acid in macrophages. *Brazilian Journal of Pharmacology* 117, 1163–1170.

Kengatharan, K.M., De Kimpe, S., Robson, C., Foster, S.J. and Thiemermann, C. (1998) Mechanism of Gram-positive shock: identification of peptidoglycan and lipoteichoic acid moieties essential in the induction of nitric oxide synthase, shock, and multiple organ failure. *Journal of Experimental Medicine* 188, 305–315.

Kerr, A.R., Wei, X.Q., Andrew, P.W. and Mitchell, T.J. (2004) Nitric oxide exerts distinct effects in local and systemic infections with *Streptococcus pneumoniae*. *Microbial Pathogenesis* 36, 303–310.

Kerwin, J.F. Jr, Lancaster, J.R. Jr and Feldman, P.L. (1995) Nitric oxide: a new paradigm for second messengers. *Journal of Medical Chemistry* 38, 4343–4362.

Kidd, S.P., Potter, A.J., Apicella, M.A., Jennings, M.P. and McEwan, A.G. (2005) NmlR of *Neisseria gonorrhoeae:* a novel redox responsive transcription factor from the MerR family. *Molecular Microbiology* 57, 1676–1689.

Kidd, S.P., Jiang, D., Jennings, M.P. and McEwan, A.G. (2007) A glutathione-dependent alcohol dehydrogenase (AdhC) is required for defense against nitrosative stress in *Haemophilus influenzae*. *Infection and Immunity* 75, 4506–4513.

Kim, J.H., Seo, H., Han, S.H., Lin, J., Park, M.K., Sorensen, U.B. and Nahm, M.H. (2005) Monoacyl lipoteichoic acid from pneumococci stimulates human cells but not mouse cells. *Infection and Immunity* 73, 834–840.

Kim, S.O., Orii, Y., Lloyd, D., Hughes, M.N. and Poole, R.K. (1999) Anoxic function for the *Escherichia coli* flavohaemoglobin (Hmp): reversible binding of nitric oxide and reduction to nitrous oxide. *Febs Letters* 445, 389–394.

Kim, Y.S. and Tauber, M.G. (1996) Neurotoxicity of glia activated by Gram-positive bacterial products depends on nitric oxide production. *Infection and Immunity* 64, 3148–3153.

Kroncke, K.D. and Carlberg, C. (2000) Inactivation of zinc finger transcription factors provides a mechanism for a gene regulatory role of nitric oxide. *FASEB Journal* 14, 166–173.

Kroncke, K.D., Suschek, C.V. and Kolb-Bachofen, V. (2000) Implications of inducible nitric oxide synthase expression and enzyme activity. *Antioxidant and Redox Signalling* 2, 585–605.

Lancaster, J.R. Jr and Hibbs, J.B. Jr (1990) EPR demonstration of iron-nitrosyl complex formation by cytotoxic activated macrophages. *Proceedings of the National Academy of Sciences of the United States of America* 87, 1223–1227.

Leenen, P.J., Canono, B.P., Drevets, D.A., Voerman, J.S. and Campbell, P.A. (1994) TNF-alpha and IFN-gamma stimulate a macrophage precursor cell line to kill *Listeria monocytogenes* in a nitric oxide-independent manner. *Journal of Immunology* 153, 5141–5147.

Lehnardt, S., Henneke, P., Lien, E., Kasper, D.L., Volpe, J.J., Bechmann, I., Nitsch, R., Weber, J.R., Golenbock, D.T. and Vartanian, T. (2006) A mechanism for neurodegeneration induced by group B streptococci through activation of the TLR2/MyD88 pathway in microglia. *Journal of Immunology* 177, 583–592.

Lehnardt, S., Wennekamp, J., Freyer, D., Liedtke,

C., Krueger, C., Nitsch, R., Bechmann, I., Weber, J.R. and Henneke, P. (2007) TLR2 and caspase-8 are essential for group B Streptococcus-induced apoptosis in microglia. *Journal of Immunology* 179, 6134–6143.

Leib, S.L., Kim, Y.S., Black, S.M., Tureen, J.H. and Tauber, M.G. (1998) Inducible nitric oxide synthase and the effect of aminoguanidine in experimental neonatal meningitis. *Journal of Infectious Diseases* 177, 692–700.

Liew, F.Y. (1994) Regulation of nitric oxide synthesis in infectious and autoimmune diseases. *Immunology Letters* 43, 95–98.

Liew, F.Y. (1995) Regulation of lymphocyte functions by nitric oxide. *Current Opinion in Immunology* 7, 396–399.

Liew, F.Y. and Cox, F.E. (1991) Non-specific defence mechanism: the role of nitric oxide. *Immunology Today* 12, A17–21.

Liew, F.Y., Millott, S., Parkinson, C., Palmer, R.M. and Moncada, S. (1990) Macrophage killing of Leishmania parasite *in vivo* is mediated by nitric oxide from L-arginine. *Journal of Immunology* 144, 4794–4797.

Liu, L., Hausladen, A., Zeng, M., Que, L., Heitman, J. and Stamler, J.S. (2001) A metabolic enzyme for *S*-nitrosothiol conserved from bacteria to humans. *Nature* 410, 490–494.

Long, J.P., Tong, H.H., Shannon, P.A. and DeMaria, T.F. (2003) Differential expression of cytokine genes and inducible nitric oxide synthase induced by opacity phenotype variants of *Streptococcus pneumoniae* during acute otitis media in the rat. *Infection and Immunity* 71, 5531–5540.

Lyons, C.R. (1995) The role of nitric oxide in inflammation. *Advances in Immunology* 60, 323–371.

McInnes, I.B., Leung, B.P., Field, M., Wei, X.Q., Huang, F.P., Sturrock, R.D., Kinninmonth, A., Weidner, J., Mumford, R. and Liew, F.Y. (1996) Production of nitric oxide in the synovial membrane of rheumatoid and osteoarthritis patients. *Journal of Experimental Medicine* 184, 1519–1524.

McInnes, I.B., Leung, B., Wei, X.Q., Gemmell, C.C. and Liew, F.Y. (1998) Septic arthritis following *Staphylococcus aureus* infection in mice lacking inducible nitric oxide synthase. *Journal of Immunology* 160, 308–315.

MacLean, A., Wei, X.Q., Huang, F.P., Al-Alem, U.A., Chan, W.L. and Liew, F.Y. (1998) Mice lacking inducible nitric-oxide synthase are more susceptible to herpes simplex virus infection despite enhanced Th1 cell responses. *Journal of General Virology* 79 (Pt 4), 825–830.

MacMicking, J., Xie, Q.W. and Nathan, C. (1997) Nitric oxide and macrophage function. *Annual Reviews in Immunology* 15, 323–350.

MacMicking, J.D., Nathan, C., Hom, G., Chartrain, N., Fletcher, D.S., Trumbauer, M., Stevens, K., Xie, Q.W., Sokol, K., Hutchinson, N., Chen, H. and Mudget, J.S. (1995) Altered responses to bacterial infection and endotoxic shock in mice lacking inducible nitric oxide synthase. *Cell* 81, 641–650.

McNaughton, L., Puttagunta, L., Martinez-Cuesta, M.A., Kneteman, N., Mayers, I., Moqbel, R., Hamid, Q. and Radomski, M.W. (2002) Distribution of nitric oxide synthase in normal and cirrhotic human liver. *Proceedings of the National Academy of Sciences of the United States of America* 99, 17161–17166.

Malawista, S.E., Montgomery, R.R. and van Blaricom, G. (1992) Evidence for reactive nitrogen intermediates in killing of staphylococci by human neutrophil cytoplasts. A new microbicidal pathway for polymorphonuclear leukocytes. *Journal of Clinical Investigations* 90, 631–636.

Malawista, S.E., Montgomery, R.R. and Van Blaricom, G. (1996) Microbial killing by human neutrophil cytokineplasts: similar suppressive effects of reversible and irreversible inhibitors of nitric oxide synthase. *Journal of Leukocyte Biology* 60, 753–757.

Maloney, C.G., Thompson, S.D., Hill, H.R., Bohnsack, J.F., McIntyre, T.M. and Zimmerman, G.A. (2000) Induction of cyclooxygenase-2 by human monocytes exposed to group B streptococci. *Journal of Leukocyte Biology* 67, 615–621.

Maragos, C.M., Morley, D., Wink, D.A., Dunams, T.M., Saavedra, J.E., Hoffman, A., Bove, A.A., Isaac, L., Hrabie, J.A. and Keefer, L.K. (1991) Complexes of .NO with nucleophiles as agents for the controlled biological release of nitric oxide. Vasorelaxant effects. *Journal of Medical Chemistry* 34, 3242–3247.

Marriott, H.M., Ali, F., Read, R.C., Mitchell, T.J., Whyte, M.K. and Dockrell, D.H. (2004) Nitric oxide levels regulate macrophage commitment to apoptosis or necrosis during pneumococcal infection. *Faseb Journal* 18, 1126–1128.

Marshall, H.E. and Stamler, J.S. (2001) Inhibition of NF-kappa B by S-nitrosylation. *Biochemistry* 40, 1688–1693.

Meadow, W., Rudinsky, B., Bell, A. and Hipps, R. (1995) Effects of inhibition of endothelium-derived relaxation factor on hemodynamics and oxygen utilization during group B streptococcal sepsis in piglets. *Critical Care in Medicine* 23, 705–714.

Moffat, F.L. Jr, Han, T., Li, Z.M., Peck, M.D., Jy, W., Ahn, Y.S., Chu, A.J. and Bourguignon, L.Y. (1996) Supplemental L-arginine HCl augments bacterial phagocytosis in human polymorphonuclear leukocytes. *Journal of Cellular Physiology* 168, 26–33.

Moncada, S. and Higgs, E.A. (1993) The L-arginine-nitric oxide pathway. *New England Journal of Medicine* 329, 2002–2012.

Moncada, S. and Higgs, E.A. (1991) Endogenous nitric oxide: physiology, pathology and clinical relevance. *European Journal of Clinical Investigations* 21, 361–374.

Moncada, S. and Higgs, E.A. (1995) Molecular mechanisms and therapeutic strategies related to nitric oxide. *FASEB Journal* 9, 1319–1330.

Moncada, S., Palmer, R.M. and Higgs, E.A. (1991) Nitric oxide: physiology, pathophysiology, and pharmacology. *Pharmacology Reviews* 43, 109–142.

Moore, C.M., Nakano, M.M., Wang, T., Ye, R.W. and Helmann, J.D. (2004) Response of *Bacillus subtilis* to nitric oxide and the nitrosating agent sodium nitroprusside. *Journal of Bacteriology* 186, 4655–4664.

Morath, S., Geyer, A., Spreitzer, I., Hermann, C. and Hartung, T. (2002) Structural decomposition and heterogeneity of commercial lipoteichoic acid preparations. *Infection and Immunity* 70, 938–944.

Mukhopadhyay, P., Zheng, M., Bedzyk, L.A., LaRossa, R.A. and Storz, G. (2004) Prominent roles of the NorR and Fur regulators in the *Escherichia coli* transcriptional response to reactive nitrogen species. *Proceedings of the National Academy of Sciences of the United States of America* 101, 745–750.

Mulligan, M.S., Hevel, J.M., Marletta, M.A. and Ward, P.A. (1991) Tissue injury caused by deposition of immune complexes is L-arginine dependent. *Proceedings of the National Academy of Sciences of the United States of America* 88, 6338–6342.

Munoz-Fernandez, M.A., Fernandez, M.A. and Fresno, M. (1992) Activation of human macrophages for the killing of intracellular *Trypanosoma cruzi* by TNF-alpha and IFN-gamma through a nitric oxide-dependent mechanism. *Immunology Letters* 33, 35–40.

Murray, H.W. and Teitelbaum, R.F. (1992) L-arginine-dependent reactive nitrogen intermediates and the antimicrobial effect of activated human mononuclear phagocytes. *Journal of Infectious Diseases* 165, 513–517.

Myers, J.T., Tsang, A.W. and Swanson, J.A. (2003) Localized reactive oxygen and nitrogen intermediates inhibit escape of *Listeria monocytogenes* from vacuoles in activated macrophages. *Journal of Immunology* 171, 5447–5453.

Nablo, B.J., Rothrock, A.R. and Schoenfisch, M.H. (2005) Nitric oxide-releasing sol-gels as antibacterial coatings for orthopedic implants. *Biomaterials* 26, 917–924.

Nakano, M.M. (2002) Induction of ResDE-dependent gene expression in *Bacillus subtilis* in response to nitric oxide and nitrosative stress. *Journal of Bacteriology* 184, 1783–1787.

Natarajan, G., Glibetic, M., Raykova, V., Ofenstein, J.P., Thomas, R.L. and Aranda, J.V. (2007) Nitric oxide and prostaglandin response to group B streptococcal infection in the lung. *Annual Clinical Laboratory Sciences* 37, 170–176.

Nathan, C. (1992) Nitric oxide as a secretory product of mammalian cells. *Faseb Journal* 6, 3051–3064.

Nathan, C. and Xie, Q.W. (1994a) Nitric oxide synthases: roles, tolls, and controls. *Cell* 78, 915–918.

Nathan, C. and Xie, Q.W. (1994b) Regulation of biosynthesis of nitric oxide. *Journal of Biological Chemistry* 269, 13725–13728.

Nathan, C.F. and Hibbs, J.B. Jr (1991) Role of nitric oxide synthesis in macrophage antimicrobial activity. *Current Opinion in Immunology* 3, 65–70.

Neighbors, M., Xu, X., Barrat, F.J., Ruuls, S.R., Churakova, T., Debets, R., Bazan, J.F., Kastelein, R.A., Abrams, J.S. and O'Garra, A. (2001) A critical role for interleukin 18 in primary and memory effector responses to *Listeria monocytogenes* that extends beyond its effects on interferon gamma production. *Journal of Experimental Medicine* 194, 343–354.

Neonatal Inhaled Nitric Oxide Study Group, The (1997) Inhaled nitric oxide in full-term and nearly full-term infants with hypoxic respiratory failure. *New England Journal of Medicine* 336, 597–604.

Nicholson, S., Bonecini-Almeida Mda, G., Lapa e Silva, J.R., Nathan, C., Xie, Q.W., Mumford, R., Weidner, J.R., Calaycay, J., Geng, J., Boechat, N., Linhares, C., Rom, W. and Ho, J.L. (1996) Inducible nitric oxide synthase in pulmonary alveolar macrophages from patients with tuberculosis. *Journal of Experimental Medicine* 183, 2293–2302.

Nobre, L.S., Goncalves, V.L., Vicente, J.B., Teixiera, M. and Saraiva, L.M. (2007) Flavohemoglobin is an important protein involved in nitrosative stress defence of *Staphylococcus aureus*. *Free Radical Biology and Medicine* 43, S25–S25.

Nussler, A., Drapier, J.C., Renia, L., Pied, S., Miltgen, F., Gentilini, M. and Mazier, D. (1991)

L-arginine-dependent destruction of intrahepatic malaria parasites in response to tumor necrosis factor and/or interleukin 6 stimulation. *European Journal of Immunology* 21, 227–230.

Nussler, A.K., Di Silvio, M., Billiar, T.R., Hoffman, R.A., Geller, D.A., Selby, R., Madariaga, J. and Simmons, R.L. (1992) Stimulation of the nitric oxide synthase pathway in human hepatocytes by cytokines and endotoxin. *Journal of Experimental Medicine* 176, 261–264.

Oates, J.C. and Gilkeson, G.S. (2006) The biology of nitric oxide and other reactive intermediates in systemic lupus erythematosus. *Clinical Immunology* 121, 243–250.

Ogawa, R., Pacelli, R., Espey, M.G., Miranda, K.M., Friedman, N., Kim, S.M., Cox, G., Mitchell, J.B., Wink, D.A. and Russo, A. (2001) Comparison of control of Listeria by nitric oxide redox chemistry from murine macrophages and NO donors: insights into listeriocidal activity of oxidative and nitrosative stress. *Free Radical Biology and Medicine* 30, 268–276.

Ouadrhiri, Y., Scorneaux, B., Sibille, Y. and Tulkens, P.M. (1999) Mechanism of the intracellular killing and modulation of antibiotic susceptibility of *Listeria monocytogenes* in THP-1 macrophages activated by gamma interferon. *Antimicrobial Agents and Chemotherapy* 43, 1242–1251.

Overton, T.W., Justino, M.C., Li, Y., Baptista, J.M., Melo, A.M.P., Cole, J.A. and Saraiva, L.M. (2008) Widespread distribution in pathogenic bacteria of di-iron proteins that repair oxidative and nitrosative damage to iron-sulfur centers. *Journal of Bacteriology* 190, 2004–2013.

Passalacqua, K.D., Bergman, N.H., Herring-Palmer, A. and Hanna, P. (2006) The superoxide dismutases of *Bacillus anthracis* do not cooperatively protect against endogenous superoxide stress. *Journal of Bacteriology* 188, 3837–3848.

Patel, B.A., Moreau, M., Widom, J., Chen, H., Yin, L., Hua, Y. and Crane, B.R. (2009) Endogenous nitric oxide regulates the recovery of the radiation-resistant bacterium *Deinococcus radiodurans* from exposure to UV light. *Proceedings of the National Academy of Sciences of the United States of America* 106, 18183–18188.

Paul-Clark, M.J., McMaster, S.K., Belcher, E., Sorrentino, R., Anandarajah, J., Fleet, M., Sriskandan, S. and Mitchell, J.A. (2006) Differential effects of Gram-positive versus Gram-negative bacteria on NOSII and TNFalpha in macrophages: role of TLRs in synergy between the two. *Brazilian Journal of Pharmacology* 148, 1067–1075.

Petros, A., Lamb, G., Leone, A., Moncada, S.,

Bennett, D. and Vallance, P. (1994) Effects of a nitric oxide synthase inhibitor in humans with septic shock. *Cardiovascular Research* 28, 34–39.

Pollock, J.D., Williams, D.A., Gifford, M.A., Li, L.L., Du, X., Fisherman, J., Orkin, S.H., Doerschuk, C.M. and Dinauer, M.C. (1995) Mouse model of X-linked chronic granulomatous disease, an inherited defect in phagocyte superoxide production. *Nature Genetics* 9, 202–209.

Poon, B.Y., Raharjo, E., Patel, K.D., Tavener, S. and Kubes, P. (2003) Complexity of inducible nitric oxide synthase: cellular source determines benefit versus toxicity. *Circulation* 108, 1107–1112.

Potter, A.J., Kidd, S.P., Jennings, M.P. and McEwan, A.G. (2007) Evidence for distinctive mechanisms of *S*-nitrosoglutathione metabolism by AdhC in two closely related species, *Neisseria gonorrhoeae* and *Neisseria meningitidis*. *Infection and Immunity* 75, 1534–1536.

Potter, A.J., Kidd, S.P., Edwards, J.L., Falsetta, M.L., Apicella, M.A., Jennings, M.P. and McEwan, A.G. (2009a) Esterase D is essential for protection of *Neisseria gonorrhoeae* against nitrosative stress and for bacterial growth during interaction with cervical epithelial cells. *Journal of Infectious Diseases* 200, 273–278.

Potter, A.J., Kidd, S.P., Edwards, J.L., Falsetta, M.L., Apicella, M.A., Jennings, M.P. and McEwan, A.G. (2009b) Thioredoxin reductase is essential for protection of *Neisseria gonorrhoeae* against killing by nitric oxide and for bacterial growth during interaction with cervical epithelial cells. *Journal of Infectious Diseases* 199, 227–235.

Puliti, M., von Hunolstein, C., Bistoni, F., Orefici, G. and Tissi, L. (2004) Inhibition of nitric oxide synthase exacerbates group B streptococcus sepsis and arthritis in mice. *Infection and Immunity* 72, 4891–4894.

Radi, R. (2004) Nitric oxide, oxidants, and protein tyrosine nitration. *Proceedings of the National Academy of Sciences of the United States of America* 101, 4003–4008.

Radi, R., Beckman, J.S., Bush, K.M. and Freeman, B.A. (1991) Peroxynitrite-induced membrane lipid-peroxidation – the cytotoxic potential of superoxide and nitric-oxide. *Archives of Biochemistry and Biophysics* 288, 481–487.

Raines, K.W., Kang, T.J., Hibbs, S., Cao, G.L., Weaver, J., Tsai, P., Baillie, L., Cross, A.S. and Rosen, G.M. (2006) Importance of nitric oxide synthase in the control of infection by *Bacillus anthracis*. *Infection and Immunity* 74, 2268–2276.

Raykova, V.D., Glibetic, M., Ofenstein, J.P. and

Aranda, J.V. (2003) Nitric oxide-dependent regulation of pro-inflammatory cytokines in group B streptococcal inflammation of rat lung. *Annual Clinical Laboratory Sciences* 33, 62–67.

Redmond, C., Baillie, L.W.J., Hibbs, S., Moir, A.J.G. and Moir, A. (2004) Identification of proteins in the exosporium of *Bacillus anthracis*. *Microbiology-SGM* 150, 355–363.

Reiling, N., Ulmer, A.J., Duchrow, M., Ernst, M., Flad, H.D. and Hauschildt, S. (1994) Nitric oxide synthase: mRNA expression of different isoforms in human monocytes/macrophages. *European Journal of Immunology* 24, 1941–1944.

Remer, K.A., Jungi, T.W., Fatzer, R., Tauber, M.G. and Leib, S.L. (2001) Nitric oxide is protective in listeric meningoencephalitis of rats. *Infection and Immunity* 69, 4086–4093.

Ren, B.B., Zhang, N.H., Yang, J.J. and Ding, H.G. (2008) Nitric oxide-induced bacteriostasis and modification of iron-sulphur proteins in *Escherichia coli*. *Molecular Microbiology* 70, 953–964.

Richardson, A.R., Dunman, P.M. and Fang, F.C. (2006) The nitrosative stress response of *Staphylococcus aureus* is required for resistance to innate immunity. *Molecular Microbiology* 61, 927–939.

Richardson, A.R., Libby, S.J. and Fang, F.C. (2008) A nitric oxide-inducible lactate dehydrogenase enables *Staphylococcus aureus* to resist innate immunity. *Science* 319, 1672–1676.

Ring, A., Depnering, C., Pohl, J., Nizet, V., Shenep, J.L. and Stremmel, W. (2002) Synergistic action of nitric oxide release from murine macrophages caused by group B streptococcal cell wall and beta-hemolysin/cytolysin. *Journal of Infectious Diseases* 186, 1518–1521.

Rogers, H.W., Callery, M.P., Deck, B. and Unanue, E.R. (1996) *Listeria monocytogenes* induces apoptosis of infected hepatocytes. *Journal of Immunology* 156, 679–684.

Rogstam, A., Larsson, J.T., Kjelgaard, P. and von Wachenfeldt, C. (2007) Mechanisms of adaptation to nitrosative stress in *Bacillus subtilis*. *Journal of Bacteriology* 189, 3063–3071.

Rose, F., Zeller, S.A., Chakraborty, T., Domann, E., Machleidt, T., Kronke, M., Seeger, W., Grimminger, F. and Sibelius, U. (2001) Human endothelial cell activation and mediator release in response to *Listeria monocytogenes* virulence factors. *Infection and Immunity* 69, 897–905.

Rothfork, J.M., Timmins, G.S., Harris, M.N., Chen, X., Lusis, A.J., Otto, M., Cheung, A.L. and Gresham, H.D. (2004) Inactivation of a bacterial virulence pheromone by phagocyte-derived oxidants: new role for the NADPH oxidase in host defense. *Proceedings of the National Academy of Sciences of the United States of America* 101, 13867–13872.

Rudinsky, B., Bell, A., Hipps, R. and Meadow, W. (1994) The effects of intravenous L-arginine supplementation on systemic and pulmonary hemodynamics and oxygen utilization during group B streptococcal sepsis in piglets. *Journal of Critical Care* 9, 34–46.

St Clair, E.W., Wilkinson, W.E., Lang, T., Sanders, L., Misukonis, M.A., Gilkeson, G.S., Pisetsky, D.S., Granger, D.I. and Weinberg, J.B. (1996) Increased expression of blood mononuclear cell nitric oxide synthase type 2 in rheumatoid arthritis patients. *Journal of Experimental Medicine* 184, 1173–1178.

Sakiniene, E., Bremell, T. and Tarkowski, A. (1997) Inhibition of nitric oxide synthase (NOS) aggravates *Staphylococcus aureus* septicaemia and septic arthritis. *Clinical and Experimental Immunology* 110, 370–377.

Santolini, J., Roman, M., Stuehr, D.J. and Mattioli, T.A. (2006) Resonance Raman study of *Bacillus subtilis* NO synthase-like protein: similarities and differences with mammalian NO synthases. *Biochemistry* 45, 1480–1489.

Sasaki, S., Miura, T., Nishikawa, S., Yamada, K., Hirasue, M. and Nakane, A. (1998) Protective role of nitric oxide in *Staphylococcus aureus* infection in mice. *Infection and Immunity* 66, 1017–1022.

Schneemann, M. and Schoedon, G. (2002) Species differences in macrophage NO production are important. *Nature Immunology* 3, 102–106.

Schneemann, M., Schoedon, G., Hofer, S., Blau, N., Guerrero, L. and Schaffner, A. (1993) Nitric oxide synthase is not a constituent of the antimicrobial armature of human mononuclear phagocytes. *Journal of Infectious Diseases* 167, 1358–1363.

Schopfer, F.J., Baker, P.R. and Freeman, B.A. (2003) NO-dependent protein nitration: a cell signaling event or an oxidative inflammatory response? *Trends in Biochemical Sciences* 28, 646–654.

Serbina, N.V., Salazar-Mather, T.P., Biron, C.A., Kuziel, W.A. and Pamer, E.G. (2003) TNF/iNOS-producing dendritic cells mediate innate immune defense against bacterial infection. *Immunity* 19, 59–70.

Sharma, S., Sharma, M., Roy, S., Kumar, P. and Bose, M. (2004) *Mycobacterium tuberculosis* induces high production of nitric oxide in coordination with production of tumour necrosis factor-alpha in patients with fresh active tuberculosis but not in MDR tuberculosis.

Immunology and Cellular Biology 82, 377–382.

Shatalin, K., Gusarov, I., Avetissova, E., Shatalina, Y., McQuade, L.E., Lippard, S.J. and Nudler, E. (2008) *Bacillus anthracis*-derived nitric oxide is essential for pathogen virulence and survival in macrophages. *Proceedings of the National Academy of Sciences of the United States of America* 105, 1009–1013.

Shay, A.H., Choi, R., Whittaker, K., Salehi, K., Kitchen, C.M., Tashkin, D.P., Roth, M.D. and Baldwin, G.C. (2003) Impairment of antimicrobial activity and nitric oxide production in alveolar macrophages from smokers of marijuana and cocaine. *Journal of Infectious Diseases* 187, 700–704.

Shiloh, M.U., MacMicking, J.D., Nicholson, S., Brause, J.E., Potter, S., Marino, M., Fang, F., Dinauer, M. and Nathan, C. (1999) Phenotype of mice and macrophages deficient in both phagocyte oxidase and inducible nitric oxide synthase. *Immunity* 10, 29–38.

Singh, V.K., Mehrotra, S., Narayan, P., Pandey, C.M. and Agarwal, S.S. (2000) Modulation of autoimmune diseases by nitric oxide. *Immunology Research* 22, 1–19.

Stroeher, U.H., Kidd, S.P., Stafford, S.L., Jennings, M.P., Paton, J.C. and McEwan, A.G. (2007) A pneumococcal MerR-like regulator and S-nitrosoglutathione reductase are required for systemic virulence. *Journal of Infectious Diseases* 196, 1820–1826.

Stuehr, D.J. and Marletta, M.A. (1985) Mammalian nitrate biosynthesis: mouse macrophages produce nitrite and nitrate in response to *Escherichia coli* lipopolysaccharide. *Proceedings of the National Academy of Sciences of the United States of America* 82, 7738–7742.

Stuehr, D.J., Cho, H.J., Kwon, N.S., Weise, M.F. and Nathan, C.F. (1991) Purification and characterization of the cytokine-induced macrophage nitric oxide synthase: an FAD- and FMN-containing flavoprotein. *Proceedings of the National Academy of Sciences of the United States of America* 88, 7773–7777.

Subczynski, W.K. and Wisniewska, A. (2000) Physical properties of lipid bilayer membranes: relevance to membrane biological functions. *Acta Biochimica Polonica* 47, 613–625.

Szalay, G., Hess, J. and Kaufmann, S.H. (1995) Restricted replication of *Listeria monocytogenes* in a gamma interferon-activated murine hepatocyte line. *Infection and Immunity* 63, 3187–3195.

Tanaka, T., Akira, S., Yoshida, K., Umemoto, M., Yoneda, Y., Shirafuji, N., Fujiwara, H.,

Suematsu, S., Yoshida, N. and Kishimoto, T. (1995) Targeted disruption of the NF-IL6 gene discloses its essential role in bacteria killing and tumor cytotoxicity by macrophages. *Cell* 80, 353–361.

Tavares, A.F.N., Nobre, L.S., Melo, A.M.P. and Saraiva, L.M. (2009) A novel nitroreductase of *Staphylococcus aureus* with S-nitrosoglutathione reductase activity. *Journal of Bacteriology* 191, 3403–3406.

Taylor-Robinson, A.W., Liew, F.Y., Severn, A., Xu, D., McSorley, S.J., Garside, P., Padron, J. and Phillips, R.S. (1994) Regulation of the immune response by nitric oxide differentially produced by T helper type 1 and T helper type 2 cells. *European Journal of Immunology* 24, 980–984.

Tezuka, H., Abe, Y., Iwata, M., Takeuchi, H., Ishikawa, H., Matsushita, M., Shiohara, T., Akira, S. and Ohteki, T. (2007) Regulation of IgA production by naturally occurring TNF/iNOS-producing dendritic cells. *Nature* 448, 929–933.

Thiemermann, C. (1998) The use of selective inhibitors of inducible nitric oxide synthase in septic shock. *Sepsis* 1, 123–129.

Thomas, D.D., Espey, M.G., Vitek, M.P., Miranda, K.M. and Wink, D.A. (2002) Protein nitration is mediated by heme and free metals through Fenton-type chemistry: an alternative to the NO/O2- reaction. *Proceedings of the National Academy of Sciences of the United States of America* 99, 12691–12696.

Todorovic, S., Justino, M.C., Wellenreuther, G., Hildebrandt, P., Murgida, D.H., Meyer-Klaucke, W. and Saraiva, L.M. (2008) Iron-sulfur repair YtfE protein from *Escherichia coli*: structural characterization of the di-iron center. *Journal of Biological and Inorganic Chemistry* 13, 765–770.

Tucker, N.P., Le Brun, N.E., Dixon, R. and Hutchings, M.I. (2010) There's NO stopping NsrR, a global regulator of the bacterial NO stress response. *Trends in Microbiology* 18, 149–156.

Ulett, G.C. and Adderson, E.E. (2005) Nitric oxide is a key determinant of group B streptococcus-induced murine macrophage apoptosis. *Journal of Infectious Diseases* 191, 1761–1770.

Ulett, G.C., Maclean, K.H., Nekkalapu, S., Cleveland, J.L. and Adderson, E.E. (2005) Mechanisms of group B streptococcal-induced apoptosis of murine macrophages. *Journal of Immunology* 175, 2555–2562.

Ullmann, B.D., Myers, H., Chiranand, W., Lazzell, A.L., Zhao, Q., Vega, L.A., Lopez-Ribot, J.L., Gardner, P.R. and Gustin, M.C. (2004) Inducible defense mechanism against nitric oxide in *Candida albicans*. *Eukaryote Cell* 3, 715–723.

Weaver, J., Kang, T.J., Raines, K.W., Cao, G.L.,

Hibbs, S., Tsai, P., Baillie, L., Rosen, G.M. and Cross, A.S. (2007) Protective role of *Bacillus anthracis* exosporium in macrophage-mediated killing by nitric oxide. *Infection and Immunity* 75, 3894–3901.

Weinberg, J.B. (1998) Nitric oxide production and nitric oxide synthase type 2 expression by human mononuclear phagocytes: a review. *Molecular Medicine* 4, 557–591.

Weinberg, J.B., Misukonis, M.A., Shami, P.J., Mason, S.N., Sauls, D.L., Dittman, W.A., Wood, E.R., Smith, G.K., McDonald, B., Bachus, K.E., Haney, A.F. and Granger, D.L. (1995) Human mononuclear phagocyte inducible nitric oxide synthase (iNOS): analysis of iNOS mRNA, iNOS protein, biopterin, and nitric oxide production by blood monocytes and peritoneal macrophages. *Blood* 86, 1184–1195.

Weinstein, M.P., Murphy, J.R., Reller, L.B. and Lichtenstein, K.A. (1983) The clinical significance of positive blood cultures: a comprehensive analysis of 500 episodes of bacteremia and fungemia in adults. II. Clinical observations, with special reference to factors influencing prognosis. *Reviews in Infectious Diseases* 5, 54–70.

Weller, R., Price, R.J., Ormerod, A.D., Benjamin, N. and Leifert, C. (2001) Antimicrobial effect of acidified nitrite on dermatophyte fungi, Candida and bacterial skin pathogens. *Journal of Applied Microbiology* 90, 648–652.

Wheeler, M.A., Smith, S.D., Garcia-Cardena, G., Nathan, C.F., Weiss, R.M. and Sessa, W.C. (1997) Bacterial infection induces nitric oxide synthase in human neutrophils. *Journal of Clinical Investigations* 99, 110–116.

Wink, D.A. and Mitchell, J.B. (1998) Chemical biology of nitric oxide: insights into regulatory, cytotoxic, and cytoprotective mechanisms of nitric oxide. *Free Radical Biology and Medicine* 25, 434–456.

Wink, D.A., Kasprzak, K.S., Maragos, C.M., Elespuru, R.K., Misra, M., Dunams, T.M., Cebula, T.A., Koch, W.H., Andrews, A.W., Allen, J.S. and Keefer, L.K. (1991) DNA deaminating ability and genotoxicity of nitric oxide and its progenitors. *Science* 254, 1001–1003.

Xu, W., Liu, L.Z., Loizidou, M., Ahmed, M. and Charles, I.G. (2002) The role of nitric oxide in cancer. *Cell Research* 12, 311–320.

Yarwood, J.M., McCormick, J.K. and Schlievert, P.M. (2001) Identification of a novel two-component regulatory system that acts in global regulation of virulence factors of *Staphylococcus aureus*. *Journal of Bacteriology* 183, 1113–1123.

Zembala, M., Siedlar, M., Marcinkiewicz, J. and Pryjma, J. (1994) Human monocytes are stimulated for nitric oxide release *in vitro* by some tumor cells but not by cytokines and lipopolysaccharide. *European Journal of Immunology* 24, 435–439.

Zwaferink, H., Stockinger, S., Reipert, S. and Decker, T. (2008) Stimulation of inducible nitric oxide synthase expression by beta interferon increases necrotic death of macrophages upon *Listeria monocytogenes* infection. *Infection and Immunity* 76, 1649–1656.

Part 2
Novel Gene Regulation in Response to Host Defences

5 Novel Regulation in Response to Host-generated Stresses: The MerR Family of Regulators in Pathogenic Bacteria

Stephen P. Kidd

5.1 Introduction

Bacteria have evolved specialized systems that allow them to adapt to conditions relevant to their lifestyle. These systems can include metabolic and biosynthetic pathways which are essential for their growth and survival as the environmental conditions change. Importantly, in various ecological niches, this ability correlates not only with growth but also with defence against damaging environmental stresses, such as physical and chemical agents. In pathogenic bacteria this means a requirement to include diverse mechanisms for survival within their host. Central to the bacterial response to these stresses are specialized transcription factors and pathways which can respond to discrete conditions and regulate the appropriate suite of genes. Often, these are a combination of conditions and stresses, which may be complementary to toxicity, and therefore there is an interplay and overlap between the function of particular transcription factors. Also important in the response to the environment is the associated genetics and physiology of the cell. Bacteria have specialized transcription factors, and indeed pathways or a regulatory hierarchy, which suits their capabilities, lifestyle and the environmental niche in which they exist.

Recently, it has become obvious that the cell's own metabolism has a role in its stress response as well as its maintenance of the intracellular conditions. Significantly, bacteria will therefore harbour a suite of transcription factors that is appropriate for their lifestyle and their mechanisms of growth and stress response within an ecological niche. Even specific transcription factors have evolved within particular bacterial species or strains to be able to operate correspondingly to the particular exogenous and endogenous conditions of that bacterium. One particular family which fits this evolutionary theme is the MerR family of bacterial transcription factors. It is interesting to relate the role of these transcription factors to their mode of action. Bacterial transcription factors can be classed discretely into families based on the sequence or structural homologies, their function and the specific elements of their mechanisms of activation or repression.

5.2 The MerR Family: A Traditional View

5.2.1 A growing family of transcription factors

The method of classification of bacterial transcription factors into families depends on the motifs or domains which are present in a certain class. Bacterial transcription factors predominantly bind DNA through a Helix-turn-Helix (HTH) motif. In those that do contain a HTH motif, proteins can be grouped discretely (Huffman and Brennan, 2002; Rigali *et al.*, 2004; Aravind *et al.*, 2005). The MerR family are those proteins which contain the homologous N-terminal HTH motif (HTH_MerR domain) (Brown *et al.*, 2003). The sensing domain or effector domain of these proteins is in the C-terminal of the protein, and this dissimilar domain is the region that determines the environmental signal to which it responds (this domain is greatly variable among the family but the common and known features, and therefore the associated mode of action of this family of transcription factors, are discussed in the next section).

Proteins in the MerR family include, firstly, the numerous versions from the various sources (transposon, plasmid and chromosomal) of MerR itself. This is the regulator of the mercuric ion resistance operon (similar, but not always the same, are the organo-mercuric resistance systems) (Lund and Brown, 1987). There are then a range of metal ion-sensing MerR-like regulators (proteins which respond to nickel, copper, zinc, lead, cobalt or cadmium), regulators of multi-drug efflux pumps and specific regulators that sense oxidative stress, osmotic stress, pH and temperature changes, or environmental signals for phenotypic changes (such as the curli phenotype and nodulation), as well as sensing nutrient starvation and controlling nitrogen metabolism. There are still others that have been studied but their partner signal has not been identified (such as NmlR, which is known to be present in various pathogenic bacteria and which will be discussed in detail in a later section in this chapter). What is obvious is that since MerR itself was first identified and characterized in the late 1980s (Lund and Brown, 1987, 1989a,b), and classified within its own 'rogue' family together with a small number of other regulators (the MerR family at that time included only BltR, BmrR, Mta, TipAL and SoxR), this family is now known to be functional in many physiological and stress-related roles of bacterial cells and is present in many bacterial species (in 2003 it was shown to be present in 80% of the genomes which were then available; Brown *et al.*, 2003). Their role is greatly diverse.

Depending on which database is interrogated will determine the proteins which are identified that contain the specific domain or motif associated with the HTH-MerR family: PR0000551 (there are 9888 motifs with this Pfam group); the Pfam PF00376 has 8943 hits; and the SMART database (Schultz *et al.*, 1998; Letunic *et al.*, 2009) has the group SM00422 (MerR family), which has 3043 proteins. The SwissProt database contains 1022 proteins with 'MerR' within the annotation. A research group has applied a recently developed bioinformatics tool, *Provalidator*, for grouping proteins into families based on a conventional profile and has constructed a complete profile of the MerR family using all available sequence data (Molina-Henares *et al.*, 2009). They were able to confirm proteins that were part of the MerR family and thereby included or excluded those considered to be false positives or negatives or redundant (this was from the databases they used: Pfam and TrEmbl). They identified about 4500 proteins (in August 2008) that were included in the MerR family of transcription factors. It is not in the remit of this chapter to discuss the variations in or methods of classifying bacterial proteins or to argue to which family a transcription factor may belong, but this does highlight the incredible number of MerR-like regulators, mostly uncharacterized. It is interesting to note, then, that there is no correlation in the genome size to the number of MerR-like regulators it possesses, but rather an increase in the numbers of these proteins in those bacteria that live under changing environmental conditions or diverse stresses. Environmental bacterial isolates which live in niches that undergo significant changes and which are exposed to various stresses do contain numerous MerR-like regulators; the soil microbes such as *Streptomyces coelicolor* contain 34 MerR-like regulators and some *Bacillus* sp. and *Burkholderia* sp. have in the range of 20–30 members (Molina-Henares *et al.*, 2009). This is also highlighted by the variations in the number of MerR-like regulators present between species of particular genera; for example, in *Clostridium*,

C. difficile has more than 10 MerR-like regulators compared to *C. perfringens*, which has only a single copy. Most of these MerR-like regulators have not been studied.

As mentioned, the original MerR family contained MerR, SoxR (a superoxide responsive regulator that regulates SoxS, which is a global regulator important in defence against oxidative stress; see Imlay and Hassett, Chapter 1, this volume) and the multi-drug efflux pump regulators, mainly those studied from *B. subtilis*: TipL, BmrR and BltR. These respond to exogenous toxins and regulated genes, which are required for the direct detoxification of the toxin/toxic chemical. From our recent broad analysis of proteins that include the MerR domain/motif (Table 5.1; this is representative of only some of the known and characterized MerR-like regulators, with an overview of the un-characterized proteins), it is interesting to note the diversity of functions that can now be predicted as being regulated by MerR-like regulators.

The C-terminal domains of these un-characterized (and novel) MerR-like proteins, which is the effector domain, contain a Pfam or CDD motifs which represent a widespread if putative array of functions for this family; these include sensing vitamins, iron, thiol stress, temperature and reactive aldehydes, as well as, unusually, containing enzymatic functions such as transposition, DNA polymerase, polysaccharide biosynthesis and aldo/keto reductase activity. Many of these functions are not merely a response to an exogenous stress or specific toxin, and therefore simply related to the regulation of detoxification processes, but are also associated with physiological and metabolic processes and maintaining growth. This could perhaps especially be under altered environmental conditions. An interesting example, which will not be discussed in detail in this chapter, is MerR-like regulator GlnR (and similarly, TnrA; Wray *et al.*, 2001; Zalieckas *et al.*, 2006; Wray and Fisher, 2007, 2008), which is involved in nitrogen metabolism and responds to nitrogen starvation or limited glutamate, glutamine and ammonium. This is linked directly to the bacterial cell's central metabolism (the TCA

cycle). GlnR is present in several bacteria, most studied in *B. subtilis* and *Staphylococcus aureus* (Chen *et al.*, 2010a,b; Wray and Fisher, 2007, 2008; Fisher and Wray, 2008; Somerville and Proctor, 2009), but it is also present in *Enterococcus faecium*, *Lactobacillus lactis* and *Streptococcus pneumoniae*, as well as others. The mode of action of GlnR varies from the other MerR-like regulators that have been studied. Its function is obviously a response to extracellular stress (nutrient stress, or starvation), but is also tightly linked to the metabolism of the cell – the intracellular processes determining the cell's response to an extracellular 'stress'.

The recent MerR-like regulators that have been characterized also highlight this shift from that which has been defined for the early MerR-like regulators (a toxin/toxic-chemical-then-detoxification pathway) to those which are required for cellular functions but still linked to stress response; as well as being regulators that feature directly as part of metabolism or the initiation of a broader physiological outcome, and thereby permitting a lifestyle for stressful conditions. This is certainly the case for those studies of pathogenic bacteria from about 2005, linking the presence of exogenous stress or stresses, maintaining intracellular conditions and selecting the correct metabolism for the conditions. These are thereby a key part of the bacterium's pathogenesis.

In the context of this chapter, the focus will be on the MerR-like regulators that fit this theme, but it is at least worth mentioning a number of the very different or novel MerR-like regulators that have been identified as part of wider or global analyses, in both pathogenic and non-pathogenic bacteria. These are some current examples and act exactly as that, examples of the diversity and growth of our understanding of the MerR family of transcription factors and their importance in the stress response and physiology of bacteria. *Pseudomonas aeruginosa* is found in a range of environments (this includes growth in soil, marine environments and on human tissue, where it acts as an opportunistic pathogen) and its versatility can be attributed to a unique physiology and its ability to adapt. A key part of this is its

Table 5.1. Representative members of the MerR family of bacterial transcription factors, highlighting the novel and uncharacterized proteins of this family.

Example protein	Size (aa)	Effector domain (Pfam)	Example species	Reference
CueR	135	$Cu^{I/II}$ – $Cu^{I/II}$ homeostasis	*Escherichia coli* and many others	Outten *et al.*, 2000; Stoyanov *et al.*, 2001
ZntR	135	Zn^{II} – Zn^{II} homeostasis	*E. coli* and many others	Brocklehurst *et al.*, 1999
SoxR	154	Superoxide – oxidative stress response	*E. coli* and many others	Nunoshiba and Demple, 1994
NmlR	130	Unknown – ROS/RNS stress response	Pathogenic *Neisseria* and others	Kidd *et al.*, 2005, 2007; Stroeher *et al.*, 2007
BltR	273	Many dyes – multi-drug efflux	*Bacillus subtilis*	Ahmed *et al.*, 1994
GlnR	135	Glutamine – nitrogen metabolism	*B. subtilis*	Fisher and Wray, 2008
Q9HKN4_THEAC; 126 similar proteins	183	Resolvase domain	*Thermoplasma acidophilum*	
Q9KMQ6_VIBCH; 81 similar proteins	392	Methyltransferase_11 domain	*Vibrio cholerae*	
Q8YZW0_ANASP; 70 similar proteins	290	AraC_E: DNA gyrase inhibitor domain	*Anabaena* sp.	
Q6HG46_BACHK; 70 similar proteins	135	Med9: in yeast, mediator of RNA polymerase II formation	*B. thuringiensis*	
Q9F2R2_STRCO; 69 similar proteins	238	MerR: two MerR_HTH domains	*Streptomyces coelicolor*	
Q93KR7_YEREN; 44 similar proteins	388	CbiA: Cobyrinic acid A,C-diamide synthase, B12 synthesis	*Yersinia enterocolitica*	
Q50900_MYXXA; 35 similar proteins	299	B12-binding	*Myxococcus xanthus*	
Q9X8Y5_STRCO; 20 similar proteins	135	TOBE; transport domain, molybdenum or sulfate	*S. coelicolor*	
Q92V63_RHIME; 14 similar proteins	276	Cupin_2 domain	*Rhizobium meliloti*	
Q9RJZ0_STRCO; 12 similar proteins	386	DNA_pol3_beta_2 domain	*S. coelicolor*	
B9B4B9_9BURK; 8 similar proteins	410	Citrate synthase domain	*Burkholderia multivorans*	
C2DFM0_ENTFA; 5 similar proteins	258	Glyoxalase	*Enterococcus faecalis*	
B3DW85_METI4; 5 similar proteins	169	Transposase_35	*Methylacidiphilium infernorum*	
Q8YSC7_ANASP; 3 similar proteins	368	CbiC; Precorrin-8X methylmutase, B12 synthesis	*Anabaena* sp.	
B7HIC3_BACC4	126	EsxB domain; ESAT-6-like proteins	*B. cereus*	
A6NT71_9BACE	318	Isochorismatase domain	*Bacteriodes capillosus*	
B8FJQ3_DESAA	418	CapD, polysaccharide biosynthesis	*Desulfatibacillum alkenivorans*	
B1MAQ6_MTCA9	315	Redoxin; thiol antioxidant protein domain	*Mycobacterium* sp.	
A8LB94_FRASN	491	Aldo_Keto_reductase domain	*Frankia* sp.	

complex network of transcriptional pathways. One of the factors that allow *P. aeruginosa* to survive in diverse environments is its switch from a planktonic (free-living) to a biofilm lifestyle. In the case of cystic fibrosis patients, *P. aeruginosa* infection and its formation of biofilms in the lung are a main factor in the disease outcome. However, variations do exist between strains. In strain PAO1, many genes are up- and down-regulated during its confluent biofilm lifestyle. A novel MerR-like regulator (PA2718) was one of a few regulators that was up-regulated (Waite *et al.*, 2005). The gene/s that can be predicted to be the targets for this regulator do not seem to change their expression. It may be that PA2718 is part of a larger regulatory hierarchy, itself being regulated under as yet unindentified conditions by higher control systems, such as an alternative sigma factor. It is worth noting that also upregulated at this stage of biofilm development by *P. aeruginosa* are genes with the GGDEF motif; these proteins are known to function in the biosynthesis of the signalling molecule bis-(3′-5′)-cyclic di-guanosine monophosphate (c-di-GMP). In *Escherichia coli* and *Salmonella* sp., c-di-GMP is associated with a particular regulatory cascade which acts as the switch to a biofilm formation, and these are central to pathogenesis of these bacteria. Part of this regulatory cascade is MlrA (a MerR-like regulator that is discussed later in this chapter) and, while there is limited homology of MlrA to PA2718, it could be that, similar to MlrA, PA2718 is a central part of the biofilm formation under particular stresses.

Also in *P. aeruginosa* (as well as 40 other bacterial species in the γ- and β-proteobacterium), the major regulator of the genes which encode the proteins for the utilization of branched-chained amino acids (that is isoleucine, leucine and valine) is a regulator of the MerR family (this is called LiuR in *P. aeruginosa*) (Kazakov *et al.*, 2009). Its exact response and indeed cellular role has not been investigated.

In Group B Streptococcus (GBS) different MerR-like regulators were identified as part of the global response to low pH (SAG0427) (Santi *et al.*, 2009) and then a MerR-like regulator (gbs1793) was part of those genes that were upregulated when the bacterium entered the blood (Mereghetti *et al.*, 2008). In a global study of the genes expressed differentially in *Actinobacillus pleuropneumoniae* (a pathogen of the lungs of swine) when exposed to bronchoalveoloar fluid, a novel MerR-like regulator was downregulated (Lone *et al.*, 2009). As a final example, and one novel MerR-like regulator that has been studied to some degree, is PigR, a regulator from the intracellular pathogen *Francisella tularensis* (Charity *et al.*, 2009). PigR regulates the expression of two other regulators and these control numerous virulence factors. All three regulators are essential for survival in a macrophage. PigR is a truncated MerR-like regulator, the effector domain is missing and, although it has not been shown directly, PigR does seem to respond to guanosine tetraphosphate (ppGpp), a signalling molecule produced intracellularly by many bacteria in response to starvation. PigR therefore also links the nutritional state and environmental conditions to the physiology of the cell, its stress response and the expression of genes for survival in the host.

5.2.2 The MerR mode of action

The mode of action of gene regulators can reveal some significance for the cell's required response to the environmental stimuli in which the regulator functions. The members of the MerR family do appear to function in the same manner (SoxR, which uses the oxidation of its Fe-S cluster, and GlnR, whose dimerization is unique, are the only examples of differences in the known mode of action of MerR-like regulators) (Lund and Brown, 1989a,b; Lee *et al.*, 1993; Livrelli *et al.*, 1993; Zeng *et al.*, 1998; Kulkarni and Summers, 1999; Outten *et al.*, 1999). As far as can be informative of their role within the cell, we will describe what is known of the mode of action of the MerR family regulators. MerR (which has been investigated at length and is the prototype for the mode of action for the entire family) acts as a weak repressor and then, in the presence of nM concentrations of its cognate ligand (Hg^{2+}), it activates the expression of its divergent operon, *merTPAD*. This target is the particular case of the transposon Tn*501* and its MerR-*merTPAD* Hg^{2+} resistance locus. Other MerR-regulated Hg^{2+} resistant systems contain different arrangements or combinations of transport and detoxification genes, but central throughout are: *merT* (Hg^{2+} membrane transporter, a P-type ATPase), *merP* (the periplasmic Hg^{2+} transporter/chaperone) and *merA* (a mercuric reductase).

MerR forms a homo-dimeric protein which binds to a dyad symmetry located in the promoter region for their targeted gene (P_{merT}). The promoter elements (the −35 and −10 hexamer consensus sequences for σ^{70} RNA polymerase (RNAP) promoters: TTGACA and TATAAT, respectively) for P_{merT} do align closely (4/6 and 5/6, respectively) to the canonical sequences for this class of promoter. This good homology is unusual for genes which are ultimately activated; the constitutive low activity in the absence of the regulator is usually the result of the poor −10 and −35 elements preventing RNAP initiating transcription correctly and therefore requiring the activator. However, in the P_{merT}, the promoter elements, although 'good' consensus sequences, are separated by a larger spacer region than is typical; they are 19 base pairs (bp) apart rather than the 17+/−1 bp spacer which is required for σ^{70}-RNAP promoters. This removes the −10 element from the correct region of the RNAP. MerR binds to its target dyad symmetry between the −10 and −35 of P_{merT} and, under conditions where Hg^{2+} is not present, this binding is tight and results in bending the DNA around itself. This restricts the RNAP from further contact with the −10 element and thereby results in repression (Summers, 1992). The RNAP remains bound but there is no initiation of transcription. This situation is referred to as 'active repression' in eukaryotic systems, and while the recruitment of RNAP by MerR to the promoter appears to be slightly different, it is rare in bacterial systems. The MerR-RNAP is poised as a hyper-responsive switch to the presence of sublethal concentrations of Hg^{2+} (or the particular cognate stress in the case of other MerR-like regulators). In the presence of nM concentrations of Hg^{2+}, MerR binds Hg^{2+} and

this alters its protein conformation and reduces the binding to the DNA. There is subsequently an unwinding and untwisting of the operator/promoter region of P_{merT} and correct contact with the RNAP (Lee *et al.*, 1993; Livrelli *et al.*, 1993). This results in activation (shown in Fig. 5.1). The promoter for *merR* itself (P_{merR}) fits closely to canonical σ^{70}-RNAP promoters (both in the consensus sequence for the −10 and −35 elements and the spacer region) and overlaps the P_{merT} such that, in the presence or absence of Hg^{2+}, MerR represses the expression of itself (Lund and Brown, 1989a,b; Lee *et al.*, 1993; Livrelli *et al.*, 1993; Zeng *et al.*, 1998; Kulkarni and Summers, 1999; Outten *et al.*, 1999).

5.2.3 CueR: a model for MerR regulators sensing their stimuli

In terms of the mechanism of action and cellular function, most MerR-like regulators are assumed to function the same. Some of the first MerR-like regulators that were characterized beyond the original set were those that responded to copper and zinc (CueR and ZntR, respectively). These were shown initially to be required for resistance to these metal ions (Outten *et al.*, 2000; Petersen and Møller, 2000; Stoyanov *et al.*, 2001; Osman and Cavet, 2008). A major advance in the knowledge of the response of MerR-like regulators, or specifically the metal responsive subclass, was the resolution of the protein structures of CueR and ZntR (Changela *et al.*, 2003). Interestingly, the response of CueR to copper was at the level of zeptomolar concentrations, which amounted to less than one atom per cell and presumably much less than what would be the toxic level for the cell (Changela *et al.*, 2003). Once again, the role for this MerR-like regulator therefore appears to be less as a mechanism for detoxification and more in a metabolic role for the appropriate distribution of the copper to the copper enzymes (predominantly in the membrane and in functions such as the electron transport chain). The role of copper and the sensing of copper for pathogenic bacteria is discussed elsewhere in this book (Clayton *et al.*, Chapter 11, this volume).

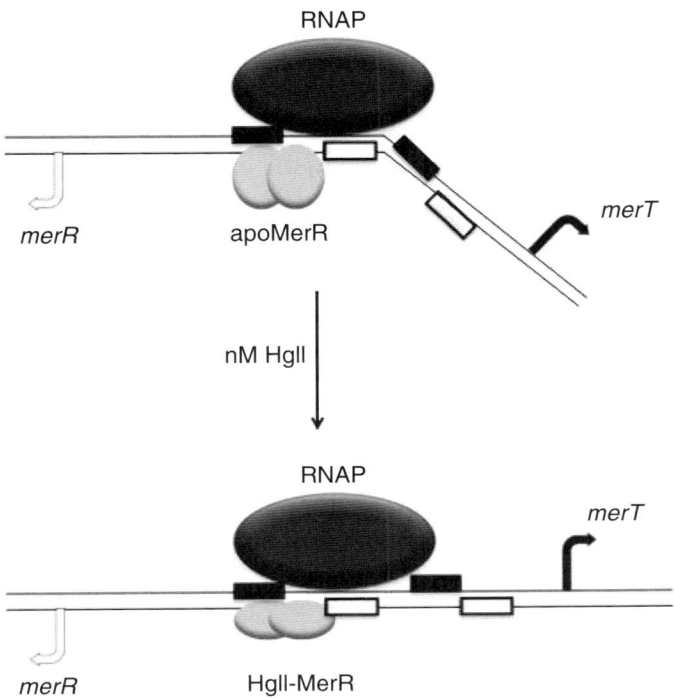

Fig. 5.1. The mode of action of the MerR transcription factors. During repression the RNA polymerase (RNAP) is bound to the target promoter (such as for the *merT*) but is unable to make contact with the −10 element of this promoter. MerR is bound in its apo-form. In the presence of Hg^{2+}, the MerR–Hg^{2+} complex now realigns the DNA so the RNAP contacts the promoter correctly; this results in activation.

5.3 The NmlR Subfamily of the MerR Family of Transcription Factors

Neisseria gonorrhoeae is a pathogen which has a major impact on health worldwide. In the females in which infections can be persistent or asymptomatic, it can be associated with activated polymorphonuclear leukocytes (PMNs) or endocervical cells. In the context of stress response, both these cell types are known to create reactive oxygen species (ROS) and reactive nitrogen species (RNS) as their bactericidal mechanisms (the physiology of this intriguing pathogen, *N. gonorrhoeae*, and its transcription factors which are important for growth within the host are discussed elsewhere in this book; see Hill, Chapter 6 and Cole, Chapter 10). What is interesting is how *N. gonorrhoeae* has

evolved a unique lifestyle to cope with its specific stresses and therefore has highly distinctive processes and transcription pathways for defence against ROS/RNS. This includes a superoxide dismutase-independent manganese-based system for quenching of superoxide and two regulators of the peroxide defence response (Tseng *et al.*, 2001; Seib *et al.*, 2006); that is, unusually, both OxyR and PerR are present in this bacterium and control different aspects of the oxidative stress response (Tseng *et al.*, 2003). Another distinctive stress response system it possesses, and which is part of the ROS/RNS response, is the novel MerR-like regulator, NmlR (*Neisseria merR-like Regulator*) (Kidd *et al.*, 2005). This regulates proteins involved in defence against both ROS and RNS. NmlR has become the prototype of a new subfamily of MerR-like regulators that is found in a range of bacterial species, and in particular is present in a number of pathogenic bacteria.

Based on the common features between the members of the MerR family, not only can MerR-like regulators themselves be identified but also predictions can be made for the genes they are regulating. The gonococcus genome has a single MerR-like regulator, *nmlR*. NmlR is 135 amino acids in length and was thought originally to be part of the metal-responsive subclass of the MerR family, but NmlR lacks the conserved cysteine residues present in MerR and other metal-responsive MerR-like regulators (that is CueR, PbrR, ZntR, ZccR). It does have cysteines, but at spatially different positions (positions 40, 54, 71, 94). A subgroup of MerR-like regulators was identified in bacterial genomes in which these cysteines were conserved.

Initial studies revealed that NmlR regulated the adjacent and divergently transcribed operon, *adhC-estD*. Further analysis of the gonococcus genome uniquely found an additional intergenic region with a MerR-like operator/promoter. These two additional genes were then shown also to be under NmlR control; these encoded a thioredoxin reductase (*trxB*) and a CPx-ATPase (*copA*) (Kidd *et al.*, 2005). The genomes of bacteria which contained NmlR-like regulators all showed that the genes that were predicted as being under the control of NmlR-like regulators were alcohol dehydrogenases or oxidoreductases. EstD is an esterase and AdhC is a zinc-dependent and glutathione (GSH)-dependent formaldehyde dehydrogenase. AdhC is part of the superfamily of zinc-containing alcohol dehydrogenases found in animals, plants, yeast and bacterial cells. Different functions have been described for these enzymes, but recently it has been argued that its common physiological role is in the degradation of the RNS *S*-nitrosoglutathione (GSNO) and through this, a role in protection against nitrosative stress (Liu *et al.*, 2001; Steverding, 2001). The GSNO-decomposing activity of AdhC requires NADH and an equal amount of GSH and GSNO. The main products of the reaction are claimed to be ammonia (NH_3)

and glutathione disulfide (GSSG) (Steverding, 2001), although experimental studies confirming this reaction pathway have not been performed. Originally, and especially in bacterial sources, research on AdhC was focused on its activity to catalyse the oxidation of *S*-hydroxy-methylglutathione (HMGSH), a chemical which was formed spontaneously by formaldehyde reacting with GSH. Through this reaction AdhC has been shown to be required for resistance to formaldehyde, which can be present as a stress either from an exogenous source, such as when it is used as a disinfectant, or as a metabolic by-product; in particular, it is formed during the oxidation of methanol by methanotrophic bacteria (Barber and Donohue, 1998; Lee *et al.*, 2002). This being said, AdhC is present in many bacteria, and not only those which metabolize methanol or which would encounter exogenous formaldehyde.

The regulation of the expression of AdhC has been characterized in a number of bacteria, including *Rhodobacter sphaeroides*, *Acinetobacter baumannii* and *Paracoccus denitrificans* (Ras *et al.*, 1995; Barber and Donohue, 1998). These bacteria are facultative methylotrophs and their ability to defend against toxic formaldehyde is central to their growth. In these bacteria there is clear evidence that the AdhC functions as an NAD^+-dependent dehydrogenase that oxidizes HMGSH. In the case of *R. sphaeroides*, expression of *adhC* in response to aldehyde is through a sensor histidine kinase-response regulator two-component system (Hickman *et al.*, 2004).

In *E. coli*, expression of *adhC* is controlled by the FrmR transcriptional repressor that has been shown to respond to formaldehyde but not GSNO (Herring and Blattner, 2004). The organization of the *adhC* locus in *S. typhimurium* shows a similar pattern of regulation. These data support the conclusion that in enteric bacteria and in facultative methylotrophs, expression of *adhC* is associated with defence against aldehyde stress. Even so, AdhC purified from *E. coli* has been shown to have much higher activity towards GSNO than formaldehyde and is argued strongly to be more involved in controlling intracellular levels of both GSNO and S-nitrosylated proteins, and this functions to protect against nitrosative challenge (Liu *et al.*, 2001). NO is a key antimicrobial element of the host's initial response to infections and this has been shown for enteric bacteria in the case of *E. coli* and *Salmonella*. But studies using these bacteria seem to indicate that they have another mechanism apart from AdhC for their detoxification (Vazquez-Torres and Fang, 2005; Bang *et al.*, 2006) (the response to NO by these enteric pathogens is discussed by Spiro, Chapter 3, this volume). In the Gram-positive model system, *B. subtilis*, the formaldehyde response included a formaldehyde dehydrogenase (AdhA)-cysteine proteinase (YraA) operon and a γ-carboxymuconolactone decarboxylase (YraC), both regulated by an NmlR-like MerR family regulator (AdhR) (Huyen *et al.*, 2009). Aldehydes or α'β-unsaturated carbonyls are toxic through their electrophilic activity against cysteine residues by thiol-S-alkylation. AdhA and YraA are functional in the detoxification of these compounds or the repair from their damage. AdhR has only one of the cysteine residues which were shown originally to be conserved through the NmlR subfamily, Cys52. Although not conclusive, the initial studies into AdhR indicate that it is alkylated at Cys52 and this generates the switch to its activator conformation.

NmlR has now been studied in detail in the pathogens *N. meningitidis*, *S. pneumoniae* and *Haemophilus influenzae*. These studies to date seem to show significant variation between systems, possibly indicating they have evolved to the bacterium's own metabolism as well as the specific stresses of their environmental niche. Perhaps highlighting this is the analysis of the Neisseria *adhC* genes. It has been shown that in *N. gonorrhoeae*, *adhC* is a pseudogene (Potter *et al.*, 2007). Despite this, research has shown that the subsequent gene in the operon, *estD*, is an essential component of *N. gonorrhoeae*'s RNS defence and presumably as a part of this, it has been shown also that *estD* is essential for its growth, infection and biofilm formation on cervical epithelial cells (Potter *et al.*, 2009b). Further to this and emphasizing the importance of the NmlR transcription pathway in this pathogen and related specifically

to its role in RNS defence, *trxB* is clearly essential for NO defence and, once again, is required for *N. gonorrhoeae* survival in cervical epithelial cells and in biofilm formation (Potter *et al.*, 2009a). In addition, the *copA* mutant of *N. gonorrhoeae* is not only more sensitive than wild-type cells to copper ions but also shows reduced ability to invade and survive in human cervical epithelial cells. However, its ability to form a biofilm on the surface of these cells is not significantly different from that of wild-type cells. Protection against copper ion toxicity by CopA is an additional and integral part of the defence system of *N. gonorrhoeae* during its intracellular colonization of cervical epithelial cells. The role of copper in the stress response of pathogenic bacteria is discussed by Clayton *et al.* in Chapter 11 of this volume.

Despite these detailed studies into the role of NmlR and its regulon in stress response and pathogenesis, the environmental signal or discrete stress to which the Neisserial NmlR responds has not been determined. Although the promoter activity for P_{AdhC} and P_{CopA} was repressed with NmlR *in trans* when under normal growth conditions with the application of environmental agents such as metal ions, O_2^- or NO-generating chemicals, there was no alteration in the NmlR repression of either P_{AdhC} or P_{CopA}. When the growth conditions were metal depleted, NmlR caused an activation (or perhaps a loss of repression) of both promoter activities. It was through the addition of zinc into metal-depleted media that there was a return to repressed promoter activity. This has led to the speculation that zinc plays a role in NmlR function. NmlR has also been shown to be an activator in the presence of diamide, a chemical which changes the intracellular redox balance by forcing the GSH pool into an oxidized state. Some redox or oxidative stress responsive proteins do bind zinc; this interaction is through cysteine residues (such as the anti-sigma factor RsrA from *S. coelicolor*; Paget *et al.*, 1998; Kang *et al.*, 1999; Li *et al.*, 2003; Paget and Buttner, 2003). Upon oxidative stress, the thiol in the cysteine residues is altered, the zinc is released and there is a conformational change in the protein. Specific biochemical studies to determine if this is the case in NmlR have not been performed.

It is interesting to note that despite the high similarity between the different NmlRs at an amino acid level, there are significant variations in its function. The subtle differences (both in protein and genetic arrangements) seem to alter the physiological role of the NmlR regulator pertinent to the cell and its environment. In the context of pathogenic bacteria, NmlR has now been characterized in *S. pneumoniae* (this NmlR$_{sp}$, similar to AdhR, has only the one cysteine, Cys52) and *H. influenzae* (NmlR$_{hi}$). Indeed, the research to date on NmlR$_{sp}$ and NmlR$_{hi}$ suggests their role is not limited to the detoxification of exogenous stresses but includes additional functions within the cell which are associated with the cell's metabolism. It has been shown that NmlR$_{sp}$ does induce *adhC* expression in the presence of either formaldehyde or methylglyoxal, despite neither *nmlR$_{sp}$* nor *adhC* mutant strains having an altered growth phenotype in the presence of these aldehydes (Potter *et al.*, 2010). The NmlR$_{sp}$ and NmlR$_{hi}$ and their related *adhC* targets have been implicated directly in the metabolism of GSNO (Kidd *et al.*, 2007; Stroeher *et al.*, 2007). In both these bacteria, not only was *adhC* required for GSNO reductase activity but also, even under normal growth conditions, a chromosomal mutation of *nmlR* (*nmlR$_{sp}$*/*nmlR$_{hi}$*) resulted in a loss of GSNO reductase activity. Given the mode of action of MerR-like regulators, these results point to these NmlR proteins being in their activator conformation under the growth conditions used for these studies (this is in rich media) and with no particular exogenous stress added. Although the GSNO reductase activity is clearly shown for these AdhC proteins, it is unlikely that GSNO is being produced from the components present or through an endogenous pathway. AdhC in humans (and other eukaryotes) is known to use multiple substrates depending on the conditions. This has been shown definitively to include both formaldehyde oxidation and GSNO reduction: in humans, these reactions are known to function concurrently. It certainly could be the case that the pathogenic AdhC proteins are required both for an endogenous pathway as well as for protection against an exogenous stress encountered in the host. It would seem that NmlR$_{sp}$ could be activated similarly to AdhR (certainly it is activated by some aldehydes), thereby responding to endogenous reactive aldehydes (which could form an adduct with GSH and be detoxified by AdhC), as well as defending against GSNO. For NmlR$_{sp}$ (and the *S. pneumoniae* AdhC) using animal model studies, it was shown that, while they (the *nmlR* and *adhC*) probably had a minor role in colonization and survival in the lung, they were essential for survival in the blood.

At the level of amino acid sequences, NmlR and NmlR$_{hi}$ are 97% identical; the few differences include the Cys at 54. The NmlR regulons and their genetic arrangements between these bacteria also seem to create the variation in their cellular function (Fig. 5.2).

In *H. influenzae*, *adhC* is functional, and while there is an *estD* (which is on the transcript with *adhC*), there is no *trxB/copA* target for NmlR$_{hi}$. In *S. pneumonaie* there is no *estD* or *trxB/copA* target and the *adhC* and *nmlRsp* are co-transcribed (they are on the same transcriptional unit, which is driven from a promoter regulated by NmlR$_{sp}$). Additionally, they are regulated together (on what appears as a separate transcript) with a zinc efflux pump (*czcD*; creating a *czcD-nmlRsp-adhC* transcript), which is controlled from an upstream promoter controlled by a TetR family transcription factor, SzcA (Kloosterman *et al.*, 2007). The induction of the expression of this longer operon is in response to zinc. The exact reason for the induction of *nmlR$_{sp}$* and *adhC* in response to zinc has not been determined. Animal model studies have not been performed with SczA and *czcD*, but it is worth noting that the level of zinc is 100 times higher in the lung than the blood. In a large-scale signature-tagged mutagenesis study to identify genes required for *S. pneumoniae* colonization of the naso-pharynx and then the lung and the blood, *adhC* seemed to be required only for infection in the lung (Hava and Camilli, 2002; Hava *et al.*, 2003). This certainly directly contradicted the studies on the virulence of *adhC* and *nmlR* in *S. pneumoniae* which used different infection types in an animal model and then different adherence studies on specific animal cell line in their conclusion that these genes were predominantly functional in the blood (Stroeher *et al.*, 2007). Subtle differences in experimental design might result in different requirements for particular transcriptional pathways, which would be emphasized further when examining the outcome of changes in a transcription factor lower down a hierarchy. This was complicated by the identification of a series of regulatory small RNAs in *S. pneumoniae*, and these were shown to regulate directly the expression of $nmlR_{sp}$–*adhC* (Tsui *et al.*, 2010). This appears to be an independent transcriptional pathway to that of SczA and, likewise, this is in response to a different signal/s, which is yet to be determined. This could explain the differences between the animal model studies; the difference in requirement for $nmlR_{sp}$-*adhC* under one set of experimental conditions could be due to using the sRNA-$nmlR_{sp}$-*adhC* pathway compared to the SczA pathway. In addition, there have been studies into the role of the pilus locus in the pathogenesis of *S. pneumoniae* and the correlated regulatory

control of this locus (Rosch *et al.*, 2008). $NmlR_{sp}$ did seem to play some role in the expression of pilus (the *rlrA* locus which encoded the *S. pnuemoniae* pilus), although this was not through a direct binding to its promoter region. The removal of $nmlR_{sp}$ caused an overexpression of pilus and the associated phenotype/s and virulence of such a strain. The regulation of the pilus was shown to require a number of regulatory pathways, once again leading to a level of overlap between signal–response pathways and the environmental conditions in play. The other regulators suggested to be involved included the metal cation responsive transcription factors, PsaR.

What this does highlight is the interplay between transcriptional pathways and, in particular, the dual or multiple nature of the cell's stress response. An environment does not necessarily and probably rarely contains a single stress. The combination of stresses may affect the level of their harm or the ability of the bacterial cell to respond. This is shown further in different studies which have identified NmlR regulators as secondary

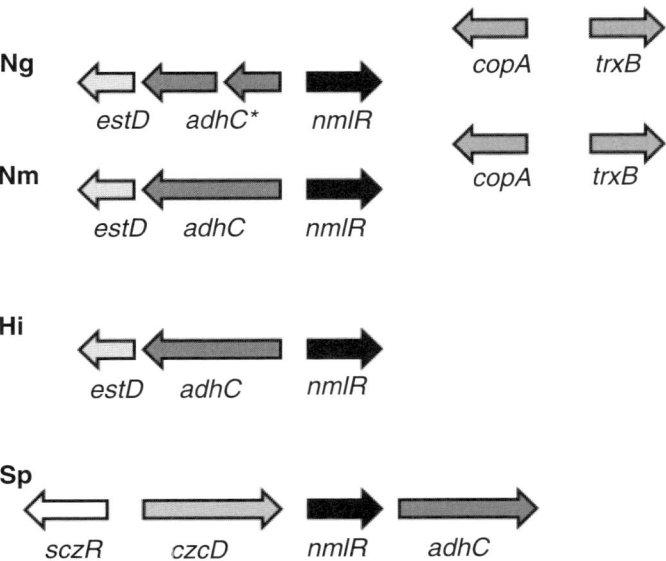

Fig. 5.2. The genetic arrangement of the characterized NmlR transcription factors. NmlR and its homologues are shown in black and the genes they regulate directly are in dark grey. In Ng the *adhC** is shown to indicate it is a pseudogene. The bacteria are annotated as: Ng, *Neisseria gonorrhoeae*; Nm, *Neisseria meningitidis*; Hi, *Haemophilus influenzae*; and Sp, *Streptococcus pneumoniae*.

factors to original regulatory or stress responses of different bacteria. As part of a transcriptional and functional examination of the role of the iron-responsive and global regulator, Fur (ferric uptake regulator) in *N. gonorrhoeae*, it was revealed that one of the transcription factors that was induced under iron-rich conditions was NmlR (Agarwal *et al.*, 2008). While these studies did include NmlR as part of the Fur-iron activated pathway, there was no direct evidence for this regulation. Under iron-rich conditions, there were four transcription factors that were induced in genomics studies. Fur binds to a consensus sequence called a Fur-box; there are some differences in this sequence between bacteria. An *N. gonorrhoeae* Fur-box has been predicted and there seems to be a Fur-box in the *nmlR* ORF. The mode of how this regulation is functioning is a question that needs to be answered to understand the cellular role of NmlR in *N. gonorrhoeae* and its relationship to iron levels. It is interesting to note that in different iron (or haem/iron)-responding analyses of *H. influenzae*, *adhC* is one of the highest genes induced in the presence of iron/haem (Whitby *et al.*, 2006, 2009). These genomics studies, firstly using the Rd KW20 strain and then other clinical isolates of *H. influenzae*, were not aiming to identify Fur-regulated genes but the iron-responsive genes. *H. influenzae* has an obligate requirement for exogenous haem under aerobic conditions; it does not have the machinery for the synthesis of haem, despite an absolute requirement for this cofactor for aerobic growth. The addition of iron/haem resulted in induction of *adhC-estD*. An explanation for this induction is not particularly obvious and is not discussed by the authors of these studies (Whitby *et al.*, 2006, 2009). It is possible that NmlR$_{hi}$ is responding to an endogenously produced and damaging by-product of its own aerobic growth (or growth with iron) and this requires *adhC-estD* for its detoxification. This is in addition to its role in the metabolism of exogenous GSNO, as discussed earlier. There does not seem to be any Fur-box to indicate that Fur is regulating *adhC-estD* directly. The induction of *adhC-estD* under 'normal' growth conditions (Kidd *et al.*, 2007) (aerobic with adequate iron/haem) is consistent with previous studies which indicate that NmlR$_{hi}$ is in its activator conformation under these conditions.

Studies into the *N. gonorrhoeae* FNR (fumarate nitrate reductase), the oxygen-sensing global regulator, revealed that compared to other bacteria, this FNR regulated a small set of genes (it activated 14 genes, predominantly dealing with anaerobic growth, and repressed 6 other genes) (Whitehead *et al.*, 2007). *nmlR* was among the genes predicted as being activated by FNR. A putative FNR-binding site was located (unusually for FNR-binding sites, this was located at 130.5 base pairs upstream of the *nmlR* start site), but direct regulation has not been shown.

The identification of NmlR in *N. gonorrhoeae* certainly has created a new understanding of bacterial response to RNS and ROS (a major shift from the previous and model systems known in other bacteria, see Imlay and Hassett, Chapter 1, Spiro, Chapter 3 and Ulett and Potter, Chapter 4, this volume). Although very similar, the slight differences in the systems in which NmlR and its regulon have now been studied (mainly other pathogenic bacteria) have shown significant shifts in physiological and cellular roles. It seems NmlR can not only play a role in the response to exogenous stresses, in particular RNS, but also is important in maintaining the intracellular redox conditions. This is linked to the particular metabolic pathways the cell employs. Particular pathways will be determined by the extracellular conditions, but also the extracellular conditions will affect the toxicity of the by-products of certain metabolic pathways. Growth with specific sugars results in the production of short-chain sugars or reactive aldehydes, for instance, and aerobically these can be toxic. Anaerobically or with different carbon/energy sources, these would not be produced. AdhC is an enzyme with the possible capacity for multiple reactions, using substrates that can form adducts with GSH: NO and aldehydes are examples. AdhC has been shown to function against GSNO, but also to be important in normal growth of some pathogenic bacteria. It is possible that this role is in the defence against the toxic effect of metabolic by-products, such as reactive aldehydes. This toxicity would be determined by extracellular conditions and it is perhaps for this reason that the NmlR-AdhC pathway seems to be linked to other environmental sensing transcriptional pathways.

5.4 HspR: A MerR-like System

As part of the MerR superfamily when it is defined through its conserved domain (CDD: clo2600) at the N-terminal HTH motif, there are numerous subgroups and among these are the HTH_HspR (cd04766), HTH_HspR-like (cd01279) and HTH_HspR_like_MBC (cd04767) classes of bacterial transcription factors. These three classes of transcription factors have been characterized, certainly initially, as heat shock protein regulators (HspR). They are present in many bacteria and have importance ecologically, environmentally and in the context of human health.

The heat shock response is an essential part of the virulence of many pathogenic bacteria. The heat shock proteins (HSPs), those proteins that are functional as a response to heat stress, repair cellular proteins which are unfolded or aggregated, and these HSPs are mainly molecular chaperones (for folding of proteins) or ATP-dependent proteases (these degrade unfolded polypeptides). Given these functions, the expression of HSPs needs to be tightly regulated; loss of this regulation or overexpression of these proteins, certainly in *E. coli*, is fatal to the cell. In several bacteria, HSPs are also induced in response to other stresses which damage protein structural formation: osmotic stress, pH variations and oxidative stress, as well being induced during stationary phase or low oxygen. For pathogenic bacteria, HSPs are obviously important in their virulence, because of their role in the bacterium's stress response, and therefore provide an ability to survive but, in addition to this stress response, some HSPs are known to function extracellularly through the direct modulation of host cell/s, and in particular those of the immune system. Although in this context as well, the overexpression or unregulated expression of HSPs

can have a deleterious outcome for the pathogen in its interaction with the host and even assist in the bacterial clearance by the host.

Generally, HspR is a repressor which binds to its DNA operator site at a consensus sequence called the HspR-associated inverted repeat (HAIR; the system is often referred to as HspR/HAIR) (Bucca *et al.*, 1995, 2000, 2003). It controls part of the heat shock response by regulating a number of genes in association with other global regulators, and these are often controlled by a sigma factor. In particular, HspR regulates the chaperones, *dnaK* and *clpB* (these are bacterial species specific). Under heat shock, HspR is no longer bound to the DNA (the HAIR operator site) and there is expression of the genes it regulates (through de-repression). In many systems, this disassociation with its cognate HAIR operator site requires DnaK (providing a feedback system). Also, in many bacterial HspR/HAIR systems, the target operator/promoter for HspR binding possesses multiple HAIR sites (at least three in some targets). This could explain the differential response seen by HspR; there are variations in the degree of its regulatory effect with a minor heat shock and then increasing in expression of its targets as the heat shock increases. HspR/HAIR is present in different bacteria; the main species in which it has been studied are the soil bacteria *Corynebacterium glutamicum* and *S. coelicolor*, the human gut commensal *Bifidobacterium* sp. (*breve* and *longum*) and the pathogenic bacteria *Helicobacter pylori*, *Campylobacter jejuni* and *Mycobacterium tuberculosis*. In *S. coelicolor*, HspR regulates the well-known heat shock-responsive chaperones, DnaK, ClpB and DnaJ, as well as the protease, Lon (Bucca *et al.*, 1995, 2000, 2003; Grandvalet *et al.*, 1997, 1999). Other stresses stimulate HspR in *S. coelicolor*, and this response overlaps with the σ^R response to oxidative stress; this particular HspR response is σ^R-independent. In the context of this chapter and the role of novel transcription factors in pathogenic bacteria, it is interesting to point out the variations in the cellular role of HspR in different pathogens. In *H. pylori* not only was HspR responsible for regulation of the conventional heat shock targets, but also a *hspR* mutant was non-motile and had minimal urease activity (urease is the key virulence factor in the pathogenesis of this pathogen) (Spohn *et al.*, 1997; Spohn and Scarlato, 1999). It is worth noting that, in most systems, the HspR overlaps or intersects with other regulatory pathways, either by regulating similar genes or being regulated by or regulating another regulator. Good examples to highlight the key elements of HspR in pathogenic bacteria are *M. tuberculosis* and *C. jejuni*.

In *M. tuberculosis* the heat shock response is dominated by molecular chaperones, and this role is in the repair of intracellular proteins. Studies have shown that many HSPs are induced during infection and they have an extracellular role through direct interaction with the host immune cell (Stewart *et al.*, 2001). *M. tuberculosis* is a pathogen that has many adaptive mechanisms to allow it to interact, respond and survive through a prolonged interaction with immune cells. Although this interaction also seems to be essential in the host recognizing the *M. tuberculosis* cells that are infecting and certainly during the chronic stages of infection, the persistent levels of particular HSPs extracellular to the bacteria (especially Hsp70) actually reduces the ability of the pathogen to survive (Stewart *et al.*, 2001). The regulation of Hsp70 expression is therefore vital and links the extracellular conditions and the ability to persist. In *M. tuberculosis* there seem to be the two regulatory systems that respond to 'heat stress' (this term can include other factors that result in activation in what are known as heat shock proteins, as discussed earlier), HspR and the HcrA transcription factor which binds to the CIRCE (controlling inverted repeat of chaperone expression) DNA operator site. This binding results in repression of gene expression; HcrA then dissociates from the CIRCE site at elevated temperatures, a process which requires GroE/S/L (genes themselves regulated by HcrA) to refold HcrA and permit HcrA/CIRCE contact (and repression) if conditions return to a non-stressed state (thus providing a feedback mechanism). HspR (chromosomal gene locus Rv0353) is on an operon with three other genes, *dnaK* (or

hsp70), *grpE* and *dnaJ*. The *M. tuberculosis* HspR also regulates *clpB*, an ATPase and *acr3* (or *hsp18*), an α-crystallin family protein and perhaps other virulence factors (Stewart *et al.*, 2002). Acr3 is known as a major antigen and is present in response to various stresses; hypoxia and RNS are examples beyond heat shock. The expression in *M. tuberculosis* in response to NO is controlled mainly by the two-component regulatory system, DevRS, and this is known to be a central pathway in the initial stages of *M. tuberculosis* response to host conditions. DnaK together with HspR forms the repressor complex which is able to bind the HAIR operator site and then, under heat stress, the DnaK is recruited away from HspR to other proteins which are denatured by the stress, thereby preventing HspR binding to the HAIR. An *hspR* mutant only affected the chronic stage of infection by *M. tuberculosis*. HspR is a repressor of the *dnaK* operon; the mutant results in overexpression of this operon (unlike in *E. coli*, where overexpression of DnaK has a toxic effect on the cells, there was only a small effect in *M. tuberculosis*) and this is thought to prime the immune system (particularly the adaptive immune response) for a more effective response to the infecting pathogen. In *M. tuberculosis* the heat shock responsive pathways overlap with ECF alternative sigma factor, σ^E.

The heat stress response in *C. jejuni* is similar to other systems studied, but there is no homology of the heat shock responsive alternative sigma factors; it has only RpoN (σ^{54}) and FliA (σ^{28}), both of which control flagella expression. It does have HcrA and HspR. The HspR has also been implicated in key stages of the bacterial pathogenesis of *C. jejuni*, its adherence and invasion of epithelial cells. Different studies are conflicting over whether flagella are also regulated directly or indirectly by HspR (Andersen *et al.*, 2005; Holmes *et al.*, 2010).

While much research remains to be done to appreciate fully the role of heat shock responsive proteins in pathogenesis of different bacteria, what is obvious although different between pathogens is that not only is this a response to heat shock but also to other stresses which are central to survival within the host. There are also direct interactions with elements in the host. Therefore, the regulation of the expression of HSPs is an important part of the pathogenesis.

5.5 Another MerR-like Regulator in Pathogenesis: The Curli Phenotype

In many environments, bacteria function with multicellular behaviour. Certainly for pathogenic bacteria, this state, and then the physiological transition back to a single cell state, is central to the stress response and the survival of the bacteria within the host. For Enterobacteriaceae, the extracellular matrices that mediate the multicellular behaviour are factors whose expression is in response to different external signals and stresses, and therefore their role is subsequently a key part of the pathogenesis of the bacterium. This includes the adhesion to host cells and in particular the formation of biofilms. Biofilms in their simplest form are bacterial communities which, through their formation of a large structure and the production of an extracellular matrix, are able to persist in an environment. Importantly, bacterial biofilms display an increased resistance to exogenous stresses (for instance, bacterial biofilms are generally thought to be 1000 times more resistant to antibiotics). Certainly for pathogenic *E. coli* and *S. enteric* serovar Typhimurium, a major proteinaceous element of this extracellular matrix is curli: these are amyloid fibres formed by the bacteria and attached to the outside of their cells. In these bacteria, curli are associated with attachment and invasion of host cells, direct interactions with proteins in the host and specifically those host elements of the immune system (Barnhart and Chapman, 2006; Pesavento *et al.*, 2008; Saldaña *et al.*, 2009; Weiss-Muszkat *et al.*, 2010). Curli are major inducers of the host inflammatory response.

The development of bacterial biofilms has been categorized into discrete steps: initial and reversible attachment, irreversible attachment, formation of extracellular structure, maturation of the biofilm into a three-dimensional structure and dispersal (Schembri *et al.*, 2002). Depending on the

bacterial species, and in some cases the particular strain, each of these steps involves specific molecular mechanisms and, as such, there have been numerous surface structures described which are important for the specific steps in bacterial biofilm development. Studies into *E. coli* biofilm formation have indicated that curli is part of the cell's ability in the attachment phases in the initial steps of biofilm formation (Barnhart and Chapman, 2006). Strains lacking curli form flat, irregular biofilm structures. The role in pathogenesis for curli is therefore both through direct interactions with various host proteins, as well as in the bacterium's ability to form a mature biofilm, and in this capacity curli provides an ability to resist various stresses of its environment and thereby persist within an anatomical niche. Indeed, curliated *E. coli* attach better to numerous human cell lines (Pesavento *et al.*, 2008; Saldaña *et al.*, 2009; Uhlich *et al.*, 2009; Weiss-Muszkat *et al.*, 2010) and invade these cells better than non-curliated bacteria. Curliated bacteria are more virulent in animal models.

The regulation, expression and function of curli is very complex and does vary significantly between strains and clinical isolates. Indeed, even the suggested role and those elements within the host to which curli has been implicated to interact is not definite; in different cases, it has been shown to interact with the host fibronectin, laminin, antigen-presenting molecules, Toll-like receptors, plasminogen and plasminogen-processing factors (Antao *et al.*, 2009). These are all constituents that are known to be important in the pathogen's ability for invasion, proliferation and spread within the host. Without argument, curli is an important virulence factor.

Much is still being resolved in regard to the physical attributes of curli, but interestingly the biochemical and structural properties of curli resemble those of eukaryotic amyloid fibres (Cegelski *et al.*, 2009). The biosynthesis of curli structures is through genes encoded on divergent operons, *csgAB* (and maybe *csgC*) and *csgDEFG* (Fig. 5.3).

CsgB is the nucleator protein which functions on CsgA (CsgB polymerizing CsgA

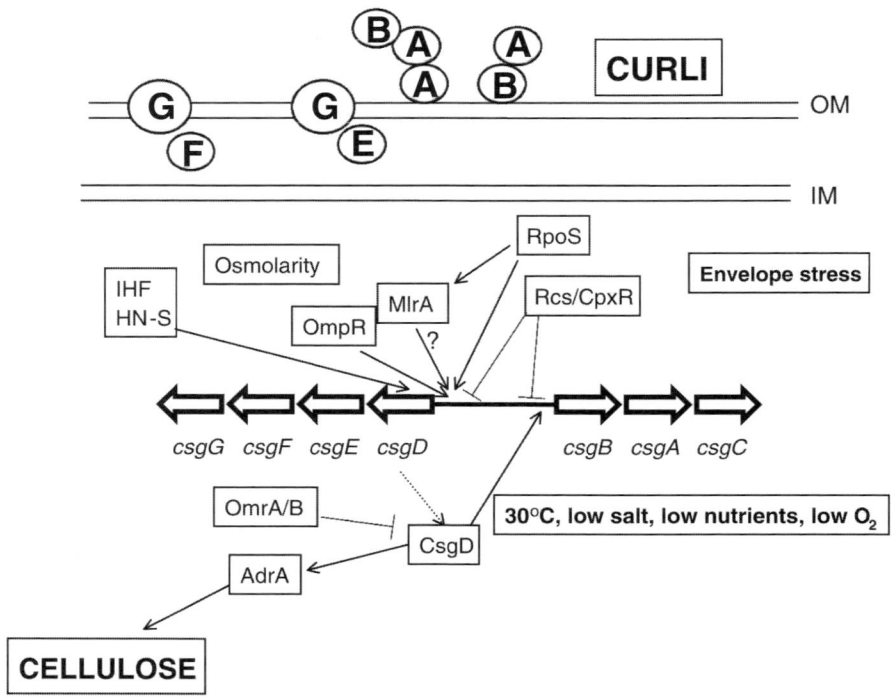

Fig. 5.3. The genetic arrangement of genes for curli production and the regulators affecting their expression. The structural genes for curli are *csgAB* and *csgEFG*; their products and cellular locations are shown as described in the text. The environmental conditions and the various regulators known to affect the expression of the curli genes are also shown.

to the outside of the cell); CsgA forms the major subunit of curli. CsgG is an outer membrane protein which seems to be the transporter for CsgA and CsgB. CsgE is a periplasmic protein which interacts with CsgG and seems to be important in the assembly of CsgA and CsgB on to the surface of the cell. CsgF is also located in the periplasm and this also interacts directly with CsgG, and while its function is known to be different from CsgE, its exact role is still unclear. CsgD is a transcriptional activator of the FixJ family of bacterial transcription factors and activates *csgAB* directly. Unlike many activators of this class of regulator, it does not regulate its own expression (the *csgDEFG* operon). Originally, and seemingly predominantly, curli production was maximal under low temperature (30°C, or less) and in stationary phase. But in some strains, curli structures are expressed at 37°C and have been shown to be expressed

or regulated in response to specific extracellular conditions such as low salt or nutrients (in particular phosphate) and microaerophilic conditions, as well as osmotic stress and membrane stress. Interestingly, these responses have been shown variably to be independent of CsgD and it is now known that regulation of curli biogenesis is through a cascade of regulatory pathways and indeed involves the interplay between transcriptional pathways (Fig. 5.3). CsgD is itself activated by the MerR-like regulator MlrA (Brown *et al.*, 2001), as well as the two-component OmpR/ EnvZ system, the Crl regulator, the general stress responsive/stationary phase sigma factor (RpoS), IHF, HN-S and is also repressed by CpxR/A, Rcs and post-transcriptionally by the sRNA OmrA and OmrB (Gerstel and Römling, 2003; Barnhart and Chapman, 2006; Weber *et al.*, 2006; Pesavento *et al.*, 2008; Saldaña *et al.*, 2009; Holmqvist *et al.*, 2010).

The *csgABC* operon is also repressed by CpxR and Rcs.

The MlrA regulator is itself activated by RpoS. MlrA does contain cysteine residues in its C-terminal domain; they do not correlate to the conserved cysteines in other MerR-like regulators and their role is not known, and indeed the environmental signal that MlrA responds to has not been determined. Given that it is regulated by RpoS, it seems reasonable to postulate that its response is perhaps to an endogenous signal associated with the cell moving into stationary phase (see below).

The array of transcription factors which affect curli expression, either directly or indirectly, consequently means there are numerous environmental signal/conditions and stresses which result in activation or repression of curli formation. CsgD seems to respond, at least in some cases, to low nutrients, temperature and oxygen levels. Osmolarity and other stresses have been shown to be important in OmpR/EnvZ activation of *csgDEFG*. Different forms of envelope stress result in Rcs/Cpx repression of *csgD* and *csgBA*. The sRNA OmrA/B (themselves regulated by OmpR) regulate *csgD* expression, repressing its translation through binding its mRNA directly (Weber *et al.*, 2006). This regulation may be linked directly to OmpR and its corresponding stress response pathway (osmotic stress), or it has been suggested that this sRNA pathway is associated with temperature. An alternative possibility is that these sRNA provide a stochastic fluctuation in *csgD* expression.

Also linked with the production of an extracellular matrix and biofilm formation and concurrently regulated with curli production is the biosynthesis of cellulose. CsgD has been shown to activate AdrA (as it is named in *Salmonella*, or is YaiC in *E. coli*), which itself is a regulator that activates the production and transport of cellulose (an extracellular factor also implicated in cell adhesion and biofilm formation). While the exact function of AdrA is unclear, it is understood to respond to the signalling molecule cyclic nucleotide bis-(3'-5')-cyclic di-guanosine monophosphate (c-di-GMP). This c-di-GMP is believed also to activate

csgD, as well as being part of the mechanism for CsgD activation of *csgABC*. The c-di-GMP is synthesized by diguanylate cyclase (DGC) from two GTP molecule proteins which contain the GGDEF domain. It (c-di-GMP) is degraded by c-di-GMP-specific phospho-diesterases (PDE), which have an EAL domain. The role of c-di-GMP as a signalling molecule is complex and varied (Weber *et al.*, 2006, Hengge, 2009). Certainly, the bio-synthesis of c-di-GMP and the associated interplay between regulatory pathways involved in the bacterial switch in lifestyles, which is largely characterized by an activation of c-di-GMP metabolism, is a rapidly evolving field of study. There is an overall hierarchy and interaction between pathways important in the cell's (based predominantly on studies in *E. coli*) switch from exponential to stationary phases of growth; concurrently a motile to adhesive cellular physiology. This switch is linked to the transcription pathways required for motility and general stress response. These pathways are directed firstly from the start of a regulatory hierarchy by specific sigma factors (σ^{70}/sFliA and σ^s), but also the complex regulator of flagella expression, FlhDC. This also controls FliZ, a regulator which itself represses the production of σ^s. σ^s represses the array of regulatory pathways involved in flagella biosynthesis but also includes the activation of MlrA which, as described earlier, activates *csgD*. σ^s, along with some of these regulators, controls the expression of the genes for c-di-GMP production, which acts as a signal for the downstream regulators of the production of curli and cellulose. Together, these provide a coordinated regulatory response to en-vironmental or growth signals which allow the bacterium to switch from its flagella-based motile and foraging lifestyle to the general stress response and non-motile curli production and biofilm formation.

Within *E. coli* there is another MerR-like regulator whose function is associated with biofilm formation and stress response, YcgE (Tschowri *et al.*, 2009). YcgE is co-transcribed with YcgF, a protein with a degenerate EAL domain and a blue-light sensing BLUF domain (blue light using FAD protein). Initial studies have shown that six genes are activated by YcgE: the polycistronic *ycgZ*, *ymgA*, *ymgB* and *ymgC* and then *yliL* and *ynaK*. All six of these genes encode small proteins (78–90 amino acids). YmgA and YmgB lead to the activation of the Rcs cascade and thereby, through this pathway, the production of curli and cellulose and biofilm formation. The function of the other genes remains unclear. YcgF responds to blue light and acts directly on YcgE (uniquely through the reaction with its EAL domain) and this in effect represses the YcgE repression. The induction of YcgF/E and their targets is in response to low temperature. It is also worth noting that *ycgZ*, *ycmA* and *ycmB* are all under the control of σ^s. This complex array of transcriptional pathways and the interplay between environmental signals highlights the necessary control of bacterial physiology and lifestyle in response to various elements of its environment. This stress response is central to the virulence of the pathogen.

5.6 Conclusions

The MerR family of bacterial transcription factors, once thought to be rare regulators of gene expression, is a growing and important family of proteins which are present in numerous bacteria with a diversity of functions. In recent years numerous members of this family have been characterized, and these respond to stresses different from those envisaged previously for this family. Importantly, they seem to be central in both the stress response and physiological function of bacterial cells. In pathogenic bacteria these roles have been shown to translate to a central and essential feature of the pathogenesis and virulence of the bacteria. From preliminary studies it has also become obvious that the diversity of the MerR family is only just being realized, and when these are characterized, the role of these regulators and their importance will become more obvious.

References

Agarwal, S., Sebastian, S., Szmigielski, B., Rice, P.A. and Genco, C.A. (2008) Expression of the Gonococcal global regulatory protein Fur and genes encompassing the Fur and iron regulon during *in vitro* and *in vivo* infection in women. *Journal of Bacteriology* 190, 3129–3139.

Ahmed, M., Borsch, C.M., Taylor, S.S., Vazquez-Laslop, N. and Neyfakh, A.A. (1994) A protein that activates expression of a multidrug efflux transporter upon binding the transporter substrates. *Journal of Biological Chemistry* 269, 28506–28513.

Andersen, M.T., Brondsted, L., Pearson, B.M., Mulholland, F., Parker, M., Pin, C., Wells, J.M. and Ingmer, H. (2005) Diverse roles for HspR in *Campylobacter jejuni* revealed by the proteome, transcriptome and phenotypic characterization of an *hspR* mutant. *Microbiology* 151, 905–915.

Antao, E.-M., Wieler, L. and Ewers, C. (2009) Adhesive threads of extraintestinal pathogenic *Escherichia coli*. *Gut Pathogens* 1, 22.

Aravind, L., Anantharaman, V., Balaji, S., Babu, M.M. and Iyer, L.M. (2005) The many faces of the helix-turn-helix domain: transcription regulation and beyond. *Fems Microbiology Reviews* 29, 231–262.

Bang, I.-S., Liu, L., Vazquez-Torres, A., Crouch, M.-L., Stamler, J.S. and Fang, F.C. (2006) Maintenance of nitric oxide and redox homeostasis by the Salmonella flavohemoglobin Hmp. *Journal of Biological Chemistry* 281, 28039–28047.

Barber, R.D. and Donohue, T.J. (1998) Function of a glutathione-dependent formaldehyde dehydrogenase in *Rhodobacter sphaeroides* formaldehyde oxidation and assimilation. *Biochemistry* 37, 530–537.

Barnhart, M.M. and Chapman, M.R. (2006) Curli biogenesis and function. *Annual Review of Microbiology* 60, 131–147.

Brocklehurst, K.R., Hobman, J.L., Lawley, B., Blank, L., Marshall, S.J., Brown, N.L. and Morby, A.P. (1999) ZntR is a Zn(II) responsive MerR-like transcriptional regulator of *zntA* in *Escherichia coli*. *Molecular Microbiology* 31, 893–902.

Brown, N.L., Stoyanov, J.V., Kidd, S.P. and Hobman, J.L. (2003) The MerR family of transcriptional regulators. *FEMS Microbiology Reviews* 27, 145–163.

Brown, P.K., Dozois, C.M., Nickerson, C.A., Zuppardo, A., Terlonge, J. and Curtiss, R. III (2001). MlrA, a novel regulator of curli (AgF) and extracellular matrix synthesis by *Escherichia coli* and *Salmonella enterica* serovar Typhimurium. *Molecular Microbiology* 41, 349–363.

Bucca, G., Ferina, G., Puglia, A.M. and Smith, C.P. (1995) The *dnaK* operon of *Streptomyces coelicolor* encodes a novel heat-shock protein which binds to the promoter region of the operon. *Molecular Microbiology* 17, 663–674.

Bucca, G., Brassington, A.M., Schönfeld, H.-J. and Smith, C.P. (2000) The HspR regulon of *Streptomyces coelicolor*: a role for the DnaK chaperone as a transcriptional co-repressor. *Molecular Microbiology* 38, 1093–1103.

Bucca, G., Brassington, A.M., Hotchkiss, G., Mersinias, V. and Smith, C.P. (2003) Negative feedback regulation of *dnaK*, *clpB* and *lon* expression by the DnaK chaperone machine in *Streptomyces coelicolor*, identified by transcriptome and *in vivo* DnaK-depletion analysis. *Molecular Microbiology* 50, 153–166.

Cegelski, L., Pinkner, J.S., Hammer, N.D., Cusumano, C.K., Hung, C.S., Chorell, E., Åberg, V., Walker, J.N., Seed, P.C., Almqvist, F., Chapman, M.R. and Hultgren, S J. (2009) Small-molecule inhibitors target *Escherichia coli* amyloid biogenesis and biofilm formation. *Nature Chemical Biology* 5, 913–919.

Changela, A., Chen, K., Xue, Y. Holschen, J., Outten, C.E., O'Halloran, T.V. and Mondrago, A. (2003) Molecular basis of metal-ion selectivity and zeptomolar sensitivity by CueR. *Science* 301, 1383–1387.

Charity, J.C., Blalock, L.T., Costante-Hamm, M.M., Kasper, D.L. and Dove, S.L. (2009) Small molecule control of virulence gene expression in *Francisella tularensis*. *Plos Pathogens* 5(10), e1000641, doi:10.1371/journal.ppat.1000641.

Chen, P.-M., Chen, Y.-Y., Yu, S.-L., Sher, S., Lai, C.-H. and Chia, J.-S. (2010a) Role of GlnR in acid-mediated repression of genes encoding proteins involved in glutamine and glutamate metabolism in *Streptococcus mutans*. *Applied and Environmental Microbiology* 76, 2478–2486.

Chen, P.-M., Chen, Y.-Y.M., Yu, S.-L., Sher, S., Lai, C.-H. and Chia, J.-S. (2010b) Role of GlnR in acid-mediated repression of genes encoding proteins involved in glutamine and glutamate metabolism in *Streptococcus mutans*. *Applied and Environmental Microbiology* 76, 2478–2486.

Fisher, S.H. and Wray, L.V. (2008) *Bacillus subtilis* glutamine synthetase regulates its own synthesis by acting as a chaperone to stabilize GlnR-DNA complexes. *Proceedings of the National Academy of Sciences of the United States of America* 105, 1014–1019.

Gerstel, U. and Römling, U. (2003) The *csgD* promoter, a control unit for biofilm formation in *Salmonella typhimurium*. *Research in Microbiology* 154, 659–667.

Grandvalet, C., Servant, P. and Mazodier, P. (1997) Disruption of *hspR*, the repressor gene of the *dnaK* operon in *Streptomyces albus* G. *Molecular Microbiology* 23, 77–84.

Grandvalet, C., Crécy-Lagard, V. d. and Mazodier, P. (1999) The ClpB ATPase of *Streptomyces albus* G belongs to the HspR heat shock regulon. *Molecular Microbiology* 31, 521–532.

Hava, D.L. and Camilli, A. (2002) Large-scale identification of serotype 4 *Streptococcus pneumoniae* virulence factors. *Molecular Microbiology* 45, 1389–1406.

Hava, D.L., LeMieux, J. and Camilli, A. (2003) From nose to lung: the regulation behind *Streptococcus pneumoniae* virulence factors. *Molecular Microbiology* 50, 1103–1110.

Hengge, R. (2009) Principles of c-di-GMP signalling in bacteria. *Nature Reviews Microbiology* 7, 263–273.

Herring, C.D. and Blattner, F.R. (2004) Global transcriptional effects of a suppressor tRNA and the inactivation of the regulator *frmR*. *Journal of Bacteriology* 186, 6714–6720.

Hickman, J.W., Witthuhn, V.C. Jr, Dominguez, M. and Donohue, T.J. (2004) Positive and negative transcriptional regulators of glutathione-dependent formaldehyde metabolism. *Journal of Bacteriology* 186, 7914–7925.

Holmes, C.W., Penn, C.W. and Lund, P.A. (2010) The *hrcA* and *hspR* regulons of *Campylobacter jejuni*. *Microbiology* 156, 158–166.

Holmqvist, E., Reimegard, J., Sterk, M., Grantcharova, N., Romling, U. and Wagner, E.G.H. (2010) Two antisense RNAs target the transcriptional regulator CsgD to inhibit curli synthesis. *EMBO Journal* 29, 1840–1850.

Huffman, J.L. and Brennan, R.G. (2002) Prokaryotic transcription regulators: more than just the helix-turn-helix motif. *Current Opinion in Structural Biology* 12, 98–106.

Huyen, N.T.T., Eiamphungporn, W., Mader, U., Liebeke, M., Lalk, M., Hecker, M., Helmann, J.D. and Antelmann, H. (2009) Genome-wide responses to carbonyl electrophiles in *Bacillus subtilis*: control of the thiol-dependent formaldehyde dehydrogenase AdhA and cysteine proteinase YraA by the MerR-family regulator YraB (AdhR). *Molecular Microbiology* 71, 876–894.

Kang, J.-G., Paget, M.S.B., Seok, Y.-J., Hahn, M.-Y., Bae, J.-B. and Kleanthous, C. (1999) RsrA, an anti-sigma factor regulated by redox change. *EMBO Journal* 18, 4292–4298.

Kazakov, A.E., Rodionov, D.A., Alm, E., Arkin, A.P., Dubchak, I. and Gelfand, M.S. (2009) Comparative genomics of regulation of fatty acid and branched-chain amino acid utilization in proteobacteria. *Journal of Bacteriology* 191, 52–64.

Kidd, S.P., Potter, A.J., Apicella, M.A., Jennings, M.P. and McEwan, A.G. (2005) NmlR of *Neisseria gonorrhoeae:* a novel redox responsive transcription factor from the MerR family. *Molecular Microbiology* 57, 1676–1689.

Kidd, S.P., Jiang, D., Jennings, M.P. and McEwan, A.G. (2007) A glutathione-dependent alcohol dehydrogenase (AdhC) is required for defense against nitrosative stress in *Haemophilus influenzae. Infection and Immunity* 75, 4506–4513.

Kloosterman, T.G., van der Kooi-Pol, M.M., Bijlsma, J.J. and Kuipers, O.P. (2007) The novel transcriptional regulator SczA mediates protection against Zn^{2+} stress by activation of the Zn^{2+}-resistance gene *czcD* in *Streptococcus pneumoniae. Molecular Microbiology* 65, 1049–1063.

Kulkarni, R.D. and Summers, A.O. (1999) MerR cross-links to the alpha, beta, and sigma(70) subunits of RNA polymerase in the preinitiation complex at the *merTPCAD* promoter. *Biochemistry* 38, 3362–3368.

Lee, B., Yurimoto, H., Sakai, Y. and Kato, N. (2002) Physiological role of the glutathione-dependent formaldehyde dehydrogenase in the methylotrophic yeast *Candida boidinii. Microbiology* 148, 2697–2704.

Lee, I.W., Livrelli, V., Park, S.J., Totis, P.A. and Summers, A.O. (1993) *In vivo* DNA-protein interactions at the divergent mercury resistance (Mer) promoters. 2. Repressor activator (MerR)-RNA polymerase interaction with merop mutants. *Journal of Biological Chemistry* 268, 2632–2639.

Letunic, I., Doerks, T. and Bork, P. (2009) SMART 6: recent updates and new developments. *Nucleic Acids Research* 37, D229–232.

Li, W., Bottrill, A.R., Bibb, M.J., Buttner, M.J., Paget, M.S.B. and Kleanthous, C. (2003) The role of zinc in the disulfide stress regulated anti-sigma factor RsrA from *Streptomyces coelicolor. Journal of Molecular Biology* 333, 461–472.

Liu, L., Hausladen, A., Zeng, M., Que, L., Heitman, J. and Stamler, J.S. (2001) A metabolic enzyme for *S*-nitrosothiol conserved from bacteria to humans. *Nature* 410, 490–494.

Livrelli, V., Lee, I.W. and Summers, A.O. (1993) *In vivo* DNA–protein interactions at the divergent mercury resistance (*mer*) promoters. 1. Metalloregulatory protein *merR* mutants. *Journal of Biological Chemistry* 268, 2623–2631.

Lone, A.G., Deslandes, V., Nash, J.H.E., Jacques, M. and MacInnes, J.I. (2009) Modulation of gene expression in *Actinobacillus pleuropneumoniae* exposed to bronchoalveolar fluid. *Plos One* 4, e6139.

Lund, P.A. and Brown, N.L. (1987) Role of the MerT and MerP gene-products of transposon Tn501 in the induction and expression of resistance to mercuric ions. *Gene* 52, 207–214.

Lund, P. and Brown, N. (1989a) Up-promoter mutations in the positively-regulated *mer* promoter of Tn50l. *Nucleic Acids Research* 17, 5517–5528.

Lund, P.A. and Brown, N.L. (1989b) Regulation of transcription in *Escherichia coli* from the Mer and MerR promoters in the transposon Tn501. *Journal of Molecular Biology* 205, 343–353.

Mereghetti, L., Sitkiewicz, I., Green, N.M. and Musser, J.M. (2008) Extensive adaptive changes occur in the transcriptome of *Streptococcus agalactiae* (Group B Streptococcus) in response to incubation with human blood. *Plos One* 3, e3143.

Molina-Henares, M.A., de la Torre, J., Godoy, P., Duque, E. and Ramos, J.L. (2009) A general profile for the MerR family of transcriptional regulators constructed using the semi-automated provalidator tool. *Environmental Microbiology Reports* 1, 518–523.

Nunoshiba, T. and Demple, B. (1994) A cluster of constitutive mutations affecting the C-terminus of the redox-sensitive SoxR transcriptional activator. *Nucleic Acids Research* 22, 2958–2962.

Osman, D. and Cavet, J.S. (2008) Copper homeostasis in bacteria. *Advances in Applied Microbiology* 65, 217–247.

Outten, C.E., Outten, F.W. and O'Halloran, T.V. (1999) DNA distortion mechanism for transcriptional activation by ZntR, a Zn(II)-responsive MerR homologue in *Escherichia coli. Journal of Biological Chemistry* 274, 37517–37524.

Outten, F.W., Outten, C.E., Hale, J. and O'Halloran, T.V. (2000) Transcriptional activation of an *Escherichia coli* copper efflux regulon by the chromosomal MerR homologue CueR. *Journal of Biological Chemistry* 275, 31024–31029.

Paget, M.S.B. and Buttner, M.J. (2003) Thiol-based regulatory switches. *Annual Review of Genetics* 37, 91–121.

Paget, M.S.B., Kang, J.-G., Roe, J.-H. and Buttner, M.J. (1998) σ^{R}, an RNA polymerase sigma factor that modulates expression of the thioredoxin system in response to oxidative stress in *Streptomyces coelicolor. EMBO Journal* 17, 5776–5782.

Pesavento, C., Becker, G., Sommerfeldt, N., Possling, A., Tschowri, N., Mehlis, A. and Hengge, R. (2008) Inverse regulatory coordination of motility and curli-mediated adhesion in *Escherichia coli. Genes and Development* 22, 2434–2446.

Petersen, C. and Møller, L.B. (2000) Control of copper homeostasis in *Escherichia coli* by a P-type ATPase, CopA, and a MerR-like transcriptional activator, CopR. *Gene* 261, 289–298.

Potter, A.J., Kidd, S.P., Jennings, M.P. and McEwan, A.G. (2007) Evidence for distinctive mechanisms of *S*-nitrosoglutathione metabolism by AdhC in two closely related species, *Neisseria gonorrhoeae* and *Neisseria meningitidis. Infection and Immunity* 75, 1534–1536.

Potter, A., Kidd, S., Edwards, J., Falsetta, M., Apicella, M., Jennings, M. and McEwan, A. (2009a) Thioredoxin reductase is essential for protection of *Neisseria gonorrhoeae* against killing by nitric oxide and for bacterial growth during interaction with cervical epithelial cells. *The Journal of Infectious Diseases* 199, 227–235.

Potter, A.J., Kidd, S.P., Edwards, J.L., Falsetta, M.L., Apicella, M.A., Jennings, M.P. and McEwan, A.G. (2009b) Esterase D is essential for protection of *Neisseria gonorrhoeae* against nitrosative stress and for bacterial growth during interaction with cervical epithelial cells. *The Journal of Infectious Diseases* 200, 273–278.

Potter, A.J., Kidd, S.P., McEwan, A.G. and Paton, J.C. (2010) The MerR/NmlR family transcription factor of *Streptococcus pneumoniae* responds to carbonyl stress and modulates hydrogen peroxide production. *Journal of Bacteriology* 192, 4063–4066.

Ras, J., Van Ophem, P., Reijnders, W., Van Spanning, R., Duine, J., Stouthamer, A. and Harms, N. (1995) Isolation, sequencing, and mutagenesis of the gene encoding NAD- and glutathione-dependent formaldehyde dehydrogenase (GD-FALDH) from *Paracoccus denitrificans*, in which GD-FALDH is essential for methylotrophic growth. *Journal of Bacteriology* 177, 247–251.

Rigali, S., Schlicht, M., Hoskisson, P., Nothaft, H., Merzbacher, M., Joris, B. and Titgemeyer, F. (2004) Extending the classification of bacterial transcription factors beyond the helix-turn-helix motif as an alternative approach to discover new cis/trans relationships. *Nucleic Acids Research* 32, 3418–3426.

Rosch, J.W., Mann, B., Thornton, J., Sublett, J. and Tuomanen, E. (2008) Convergence of regulatory networks on the pilus locus of *Streptococcus pneumoniae. Infection and Immunity* 76, 3187–3196.

Saldaña, Z., Xicohtencatl-Cortes, J., Avelino, F., Phillips, A.D., Kaper, J.B., Puente, J.L. and

Girón, J.A. (2009) Synergistic role of curli and cellulose in cell adherence and biofilm formation of attaching and effacing *Escherichia coli* and identification of Fis as a negative regulator of curli. *Environmental Microbiology* 11, 992–1006.

Santi, I., Grifantini, R., Jiang, S.-M., Brettoni, C., Grandi, G., Wessels, M.R. and Soriani, M. (2009) CsrRS regulates Group B Streptococcus virulence gene expression in response to environmental pH: a new perspective on vaccine development. *Journal of Bacteriology* 191, 5387–5397.

Schembri, M.A., Givskov, M. and Klemm, P. (2002) An attractive surface: Gram-negative bacterial biofilms. *Science: STKE* 2002, 6–12.

Schultz, J.R., Milpetz, F., Bork, P. and Ponting, C.P. (1998) SMART, a simple modular architecture research tool: identification of signalling domains. *Proceedings of the National Academy of Sciences of the United States of America* 95, 5857–5864.

Seib, K.L., Wu, H.J., Kidd, S.P., Apicella, M.A., Jennings, M.P. and McEwan, A.G. (2006) Defenses against oxidative stress in *Neisseria gonorrhoeae*: a system tailored for a challenging environment. *Microbiology and Molecular Biology Reviews* 70, 344–364.

Somerville, G.A. and Proctor, R.A. (2009) At the crossroads of bacterial metabolism and virulence factor synthesis in Staphylococci. *Microbiology and Molecular Biology Reviews* 73, 233–248.

Spohn, G. and Scarlato, V. (1999) The autoregulatory HspR repressor protein governs chaperone gene transcription in *Helicobacter pylori. Molecular Microbiology* 34, 663–674.

Spohn, G., Beier, D., Rappuoli, R. and Scarlato, V. (1997) Transcriptional analysis of the divergent *cagAB* genes encoded by the pathogenicity island of *Helicobacter pylori. Molecular Microbiology* 26, 361–372.

Steverding, D. (2001) Nitrosative stress: protection by glutathione-dependent formaldehyde dehydrogenase. *Redox Report* 6, 209–210.

Stewart, G.R., Snewin, V.A., Walzl, G., Hussell, T., Tormay, P., O'Gaora, P., Goyal, M., Betts, J., Brown, I.N. and Young, D.B. (2001) Overexpression of heat-shock proteins reduces survival of *Mycobacterium tuberculosis* in the chronic phase of infection. *Nature Medicine* 7, 732–737.

Stewart, G.R., Wernisch, L., Stabler, R., Mangan, J.A., Hinds, J., Laing, K.G., Young, D.B. and Butcher, P.D. (2002) Dissection of the heat-shock response in *Mycobacterium tuberculosis* using mutants and microarrays. *Microbiology* 148, 3129–3138.

Stoyanov, J.V., Hobman, J.L. and Brown, N.L. (2001) CueR: (YbbI) of *Escherichia coli*. *Molecular Microbiology* 39, 502–511.

Stroeher, U.H., Kidd, S.P., Stafford, S.L., Jennings, M.P., Paton, J.C. and McEwan, A.G. (2007) A pneumococcal MerR-like regulator and *S*-nitrosoglutathione reductase are required for systemic virulence. *Journal of Infectious Diseases* 196, 1820–1826.

Summers, A.O. (1992) Untwist and shout: a heavy metal-responsive transcriptional regulator. *Journal of Bacteriology* 174, 3097–3101.

Tschowri, N., Busse, S. and Hengge, R. (2009) The BLUF-EAL protein YcgF acts as a direct anti-repressor in a blue-light response of *Escherichia coli*. *Genes and Development* 23, 522–534.

Tseng, H.J., Srikhanta, Y., McEwan, A.G. and Jennings, M.P. (2001) Accumulation of manganese in *Neisseria gonorrhoeae* correlates with resistance to oxidative killing by superoxide anion and is independent of superoxide dimutasae activity. *Molecular Microbiology* 40, 1175–1186.

Tseng, H.J., McEwan, A.G., Apicella, M.A. and Jennings, M.P. (2003) OxyR acts as a repressor of catalase expression in *Neisseria gonorrhoeae*. *Infection and Immunity* 71, 550–556.

Tsui, H.-C.T., Mukherjee, D., Ray, V.A., Sham, L.-T., Feig, A.L. and Winkler, M.E. (2010) Identification and characterization of non-coding small RNAs in *Streptococcus pneumoniae* serotype 2 strain D39. *Journal of Bacteriology* 192, 264–279.

Uhlich, G.A., Gunther, N.W.I.V., Bayles, D.O. and Mosier, D.A. (2009) The CsgA and Lpp proteins of an *Escherichia coli* O157:H7 strain affect HEp-2 cell invasion, motility, and biofilm formation. *Infection and Immunity* 77, 1543–1552.

Vazquez-Torres, A. and Fang, F.C. (2005) *Nitric Oxide in Salmonella and Escherichia coli Infections. Host Responses*. Fang, F.C. and Kagnoff, M. (series eds). ASM Press, Washington, DC.

Waite, R.D., Papakonstantinopoulou, A., Littler, E. and Curtis, M.A. (2005) Transcriptome analysis of *Pseudomonas aeruginosa* growth: comparison of gene expression in planktonic cultures and developing and mature biofilms. *Journal of Bacteriology* 187, 6571–6576.

Weber, H., Pesavento, C., Possling, A., Tischendorf, G. and Hengge, R. (2006) Cyclic-di-GMP-mediated signalling within the sigmaS network of *Escherichia coli*. *Molecular Microbiology* 62, 1014–1034.

Weiss-Muszkat, M., Shakh, D., Zhou, Y., Pinto, R., Belausov, E., Chapman, M.R. and Sela, S. (2010) Biofilm formation by and multicellular behavior of *Escherichia coli* O55:H7, an atypical enteropathogenic strain. *Applied and Environmental Microbiology* 76, 1545–1554.

Whitby, P.W., VanWagoner, T.M., Seale, T.W., Morton, D.J. and Stull, T.L. (2006) Transcriptional profile of *Haemophilus influenzae*: effects of iron and heme. *Journal of Bacteriology* 188, 5640–5645.

Whitby, P., Seale, T., VanWagoner, T., Morton, D. and Stull, T. (2009) The iron/heme regulated genes of *Haemophilus influenzae*: comparative transcriptional profiling as a tool to define the species core modulon. *BMC Genomics* 10, 6.

Whitehead, R., Overton, T., Snyder, L., McGowan, S., Smith, H., Cole, J. and Saunders, N. (2007) The small FNR regulon of *Neisseria gonorrhoeae*: comparison with the larger *Escherichia coli* FNR regulon and interaction with the NarQ-NarP regulon. *BMC Genomics* 8, 35.

Wray, L.V. Jr and Fisher, S.H. (2007) Functional analysis of the carboxy-terminal region of *Bacillus subtilis* TnrA, a MerR family protein. *Journal of Bacteriology* 189, 20–27.

Wray, L.V. and Fisher, S.H. (2008) *Bacillus subtilis* GlnR contains an autoinhibitory C-terminal domain required for the interaction with glutamine synthetase. *Molecular Microbiology* 68, 277–285.

Wray, L.V. Jr, Zalieckas, J.M. and Fisher, S.H. (2001) *Bacillus subtilis* glutamine synthetase controls gene expression through a protein–protein interaction with transcription factor TnrA. *Cell* 107, 427–435.

Zalieckas, J.M., Wray, L.V. and Fisher, S.H. (2006) Cross-regulation of the *Bacillus subtilis glnRA* and *tnrA* genes provides evidence for DNA binding site discrimination by GlnR and TnrA. *Journal of Bacteriology* 188, 2578–2585.

Zeng, Q.D., Stalhandske, C., Anderson, M.C., Scott, R.A. and Summers, A.O. (1998) The core metal-recognition domain of MerR. *Biochemistry* 37, 15885–15895.

6 Stress Responses in the Pathogenic *Neisseria*: Overlapping Regulons and sRNA Regulation

Stuart A. Hill

6.1 Introduction

For the genus *Neisseria*, there are two pathogenic species, *N. gonorrhoeae* (the gonococcus: Gc) and *N. meningitidis* (the meningococcus: Mc), and these cause the human mucosal diseases gonorrhoea and bacterial meningitis, respectively. Genetically, each organism is highly similar, both at the DNA homology level as well as in the conservation of their gene complement (give or take one or two notable exceptions such as the capsule production in the meningococcus). However, each organism resides in very distinct ecological niches that require unique genetic responses in order to survive. Gonococci primarily infect mucosal surfaces in the urogenital tract, either within the male urethra, or, at the cervix within the vagina. In many patients, Gc infections elicit a pronounced inflammatory response with a significant recruitment of polymorphonuclear leukocytes (PMNs; neutrophils). In contrast, Mc tends to reside as a commensal organism in the nasopharynx. At normal carriage loads, colonization of the nasopharyngeal tissues does not appear to incite a pronounced inflammatory response. However, occasionally Mc can transcytose the mucosal epithelium, leading to disseminated infections that can progress to fulminating septicaemia and/or a concomitant infection of the meninges. Whether transcytosis is facilitated by TNF-alpha-induced apoptosis due to lipooligosaccharide (LOS) deposition on the mucosal epithelium, or through PorB induction of apoptosis of infected host cells (e.g. *N. meningitidis* strain Z2491; Kozjak-Pavlovic *et al.*, 2009), the exact mechanism is poorly understood (Deghmane *et al.*, 2009). However, a correlation does exist between the ability of hyperinvasive strains to be able to induce apoptosis, in contrast to commensal carriage strains which seem to cause less cellular damage for unknown reasons (Deghmane *et al.*, 2009). Consequently, Mc can cause infections in quite disparate environments.

In these various locales within the body the chemical and physical characteristics will change depending on the particular host environment. When Gc infect human genital mucosal tissue, the bacteria will encounter a reduced oxygen environment (pO_2 for human buccal folds is 3 mm Hg (400 Pa); Archibald and Duong, 1986), which is dramatically different when compared to Mc residing in the nasopharynx where high oxygen partial pressures are experienced. Consequently, Gc have evolved several strategies to cope with reduced oxygen partial pressures (which will be outlined in a later chapter; see Cole, Chapter 10, this volume). Despite these obvious differences in the pO_2 levels experienced by the two species, each organism can reduce nitrite, with even some Mc strains being capable of growing under anaerobic conditions (Berger, 1986). Whether this feature is a remnant of the co-evolution of the two species from a common ancestor or actually reflects a need for Mc occasionally to grow anaerobically is currently not well understood.

Besides the differences in oxygen partial pressures experienced by the bacteria, pH is also likely to differ dramatically in the various infection sites, with pH occasionally changing during the course of a single infection. The maintenance of a specific mucosal pH reflects both the composition of the resident bacterial flora in conjunction with the extent of inflammation that is provoked during an infection. The male urethra contains very few resident bacteria. Consequently, urethral pH is likely to reflect the extent of the inflammatory response induced during colonization of the mucosal tissue. This is in dramatic contrast to colonization in the vagina, where many different species of microorganisms reside (e.g. Streptococci, Lactobacilli, Peptococci and *Bacteroides* species). Because the Lactobacilli produce L-lactate, the vagina is acidic (approximately pH 5), with pH possibly varying with age and/or colonization of the tissue with other microbial species (Cohen, 1969). Therefore, for Gc transiting between the male and female urogenital tracts following sexual intercourse, these organisms are likely to encounter dramatically different pH environments in each host and will need to respond accordingly. Meningococci may also encounter acidic conditions if inflammation is induced by a heavy pharyngeal carriage load (Masson and Holbein, 1985). However, following transcytosis across the endothelium, a more benign pH environment (approximately pH 7.5) will be encountered in the bloodstream when causing septicaemia. Consequently, Mc will need to adapt to pH differences depending on the type of infection they cause. Furthermore, each organism will encounter acidic conditions when causing an intracellular infection when enclosed within a vacuole, as acidification of the endosomal

compartment is known to correlate with bacterial killing (Booth *et al.*, 2003).

Nutrient composition, as well as nutrient availability, is also going to differ dramatically at the various infection sites (Smith, 2000), where *in vivo* nutrient availability is expected to be limiting. Consequently, if nutrient availability is limiting, then *Neisseria* will always be under nutrient stress, which may lead to abnormal/alternative metabolic strategies that could influence profoundly the expression of the various virulence functions *in vivo* (Holbein *et al.*, 1979; Masson and Holbein, 1985). Thus, the behaviour of Gc and Mc, when residing in the human body, may be totally different than when observed under optimum laboratory conditions.

6.2 Chemical and Physical Stress

6.2.1 Growth under acidic conditions

As indicated above, a strong case can be made that the pathogenic *Neisseria* will encounter acidic conditions *in vivo*, especially when Gc are infecting the female urogenital tract. When Gc are growing under acidic conditions, the organism alters its physiological state completely in response to the changing pH conditions; at high pH, glucose catabolism proceeds primarily by the Entner–Doudoroff pathway in conjunction with the pentose phosphate stunt, with no indication of the involvement of the tricarboxylic acid (TCA) pathway; at low pH (pH 6), gonococci turn off the Entner–Doudoroff pathway and shift glucose catabolism through the pentose phosphate shunt as well as through the TCA cycle (Morse and Hebeler, 1978).

Structurally, Gc are more autolytic at basic pH than at low pH (Weneger *et al.*, 1977a), with the increased stability at low pH correlating with changes in the cell envelope (Weneger *et al.*, 1977b), which coincides with a greater accumulation of protein with the peptidoglycan cell envelope layer (Hebeler *et al.*, 1978; Hill and Judd, 1989). Acid stress also upregulates outer membrane protein synthesis in clinical isolates, but not in the closely related commensal species (Pettit *et al.*, 2001). For Mc, low pH in conjunction with low iron availability affects virulence (Brener *et al.*, 1981) due to changes in capsular polysaccharide levels (Masson and Holbein, 1985). Therefore, growth under different pH levels may have a profound effect on their growth and metabolism, as well as in the expression of various surface antigens. Just how *Neisseria* responds at the global molecular level to changes in pH has yet to be addressed. However, given the ease with which to conduct microarray analysis, deducing the changes in the expression profiles under varying pH growth conditions may yield incisive observations with regard to pathogenicity.

6.2.2 Physical stress – activation of the general stress response

Changes in physical parameters, such as temperature and osmotic pressure, generally lead to the accumulation of misfolded proteins, which activates the general stress response. The general stress response up-regulates the expression of various chaperone-like proteins in an effort to clear the misfolded protein log jam (Yura and Nakahigashi, 1999). When Gc are exposed to physical stress, a 63 kD protein, with homology to the Hsp60 heat shock family of proteins, is readily identifiable in stressed cells (Pannekoek *et al.*, 1992, 1995). Other stressful conditions (iron limitation, glucose deprivation, pH changes) also cause an accumulation of this particular protein, which has also been identified in serogroup B meningococci (Arakere *et al.*, 1993).

The general stress response is mediated by the RpoH sigma factor (σ^{32}) and in *E. coli* involves the upregulation of various chaperones, as well as several proteases (Yura and Nakahigashi, 1999). Some of the common proteins involved in this stress response are GroEL/GroES, DnaK, DnaJ and GrpE. The Gc RpoH regulon has been investigated by overexpressing RpoH (the lack of success in obtaining *rpoH* null mutants indicates that this sigma factor may be essential). Transcriptome analysis found that the

expression of the *dnaK*, *dnaJ* and *grpE* genes was controlled by the RpoH sigma factor and appeared to constitute a limited general stress response (Laskos *et al.*, 2004). The study also revealed an unusual aspect of the *Neisseria* RpoH-mediated stress regulon in that Gc appear to maintain preformed RpoH in the cytoplasm in readiness for the organism to respond rapidly to stress, which is unlike what is observed with other organisms (Laskos *et al.*, 2004). Whether this reflects a continuous need for rapid upregulation of the regulon due to continually encountering hostile chemical/physical environments is presently unclear. The Mc heat shock response was partially analysed using a limited DNA microarray. Again, expression of DnaJ, DnaK and GroEL/GroES was upregulated as well as several thioredoxin and glutaredoxin genes, indicating a link between the general stress and oxidative stress responses (Guckenberger *et al.*, 2002). Overall, the Mc response appeared to be broader in scope and, as such, appeared to resemble more closely the general stress response in other bacteria.

6.3 Nutrient Stress

6.3.1 The gonococcal stringent response – responding to limited nutrient availability

Due to the highly varied nutrient landscapes in which bacteria find themselves, often under severely limiting nutrient conditions, common survival mechanisms have evolved across all genera of bacteria that allow for cells to be able to adapt readily to their respective environments. A common bacterial response to nutrient stress is the establishment of specialized regulon (the stringent response) that is induced through the synthesis of unusual hyperphosphorylated guanosine nucleotides. Collectively, these effector molecules are referred to as (p)ppGpp, which includes both guanosine 3′,5′-bis(diphosphate) (ppGpp) and guanosine 3′,5′-bis(triphosphate) (pppGpp). Various conditions are able to stimulate the stringent response in bacteria and include amino acid and carbon energy source deprivation, phosphorus and nitrogen depletion, as well as abrupt temperature and pH changes (reviewed in Cashel and Gallant, 1969; Chatterji and Ojha, 2001).

The stringent response has been implicated in contributing to pathogenesis associated with both intracellular and extracellular pathogens (Godfrey et al., 2002). For two intracellular pathogens, Mycobacterium tuberculosis and Legionella pneumophila, a fully functioning stringent response is required for survival in macrophages where the organism may be deprived of various essential nutrients such as oxygen, amino acids and carbohydrates. Consequently, bacterial virulence and persistence is associated closely with the ability to mount a stringent response (Hammer and Swanson, 1999; Primm et al., 2000; Bachman and Swanson, 2001; Zusman et al., 2002; Dahl et al., 2003). The ability to activate the stringent response regulon has also been shown to be important with respect to the virulence properties of several extracellular pathogens such as Streptococcus pyogenes (Steiner and Malke, 2001), Pseudomonas aeruginosa (van Delden et al., 2001) and Staphylococcus aureus (Cassels et al., 1995; Gentry et al., 2000), as well as with Borrelia burgdorferi infections (Yang et al., 2000; Hubner et al., 2001; Bugrysheva et al., 2002; Revel et al., 2002). Therefore, nutrient stress and the ability of a pathogen to be able to deal with it appears to have significant effects with respect to the observed pathogenicity associated with a particular disease.

Generally, two genes are involved in establishing the stringent response: relA, which encodes a ribosome-associated (p)ppGpp synthetase (PSI) that responds primarily to amino acid deprivation (Fehr and Richter, 1981); and spoT, which encodes a cytosolic (p)ppGpp synthetase (PSII) (Fehr and Richter, 1981). Therefore, when unfavourable conditions are encountered, the combined action of these two enzymes leads to a dramatic increase in (p)ppGpp levels in the cell, which leads to the activation of the stringent response regulon. Both pathogenic Neisseria species possess the relA and spoT genes (Parkhill et al., 2000; Tettelin et al.,

2000; Fisher et al., 2005). In E. coli the omega subunit of RNA polymerase (the rpoZ gene, which resides within the spoT operon) is required for mounting and maintaining the stringent response (Chatterji et al., 2007). Similarly, in Gc the rpoZ gene is also located in the spoT operon, and presumably this sigma factor will also modulate the Neisserial stringent response (unpublished observations).

The adaptation to nutrient deprivation is a pleiotrophic response that effectively shuts down macromolecular synthesis and cellular metabolism (negative regulation, e.g. rRNA and tRNA synthesis; Cashel et al., 1996) in conjunction with upregulating several specific survival systems (positive regulation, e.g. induction of the RpoS regulon, Gentry et al., 1993; and the RpoH heat shock regulon, Grundy et al., 1994). Several of the responses to (p)ppGpp accumulation are mediated through the use of alternative sigma factors such as σ^S and σ^{54}, which interestingly the pathogenic Neisseria lack. None the less, Gc show a pronounced accumulation of (p)ppGpp when starved of the amino acid serine when grown in chemically defined medium (Fisher et al., 2005). Therefore, if Gc are growing under nutrient deprivation on an inflamed urogenital mucosa, then a robust elaboration of (p)ppGpp molecules may influence the physiology of the organism profoundly. Therefore, transcriptome analysis should reveal the extent of the response when Gc are growing under nutrient stress. For Gc, the ability to engage in a stringent response appears to be important as N. gonorrhoeae relA mutants present a severe growth defect, which can be alleviated by preventing DNA uptake into the cell (i.e. N. gonorrhoeae relA pilT mutants present a wild-type growth phenotype; unpublished observations). When the Gc stringent response is compared to E. coli's, it would appear that Gc engage in the more robust response (Fig. 6.1). Whether this indicates that Gc have evolved to exploit more fully this stress response mechanism due to spending considerable time in an unfavourable environment is open to conjecture and further experimentation.

(p)ppGpp turnover in Gc is examined in

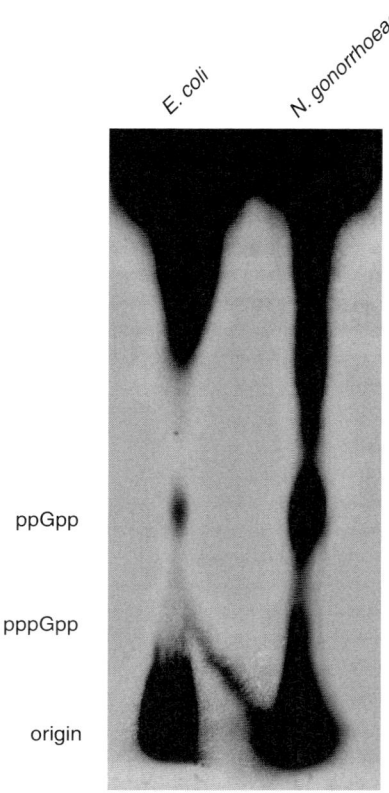

Fig. 6.1. Stringent response comparison between *N. gonorrhoeae* and *E. coli.* Both organisms were grown initially in fully complemented minimal medium (M9 and WSJM) containing [^{32}P]-orthophosphate to an optical density of 0.3 OD_{540nm}. The cells were then transferred to minimal medium without the amino acid serine, yet with the stringent response being optimally induced through the addition of sodium hydroxamate to the culture. The culture then was allowed to grow for a further 30 min prior to extraction of the labelled nucleotides. The labelled nucleotides were fractionated by using thin-layer chromatography on polyethyleneimine-cellulose plates using 1.5 M KH_2PO_4 (pH 3.4)

Fig. 6.2, where it can be seen that different cations in the medium can have a pronounced effect on turnover: when Mg^{2+} ions are present, (p)ppGpp levels appear to remain constant; when Mn^{2+} ions replace the Mg^{2+} ions in the gonococcal growth medium, (p)ppGpp hydrolysis occurs rapidly (within

5 min) in wild-type cultures, whereas hydrolysis is abrogated in a *spoT::ermC* insertional mutant. Therefore, the gonococcal *spoT* phosphatase activity requires Mn^{2+} uptake into the cell.

Of the various Neisserial stress responses, the stringent response has received the least amount of investigation. Given that several species of bacteria are unable to survive in an intracellular environment (Hammer and Swanson, 1999; Primm *et al.*, 2000; Bachman and Swanson, 2001; Zusman *et al.*, 2002; Dahl *et al.*, 2003; Zhou *et al.*, 2008; Dean *et al.*, 2009), nor are able to withstand acid stress without being able to engage in an active stringent response (Rallu *et al.*, 1996, 2000; Mouery *et al.*, 2006), then it may prove to be highly informative to determine how the pathogenic *Neisseria* utilize this stress regulon to moderate their pathogenic potential.

6.4 Oxidative Stress

It has become increasingly apparent that the two pathogenic *Neisseria* species differ dramatically in their mechanisms with which to cope with oxidative stress, with the mechanistic differences being attributed to the radically different lifestyles enjoyed by each organism within their unique environmental niches (Seib *et al.*, 2004). Oxidative stress occurs due to the exposure of the bacterium to a variety of oxygen- or nitrogen-based oxidants such as superoxide ions ($O_2^{·-}$), hydrogen peroxide (H_2O_2), hydroxyl radicals ($OH^·$), nitric oxide (NO) and peroxynitrite ions ($ONOO^-$). Such oxidants arise, either as:

1. Metabolic by-products during respiration;
2. Reactive intermediates that are formed during the innate immune response (i.e. PMN killing schemes);
3. A negative consequence of iron homeostasis; or
4. A consequence of co-colonization with other radical-producing organisms (e.g. Gc co-inhabiting the vaginal tract with lactobacilli, which produce H_2O_2) (Touati, 2000; Stohl *et al.*, 2005; Seib *et al.*, 2006).

Therefore, oxidative damage may occur either from exo- or endogenously produced

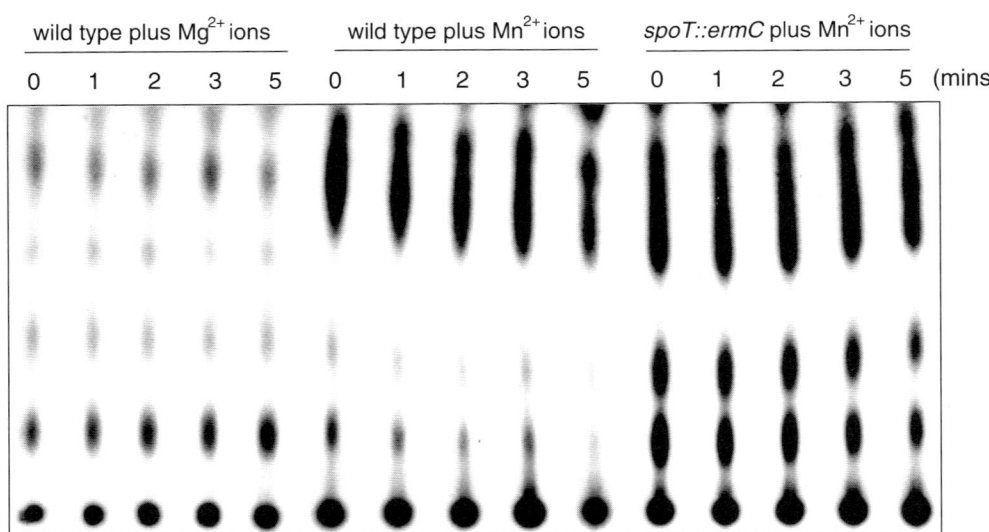

Fig. 6.2. The gonococcal stringent response. Gc are grown initially in fully complemented WSJM minimal medium containing [^{32}P]-orthophosphate to an optical density of 0.3 OD_{540nm} (Fisher *et al.*, 2005). The cells are then transferred to minimal medium lacking the amino acid serine for 15 min (in order to induce a stringent response); serine was added to the culture (final concentration 500 μg/ml) with samples removed at 1, 2, 3 and 5 min intervals thereafter. The labelled nucleotides are fractionated by using thin-layer chromatography on polyethyleneimine-cellulose plates using 1.5 M KH_2PO_4 (pH 3.4).

oxidants. Without protective mechanisms being employed to reduce the oxidant chemically, exposure to such oxidants can lead to DNA damage (e.g. chromosome breaks, an increase in spontaneous mutations) and protein inactivation, as well as membrane dysfunction due to lipid damage. This chapter shall focus primarily on detoxification of reactive oxygen species.

6.4.1 Oxidant protection – reducing the power of the radical

Initially, it appeared that Gc had evolved oxidative defensive strategies to detoxify H_2O_2 through the expression of high levels of catalase and peroxidase (Archibald and Duong, 1986). Because H_2O_2 can diffuse freely across membranes, then exogenously produced H_2O_2 within the urogenital tract, either from activated PMNs or from aerobically respiring Lactobacilli, theoretically could damage internal macromolecules. In *E. coli*,

H_2O_2 in the cytoplasm is sensed by OxyR by using a redox-sensing mechanism (Storz *et al.*, 1990), with the oxidized form of the protein causing the upregulation of catalase (*katA*) expression, which catalyses the reduction of H_2O_2. In contrast to *E. coli*, Gc OxyR negatively regulates *katA* expression (Tseng *et al.*, 2003), yet H_2O_2 treatment of Gc can still cause an approximate 50-fold increase in *katA* activity (Grifantini *et al.*, 2004; Stohl *et al.*, 2005), which appears to correlate with the observation that *katA* mutants show a heightened sensitivity to H_2O_2 exposure (Soler-Garcia and Jerse, 2004). Overall, exposure of Gc to H_2O_2 upregulates a total of 75 genes (Stohl *et al.*, 2005).

This apparent conundrum has been partially explained by OxyR studies in *N. meningitidis*, where the redox state of the OxyR protein following H_2O_2 exposure allows OxyR to regulate *katA* expression tightly; when in the reduced form, OxyR negatively regulates *katA*, whereas in the oxidized form the protein serves as an activator (Ieva *et al.*, 2008). Gonococci also possess other proteins in the cytoplasm with which to detoxify H_2O_2. Transcriptome analysis revealed a role for a putative Zn^{2+}-metalloprotease, as well as a role for an unknown hypothetical protein (NGO554), with each protein providing protection against high levels of H_2O_2 (Stohl *et al.*, 2005). Other cytoplasmic reducing systems have also been demonstrated and include a Bacterioferritin system (Chen and Morse, 1999) and an Azurin system (Wu *et al.*, 2005), as well as a coupled glutathione reductase/peroxidoxin system (Moore and Sparling, 1996; Seib *et al.*, 2006). Gonococci also protect themselves externally from H_2O_2-induced oxidative damage using an MsrA/B system, which is located in the outer membrane (Skaar *et al.*, 2002); cytochrome *c* peroxidase, which is located in the cytoplasmic membrane yet faces towards the periplasmic space (Turner *et al.*, 2003); and a periplasmic-located thioredoxin-like protein (Achard *et al.*, 2009).

With all of these multiple H_2O_2 detoxification systems being available, it was thought initially that Gc lacked the enzyme superoxide dismutase (SOD) which was required to reduce superoxide ($O_2^{\cdot-}$) ions to H_2O_2. This belief was based primarily on the high degree of sensitivity that Gc demonstrate when exposed to plumbagin, which caused high intracellular $O_2^{\cdot-}$ levels (Archibald and Duong, 1986). Many bacteria possess multiple SOD enzymes. However, gonococcal genome analysis indicates only a single, cytoplasmically located Fe-dependent SodB, with the other cytoplasmically located SOD (the Mn-dependent SodA) and the periplasmically located Cu/Zn SodC being absent (Tseng *et al.*, 2001). Consequently, it was assumed that SodB would reduce any superoxide ions that arose in the cytoplasm. However, *N. gonorrhoeae sodB* mutants still presented a wild-type phenotype when exposed to superoxide, indicating the presence of an alternative system (Tseng *et al.*, 2001). This secondary system was identified by applying an observation initially made with *Lactobacillus plantarum*, where supplementation of the growth medium with Mn^{2+} ions provided protection against superoxide toxicity (Archibald and Fridovich, 1981 a,b).

Similarly, adding Mn^{2+} to the Gc growth medium also protected against superoxide damage (Tseng *et al.*, 2001). Moreover, Gc mutants that were unable to take up Mn^{2+} from the environment (*mtnC* mutants) were found to be highly susceptible to oxidative killing following either $O_2^{\cdot-}$ or H_2O_2 exposure (Tseng *et al.*, 2001; Seib *et al.*, 2004). Therefore, the presence of Mn^{2+} ions in the cytoplasm protects against oxidative damage via a catalase-independent mechanism (Seib *et al.*, 2004). Subsequently, an Mn^{2+}-dependent oxidative stress regulator, PerR, was then identified (Wu *et al.*, 2006). When *N. gonorrhoeae perR* mutants were examined, the PerR regulon was identified, where it was also shown that the ability to engage the PerR regulon was apparently essential *in vivo* as such mutants were more susceptible to intracellular killing using a human cervical tissue culture model system (Wu *et al.*, 2006). In addition, Mn^{2+} availability also moderates phosphate metabolism as Mn^{2+} ions serve as a cofactor for an inorganic pyrophosphatase, which strikingly mirrors the Mn^{2+} requirement for SpoT phosphatase activity (Fig. 6.2; Wu *et al.*, 2010). Consequently, the availability of Mn^{2+} ions in the cytoplasm may couple the stringent and oxidative stress responses.

Despite this apparent plethora of oxidative defence mechanisms deployed by Gc, a recent study using a complete set of mutants that were deficient in all of the oxidative stress defences presented some perplexing observations, notably that each mutant strain showed no greater susceptibility to oxidative stress than the isogenic wild type when survival assays were performed with adherent PMNs, which was a scenario that was believed to mimic *in vivo* killing (Seib *et al.*, 2005). Consequently, it was speculated that these observations indicated that Gc might experience significant oxidative stress from sources other than from the inflammatory response, which required the maintenance of all these disparate reducing systems. However, the source of this 'alternative' oxidative stress remains to be elucidated.

Historically, Mc have always been thought to respond to oxidative stress somewhat differently because *sodC* (which encodes the periplasmically located Cu/Zn SodC) is present in these bacteria and *sodC* mutants are 1000-times more sensitive to exogenously produced $O_2^{.-}$ than the isogenic wild-type strain (Wilks *et al.*, 1998). Moreover, *N. meningitidis sodC* mutants were also shown to be less virulent in a mouse model for pathogenesis (Wilks *et al.*, 1998). However, these observations are controversial (Seib *et al.*, 2004), because *N. meningitidis mntC* mutants still remain susceptible to oxidative killing (as do *N. meningitidis sodB* mutants). Therefore, Mc do not appear to show the Mn^{2+} protective effect (Seib *et al.*, 2004). Also, Mc lack the periplasmic-facing cytochrome *c* peroxidase (Seib *et al.*, 2004). Consequently, it has been proposed that Gc have evolved multiple mechanisms to cope with chronic oxidative stress in the low pH environment of the female genital tract using primarily the unusual Mn^{2+} redox mechanism in conjunction with the other mechanistic schemes to detoxify H_2O_2, whereas Mc has adapted to the high pO_2 environment of the nasopharynx by using SOD to detoxify $O_2^{.-}$ due to its greater abundance on these tissues (Seib *et al.*, 2004, 2006).

6.4.2 Iron homeostasis – forging a double-edged sword

All living systems require iron for growth due to its involvement in many enzyme systems that utilize its potent redox potential. However, too much iron in the presence of oxygen can be lethal unless the necessary defensive schemes are in place to combat the production of oxygen radicals (see Imlay and Hassett, Chapter 1, this volume). Oxygen radicals arise due to ferrous ions reacting with hydrogen peroxide via the Fenton reaction:

$$Fe^{2+} + H_2O_2 \rightarrow Fe^{3+} + OH^. + OH^- \qquad (6.1)$$

Consequently, when growing *in vivo*, bacteria must scavenge iron from host iron–protein complexes without disrupting iron homeostasis within the cell. If iron homeostasis becomes uncoupled, then this will lead to oxidative stress (and potentially cell death),

which will therefore require the upregulation of oxidative defence mechanisms, as well as tighter controls on iron metabolism (Touati, 2000).

Because iron is essential for bacterial growth, and the free iron concentration in the human host varies between 10^{-9} and 10^{-18} M, then specific mechanisms must be deployed to obtain the necessary iron. Humans sequester iron in the form of iron–protein complexes such as transferrin, lactoferrin, haemoglobin, haptoglobin and ferritin, with each iron–protein complex being associated with distinct sites within the body. For example, transferrin is present in serum, as well as being associated with serous exudates; lactoferrin is found on various mucosal tissues, whereas haemoglobin is a major component of erythrocytes that are located in the bloodstream. Many bacteria scavenge iron from the host through the production of siderophores. However, the pathogenic *Neisseria* do not possess such molecules. Therefore, depending on the type of infection and its location, the availability of a specific iron source is likely to change. All Gc are able to scavenge iron from human transferrin as well as from haemoglobin, with only a subset of strains obtaining iron from lactoferrin (Gray-Owen and Schryvers, 1996; Cornelissen *et al.*, 1998). Furthermore, the Gc transferrin receptor is unique in that it is only able to recruit iron from human transferrin and not from other mammalian transferrin complexes. Indeed, the importance of the gonococcal transferrin receptor is well demonstrated in human experimental challenge studies where gonococcal transferrin receptor mutants are non-infectious (Cornelissen *et al.*, 1998).

Meningococci also possess similar iron-scavenging systems and respond similarly *in vitro* to limiting iron conditions. However, depending on the type of Mc infection, the organism will be exposed to different sets of iron sources. Recently, it was speculated that the specific iron donor molecules might serve as niche indicators for Mc. Thereby, the iron donor molecule that is present would then activate specific responses within the bacterial cell to facilitate growth in the environmental niche that was indicated by that particular iron donor (Jordan and Saunders, 2009). Specific transcriptional responses were observed in gene expression pathways depending on which iron source was used for *in vitro* growth; when growth was facilitated by the presence of haemoglobin, the expression profile response indicated an adaptation to growth as though the organism was growing within the bloodstream, as defence mechanisms against serum killing were upregulated; when growth was supported by lactoferrin (which indicates a mucosal existence), then elevated oxidative stress responses were observed. Therefore, the host iron–protein complexes may serve as niche indicators for pathogenic bacteria.

In iron-starvation studies for both species, multiple genes are activated or repressed in response to the availability of iron in the growth medium (Grifantini *et al.*, 2003; Ducey *et al.*, 2005). For an *N. meningitidis* serogroup B strain (MC58), 233 genes were affected by the lack of iron and covered a broad array of genes involved in iron metabolism (Grifantini *et al.*, 2003). Of these genes, approximately 50% possessed a Fur-binding domain (see below). Interestingly, limiting iron availability appeared to control the expression of several oxidative stress response genes, such as *katA*, *sodB* and the NMB1436-38 operon (Delany *et al.*, 2004). A similar overlap between iron homeostasis and oxidative stress was also observed when Gc were exposed to H_2O_2, with 15 genes involved in iron homeostasis (including *fur*) being upregulated (Sebastian *et al.*, 2002; Stohl *et al.*, 2005). Similarly, iron homeostasis and nutrient stress responses via the stringent response may also overlap as iron availability influences the expression of *secY*, which controls the expression of numerous ribosomal proteins (Shaik *et al.*, 2007), down-regulation of ribosomal protein synthesis being one of the hallmarks of an induced stringent response. For Gc, 203 genes were affected by the absence of iron (Ducey *et al.*, 2005). However, only 12 genes corresponded to those affected in Mc. Consequently, as local environments are going to be highly disparate with respect to iron availability, this feature has probably driven each

organism down a different evolutionary path (Ducey *et al.*, 2005).

6.4.3 FUR – putting a wrap on stress

Many Gram-negative bacteria coordinate iron homeostasis by using the Fur regulatory protein (Ferric uptake regulator). Fur is a small protein (approximately 15 kD) that binds ferrous ions. When bound to Fe^{2+}, a Fur dimer forms which is the active form of the protein. Upon dimer formation, the Fur complex then binds to a specific sequence (Fur boxes; consensus sequence 5′-GATAATGATAATCATTATC-3′) located on the chromosome. Invariably, Fur boxes are associated with promoters and Fur generally acts as a negative regulator. However, in some species of bacteria, Fur can activate certain genes (Delany *et al.*, 2001).

With Gc, the isolation of *fur* null mutants has proven to be difficult. A *fur* missense mutant was obtained using manganese selection and revealed that many iron-induced genes were apparently regulated by Fur protein (Thomas and Sparling, 1996). A more recent study, which incorporated several global analytic strategies to detect chromosomal regions that bound Fur *in vivo*, showed that the Gc Fur regulon apparently involved a cascade of events, mediated by multiple regulators, many of which were under indirect control by Fur protein (Jackson *et al.*, 2010). This demonstration of a cascade effect may explain partially the observed non-Fur regulation of genes under iron-limiting conditions.

Greater success has been forthcoming in obtaining *N. meningitidis fur* null mutants, allowing the Fur regulon to be identified, with 83 genes being controlled directly by Fur protein (Delany *et al.*, 2006). Again, there appeared to be overlap between the various stress responses because the heat shock regulon was induced in the *fur* mutant indirectly. Unusually, Fur can function as a positive and negative regulator in Mc (Delany *et al.*, 2004). Of the specific genes regulated directly by Fur, several have interesting characteristics. Fur serves as a positive regulator of *pan1*, *norB* and *nuoA* genes, the expression of which allows the organism to respond to aerobic as well as anaerobic conditions in an iron-responsive fashion. Also, Fur binding upregulates expression of the NMB1436-38 operon which protects against oxidative stress; mutants lacking this operon show a greater susceptibility to PMN killing, as well as being attenuated for virulence in a mouse infection model (Delany *et al.*, 2004). In contrast, Fur regulates the *tbpB* gene negatively (which encodes the transferrin-binding protein) under iron-replete conditions. With the delineation of the Fur regulon, this should allow a more rigorous analysis of the various overlapping regulons to be undertaken.

6.5 sRNA Regulation and the Stress Responses

Short interfering RNAs (siRNA) have been shown to play a prominent role in regulating gene expression in eukaryotes (Wu and Belasco, 2008). In recent years, similar non-coding RNAs have been identified in bacteria, where they have been shown to stimulate mRNA degradation and stabilize certain mRNAs within the cell, as well as regulate translation of a specific message. In each case, complementary binding occurs between the small RNA (sRNA) molecule and a specific mRNA, with this interaction being facilitated by the RNA chaperone, Hfq protein. Indeed, for *E. coli* several stress response mechanisms (Fe^{2+} limitation, oxidative stress, temperature stress) utilize sRNA regulation to mitigate the effects of the various stressors (Gottesman, 2005). Following the initial observations in *E. coli*, a concerted effort was undertaken to find sRNA regulation in other organisms, especially in those bacteria that caused disease. The importance of sRNA regulation on pathogenesis became immediately apparent when it was shown that *Vibrio cholerae hfq* mutants were attenuated for virulence in an infant rat animal model (Ding *et al.*, 2004). Subsequently, studies with various *Vibrio* spp. have shown that sRNAs regulate quorum sensing (Lenz *et al.*, 2004; Hammer and Bassler, 2007; Svenningsen *et*

al., 2009) and are involved in the Fur regulon (Davis *et al.*, 2005), as well as promoting the formation of outer membrane vesicles in response to environmental stimuli (Song *et al.*, 2008; Song and Wai, 2009). Likewise, sRNA regulation appears to operate in several other human pathogens, most notably *P. aeruginosa* (Livny *et al.*, 2006), *Salmonella* (Sittka *et al.*, 2009; Vogel, 2009), *M. tuberculosis* (Arnvig and Young, 2009), *S. pneumoniae* (Tsui *et al.*, 2010) and Enterohaemorraghic *E. coli* (Shakhnovich *et al.*, 2009), as well as with group A streptococci (Perez *et al.*, 2009).

Studies on sRNA regulation of gene expression in the *Neisseria* are currently in their infancy. A major reason for this lack of enquiry is that the identification of non-coding RNA is extremely difficult (Livny *et al.*, 2006; Cao *et al.*, 2009; Sharma and Vogel, 2009; Sittka *et al.*, 2009). The first demonstration of sRNA regulation in *Neisseria* came through a bioinformatics approach that mirrored work originally performed in *E. coli* (Mellin *et al.*, 2007). This approach identified a single sRNA (NrrF; Neisserial regulatory RNA responsive to iron) whose expression was shown to be dictated by iron availability. Therefore, *N. meningitidis*, like *E. coli* and *V. cholerae* (Masse and Gottesman, 2002; Davis *et al.*, 2005), use sRNA in response to changing iron availability; in this case, transcription of the *nffR* gene is repressed by Fur protein under iron-replete conditions (Mellin *et al.*, 2007). Expression of NrrF RNA was also shown to regulate *sdhA* and *sdhC*, which are two iron-containing polypeptides that are part of the TCA cycle enzyme succinate dehydrogenase, which again mirrors the situation with *E. coli* (Masse and Gottesman, 2002). Subsequently, it was shown that NrrF RNA directly mediated Fur-dependent regulation of succinate dehydrogenase by utilizing the RNA chaperone Hfq; Hfq delivery of NrrF RNA to the *sdhCDAB* message caused rapid degradation of the transcript, which correlated with the positive Fur-dependent regulation (Metruccio *et al.*, 2009). Consequently, iron homeostasis in *Neisseria* involves sRNA regulation, as observed in other species.

When a proteomic approach was employed using *N. meningitidis hfq* mutants,

28 proteins were shown to be affected, most notably *sodB* (encoding the Fe-Mn superoxide dismutase protein) and *sodC* (encoding the Cu-Zn superoxide dismutase protein). Therefore, this study established a link between sRNA regulation and the oxidative stress responses (see Fig. 6.3; Pannekoek *et al.*, 2009).

The link between stress and sRNA regulation was further augmented when the heat stress chaperonin gene, *groEL*, was shown to be upregulated in the *N. meningitidis hfq* mutant, as was the oxidative stress responder Peroxiredoxin (*prx*) (Fantappie *et al.*, 2009). Besides sRNA coupling to oxidative stress, iron homeostasis was also affected in the *N. meningitidis hfq* mutant, as lactoferrin-binding protein B (*lbpB*) and the iron ABC transporter (*fbpA*) were also upregulated. Thus, Hfq apparently plays a key role in modulating the stress responses in Mc (Fantappie *et al.*, 2009). And, as has been observed for *V. cholerae* (Ding *et al.*, 2004), *N. meningitidis hfq* mutants are also attenuated for virulence using an animal model (Fantappie *et al.*, 2009).

A similar set of observations was also obtained for *N. gonorrhoeae hfq* mutants where slightly more genes appeared to be regulated differentially; 369 in total, with 202 upregulated and 167 downregulated (Dietrich *et al.*, 2009). As observed with Mc, Hfq appears to be involved in modulating various stress responses by being a positive expression factor for several chaperonin genes, as well as several genes involved in iron homeostasis. Exploring this latter effect further showed that the *nrrF* transcript was also a Fur-repressed sRNA (Ducey *et al.*, 2009). Therefore, given that very few studies have examined the effects of sRNA regulation in the pathogenic *Neisseria*, a consensus seems to be emerging that indicates that sRNA production and regulation of the various stress responses are intimately coupled.

6.6 Interaction and Cooperation of the Various Stress Responses

As *in vivo* growth conditions appear to be suboptimal for most organisms that cause infectious disease, it is highly likely that some overlap exists between the various stress

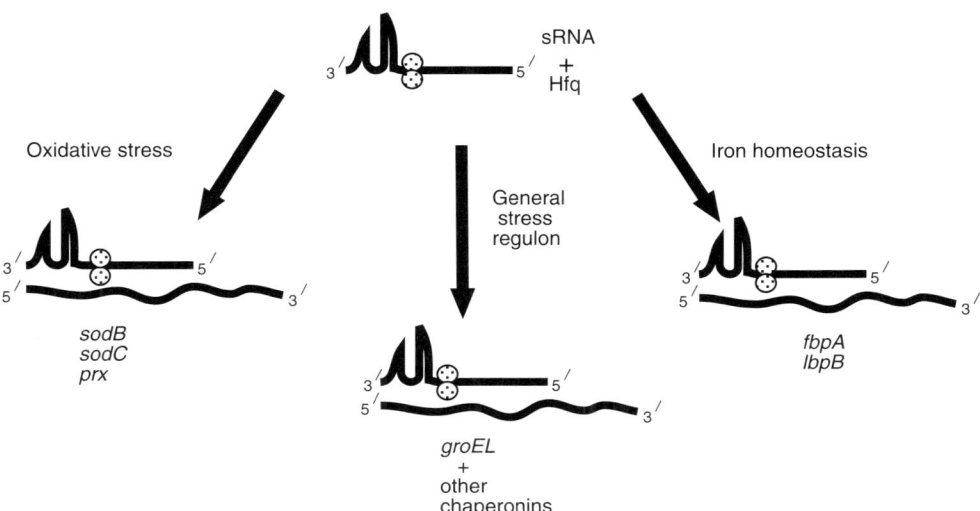

Fig. 6.3. Overview of sRNA regulation of the stress responses. sRNA, in conjunction with the RNA chaperone Hfq, bind to its complementary sequence in mRNA and regulate either the translation of the message or message stability. The involvement of sRNA regulation in the various stress regulons is shown.

responses. Such an overlap is well documented in *Salmonella, E. coli, P. aeruginosa* and *Bacillus subtilis* where H_2O_2 exposure not only stimulates the oxidative stress response but also activates the general stress response mediated by the heat shock chaperones (Morgan *et al.*, 1986; Zheng *et al.*, 2001; Mostertz *et al.*, 2004; Palma *et al.*, 2004). Likewise, for *Enterococcus faecalis*, the oxidative stress response appears to be coupled with the stringent response, as *relA* mutants are more sensitive to the effects of H_2O_2 than wild-type cultures (Yan *et al.*, 2009). Along these lines, the ability of *Lactococcus lactis* to engage in a stringent response is also required for dealing with acid stress (Rallu *et al.*, 1996, 2000). Acid stress is also encountered following uptake into cells and, for *Francisella novicida, Helicobacter pylori* and *L. pnuemophila*, intracellular survival requires the ability to mount a stringent response (Hammer and Swanson, 1999; Zhou *et al.*, 2008; Dean *et al.*, 2009). Therefore, it should come as no surprise to see similar overlaps between the various stress regulons in the pathogenic *Neisseria* (outlined in Fig. 6.4).

Transcriptome analysis of Gc when exposed to H_2O_2 revealed an overlap between the oxidative and the general stress responses, as well as demonstrating the upregulation of several genes involved in iron homeostasis (mainly those used for iron acquisition) (Stohl *et al.*, 2005). Similarly for Mc, iron availability, in conjunction with Fur activation, was shown to correlate with the expression of the general stress regulon where, at low activated Fur concentrations, the heat shock response was upregulated. However, somewhat surprisingly, control of the heat shock regulon did not appear to respond to changing iron concentrations (Delany *et al.*, 2006). Iron availability also affected the regulation of *secY*, whose homologue in *E. coli* was found in a gene cluster that contained numerous genes encoding ribosomal proteins (Shaik *et al.*, 2007). SecY controls the expression of ribosomal proteins, among other things. Therefore, iron availability may also intersect with the Neisserial stringent response, as downregulation of ribosomal protein synthesis is a hallmark of activation of this response. The observation that acid stress causes the upregulation of GroEL clearly links acid stress to the general stress response (Pannekoek *et al.*, 1992; Arakere *et al.*, 1993). Therefore, if gonococci mediate acid stress as do Lactobacilli, then exposure of *Neisseria* to acidic conditions may also activate the stringent response. An overlap between the stringent response and the oxidative stress response and phosphate metabolism also appears to be hinted at with the requirement

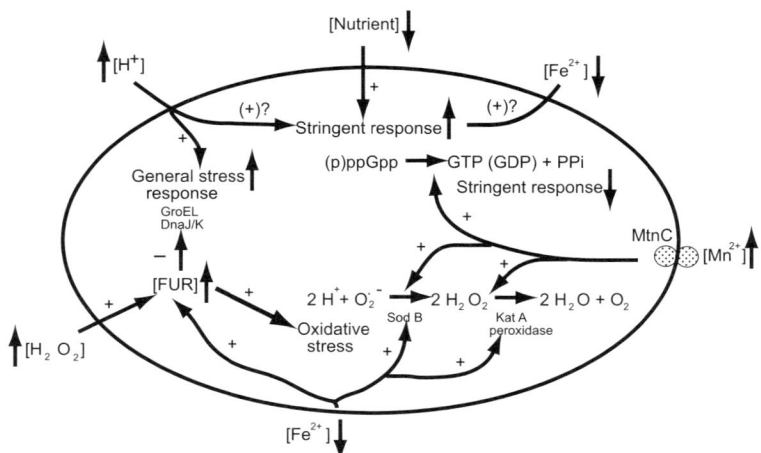

Fig. 6.4. Overlap of the various stress regulons in the pathogenic *Neisseria*. Overlap of the various stress regulons is indicated as either a positive or negative interaction. Question marks indicate conjecture discussed in the text.

for Mn^{2+} ions for hydrolysis of (p)ppGpp and its role as a cofactor for an inorganic pyrophosphatase (Wu *et al.*, 2010). A further coupling of the two responses also seems to be evident by the observation that a homologue to *E. coli's* glutaredoxin B gene (*grxB*) resides immediately upstream of the Gc *relA* gene (unpublished observations), with the putative Gc *grxB* being transcribed divergently to *relA* using overlapping promoters (also, a prominent IHF-binding site is centrally located between the two promoters; unpublished observations). Therefore, if control of *grxB* expression in *Neisseria* is similar to that observed in *E. coli*, then (p) ppGpp availability (in conjunction with IHF binding) may provide GrxB protection against oxidants (Potamitou *et al.*, 2002).

As more examples of transcription profiling become available on the effects of various stressors on *Neisseria* physiology, further overlaps are expected to be found in the future. From the preceding discussion it would seem that previously unknown overlaps might exist between the stringent response and the oxidative and acid responses. Therefore, perhaps by determining how the pathogenic *Neisseria* deal physiologically with acid and nutrient stress may provide considerable insight into their pathogenicity. Moreover, such analysis will illustrate further how they cope with stress and may provide a more detailed picture of the overlapping stress responses.

References

Achard, M.E.S., Hamilton, A.J., Dankowski, T., Heras, B., Schembri, M.S., Edwards, J.L., Jennings, M.P. and McEwan, A.G. (2009) A periplasmic thioredoxin-like protein plays a role in defence against oxidative stress in *Neisseria gonorrhoeae*. *Infection and Immunity* 77, 4934–4939.

Arakere, G., Kessel, M., Nguyen, N. and Frasch, C.E. (1993) Characterization of a stress protein from Group B *Neisseria meningitidis*. *Journal of Bacteriology* 175, 3664–3668.

Archibald, F.S. and Duong, M.-N. (1986) Superoxide dismutase and oxygen toxicity defences in the genus *Neisseria*. *Infection and Immunity* 51, 631–641.

Archibald, F.S. and Fridovich, I. (1981a) Manganese and defenses against oxygen toxicity in *Lactobacillus plantarum*. *Journal of Bacteriology* 145, 442–451.

Archibald, F.S. and Fridovich, I. (1981b) Manganese, superoxide dismutase, and oxygen tolerance in some lactic acid bacteria. *Journal of Bacteriology* 146, 928–936.

Arnvig, K.B. and Young, D.B. (2009) Identification of small RNAs in *Mycobacterium tuberculosis*. *Molecular Microbiology* 73, 397–408.

Bachman, M.A. and Swanson, M.S. (2001) RpoS co-operates with other factors to induce *Legionella pneumophila* virulence in the stationary phase. *Molecular Microbiology* 40, 1201–1214.

Berger, U. (1986) Nitrite reduction related to serogroups in *Neisseria meningitidis*. *Zentralblatt fuer Bakteriologie, Mikrobiologie und Hygiene (Reihe A)* 261, 140–146.

Booth, J.W., Telio, D., Liao, E.H., McCaw, S.E., Matsuo, T., Grinstein, S. and Gray-Owen, S.D. (2003) Phosphyatidylinositol 3-kinases in carcinoembryonic antigen-related cellular adhesion molecule-mediated internalization of *Neisseria gonorrhoeae*. *Journal of Biological Chemistry* 278, 14037–14045.

Brener, D., DeVoe, I.W. and Holbein, B.E. (1981) Increased virulence of *Neisseria meningitidis* after *in vitro* iron-limited growth and low pH. *Infection and Immunity* 33, 59–66.

Burgrysheva, J., Dobrikova, E.Y., Godfrey, H.P., Sartakova, M.L. and Cabello, F.C. (2002) Modulation of *Borrelia burgdorferi* stringent response and gene expression during extracellular growth with tick cells. *Infection and Immunity* 70, 3061–3067.

Cao, Y., Zhao, Y., Cha, L., Ying, X., Wang, L., Shao, N. and Li, W. (2009) sRNATarget: a web server for prediction of bacterial sRNA targets. *Bioinformation* 3, 364–366.

Cashel, M. and Gallant, J. (1969) Two compounds implicated in the function of the RC gene of *Escherichia coli*. *Nature* 221, 838–841.

Cashel, M., Gentry, D.R., Hernandez, V.J. and Vinella, D. (1996) The stringent response. In: Neidhardt, F.C., Curtis, R., Ingraham, J.L., Lin, E.C.C., Low, K.B., Magasanik, B., Resnikoff, W.S., Riley, M., Schaechter, M. and Umbarger, H.E. (eds) *Escherichia coli and Salmonella: Cellular and Molecular Biology*, Vol. 1, 2 vols. ASM Press, Washington, DC, pp. 1458–1496.

Cassels, R., Oliva, B. and Knowles, D. (1995) Occurrence of the regulatory nucleotides ppGpp and pppGpp following induction of the stringent response in *Staphylococcus aureus*. *Journal of Bacteriology* 177, 5161–5165.

Chatterji, D. and Ojha, A.K. (2001) Revisiting the stringent response, ppGpp and starvation signaling. *Current Opinion in Microbiology* 4, 160–165.

Chatterji, D., Ogawa, Y., Shimada, T. and Ishihama, A. (2007) The role of the omega subunit of RNA polymerase in expression of the *relA* gene in *Escherichia coli*. *FEMS Microbiology Letters* 267, 51–55.

Chen, C.Y. and Morse, S.A. (1999) *Neisseria gonorrhoeae* bacterioferritin: structural heterogeneity, involvement in iron storage and protection against oxidative stress. *Microbiology* 145, 2967–2975.

Cohen, L. (1969) Influence of pH on vaginal discharges. *British Journal of Venereal Disease* 45, 241–245.

Cornelissen, C.N., Kelley, M., Hobbs, M.M., Anderson, J.E., Cannon, J.G., Cohen, M.S. and Sparling, P.F. (1998) The transferrin receptor expressed by gonococcal strain FA1090 is required for experimental infection of human male volunteers. *Molecular Microbiology* 27, 611–616.

Dahl, J.L., Kraus, C.N., Boshoff, H.I., Doan, B., Foley, K., Avarbock, D., Kaplan, G., Mizrahi, V., Rubin, H. and Barry, C.E. (2003) The role of RelMtb-mediated adaptation to stationary phase in long-term persistence of *Mycobacterium tuberculosis* in mice. *Proceedings of the National Academy of Sciences of the United States of America* 100, 10026–10031.

Davis, B.M., Quinones, M., Pratt, J., Ding, Y. and Waldor, M.K. (2005) Characterization of the small untranslated RNA RyhB and its regulon in *Vibrio cholerae*. *Journal of Bacteriology* 187, 4005–4014.

Dean, R.E., Ireland, P.M., Jordan, J.E., Titball, R.W. and Oyston, P.C. (2009) RelA regulates virulence and intracellular survival of *Francisella novicida*. *Microbiology* 155, 4104–4113.

Deghmane, A.-E., Veckerle, C., Giorgini, D., Hong, E., Ruckly, C. and Taha, M.-K. (2009) Differential modulation of TNF-alpha-induced apoptosis by *Neisseria meningitidis*. *PLoS Pathogens* 5, e1000405.

Delany, I., Spohn, G., Rappuoli, R. and Scarlato, V. (2001) The Fur repressor controls transcription of iron-activated and -repressed genes in *Helicobacter pylori*. *Molecular Microbiology* 42, 1297–1309.

Delany, I., Rappuoli, R. and Scarlato, V. (2004) Fur functions as an activator and as a repressor of putative virulence genes in *Neisseria meningitidis*. *Molecular Microbiology* 52, 1081–1090.

Delany, I., Grifantini, R., Bartolini, E., Rappuoli, R. and Scarlato, V. (2006) Effect of *Neisseria meningitidis* Fur mutations on global control of gene transcription. *Journal of Bacteriology* 188, 2483–2492.

Dietrich, M., Munke, R., Gottschald, M., Ziska, E., Boettcher, J.P., Mollenkopt, H. and Friedrich, A. (2009) The effect of Hfq on global gene expression and virulence in *Neisseria gonorrhoeae*. *FEBS Journal* 276, 5507–5520.

Ding, Y., Davis, B.M. and Waldor, M.K. (2004) Hfq is required for *Vibrio cholerae* virulence and down-regulates sigma expression. *Molecular Microbiology* 53, 345–354.

Ducey, T.F., Carson, M.B., Orvis, J., Stintzi, A.P. and Dyer, D.W. (2005) Identification of the iron-responsive genes of *Neisseria gonorrhoeae* by microarray analysis in defined medium. *Journal of Bacteriology* 187, 4865–4874.

Ducey, T.F., Jackson, L., Orvis, J. and Dyer, D.W. (2009) Transcript analysis of *nrrF*, a Fur repressed sRNA of *Neisseria gonorrhoeae*. *Microbial Pathogenesis* 3, 166–170.

Fantappie, L., Metruccio, M.M.E., Seib, K.L., Oriente, F., Catocci, E., Ferlicca, F., Giulianai, M.M., Scarlato, V. and Delany, I. (2009) The RNA chaperone Hfq is involved in stress response and virulence in *Neisseria meningitidis* and is a pleiotropic regulator of protein expression. *Infection and Immunity* 77, 1842–1853.

Fehr, S. and Richter, D. (1981) Stringent response of *Bacillus stearothermophilus*: evidence for the existence of two distinct guanosine $3',5'$-polyphosphate synthetases. *Journal of Bacteriology* 145, 68–73.

Fisher, S.D., Reger, A.D., Baum, A. and Hill, S.A. (2005) *relA* alone appears essential for (p) ppGpp production when *Neisseria gonorrhoeae* encounters nutritional stress. *FEMS Microbiology Letters* 248, 1–8.

Gentry, D.R., Hernandez, V.J., Nguyen, D.E., Jensen, D.B. and Cashel, M. (1993) Synthesis of the stationary phase specific sigma factor, σ^S is positively regulated by ppGpp. *Journal of Bacteriology* 175, 7982–7989.

Gentry, D., Tong, L., Rosenberg, M. and McDevitt, D. (2000) The *rel* gene is essential for *in vitro* growth of *Staphylococcus aureus*. *Journal of Bacteriology* 182, 4995–4997.

Godfrey, H.P., Bugrysheva, J.V. and Cabello, F.C. (2002) The role of the stringent response in the pathogenesis of bacterial infections. *Trends in Microbiology* 10, 349–351.

Gottesman, S. (2005) Micros for microbes: non-coding regulatory RNAs in bacteria. *Trends in Genetics* 7, 399–404.

Gray-Owen, S. and Schryvers, A.B. (1996) Bacterial

transferrin and lactoferrin receptors. *Trends in Microbiology* 4, 185–191.

Grifantini, R., Sebastian, S., Frigimelica, E., Draghi, M., Bartolini, E., Muzzi, A., Rappuoli, R., Grandi, G. and Genco, C.A. (2003) Identification of iron-activated and -repressed Fur-dependent genes by transcriptome analysis of *Neisseria meningitidis* groupB. *Proceedings of the National Academy of Sciences of the United States of America* 100, 9542–9547.

Grifantini, R., Frigimelica, E., Delany, I., Bartolini, E., Giovinazzi, S., Balloni, S., Agarwal, S., Galli, G., Genco, C. and Grandi, G. (2004) Characterization of a novel *Neisseria meningitidis* Fur and iron-regulated operon required for protection from oxidative stress: utility of DNA microarray in the assignment of the biological role of hypothetical genes. *Molecular Microbiology* 54, 962–979.

Grundy, F.J., Rollins, S.M. and Henkin, T.M. (1994) Interaction between the acceptor end of tRNA and the T box stimulates antitermination in the *Bacillus subtilis tyrS* gene: a new role for the discriminator base. *Journal of Bacteriology* 176, 4518–4526.

Guckenberger, M., Kurz, S., Aepinus, C., Theiss, S., Haller, S., Leimbach, T., Panzner, U., Weber, J., Paul, H., Unkmeir, A., Frosch, M. and Dietrich, G. (2002) Analysis of the heat shock response of *Neisseria meningitidis* with cDNA- and oligonucleotide-based DNA microarrays. *Journal of Bacteriology* 184, 2546–2551.

Hammer, B.K. and Bassler, B.L. (2007) Regulatory small RNAs circumvent the conventional quorum sensing pathway in pandemic *Vibrio cholerae*. *Proceedings of the National Academy of Sciences of the United States of America* 104, 11145–11149.

Hammer, B.K. and Swanson, M.S. (1999) Co-ordination of *Legionella pneumophila* virulence with entry into stationary phase by ppGpp. *Molecular Microbiology* 33, 721–731.

Hebeler, B.H., Morse, S.A., Wong, W. and Young, F.E. (1978) Evidence for peptidoglycan-associated protein(s) in *Neisseria gonorrhoeae*. *Biochemica Biophysica Research Communications* 81, 1011–1017.

Hill, S.A. and Judd, R.C. (1989) Identification and characterization of peptidoglycan-associated proteins in *Neisseria gonorrhoeae*. *Infection and Immunity* 57, 3612–3618.

Holbein, B.E., Jericho, K.W.F. and Likes, G.C. (1979) *Neisseria meningitidis* infection in mice: influence of iron, variations in virulence among strains, and pathology. *Infection and Immunity* 24, 545–551.

Hubner, A., Yang, X., Nolen, D.M., Popova, T.G., Cabello, F.C. and Norgard, M.V. (2001)

Expression of *Borrelia burgdorferi* OspC and DbpA is controlled by a RpoN-RpoS regulatory pathway. *Proceedings of the National Academy of Sciences of the United States of America* 98, 12724–12729.

Ieva, R., Roncarati, D., Metruccio, M.M., Seib, K.L., Scarlato, V. and Delany, I. (2008) OxyR tightly regulates catalase expression in *Neisseria meningitidis* through both repression and activation mechanisms. *Molecular Microbiology* 70, 1152–1165.

Jackson, L.A., Ducey, T.F., Day, M.W., Zaitshik, J.B., Orvis, J. and Dyer, D.W. (2010) Transcriptional and functional analysis of the *Neisseria gonorrhoeae* Fur regulon. *Journal of Bacteriology* 192, 77–85.

Jordan, P.W. and Saunders, N.J. (2009) Host iron binding proteins acting as niche indicators for *Neisseria meningitidis*. *PLoS One* 4, e5198.

Kozjak-Pavlovic, V., Dian-Lothrop, E.A., Meinecke, M., Kepp, O., Ross, K., Rajalingam, K., Harsman, A., Hauf, E., Brinkmann, V., Gunther, D., Herrmann, I., Hurwitz, R., Rassow, J., Wagner, R. and Rudel, T. (2009) Bacterial Porin disrupts mitochondrial membrane potential and sensitizes host cells to apoptosis. *PLoS Pathogens* 5, e1000629.

Laskos, L., Ryan, C.S., Fyfe, J.A. and Davies, J.K. (2004) The RpoH-mediated stress response in *Neisseria gonorrhoeae* is regulated at the level of activity. *Journal of Bacteriology* 186, 8443–8452.

Lenz, D.H., Mok, K.C., Lilley, B.N., Kulkarni, R.V., Wingreens, N.S. and Bassler, B.L. (2004) The small RNA chaperone Hfq and multiple small RNAs control quorum sensing in *Vibrio harveyi* and *Vibrio cholerae*. *Cell* 118, 69–82.

Lim, K.H., Jones, C.E., vanden Hoven, R.N., Edwards, J.L., Falsetta, M.L., Apicella, M.A., Jennings, M.P. and McEwan, A.G. (2008) Metal binding specificity of the MntABC permease of *Neisseria gonorrhoeae* and its influence on bacterial growth and interaction with cervical cells. *Infection and Immunity* 76, 3569–3576.

Livny, J., Brencic, A., Lory, S. and Waldor, M.K. (2006) Identification of 17 *Pseudomonas aeruginosa* sRNAs and prediction of sRNA-encoding genes in 10 diverse pathogens using bioinformatic tool sRNAPredict2. *Nucleic Acids Research* 34, 3484–3493.

Masse, E. and Gottesman, S. (2002) A small RNA regulates the expression of genes involved in iron metabolism in *Escherichia coli*. *Proceedings of the National Academy of Sciences of the United States of America* 99, 4620–4625.

Masson, L. and Holbein, B.E. (1985) Influence of nutrient limitation and low pH on Serogroup B

Neisseria meningitidis capsular polysaccharide levels: correlation with virulence for mice. *Infection and Immunity* 47, 465–471.

Mellin, J.R., Goswami, S., Grogan, S., Tjaden, B. and Genco, C.A. (2007) A novel Fur- and iron-regulated small RNA, NrrF, is required for indirect Fur-mediated regulation of the *sdhA* and *sdhC* genes in *Neisseria meningitidis*. *Journal of Bacteriology* 189, 3686–3694.

Metruccio, M.M.E., Fantappie, L., Serrto, D., Muzzi, A., Roncarati, D., Donati, C., Scarlato, V. and Delany, I. (2009) The Hfq-dependent small non-coding RNA NrrF directly mediates Fur-dependent positive regulation of succinate dehydrogenase in *Neisseria meningitidis*. *Journal of Bacteriology* 191, 1330–1342.

Moore, T.D. and Sparling, P.F. (1996) Interruption of the *grxA* gene increases the sensitivity of *Neisseria meningitidis* to paraquat. *Journal of Bacteriology* 178, 4301–4305.

Morgan, R.W., Christman, M.F., Jacobson, F.S., Storz, G. and Ames, B.N. (1986) Hydrogen peroxide-inducible proteins in *Salmonella typhimurium* overlap with heat shock and other stress proteins. *Proceedings of the National Academy of Sciences of the United States of America* 83, 8059–8063.

Morse, S.A. and Hebeler, B.H. (1978) Effect of pH on the growth and glucose metabolism of *Neisseria gonorrhoeae*. *Infection and Immunity* 21, 87–95.

Mostertz, J., Scharf, C., Hecker, M. and Homuth, G. (2004) Transcriptome and proteome analysis of *Bacillus subtilis* gene expression in response to superoxide and peroxide stress. *Microbiology* 150, 497–512.

Mouery, K., Rader, B.A., Gaynor, E.C. and Guillemin, K. (2006) The stringent response is required for *Helicobacter pylori* survival of stationary phase, exposure to acid and aerobic shock. *Journal of Bacteriology* 188, 5494–5500.

Palma, M., DeLuca, D., Worgall, S. and Quadri, L.E. (2004) Transcriptome analysis of the response of *Pseudomonas aeruginosa* to hydrogen peroxide. *Journal of Bacteriology* 186, 248–252.

Pannekoek, Y., van Putten, J.P. and Dankert, J. (1992) Identification and molecular analysis of a 63-Kilodalton stress protein from *Neisseria gonorrhoeae*. *Journal of Bacteriology* 174, 6928–6937.

Pannekoek, Y., Dankert, J. and van Putten, J.P. (1995) Construction of recombinant neisserial Hsp60 proteins and mapping antigenic determinants. *Molecular Microbiology* 15, 277–285.

Pannekoek, Y., Huis in 't Veld, R., Hopman, C.T.P.,

Langerak, A.A.J., Speijer, D. and van der Ende, A. (2009) Molecular characterization and identification of proteins regulated by Hfq in *Neisseria meningitidis*. *FEMS Microbiology Letters* 294, 216–224.

Parkhill, J., Achtman, M., James, K.D., Bentley, S.D., Churcher, C., Klee, S.R., Morelli, G., Basham, D., Brown, D., Chillingworth, T., Davies, R.M., Davis, P., Devlin, K., Feltwell, T., Hamlin, N., Holroyd, S., Jagels, K., Leather, S., Moule, S., Mungall, K., Quail, M.A., Rajandream, M.-A., Rutherford, K.M., Simmonds, M., Skelton, J., Whitehead, S., Spratt, B.G. and Barrell, B.G. (2000) Complete DNA sequence of a serogroup A strain of *Neisseria meningitidis* Z2491. *Nature* 404, 502–506.

Perez, N., Trevino, J., Liu, Z., Ho, S.C., Babitke, P. and Sumby, P. (2009) A genome-wide analysis of small regulatory RNAs in the human pathogen group A *Streptococcus*. *PLoS One* 4, e7668.

Pettit, R.K., Whelan, T.M. and Woo, K.S. (2001) Acid stress upregulated outer membrane proteins in clinical isolates of *Neisseria gonorrhoeae*, but not most commensal *Neisseria*. *Canadian Journal of Microbiology* 47, 871–876.

Potamitou, A., Neibauer, P., Holmgren, A. and Vlamis-Gardikas, A. (2002) Expression of *Escherichia coli* Glutaredoxin 2 is mainly regulated by ppGpp and sigma S. *Journal of Biological Chemistry* 277, 17775–17780.

Primm, T.P., Andersen, S.J., Mizrahi, V., Avarbock, D., Rubin, H. and Barry, C.E. III (2000) The stringent response of *Mycobacterium tuberculosis* is required for long-term survival. *Journal of Bacteriology* 182, 4889–4898.

Rallu, F., Gruss, A. and Maguin, E. (1996) *Lactococcus lactis* and stress. *Antonie van Leeuwenhoek* 70, 243–251.

Rallu, F., Gruss, A., Ehrlich, S.D. and Maguin, E. (2000) Acid- and multistress-resistant mutants of *Lactococcus lactis*: identification of intracellular stress signals. *Molecular Microbiology* 35, 517–528.

Revel, A.T., Talaat, A.M. and Norgard, M.V. (2002) DNA microarray analysis of differential gene expression in *Borrelia burgdorferi*, the lyme disease spirochete. *Proceedings of the National Academy of Sciences of the United States of America* 99, 1562–1567.

Sebastian, S., Agarwal, S., Murphy, J.R. and Genco, C.A. (2002) The gonococcal *fur* regulon: identification of additional genes involved in major catabolic, recombination and secretory pathways. *Journal of Bacteriology* 184, 3965–3974.

Seib, K.L., Tseng, H.-J., McEwan, A.G., Apicella,

M.A. and Jennings, M.P. (2004) Defenses against oxidative stress in *Neisseria gonorrhoeae* and *Neisseria meningitidis*: distinctive systems for different lifestyles. *Journal of Infectious Diseases* 190, 136–147.

Seib, K.L., Simons, M.P., Wu, H.-J., McEwan, A.G., Nauseef, W.M., Apicella, M.A. and Jennings, M.P. (2005) Investigation of oxidative stress defenses of *Neisseria gonorrhoeae* by using a human polymorphonuclear leukocyte survival assay. *Infection and Immunity* 73, 5269–5272.

Seib, K.L., Wu, H.-J., Kidd, S.P., Apicella, M.A., Jennings, M.P. and McEwan, A.G. (2006) Defenses against oxidative stress in *Neisseria gonorrhoeae*: a system tailored for a challenging environment. *Microbiology and Molecular Biology Reviews* 70, 344–361.

Shaik, Y.B., Grogan, S., Davey, M., Sebastian, S., Goswami, S., Szmigielski, B. and Genco, C.A. (2007) Expression of the iron-activated *nspA* and *secY* genes in *Neisseria meningitidis* Group B by Fur-dependent and -independent mechanisms. *Journal of Bacteriology* 189, 663–669.

Shakhnovich, E.A., Davis, B.M. and Waldor, M.K. (2009) Hfq negatively regulates type III secretion in EHEC and several other pathogens. *Molecular Microbiology* 74, 347–363.

Sharma, C.M. and Vogel, J. (2009) Experimental approaches for the discovery and characterization of regulatory small RNA. *Current Opinion in Microbiology* 12, 536–546.

Sittka, A., Sharma, C.M., Rolle, K. and Vogel, J. (2009) Deep sequencing of Salmonella RNA associated with heterologous Hfq proteins *in vivo* reveals small RNAs as a major target class and identifies RNA processing phenotypes. *RNA Biology* 6, 266–275.

Skaar, E.P., Tobiason, D.M., Quick, J., Judd, R.C., Weissbach, H., Etienne, F., Brot, N. and Seifert, H.S. (2002) The outer membrane localization of the *Neisseria gonorrhoeae* MsrA/B is involved in survival against reactive oxygen species. *Proceedings of the National Academy of Sciences of the United States of America* 99, 10108–10113.

Smith, H. (2000) Questions about the behavior of bacterial pathogens *in vivo*. *Philosophical Transactions of the Royal Society of London Biological Sciences* 355, 551–564.

Soler-Garcia, A.A. and Jerse, A.E. (2004) A *Neisseria gonorrhoeae* catalase mutant is more sensitive to hydrogen peroxide and paraquat, an inducer of toxic oxygen radicals. *Microbial Pathogenesis* 37, 55–63.

Song, T. and Wai, S.N. (2009) A novel sRNA that modulates virulence and environmental fitness of *Vibrio cholerae*. *RNA Biology* 6, 254–258.

Song, T., Mika, F., Lindmark, B., Liu, Z., Schild, S., Bishop, A., Zhu, J., Camilli, A., Johansson, J., Vogel, J. and Wai, S.N. (2008) A new *Vibrio cholerae* sRNA modulates colonization and affects release of outer membrane vesicles. *Molecular Microbiology* 70, 100–111.

Steiner, K. and Malke, H. (2001) Independent amino acid starvation response network of *Streptococcus pyogenes*. *Journal of Bacteriology* 183, 7354–7364.

Stohl, E.A., Criss, A.K. and Seifert, H.S. (2005) The transcriptome response of *Neisseria gonorrhoeae* to hydrogen peroxide reveals genes with previously uncharacterized roles in oxidative damage protection. *Molecular Microbiology* 58, 520–532.

Storz, G., Tartaglia, L.A. and Ames, B.N. (1990) Transcriptional regulator of oxidative stress-inducible genes: direct activation by oxidation. *Science* 248, 189–194.

Svenningsen, S.L., Tu, K.C. and Bassler, B.L. (2009) Gene dosage compensation calibrates four regulatory RNAs to control *Vibrio cholerae* quorum sensing. *EMBO Journal* 28, 429–439.

Tettelin, H., Saunders, N.J., Heidelberg, J., Jeffries, A.C., Nelson, K.E., Eisen, J.A., Ketchum, K.A., Hood, D.W., Peden, J.F., Dodson, R.J., Nelson, W.C., Gwinn, M.L., DeBoy, R., Peterson, J.D., Hickey, E.K., Haft, D.H., Salzberg, S.L., White, O., Fleischmann, R.D., Dougherty, B.A., Mason, T., Ciecko, A., Parksey, D.S., Blair, E., Cittone, H., Clark, E.B., Cotton, M.D., Utterback, T.R., Khouri, H., Qin, H., Vamathevan, J., Gill, J., Scarlato, V., Masignani, V., Pizza, M., Grandi, G., Sun, L., Smith, H.O., Fraser, C.M., Moxon, E.R., Rappuoli, R. and Venter, J.C. (2000) Complete genome sequence of *Neisseria meningitidis* serogroup B strain MC58. *Science* 287, 1809–1815.

Thomas, C.E. and Sparling, P.F. (1996) Isolation and analysis of a *fur* mutant of *Neisseria gonorrhoeae*. *Journal of Bacteriology* 178, 4224–4232.

Touati, D. (2000) Iron and oxidative stress in bacteria. *Archives of Biochemistry and Biophysics* 373, 1–6.

Tseng, H.-J., Srikhanta, Y., McEwan, A.G. and Jennings, M.P. (2001) Accumulation of manganese in *Neisseria gonorrhoeae* correlates with resistance to oxidative killing by superoxide anion and is independent of superoxide dismutase activity. *Molecular Microbiology* 40, 1175–1186.

Tseng, H.-J., McEwan, A.G., Apicella, M.A. and Jennings, M.P. (2003) OxyR acts as a repressor of catalase expression in *Neisseria gonorrhoeae*. *Infection and Immunity* 71, 550–556.

Tsui, H.C., Mukherjee, D., Ray, V.A., Sham, L.T., Feig, A.L. and Winkler, M.E. (2010) Identification and characterization of non-coding small RNAs in *Streptococcus pneumoniae* serotype 2 strain D39. *Journal of Bacteriology* 192, 264–279.

Turner, S., Reid, E., Smith, H. and Cole, J. (2003) A novel cytochrome *c* peroxidase from *Neisseria gonorrhoeae*: a lipoprotein from a Gram-negative bacterium. *Biochemistry Journal* 373, 865–873.

van Delden, C., Comte, R. and Bally, A.M. (2001) Stringent response activates quorum sensing and modulates cell density-dependent gene expression in *Pseudomonas aeruginosa*. *Journal of Bacteriology* 183, 5376–5384.

Vogel, J. (2009) A rough guide to the non-coding RNA world of *Salmonella*. *Molecular Microbiology* 71, 1–11.

Weneger, W.S., Hebeler, B.H. and Morse, S.A. (1977a) Cell envelope of *Neisseria gonorrhoeae*: relationship between autolysis in buffer and the hydrolysis of peptidoglycan. *Infection and Immunity* 18, 210–219.

Weneger, W.S., Hebeler, B.H. and Morse, S.A. (1977b) Cell envelope of *Neisseria gonorrhoeae*: penicillin enhancement of peptidoglycan hydrolysis. *Infection and Immunity* 18, 717–725.

Wilks, K.E., Dunn, K.L.R., Farrant, J.L., Reddin, K.M., Gorringe, A.R., Langford, P.R. and Kroll, J.S. (1998) Periplasmic superoxide dismutase in meningococcal pathogenicity. *Infection and Immunity* 66, 213–217.

Wu, H.-J., Seib, K.L., Edwards, J.L., Apicella, M.A., McEwan, A.G. and Jennings, M.P. (2005) Azurin of pathogenic *Neisseria* spp. is involved in defense against hydrogen peroxide and survival within cervical epithelial cells. *Infection and Immunity* 73, 8444–8448.

Wu, H.-J., Seib, K.L., Srikhanta, Y.N., Kidd, S.P., Edwards, J.L., Maguire, T.L., Grimmond, S.M., Apicella, M.A., McEwan, A.G. and Jennings, M.P. (2006) PerR controls Mn-dependent resistance to oxidative stress in *Neisseria gonorrhoeae*. *Molecular Microbiology* 60, 401–416.

Wu, H.-J., Seib, K.L., Srikhanta, Y.N., Edwards, J., Kidd, S.P., Maguire, T.L., Hamilton, A., Pan, K.-T., Hsiao, H.-H., Yao, C.-W., Grimmond, S.M., Apicella, M.A., McEwan, A.G., Wang, A.H.-J. and Jennings, M.P. (2010) Manganese regulation of virulence factors and oxidative stress resistance in *Neisseria gonorrhoeae*. *Journal of Proteomics* 73, 899–916.

Wu, L. and Belasco, J.G. (2008) Let me count the ways: mechanisms of gene regulation by miRNAs and siRNAs. *Molecular Cell* 29, 1–7.

Yan, X., Zhao, C., Budin-Verneuil, A., Hartke, A., Rince, A., Gilmore, M.S., Auffray, Y. and Pichereau, V. (2009) The (p)ppGpp synthetase RelA contributes to stress adaptation and virulence in *Enterococcus faecalis* V583. *Microbiology* 155, 3226–3237.

Yang, X., Goldberg, M.S., Popova, T.G., Schoeler, G.B., Wikel, S.K., Hagman, K.E. and Norgard, M. (2000) Independence of environmental factors influencing reciprocal patterns of gene expression in virulent *Borrelia burgdorferi*. *Molecular Microbiology* 37, 1470–1479.

Yura, T. and Nakahigashi, K. (1999) Regulation of the heat shock response. *Current Opinion in Microbiology* 2, 153–158.

Zheng, M., Wang, X., Templeton, L.J., Smulski, D.R., LaRossa, R.A. and Storz, G. (2001) DNA microarray-mediated transcriptional profiling of the *Escherichia coli* response to hydrogen peroxide. *Journal of Bacteriology* 183, 4562–4570.

Zhou, Y.N., Coleman, W.G., Yang, Z., Yang, Y., Hodgson, N., Chen, F. and Jin, D.J. (2008) Regulation of cell growth during serum starvation and bacterial survival in macrophages by the bifunctional enzyme SpoT in *Helicobacter pylori*. *Journal of Bacteriology* 190, 8025–8032.

Zusman, T., Ohad, G.M. and Segal, G. (2002) Characterization of a *Legionella pneumophila* relA insertion mutant and roles of RelA and RpoS in virulence gene expression. *Journal of Bacteriology* 184, 67–75.

Part 3
Acid Stress, pH Control and Survival in the Human Host

7 Acid Survival Mechanisms of Bacterial Pathogens of the Digestive Tract

Hanan Gancz and D. Scott Merrell*

7.1 Introduction

Bacteria are versatile microorganisms that have cleverly evolved to sense and respond to their environment. These fascinating life forms have managed to colonize virtually every square inch of the planet and are able to utilize numerous sources of nutrients to support their survival. Moreover, as a whole, bacteria have developed mechanisms to withstand sudden dramatic changes in many aspects of their habitats: nutrient availability, temperature, osmotic pressure, pH, etc. This propensity for adaptation is particularly true of commensals and bacterial pathogens which occupy niches within the human body where the environment is often unpleasant from the microbial point of view. Numerically speaking, the most abundant bacterial population associated with humans is found within the digestive tract. In this tumultuous environment, where nutrient availability, osmotic pressure, ion composition and other critical factors are subject to almost instantaneous change, there is one major hurdle that all inhabitants of this site have to overcome; regardless of their prime site of colonization, whether it is the oral cavity or the large intestine, the bacteria of the digestive tract all face acid stress at some point in their lifetime. For some, the stress will come as they try to reach their primary site of colonization, and for others the stress will become evident as they attempt to disseminate and infect a new host. Herein, we will examine the strategies different enteric bacterial pathogens have adopted to withstand acid stress.

7.2 Health Burden of Bacterial Pathogens of the Digestive Tract

The bacterial pathogens of the digestive tract, although encompassing only a small subset of the entire bacterial population that can be present within this site (Bik *et al.*, 2006; Manson *et al.*, 2008; Stecher and Hardt, 2008; Qin *et al.*, 2010), are of particular significance since they are associated with considerable worldwide morbidity and mortality. Globally, the World Health Organization (WHO) estimates that there are approximately 4 billion cases of diarrhoea each year (WHO, 2008, 2009a,b). Although it is presumed that most of these cases are caused by viral infections, the contribution of bacterial infections to the overall number is considerable. For instance, the WHO estimates that 5–14% of all diarrhoeal cases are caused by *Campylobacter* spp., which are currently the leading cause of bacterial-associated diarrhoea worldwide (WHO, 2000, 2008).

* Corresponding author.

© CAB International 2011. *Stress Response in Pathogenic Bacteria*
(ed. S.P. Kidd)

135

While pathogens like *Campylobacter* spp. are a global problem, other types of bacterial infections and their associated diseases show geographic differences in their prevalence.

This is the case for cholera, which is caused by infection with *Vibrio cholerae*. Cholera is a significant health problem for affected individuals since the rapid loss of fluids and electrolytes can lead to life-threatening consequences (Kaper *et al.*, 1995). In 2008 ~190,000 cholera cases and ~5000 resulting deaths were reported officially to the WHO (Anon., 2009a). These reports came from countries in Africa, Latin America, Asia, Europe and Oceania. Conversely, only a tiny fraction of the reported cholera cases were from developed territories and states (Anon., 2008a, 2009a). Thus, the health burden attributed to cholera in the developing world is substantially elevated in comparison to the developed nations.

In the USA the incidence of enteric infections is monitored by FoodNet, which is the principal foodborne disease component of the Centers for Disease Control (CDC) Emerging Infections Program. In 2008 there were 17,446 laboratory-confirmed bacterial-associated diarrhoeal disease cases that were reported by ten US states (Anon., 2009b). Of these cases, 95% were associated with three pathogenic species: *Salmonella* spp. (43%), *Campylobacter* spp. (34%) and *Shigella* spp. (18%). The remainder of the cases were attributed (in descending order) to infections caused by Shiga toxin-producing *Escherichia coli* (STEC), *Yersinia* spp., *Listeria* spp. and *Vibrio* spp. The overall incidence of these infections has been amazingly stable from year to year and did not change significantly during 2005–2008, in spite of differences in the geographic incidence and the age distribution of infected individuals (Anon., 2009b). Although the current reported rate of food-related bacterial infection in the USA is around 36 per 100,000 people, it is believed that many more cases go undiagnosed or unreported (Anon., 2009b). For example, salmonellosis incidence may be 30 or more times greater than reported (Gold and Eisenstein, 2000). Thus, based on the incidence for each disease and the hypothesized under-reporting rate, we estimate that over 3 million people in the USA are affected each year by campylobacteriosis, salmonellosis and shigellosis. This staggering rate of bacterial infections of the digestive tract (estimated >1000 per 100,000 per year) in a developed nation clearly indicates the need for a better understanding of how these pathogens evade host defences and survive within the human body.

7.3 Acid Stress, Acid Tolerance and Other Definitions

Despite the fact that diverse gastrointestinal (GI) niches experience many different stressors, one of the first challenges bacteria encounter on entering the GI tract is the acid barrier. The acidic environment of the stomach is arguably the most efficient of the host's non-specific lines of defence and is created by the activity of parietal cells, which are found in the gastric glands that are located within the gastric pits (Dimaline and Varro, 2007). The production of acid by these cells is regulated through a complex feedback system of signals and hormones (Dimaline and Varro, 2007; Schubert, 2008). In the stomach, acid production can result in lowering of the pH of the gastric content to a pH of 1.5 (Ayazi *et al.*, 2009). Thus, bacteria that colonize this site or pass through it need to be able to withstand and cope with the deleterious effects this harsh environment can impose (Fig. 7.1).

However, it should be noted that for transient bacteria, the need to withstand low pH and acid stress does not end when they enter the duodenum, which is the next stop in the digestive tract. The production and presence of volatile organic acids in the lower digestive tract (intestine and faeces) pose yet another stressor for those microorganisms that colonize the intestine or need to pass through it to be shed and infect another host (Fig. 7.1). In order to better understand the different survival mechanisms employed by bacteria, it is important to define clearly what constitutes stress. Therefore, herein, we will adhere to Booth's general definition of stress

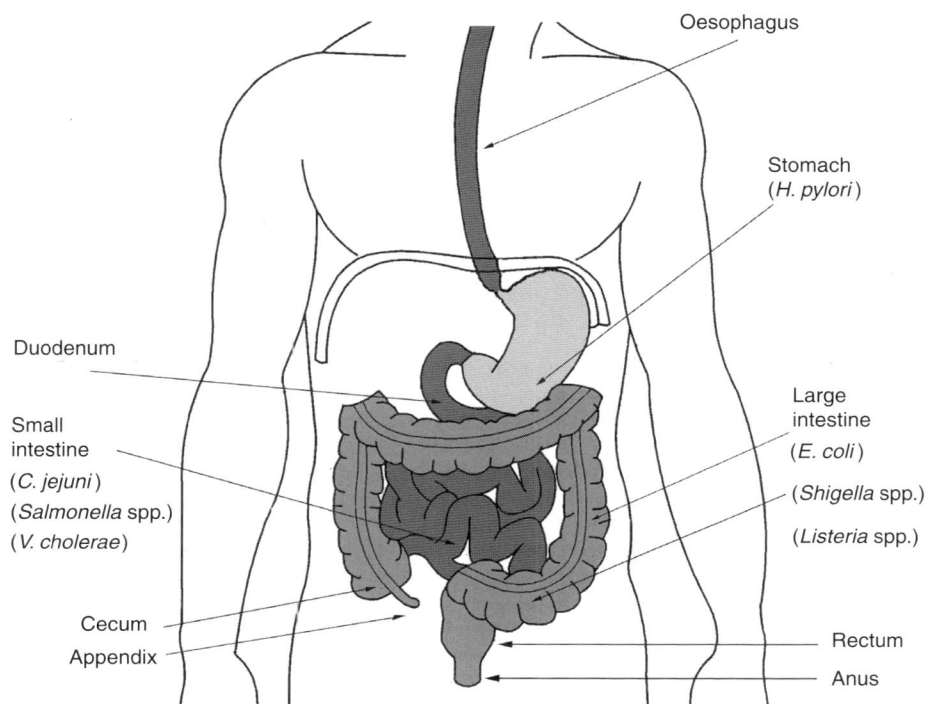

Fig. 7.1. Schematic diagram of the gastrointestinal tract illustrating the gross anatomy and initial site of colonization of particular pathogenic bacterial species. The gastrointestinal tract serves as a colonization niche for a number of bacterial pathogens. Each of these species must overcome acid stress in order to reach and subsequently thrive in its preferred section of the GI tract. Discussed in this chapter are *H. pylori* colonization of the stomach, *C. jejuni*, *Salmonella* spp. and *V. cholerae* colonization of the small intestine, and *E. coli*, *Shigella* spp. and *Listeria* spp. colonization of the large intestine. Some of these bacterial species can also be found elsewhere in the GI tract throughout the colonization/infection process.

in our discussion: '*Stress* is any change in the genome, proteome or environment that imposes either reduced growth or survival potential. Such changes lead to attempts by a cell to restore a pattern of metabolism that either fits it for survival or for faster growth' (Booth, 2002). In this context we will define acid stress as 'a lowering of the bacterial cell's environmental pH as to induce reduced growth or threaten the cell's survival'. Thus, this definition covers HCl stress imposed in the stomach, the stress imposed by short-chain fatty acids in the intestine and the acidic environment found within phagolysosomes in macrophages.

Years of research into the survival mechanisms employed by bacteria have demonstrated the existence of two over-lapping acid response mechanisms in some bacteria (Audia *et al.*, 2001). Due to this, it is important to understand the difference between the two broad systems. Many bacteria express a constitutive level of stress proteins that enable them to withstand moderate to severe acid stress. These proteins/systems constitute the acid resistance (AR) portion of the bacterial acid stress response. The existence of AR systems was identified originally by comparing the ability of a number of bacterial species to survive acid challenge as a function of the composition of the challenge medium (Foster and Hall, 1990; Gorden and Small, 1993; Small *et al.*, 1994; Benjamin and Datta, 1995). For example, in *E. coli* when the medium contained amino acids, the bacteria showed a remarkable ability to withstand the acidic environment in comparison to their survival in amino acid-deprived medium (Lin *et al.*, 1995, 1996; Hersh *et al.*, 1996). On top of this basal level of AR, other inducible systems exist that are able to facilitate bacterial survival under even more extreme conditions. These inducible systems have been categorized as the acid tolerance response (ATR), which has been defined broadly as the ability of the bacterial cell to withstand extreme acid challenge after exposure to intermediate to mild acid stress (Gale and Epps, 1942; Goodson and Rowbury, 1991; Raja *et al.*, 1991; Small *et al.*, 1994). As a whole, the combined contribution of AR and ATR determine the ability of bacteria to withstand acid stress, and the identity and importance of many of the factors that constitute the bacterial acid stress response (ASR) have been elucidated through the study of these two acid resistance systems.

7.4 Uniqueness of Acid Stress as Compared to Other Stresses

A unique feature of the ASR is its ability to confer protection from other stressors (Raja *et al.*, 1991; Bearson *et al.*, 1997; Kwon and Ricke, 1998; Koga *et al.*, 1999; Greenacre and Brocklehurst, 2006). For example, in *Salmonella enterica* serovar Typhimurium, induction of the ATR results in cross-protection against heat stress, oxidative stress and other forms of stress (Leyer and Johnson, 1993; Greenacre and Brocklehurst, 2006; Xu *et al.*, 2008). However, this phenomenon seems to be a one-way street; the ATR enables bacteria to withstand other forms of stress, but adaptation to other stressors typically does not confer acid resistance (Rodriguez-Romo and Yousef, 2005). One might conclude from these observations that the ATR induces unique components, or quantities of components, that are multifunctional and confer resistance to other forms of stress. Conversely, the same

components, or quantities of components, are not induced in a sufficient nature by the other stressful environments as to be able to confer acid resistance. While the logic for this lack of bidirectional crosstalk is not immediately evident, the answer likely lies in the multi-faceted nature of acid's effect on bacterial processes and molecules that are essential for survival.

The damaging effect of acid exposure (Fig. 7.2, panel A) can be viewed from the biochemical perspective; most cellular processes depend on enzymatic reactions, which in turn depend on the enzymatic activity of the proteins carrying them out. Most, if not all, of these proteins and reactions are pH sensitive. Thus, the narrow pH range in which enzymatic activity can take place effectively limits the boundaries of tolerable shifts in pH in the cell's cytoplasm. Additionally, changes in pH can lead to global protein denaturation due to changes in the ionization state of side groups in the protein, membrane damage and depurination, and global DNA damage (Booth *et al.*, 2002; Sachs *et al.*, 2005). Thus, any sudden drastic changes in the internal pH of a cell would result in a global shutdown of protein synthesis, ATP production and many of the repair processes that prevent accumulation of life-threatening damage to the cell (Booth *et al.*, 2002). Because of this, bacteria have developed numerous mechanisms to aid in pH homeostasis and to control changes in the cytoplasmic pH (Fig. 7.2, panels A and B). These can range from high basal levels of acid stress proteins (ASP) that aid in AR (Lin *et al.*, 1995) to elaborate coordination of the acid response through small RNA (sRNA) molecules (Opdyke *et al.*, 2004; Jin *et al.*, 2009). In general, we can divide the employed mechanisms into the following two major groups:

1. Confront acid-induced damage to cellular components – low pH and volatile fatty acids cause damage to critical components of the cell's exterior and interior. For example, acid-induced depurination of genomic DNA and pH-mediated denaturation and aggregation of proteins have been studied extensively (Jeong *et al.*, 2008; Malki *et al.*, 2008). The cell combats this stress by applying cellular repair

machinery, which is aimed at reversing acid's ill effects. For example, molecular chaperones (Fig. 7.2, panel A) protect and salvage misfolded and aggregated proteins (Malki et al., 2008). These repair mechanisms work together with the acid-resistance tools mentioned below to enable the survival of the bacteria.

2. Combat and neutralize excess protons – by actively reducing the rate of proton diffusion into the cell and utilizing various enzymatic systems, bacteria can alter/maintain their cytoplasmic pH. For example, several bacteria use amino acid decarboxylase systems to consume excess protons (Foster and Moreno, 1999; Sachs et al., 2005; Torres 2009). Similarly, Helicobacter pylori uses the enzyme urease (Fig. 7.2, panel B) to break down urea to the same end result (Dunn and Phadnis, 1998; Sachs et al., 2005). As discussed below, the latter is a key factor in H. pylori's successful colonization of the stomach.

7.5 The Host and the Bacteria of the Digestive Tract: A Continuous Cycle of Shedding, Environmental Persistence and Reinfection

7.5.1 The stomach – the great acid barrier

Strategically located in the upper section of the alimentary canal, the stomach is the first part of the GI tract to come in contact with ingested food for a substantial period of time. From a purely digestive standpoint, the stomach is critical to the efficient processing of nutrients due to its role in mechanical and chemical breakdown of food; secretion of hydrochloric acid (HCl) and digestive enzymes aid in protein digestion and facilitate nutrient absorption. However, beyond this digestive role, the idea that the stomach acts as a barrier to infection took root in the early part of the 20th century (Giannella et al., 1972). In the ensuing decades, the role of gastric acid as a barrier against infection has been studied intensely (Giannella et al., 1972; Sarker and Gyr, 1992; Dinsmore et al., 1997; Smith, 2003). Additionally, this topic has been addressed more recently using state-of-the-art tools that have reaffirmed the 'gastric barrier' concept

and have shown the effectiveness of low pH in eliminating bacterial pathogens in ingested food (Rao et al., 2006; Tennant et al., 2008). Included among these studies are the creations of transgenic mice that are unable to create an acidic environment in the stomach (Tennant et al., 2008) and new studies into the function of salivary nitrite as a bactericidal agent that functions in the acidic environment of the stomach (Rao et al., 2006). Overall, the widespread use of an 'acid barrier' by vertebrates, mammals, reptiles, etc. indicates the importance of the gastric barrier in species evolution (Koelz, 1992; Pohl et al., 2008).

In thinking about the gastric acid barrier and the pH stress encountered by different pathogenic organisms, one would be remiss to forget the fact that the environment of the stomach is constantly fluctuating. While secretion of HCl can lead to a pH of 1.5 in the stomach lumen (Dimaline and Varro, 2007), during ingestion of a food bolus the pH of the stomach's content can rise to a pH of 6. During the time the food remains in the stomach, acid secretion and mixing causes a steady drop in the pH. Thus, ingested pathogens face a significant change in pH over the course of an abbreviated time. Additionally, one should also be aware that some pathogens use fluids as their vehicle into the body (V. cholerae), while others are associated with specific solid/semi-solid foods (Salmonella spp.). Thus, the specific carrier (liquid/solid) might play a role in how fast the specific bacteria are exposed to severe acid stress and in the length this stress is imposed (Waterman and Small, 1998). Moreover, different host factors can affect the overall acidity encountered by bacterial pathogens. For instance, malnourished individuals routinely suffer from achlorhydria, where decreased acid secretion results in a higher than normal stomach pH (Shahinian et al., 2000; Williams and McColl, 2006; Windle et al., 2007). Thus, host and environmental factors markedly affect the ultimate pH stress that different bacterial pathogens encounter in the stomach. The remainder of this chapter will represent a linear view of the GI tract and broadly will address the specific mechanisms employed by a diverse number of GI pathogens to survive acid stress.

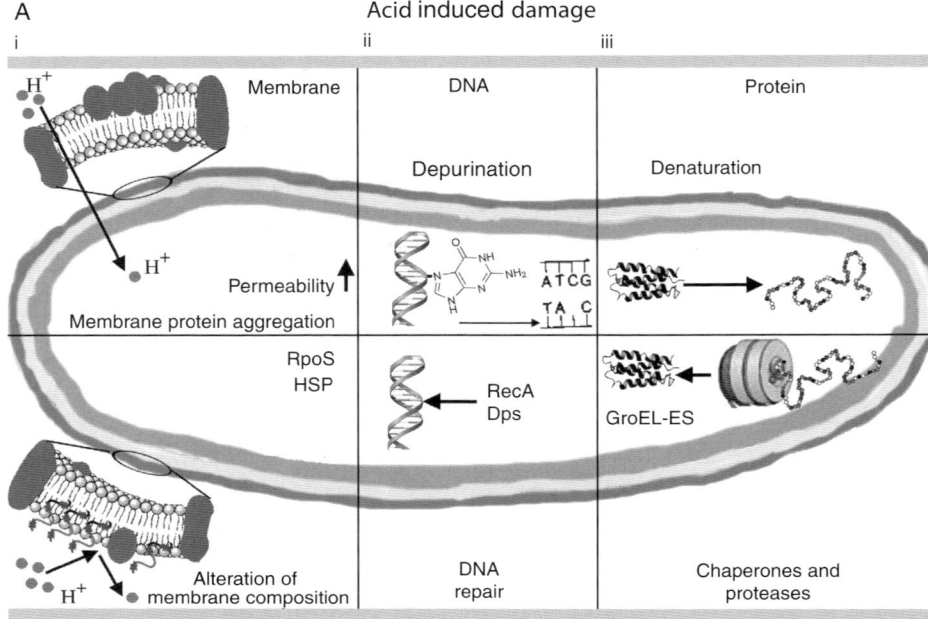

A Acid induced damage

i　　　　ii　　　　iii

Membrane　　DNA　　Protein

Depurination　　Denaturation

Permeability ↑

Membrane protein aggregation

RpoS
HSP

RecA
Dps

GroEL-ES

Alteration of
membrane composition

DNA
repair

Chaperones and
proteases

Acid stress response

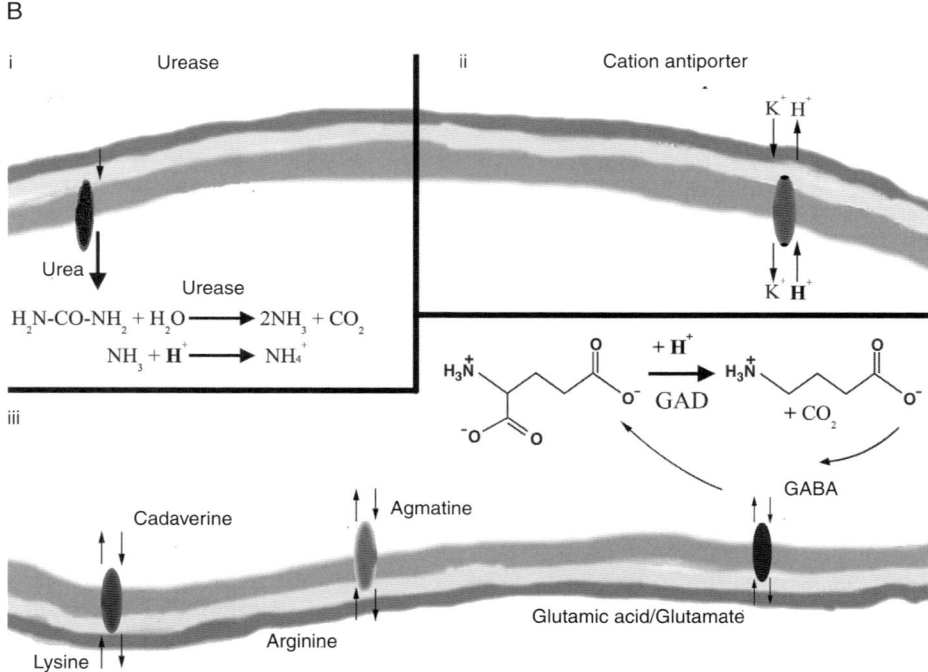

B

i　　Urease　　　　ii　　Cation antiporter

K^+ H^+

K^+ H^+

Urea

Urease

$$H_2N\text{-}CO\text{-}NH_2 + H_2O \longrightarrow 2NH_3 + CO_2$$

$$NH_3 + H^+ \longrightarrow NH_4^+$$

iii

$+ H^+$

GAD

$+ CO_2$

GABA

Cadaverine

Agmatine

Glutamic acid/Glutamate

Lysine

Arginine

Amino acid decarboxylase

Fig. 7.2. Schematic diagram of acid-induced damage and acid-resistance mechanisms utilized by enteric bacteria to overcome acid stress. Panel A: illustration of specific areas of acid-induced damage and examples of corresponding repair mechanisms that enable bacterial survival. (i) Acid stress induces changes in the outer membrane. (Top) As a result of the acidic environment, the permeability of the membrane can be altered, which results in the influx of protons. Additionally, the change in charge can result in membrane protein aggregation and dysfunction. (Bottom) RpoS regulates expression of genes that function to alter the membrane composition, while HSPs mediate disaggregation of damaged membrane proteins. (ii) Acid stress induces damage to DNA. (Top) As a result of the acidic environment, depurination can be induced; a purine is lost from the DNA backbone. The depurination of a guanine base and the subsequent lesion in the DNA sequence are illustrated. (Bottom) DNA repair is accomplished by RecA, while the DNA is protected from further damage by Dps. (iii) Acid stress induces changes in protein conformation. (Top) As a result of the acidic environment, the tertiary and quaternary structure of proteins can be affected. This altered structure can result in the loss of function of structural and enzymatic proteins as well as aggregation of proteins due to exposure of hydrophobic regions. (Bottom) Molecular chaperone systems, like the GroEL-GroES system, function to induce the proper refolding of misfolded proteins and prevent and reverse protein aggregation. Panel B: illustration of specific mechanisms used by bacterial pathogens to deal with intracellular proton accumulation. (i) Urea is pumped into the cell by a dedicated transporter and the enzyme urease breaks down the molecule to produce ammonia. In the neutral/acidic cytoplasm, the ammonia molecule is converted to an ammonium cation and consumes a proton in the process. Carbon dioxide formed during the urease reaction diffuses out of the cell. (ii) Proton transport is accomplished by dedicated transporters which exchange excess protons continuously for other, less harmful, positively charged ions. (iii) The amino acid decarboxylases consume protons in the cytoplasm by decarboxylating specific amino acid substrates. Common specific amino acid systems include the arginine, lysine and glutamic acid (glutamate) systems, which produce agmatine, cadaverine and GABA, respectively. Each system exchanges the specific amino acid and its catabolic product across the membrane through a dedicated antiporter. The detailed enzymatic reaction whereby glutamic acid is converted to GABA is shown.

7.5.2 Impossible mission: do not just survive but thrive – the amazing tale of *Helicobacter pylori*

The stomach is clearly a hostile environment and for many years was considered by most to be a sterile environment (Giannella *et al.*, 1972). Due to the relatively short time that incoming nutrients remain in the stomach, the low pH, the presence of strong proteases and the notion that the stomach functions as a barrier against infecting bacteria, the proposition that the stomach could be the site of colonization for any bacterial species was almost inconceivable. However, this all changed with the isolation and character-ization of *H. pylori* in 1982 by Marshall and Warren (Warren and Marshall, 1983; Marshall and Warren, 1984), who later received the Nobel Prize for their discovery (Anon., 2005). It is now understood that *H. pylori* persistently infects ~50% of the world's population and is the causative agent of most duodenal and gastric ulcers (Go, 2002; Basso and Plebani, 2004). Additionally, *H. pylori* is classified as a

type I carcinogen (Anon., 1994); infection is implicated in the increased risk of gastric cancer development (Asaka *et al.*, 2010). In order to succeed in the gastric environment, *H. pylori* needs not only to cope with transient acid stress, but also actually to face it repeatedly, which is a feature that calls for unique acid survival mechanisms. To this end *H. pylori* has developed a robust arsenal that enables it to thrive in this otherwise hostile environment. Perhaps the best studied of these components is the NikR-regulated urease enzyme system, which helps to buffer the bacterial cytoplasm and microenviron-ment via the import and breakdown of urea to ammonia and carbon dioxide (Sachs *et al.*, 2005). A detailed chapter regarding urease is provided by Chivers elsewhere in this book (Chivers, Chapter 8, this volume). Therefore, in this chapter, we will discuss non-urease systems and non-NikR-dependent regulatory mechanisms that facilitate *H. pylori*'s survival during acid stress.

Although it is clear that a major part of *H. pylori*'s ability to withstand acid challenge *in vivo* depends on urease activity and resultant ammonia production, it is likely no surprise that the bacterium has also developed urease-independent 'back-up plans' to deal with situations where urea is not readily available or where there is a lack of available nickel ions; nickel serves as a critical cofactor to enable urease activity. For example, *H. pylori* possesses an arginase enzyme, RocF (McGee and Mobley, 1999), which enables it to hydrolyse arginine to produce urea. In addition, *H. pylori* possesses alternative systems to produce ammonia from sources other than urea. The *H. pylori* genome encodes two aliphatic amidase paralogues: AmiE (Skouloubris *et al.*, 1997) and AmiF (Skouloubris *et al.*, 2001). These enzymes catalyse the hydrolysis of amides to organic acids and ammonia (Skouloubris *et al.*, 2001).

In addition to ammonia-producing enzymes, urease-independent acid adaptation of *H. pylori* has been studied through the use of genetics and proteomics in the absence of exogenous urea. For example, a screen of 1250 random *H. pylori* mutants identified 10 loci that were suggested as being involved in *H. pylori*'s growth at low pH but were not urease related (Bijlsma *et al.*, 2000). Subsequent work from several groups identified other such factors through proteomics approaches. For instance, of the 36 proteins related to acid stress identified by Shao *et al.*, each could be grouped into the following functional categories: molecular chaperones, ammonia production, energy metabolism and outer membrane proteins (Shao *et al.*, 2008). In addition, study of acid stress in a urea transport deficient mutant (*ΔureI*), which is deprived of the ability to produce ammonia from urea, revealed three induced proteins (Toledo *et al.*, 2002). Finally, a recent analysis of protein expression under acid stress conditions in the absence of urea revealed several proteins with altered patterns of expression (Huang *et al.*, 2010). Among these proteins were the neutrophil-activating protein A (NapA), which is a homologue of *E. coli* Dps, and the non-haem iron-containing ferritin, Pfr, which is encoded by HP0653. Subsequent experiments have revealed that Pfr can switch from an iron-storage protein to

a DNA-binding/protection factor under *in vitro* conditions (Huang *et al.*, 2010). The discovery that Pfr can change its biochemical properties to assume new functions that confer protection against acid-induced damage when the internal pH drops below a certain point may hint to yet another twist on the theme of chaperones providing protection against acid stress. NapA is also known to exhibit DNA binding when free iron is available (Wang *et al.*, 2006). Thus, since iron solubility increases at lower pH, this may explain the increased prevalence of this protein under low pH conditions. Overall, the identification of these two metal-binding proteins as conditional DNA protection molecules highlights the importance of combating DNA damage associated with acid stress.

H. pylori acid stress-mediated gene regulation is carried out by several regulatory systems. Despite the fact that this bacterium lives within the fluctuating environment of the stomach, *H. pylori* encodes a paucity of classic two-component regulatory systems which are typically used to adapt to various environmental stresses (Scarlato *et al.*, 2001). These two-component signal transduction systems (TCST) are usually composed of two protein members; one is the histidine kinase (HK) sensor which is anchored to the cell membrane, where it changes conformation upon receiving an extracellular signal. This signal activates the HK and enables the second component of the TCST, the response regulator (RR), to translate the external signal into a transcriptional response inside the cell (Parkinson, 1993; Stock *et al.*, 2000). Of the few of these systems that are found in *H. pylori*, at least one seems to be dedicated to sensing pH stress; the ArsS-ArsR system senses environmental pH through the periplasmic domain of the histidine kinase – ArsS (HP0165), which autophosphorylates and initiates a phosphorylation cascade that results in phosphorylation of the response regulator – ArsR (HP0166), which enables the bacterium to sense and respond to acid directly (Pflock *et al.*, 2006; Wen *et al.*, 2007; Loh *et al.*, 2010). In addition to genes in the urease cluster, discussed by Chivers elsewhere in this book (Chivers, Chapter 8, this volume), this system regulates genes that encode other major players in the response to acid stress: *rocF*, which encodes arginase, and *amiE* and *amiF*, which encode aliphatic amidases (Pflock *et al.*, 2006). Thus, in *H. pylori* the regulation of acid-responsive genes is mediated primarily by the ArsRS two-component system.

In addition to the ArsRS system, two major *H. pylori* metal ion regulators, the nickel regulatory protein, NikR, and the ferric uptake regulator, Fur, play a role in acid stress response. The *nikR* regulon is discussed by Chivers in Chapter 8 of this volume. However, it is worth noting that the NikR regulon partially overlaps the Fur regulon (Delany *et al.*, 2005) and that the two proteins seem to work in concert (Danielli *et al.*, 2009). Fur is a global regulator that controls intracellular iron concentration by acting as an iron sensor; when iron is abundant in the cell, the metal ion binds to Fur and causes a conformational change that enables the protein to bind to specific DNA sequences known as Fur-boxes (Whitmire *et al.*, 2007). These Fur-boxes are usually found in the promoter elements of regulated genes and Fur binding prevents transcription of the gene. Many Fur-regulated genes encode proteins that are involved in iron uptake and metabolism, and this function is conserved among many Gram-negative species (Carpenter *et al.*, 2009). The impetus to control intracellular iron levels lies in the potential for oxidative stress, which can occur due to the formation of hydroxyl radicals by the Fenton reaction (Meneghini, 1997). The initial finding that Fur functions to aid in acid acclimation came from *in vitro* work that demonstrated that a *H. pylori* Δ*fur* mutant was unable to grow on exposure to acid stress (Bijlsma *et al.*, 2002). This fact seems to be important *in vivo* since a Δ*fur* mutant is also compromised for growth/survival in a mouse model (Contreras *et al.*, 2003) and gerbil model of colonization (Gancz *et al.*, 2006; Miles *et al.*, 2010). Later work that compared the transcriptional response of wild-type *H. pylori* to an isogenic Δ*fur* mutant during acid shock revealed ~100 genes that were regulated aberrantly in the mutant strain (Gancz *et al.*, 2006). The genes encode for acid stress-related proteins such as the response regulator, ArsR,

the aliphatic amidase, AmiE, HP0869, which is involved in nickel transport to target enzymes, and other genes known to be involved in colonization. The scope of the Fur regulon under acidic conditions likely accounts for the *in vivo* colonization defect observed in the animal models (Contreras *et al.*, 2003; Gancz *et al.*, 2006).

Although lacking many of the systems that are found in other species to facilitate survival during acid stress, *H. pylori* encodes for and relies on a complex arsenal of acid-deferring systems. The main acid combating burden falls on the urease enzyme and its supporting infrastructure of nickel ion transporters and storage proteins, and urea transporters. However, alternate survival strategies exist in the form of urea-producing enzymes, nickel-independent ammonia-producing enzymes, DNA protection molecules, etc. Additionally, the complex orchestration of the response to acid relies on dedicated regulators that achieve an intimate crosstalk between metal ion availability and acid stress response. Taken together, this diverse array of survival mechanisms helps to ensure survival of *H. pylori* and plays a crucial role in the success of *H. pylori* as a pathogen.

7.5.3 The small and large intestine

Campylobacter jejuni and Campylobacter coli

C. jejuni and *C. coli* are Gram-negative, curved, motile bacilli that are similar in appearance to *H. pylori* (Sahay *et al.*, 1995) and belong to the larger genus of *Campylobacter*. As a whole, *Campylobacter* spp. are one of the two most common causes of bacterial diarrhoea in the USA (Anon., 2008a, 2009b) and probably worldwide (WHO, 2000, 2009a,b); the CDC estimates that more than one million Americans are affected by *Campylobacter* spp. yearly (Anon., 2009b). Infection occurs through ingestion of contaminated food (especially chicken) and water (Kasper, 1988). After passing through the stomach, the bacteria colonize the small intestine and colon (Russell *et al.*, 1989) and cause inflammation, which initially is limited to the small bowel but later affects the colon and the rectum (Janssen *et al.*, 2008). Given that *C. jejuni* and *C. coli* are arguably the best studied of the *Campylobacter* spp., we will confine our discussion to these organisms. For recent reviews of general survival mechanisms of *C. jejuni* in response to abiotic stress conditions and its molecular biology and pathogenesis, see Jackson *et al.* (2009) and Young *et al.* (2007), respectively.

Based on the route of infection and site of colonization, *C. jejuni* as well as other *Campylobacter* spp. must be able to survive acid stress imposed by exposure to HCl as well as to short-chain fatty acids. Despite this need, many of the elements that have been shown to play a role in conferring protection against acid in other bacterial species are not evident in the genome of this organism (Reid *et al.*, 2008b; Ma *et al.*, 2009). For instance, unlike other pathogens that use similar routes of infection, *C. jejuni* does not encode obvious homologue of RpoS or the acid-responsive amino acid decarboxylase systems (discussed below), which are crucial for pH survival of many bacterial species (Sachs *et al.*, 2005). Additionally, unlike its close relative, *H. pylori*, *C. jejuni* does not encode a urease enzyme (Reid *et al.*, 2008b). Furthermore, several *in vitro* studies have shown significant variation among *Campylobacter* spp. in their ability to withstand acid stress (Chaveerach *et al.*, 2003; Ma *et al.*, 2009); all of the *C. jejuni* or *C. coli* strains examined were killed rapidly at a pH of 4 (Chaveerach *et al.*, 2003). Moreover, pre-exposure to low pH does not result in a classical ATR in the preadapted cells (Ma *et al.*, 2009). Thus, while a full understanding of how *Campylobacter* spp. combat acid stress is clearly lacking, some possible mechanisms are evident and some insights have been gained through transcriptional profiling and mutagenesis.

One mechanism that has been suggested to support transit of *C. jejuni* through the acid stress of the stomach is the reported ability of the bacteria to enter into a viable but non-culturable (VBNC) state upon acid stress (Chaveerach *et al.*, 2003). As implied by the name, VBNC represents a physiological state where the bacteria are no longer culturable *in vitro* by typical methods but remain viable as assessed by other techniques. It is proposed that these VBNC cells have the potential to re-grow if given the proper conditions, i.e. in the complex environment found within a host. Thus, it has been suggested that the VBNC state is a persistence mechanism of several pathogens (Oliver, 2009). A recent study linked production of polyphosphate in *C. jejuni* to the ability to enter the VBNC state; the percentage of cells entering the VBNC state upon shock was decreased in a polyphosphate kinase 1 null mutant (Gangaiah *et al.*, 2009). Polyphosphate kinase 1 appears to be an important player in survival, stress response, host colonization and virulence in many diverse bacterial species (Brown and Kornberg, 2004, 2008). For instance, in addition to polyphosphate involvement in acid stress in *C. jejuni*, it has also been suggested to play a role in *S. typhimurium* resistance to organic acid stress (Price-Carter *et al.*, 2005).

To date, only one extensive mutagenesis study and one major transcriptional profiling study have been conducted to identify *C. jejuni* genes that play a role in growth at low pH and/or low pH survival (Reid *et al.*, 2008a,b). Of the 2577 viable mutants examined in the mutagenesis study, 86 showed an acid-sensitive phenotype. These mutations were mapped to genes involved in chemotaxis and motility, phosphate acquisition, outer membrane protein production, amino acid biosynthesis and transport, metabolism and bioenergetics, DNA restriction/modification and repair and lipooligosaccharide (LOS) and capsular polysaccharide (CPS) biosynthesis and expression. The definitive role of these genes in acid stress survival remains to be shown, since 'clean' mutation construction and complementation analyses have not been conducted. However, despite this caveat, most of these genes did not exhibit marked change in transcription during growth at low pH; only 27 of the genes showed marked difference in transcription upon acid shock (Reid *et al.*, 2008b). This is likely due to the ability of mutagenesis to identify genes that are required for growth at low pH, even if those genes are not expressed differentially. Interestingly, under both growth and shock conditions, the oxidoreductase subunit genes (Cj0414–Cj0415) showed an increase in expression and were identified in the mutagenesis study (Reid *et al.*, 2008b). The homologue of this oxidoreductase in *H. pylori* also showed increased expression upon acid shock and was shown to be regulated by Fur (Gancz *et al.*, 2006). This may imply a unique role for this enzyme in confronting acid stress and a conservation of this function between these two related species.

In the transcriptional profiling study, genes involved in the *in vitro* and *in vivo* responses to acid were identified (Reid *et al.*, 2008a,b). Microarray analysis of the transcriptional response during *in vitro* exposure to acid shock was integrated with microarray data from a short exposure of the bacteria to *in vivo* conditions. This approach led to the identification of a highly upregulated gene cluster, which encoded heat shock proteins, chemotaxis components and nitrosative stress response factors (Reid *et al.*, 2008a). Partial validation of the importance of some of these systems to acid stress survival subsequently came from mutational analysis; disruption of the heat shock protein, encoded by *clpB*, caused a reduction in the *in vitro* acid survival rate. However, deletion of two additional heat shock protein genes, *hrcA* or *hspR*, had no marked effect (Reid *et al.*, 2008a). This finding suggests that the transcriptome changes identified by microarray do not always correlate specifically to essential genes for bacterial fitness under a particular growth condition.

In spite of the considerable worldwide contribution to human morbidity and mortality and the wealth of information we now have, our understanding of the ability of *Campylobacter* spp. to avoid the deleterious effects of acid exposure is yet in its infancy. Clearly, the transcriptional and mutational studies that have been conducted are steps towards broadening our understanding of shared and unique mechanisms that play a role in *C. jejuni*'s acid stress response. However, further studies are needed to identify the specific role that individual genes play in survival, as well as to understand the intricacies of this pathogen's ability to survive acid stress throughout its complex life cycle.

Salmonella spp.

Salmonella species and subspecies are a versatile family of neutralophilic, Gram-negative, facultative intracellular pathogens (Boyle *et al.*, 2007). Infection usually is associated with ingestion of faecal-contaminated food or water (Darwin and Miller, 1999; Winfield and Groisman, 2003). Many domesticated animals, fowl, pets and even humans have been shown to carry the bacteria without signs of disease (Winfield and Groisman, 2003). The prevalence in the immediate environment combined with a moderate infectious dose of ~10^5 organisms creates numerous opportunities for infection (Darwin and Miller, 1999; Winfield and Groisman, 2003). Indeed, salmonellosis rates are second only to *Campylobacter* in terms of GI-related infections (Anon., 2008b). Furthermore, this diarrhoeal illness, which results in abdominal cramping and occasionally bloody diarrhoea (Darwin and Miller, 1999), is the second leading cause of hospitalization among bacterial infections of the GI tract (Anon., 2008b). After passing through the stomach and reaching the small intestine, *Salmonella* spp. penetrate the intestinal mucosa and accumulate in the lymph nodes, where they multiply and disseminate to the spleen and liver (Sivula *et al.*, 2008). It has been speculated that this enterobacterium senses pH, in combination with temperature and concentrations of particular ions, and uses these as signals that identify entrance into and location within the host (Rychlik and Barrow, 2005).

In *S. typhimurium*, a number of elaborate systems have been identified that respond to acid stress, and it has been suggested these systems have important implications in cycles of infection and shedding (Ricke, 2003; Altier, 2005). Numerous studies have shown that *S. typhimurium* is a good model organism to examine adaptive systems that lead to increased acid resistance in bacteria exposed to mild acid stress (ATR). The activity of these systems requires *de novo* protein synthesis and is seen when bacterial cells are adapted to moderate acid stress prior to a challenge with a more extreme acid stress (Foster and Hall, 1990, 1991); preadapted cells exhibit a

100–1000 fold better survival as compared to their unadapted counterparts (Lin *et al.*, 1995). In this pathogen, the ATR system is amazingly complex and components employed by logarithmically growing cells are different from the ones exploited by stationary phase bacteria (Lee *et al.*, 1994). In fact, the two systems show poor overlap, with only ~10% identity in the proteins identified in the ATR for each growth phase (Lee *et al.*, 1994). Furthermore, it is important to note that the ATR does not protect all *Salmonella* subspecies and serovars to the same extent. For example, in a recent study by Joerger *et al.* (2009), which aimed to understand the conversion of chicken isolated *S. enterica* to the Kentucky serovar in the USA, the author found marked differences between serovars in relation to their ability to survive acid stress and to mount an ATR (Joerger *et al.*, 2009). These differences may provide a competitive edge to specific subtypes populating particular environmental niches, as well as have subsequent implications for the severity of infection in humans.

As implied above, in *S. typhimurium* a complex, multi-component response to acid stress takes place. This response encompasses dozens of proteins. For the sake of brevity, we will confine our discussion to the regulatory proteins that control the expression pattern of one or more targeted cellular processes. Indeed, several regulatory proteins have been found to be involved in the acid stress response. These include two networks that function as TCST systems: PhoPQ (Adams *et al.*, 2001; Groisman, 2001) and OmpR/EnvZ (Bang *et al.*, 2000, 2002). Additionally, three other regulatory proteins are involved in acid stress response: Fur (Foster and Hall, 1992) and RpoS (Bearson *et al.*, 1996), which have been studied extensively as well as RpoE, which is a recently identified member of the regulatory proteins involved in the acid response (Muller *et al.*, 2009).

The PhoPQ TCST is found in *Salmonella* spp., *E. coli*, *Shigella* spp. and *Yersinia* spp. (Groisman, 2001). Within this system, the HK is encoded by *phoQ* and the RR is encoded by *phoP*. The HK, PhoQ, senses environmental Mg^{2+} and Ca^{2+} concentrations [for a detailed

review of this sensor and the mechanism of phagosome signal detection see Prost and Miller, 2008]. When the concentration of these ions drops to the microM range, PhoQ phosphorylates the RR, PhoP. Upon phosphorylation, PhoP promotes the transcription of ~40 genes and negatively controls expression of another set of genes (Adams *et al.*, 2001; Groisman, 2001; Prost and Miller, 2008). The genes regulated by PhoPQ are involved mainly in Mg^{2+} transport (Soncini and Groisman, 1996), resistance to antimicrobial peptides (Guina *et al.*, 2000) and pathogenicity (Miller *et al.*, 1989; Alpuche Aranda *et al.*, 1992). However, the PhoPQ system has been shown to be involved in tolerance to inorganic acid stress (HCl) but not organic acid stress (fatty acids). Furthermore, PhoP mutants failed to induce known critical ASPs (Bearson *et al.*, 1998). Interestingly, recent findings indicate that the PhoPQ signalling cascade is targeted by nitric oxide (NO), which serves as part of the innate gastric defence and is produced in the acidic environment of the stomach. Produced NO inhibits induction of an ATR in rapidly growing bacteria (Bourret *et al.*, 2008).

In a similar manner to the TCST PhoPQ regulation, the OmpR/EnvZ TCST controls expression of genes in response to osmolarity (Rychlik and Barrow, 2005). In this system EnvZ acts as the HK and OmpR acts as the RR and regulates expression of target genes (Taylor *et al.*, 1981). Among OmpR regulated genes, *ompC* encodes a major outer membrane protein that is associated with high osmolarity conditions. OmpC shows increased transcription during low pH exposure, and this increase is dependent on OmpR (Foster *et al.*, 1994). Furthermore, *ompR* null mutants are defective in their ability to mount a stationary phase ATR (Bang *et al.*, 2000, 2002). Interestingly, neither *ompC* nor *envZ* are essential to mount the stationary phase ATR (Bang *et al.*, 2000). This suggests the existence of an alternate EnvZ-independent mechanism of OmpR regulation of the ATR. Thus, in *S. typhimurium* and other *Salmonella* spp., TCST systems play a pivotal role in the proper response to acid stress through the regulation of the expression of a large number of genes involved in repairing acid-induced damage and promoting survival.

Other regulatory factors involved in the *S. typhimurium* acid stress response include Fur, RpoS and RpoE. In *S. typhimurium*, as in other bacteria, *fur* mutations have been shown to negatively impact acid adaptation *in vitro* (Foster, 1991; Foster and Hall, 1991, 1992). Exposure of acid-adapted exponential phase cultures to filter sterilized stomach contents from pigs demonstrated that the *fur* mutant showed a greater than 2-log defect in its survival in comparison to the wild type (Bearson *et al.*, 1996). Since Fur controls iron uptake and storage, and since low pH affects metal ion solubility, the defect of the *fur* mutant may be explained by increased iron uptake and the subsequent formation of oxidative stress during periods of intracellular subnormal pH (Rychlik and Barrow, 2005). However, in *S. typhimurium* the role of Fur may also be iron-independent since a *fur* mutation that converts histidine 90 to an arginine eliminates iron regulation but does not affect acid tolerance (Bearson *et al.*, 1996; Hall and Foster, 1996). This suggests iron-independent regulation of acid resistance by *apo*-Fur in this organism.

The stationary phase sigma factor, RpoS, is also involved in the ATR (Bearson *et al.*, 1996) and is needed to sustain this stress response (Lee *et al.*, 1994). As an alternative sigma factor, RpoS binds to the core RNA polymerase and directs the transcription of a specific subset of genes. In *S. typhimurium*, RpoS is important for virulence (Fang *et al.*, 1992; Coynault *et al.*, 1996) and in the starvation response (O'Neal *et al.*, 1994). For an in-depth review of RpoS function and regulation in *S. typhimurium* and *E. coli*, see Rychlik and Barrow (2005) and Dong and Schellhorn (2010). In a similar fashion to RpoS, the alternative sigma factor, RpoE, regulates genes in response to perturbations in the outer membrane and periplasm (Missiakas and Raina, 1998). Activation of RpoE occurs as a complex cascade of events that involves degradation of the anti-sigma factor, RseA, by the protease, DegS. In *E. coli*, RpoE is essential, but this is not the case in *S. typhimurium*. This fact enabled the investi-

gation of the role of RpoE in pathogenesis (Testerman *et al.*, 2002). Interestingly, the *S. typhimurium rpoE* null mutant was less virulent than the *degS* null mutant, which should be unable to degrade RseA and was required for RpoE activation (Rowley *et al.*, 2005). This finding led to the examination of alternative mechanisms of RpoE activation during the interaction of *S. typhimurium* with its host (Muller *et al.*, 2009). Since virulence factors of *S. typhimurium* are induced by acid (Rathman *et al.*, 1996), the role of RpoE in acid stress was explored (Muller *et al.*, 2009); RpoE facilitates growth at low pH and is essential for mounting an efficient ATR *in vitro*. RpoE also promotes *S. typhimurium* replication in the acidic environment found within phagolysosomes inside macrophages (Muller *et al.*, 2009). These findings indicate the intimate role of sigma factors, as well as other regulatory factors, in alteration of the transcriptome in response to acid stress.

In addition to the complex regulatory network controlling the acid stress response of *S. typhimurium*, several of the common acid stress phenotypes and mechanisms that are reviewed elsewhere in this chapter are found in *Salmonella* spp. For example, similar to what is observed in other species, the ATR supports protection against alternative forms of stress, such as salt stress (Greenacre and Brocklehurst, 2006). Other similarities include the presence of the arginine decarboxylase and lysine decarboxylase systems in *Salmonella* spp. (Foster, 1993; Audia *et al.*, 2001; Alvarez-Ordonez *et al.*, 2010). These systems import amino acids, modify them and consume an internal proton in the process. Emergency pumps, which are membrane-bound proton pumps such as the K^+/H^+ and Na^+/H^+ antiporters (Fig. 7.2, panel B), are also involved in pH homeostasis (Foster, 1993; Audia *et al.*, 2001). Finally, bacterial membranes constitute the first physical barrier against acid stress. Thus, it is no surprise that many species, including *S. typhimurium* and *E. coli*, modify their membrane composition in response to acid stress. Indeed, the ability to synthesize cyclopropane fatty acids (CFAs), which are a major component of membrane phospholipids, is a major contributor to acid stress resistance in *S. typhimurium* (Kim *et al.*, 2005), as well as in *E. coli* (Brown *et al.*, 1997; Chang and Cronan, 1999). Following the observation of a strong correlation between the resistance of various wild-type *E. coli* strains to drastic decreases in pH and the level of CFA present in cell membrane phospholipids (Brown *et al.*, 1997), transcription of *cfa* was monitored and was shown to increase upon mild acid exposure (Chang and Cronan, 1999). Deletion of the *cfa* gene in *S. typhimurium* resulted in increased acid sensitivity, which was restored partially upon *cfa* complementation. Furthermore, it was shown that *cfa* expression was regulated by RpoS (Chang and Cronan, 1999; Kim *et al.*, 2005).

Since *Salmonella* spp. are a major cause of hospitalization due to food poisoning and are a constant cause of morbidity and mortality worldwide, understanding this pathogen's strategies to withstand acid stress is crucial to our understanding of the threat it poses. Clearly, *Salmonella* spp. employ a versatile arsenal of acid damage control mechanisms, as well as a multitude of processes that are aimed at alleviation of acid stress through consumption of protons. Although energetically expensive, this strategy allows these bacteria to survive on their way to their prime colonization site.

Escherichia coli and Shigella spp.

E. coli is a commensal that exists symbiotically within the host gut (Tenaillon *et al.*, 2010). However, during the process of evolution, this commensal bacterium, which is found both in humans and in non-human vertebrates, gave rise to pathogenic clones. For example, *E. coli* belonging to the serotype O157:H7 are virulent to humans but are considered commensals in cattle, from which they are shed into the environment (Callaway *et al.*, 2009). These bacteria are associated with severe diarrhoeal disease that can lead to the development of haemolytic uraemic syndrome in infected individuals (Gyles, 2007). Infection occurs through ingestion of contaminated hamburger meat, apple cider and fruits and vegetables (Madoff and Kasper, 2004). Accordingly, *E. coli* O157:H7 has been

associated with several prominent food-poisoning outbreaks in the USA and around the world (Rangel *et al.*, 2005). An estimated 73,000 cases of *E. coli* O157:H7 infections are reported in the USA annually (Conte, 2002).

Shigella spp. arose from the *E. coli* lineage (Yang *et al.*, 2007) and share many of the *E. coli* genes (Bhagwat and Bhagwat, 2008). However, *Shigella* spp. have a greater public health impact; they are the third most common cause of foodborne bacterial infection in the USA (Anon., 2009b). Most shigellosis cases are mild and present transient clinical symptoms. However, infection can result in inflammation of the intestinal mucosa, with severe watery or bloody diarrhoea (Sansonetti, 2001). Children in childcare centres (Anon., 2004) and international travellers (Gupta *et al.*, 2004) show increased risk of infection. Infection with either one of the four *Shigella* species that affect humans (*S. dysenteriae*, *S. flexneri*, *S. boydii* and *S. sonnei*) can result in disease. However, the most severe symptoms are associated with *S. dysenteriae*. The infectious dose needed to cause illness by these microorganisms is as low as 10–500 cells, and this fact is likely due to their amazing ability to overcome the stomach acid barrier and survive and thrive in the small intestine.

In addition to *Shigella* spp. and *E. coli* O157:H7, other pathogenic strains of *E. coli* are of public health concern (Huang *et al.*, 2006; Kaur *et al.*, 2010), and thus there is a need to understand the ability of these species to withstand acid stress during their infectious cycle. Part of this has been accomplished through the extensive study of laboratory strains of *E. coli* that have been exposed to drastic downshifts in pH (Goodson and Rowbury, 1989; Benjamin and Datta, 1995). Those studies led to the identification of three main areas, each of which was necessary but not sufficient, to confer protection against acid stress. The first area is the capacity for nutrient acquisition for the subsequent ability to generate energy. Second is the ability to maintain pH homeostasis within the cytoplasm. Third is the ability to confer protection to proteins and DNA or to repair the damage caused by exposure to acidic conditions. Here, we will mention briefly previously reviewed mechanisms and focus our attention on interesting new venues of the latest research. For dated in-depth reviews of genes involved in acid resistance and mechanisms of resistance, see the fine reviews by Foster and co-workers (Bearson *et al.*, 1997; Foster and Moreno, 1999; Audia *et al.*, 2001; Richard and Foster, 2003; Foster, 2004).

A global picture of the response of *E. coli* to acid shock has been obtained through the use of microarray-based transcriptional analysis (Arnold *et al.*, 2001; Tucker *et al.*, 2002; Kannan *et al.*, 2008; House *et al.*, 2009) and 2-D gel electrophoresis (Kirkpatrick *et al.*, 2001; Stancik *et al.*, 2002; Paul and Hirshfield, 2003). For example, microarray analysis of acid-shocked cells showed 630 genes whose expression was increased and 586 genes whose expression was decreased (Kannan *et al.*, 2008). Genes showing increased expression include amino acid decarboxylase components (*cadA*, *adiY*, *gadA*), which are discussed below, succinate dehydrogenase and members of the Gad, Fur and Rcs regulons (Kannan *et al.*, 2008). These global types of approaches have also led to a better understanding of the role of specific proteins in the acid stress response. This is especially true in situations where particular mutations result in pleiotropic phenotypes that include acid sensitivity. For instance, the histone-like protein, HU, which is a small DNA-binding protein that is associated with the bacterial nucleoid, was shown to be involved in proper regulation of ~8% of *E. coli*'s genes (Oberto *et al.*, 2009). A subclass of the HU-dependent genes was those involved in acid stress (Oberto *et al.*, 2009).

Some of the best-studied systems used to combat acid stress in *E. coli* are the amino acid decarboxylase systems (Foster and Moreno, 1999). As mentioned earlier, these systems consume an internal proton during the decarboxylation of an amino acid (Fig. 7.2, panel B). In parallel, they use a unique antiporter to import the amino acid into the cell in exchange for the decarboxylation product. Thus, they produce a net decrease in proton concentration in the cytoplasm (Sachs *et al.*, 2005) and an increase in the net pH. These systems are often coupled with an acid-induced chloride channel to prevent accumulation of net negative charge within

the cell (Iyer *et al.*, 2002, 2003). The three amino acid decarboxylase systems studied include the glutamic acid, the arginine and the lysine systems. In *E. coli* the glutamic acid decarboxylase (GAD) system is composed of the gene products of 11 genes, which have been reviewed extensively earlier (Foster, 2004; Sachs *et al.*, 2005). In this system *gadE* functions as a transcriptional activator of the GAD structural genes, and thus controls the ATR mediated by this system. However, in a microarray study of the *gadE* regulon in *E. coli* O157:H7, it became obvious that *gadE* had a role in regulation of virulence, as well as in ATR; the *gadE* mutant showed increased expression of 19 virulence genes encoded in the locus of the enterocyte effacement (LEE) (Vanaja *et al.*, 2010). The finding that an acid-resistance system also controls expression of virulence genes exemplifies the adaptation of virulent strains to their environment and their need to re-wire intrinsic regulatory networks to accommodate horizontally acquired virulence genes. It is of great interest that these virulence genes have co-opted a regulatory factor to indicate presence in the host environment by sensing a drop in pH.

In *E. coli* the arginine decarboxylase system functions in parallel to the GAD system. Which system functions at a given time depends on amino acid availability and the internal pH. The mode of action of the arginine system is similar to that of the GAD system. However, the arginine-dependent enzyme has a higher pH optima, which leads to higher internal pH when this system is active (Foster, 2004). Interesting structure function insight into the pH dependency of this system has come from recent studies. The arginine decarboxylase structure is a pentamer of homodimers (Andrell *et al.*, 2009). The abundance of acidic surface residues prevents inactive homodimers from associating into active decamers at normal pH. However, acidic conditions favour the assembly of the active decamer. This implies that this enzyme's activity is modulated by the immediate environmental pH and that it will be active only if the pH is low enough. Interestingly, the structure of the 'virtual proton pump' that imports L-arginine and exports agmatine and is anchored in the inner membrane (encoded by *adiC*) is also available (Fang *et al.*, 2009; Gao *et al.*, 2009, 2010). Thus, in addition to the biochemical, genetic and physiological knowledge we possess already, we now have the addition of new crystal structures to shed light on the actual workings of these systems. This general approach is also being applied to the other amino acid decarboxylase systems (Alexopoulos *et al.*, 2008) and may help explain the differences between the systems and the preference of certain bacterial species for specific decarboxylases in their ATR arsenal.

Vibrio cholerae and other Vibrionaceae

V. cholerae is a free-living bacterium and an opportunistic human pathogen. It is believed that *V. cholerae* perceives the need to transition from an estuarine lifestyle to a lifestyle adequate for survival within the human host by sensing temperature, pH and other factors in its immediate environment (Krukonis and Dirita, 2003). Among the bacteria discussed in this chapter, *V. cholerae* has the highest infectious dose needed to cause disease; ~10^{11} bacteria have been shown to be required in human volunteer studies (Kaper *et al.*, 1995). However, the required dose can be reduced greatly by buffering of the stomach pH by the addition of bicarbonate (Kaper *et al.*, 1995). Moreover, the required dose is dependent on the physiological state of the bacteria since the ID_{50} needed to overcome the hostile environment of the stomach can be reduced substantially for bacteria that are in a 'hyperinfectious' state (Merrell *et al.*, 2002a; Butler *et al.*, 2006). Because of this, there is great interest in understanding which environmental conditions can facilitate a lower infectious dose, as well as how this organism is able to survive the acid stress in the stomach and beyond.

To address these issues and other biological questions, researchers previously used recombinase-based *in vivo* expression technology (RIVET) and signature-tagged mutagenesis (STM) to identify genes related to successful infection (Chiang and Mekalanos, 1998; Merrell and Camilli, 1999;

Merrell *et al.*, 2002b). Not surprisingly, some of the identified genes were shown subsequently to be important for *V. cholerae* to mount an ATR. Herein, we will discuss two of these genes: *cadA*, which has been shown to be important to surviving HCl stress similar to that found in the stomach (Merrell and Camilli, 1999), and *toxR*, which has been found to mediate the ATR specifically to organic acids (Merrell and Camilli, 1999).

cadA is part of a three-gene operon which encodes the lysine-cadaverine antiporter (CadB), the lysine decarboxylase (CadA) and a regulatory protein (CadC). CadA functions as a lysine decarboxylase that catalyses the decarboxylation of lysine to cadaverine (Auger and Bennett, 1989; Neely *et al.*, 1994). Conversely, CadB functions as a lysine-cadaverine antiporter, which mediates the uptake of lysine from the medium and the excretion of cadaverine that is produced as a result of CadA enzymatic activity (Neely *et al.*, 1994). Like the decarboxylases discussed in the previous section, this reaction consumes protons, and thus increases the intracellular pH (Foster, 2004). In *V. cholerae*, induction of *cadA* occurs at a pH of 5.5–6.0 depending on the growth medium (Merrell and Camilli, 1999). Disruption of *cadA* results in bacteria that are impaired in their ability to mount a robust ATR against inorganic acid stress (HCl) as well as organic acid stress (Merrell and Camilli, 1999). *cadA* and *cadB* transcription is controlled by CadC, which is a member of the ToxR-like family of regulators whose control of the promoter elements of the *cad* locus has been studied in detail (Merrell and Camilli, 2000), as well as in other *Vibrionaceae cad* systems (Kuper and Jung, 2005). In other *Vibrionaceae* it has been shown that in addition to the role that cadaverine has in neutralizing the acidic intracellular environment, it also appears to scavenge superoxide radicals (Kang *et al.*, 2007). *cadAB* are induced at lower pH in *V. vulnificus* even in the absence of a functional CadC, albeit to a lower extent (Kim *et al.*, 2006a). This induction seems to be dependent on SoxR direct binding to the *cadAB* promoter (Kim *et al.*, 2006a). In *E. coli* anaerobic respiration results in significant induction of *cadAB* transcription (Kuper and Jung, 2005).

In *V. cholerae* it seems that the regulatory circuits controlling the activation of the response to organic acids differ from those that control the ATR to inorganic acids. Studies showing the involvement of ToxR, a global regulator of *V. cholerae* virulence, in mounting the ATR revealed that ToxR was not involved in the response to inorganic acids but played a role in organic acid-induced ATR. Surprisingly, this role was independent of ToxT, the downstream transcriptional activator of many genes in the ToxR regulon (Merrell and Camilli, 1999). ToxR is an inner membrane-spanning protein that senses and transmits signals from the environment (Miller *et al.*, 1987). Originally identified due to its critical role in regulation of *V. cholerae* virulence factors, it is clear that ToxR controls expression of a larger subset of genes (Peterson and Mekalanos, 1988; Bina *et al.*, 2003). Mutation of *toxR* was shown to abrogate specifically the ability of *V. cholerae* to undergo an organic acid ATR, but had no effect on the response to HCl (Merrell and Camilli, 1999). Despite the fact that ToxR controls the expression of a large number of genes, this defect seems to be due to misregulation of expression of the outer membrane porin, OmpU; ectopic expression of OmpU from an inducible promoter was sufficient to bypass the ToxR organic ATR defect (Merrell *et al.*, 2001). While the exact mechanism by which OmpU restores the resistance to organic acid stress in a *toxR* background is unknown, two possible explanations have been suggested. First, it is possible that in the absence of OmpU, the membrane becomes more permeable to organic acids. Second, it is possible that OmpT, which is expressed aberrantly in the *toxR* mutant, enables unwanted organic acid transport into the cytoplasm (Merrell *et al.*, 2001).

In addition to *V. cholerae*, other *Vibrionaceae* are associated with human diarrhoea. For example, in Japan 400 cases of food poisoning associated with *V. parahaemolyticus* were reported in 2000. Similar to *V. cholerae*, *V. parahaemolyticus* is an opportunistic human pathogen that causes diarrhoea and infects its host through the consumption of raw fish and shellfish (Tanaka *et al.*, 2008). In Japan the consumption of contaminated mildly acidic sushi has been correlated to outbreaks of *V. parahaemolyticus* (Shimada and Arakawa, 2000; Tanaka *et al.*, 2008) and it has been speculated that this slight acidification may induce an ATR prior to the bacteria confronting the extreme acidic environment of the stomach (Tanaka *et al.*, 2008). Although various *V. parahaemolyticus* strains show differences in the overall efficiency of their ability to mount an ATR, survival upon acid challenge is consistently two logs higher in strains that have induced the ATR (Tanaka *et al.*, 2008). As with *V. cholerae* (Merrell and Camilli, 1999), *V. parahaemolyticus* appears to use the Cad system since the presence of lysine in the medium improves the survival of ATR adapted cells (Tanaka *et al.*, 2008).

Though it is apparent that *V. cholerae* and other *Vibrionaceae* occupy an environmental niche, they are opportunistic pathogens that are able to colonize the human GI tract successfully (Kaper *et al.*, 1995). Generally, these bacteria will need large numbers (and luck) to survive the stomach barrier successfully. However, since they are able to mount an ATR that enables them to become relatively acid tolerant, the *Vibrionaceae* may become hyperinfectious and survive the stomach successfully to colonize the more hospitable sites found further down the alimentary canal (Merrell *et al.*, 2002a; Butler *et al.*, 2006).

Listeria spp.

The Gram-positive, intracellular bacterium, *Listeria monocytogenes*, is the causative agent of listeriosis and typically infects individuals via contaminated food products (Gandhi and Chikindas, 2007). Though the incidence of listeriosis is infrequent, the disease results in a high fatality rate (Drevets and Bronze, 2008) and affects pregnant women disproportionately (Jamieson *et al.*, 2006). Currently, it appears that the ability to survive acid stress is important for *L. monocytogenes* virulence since the percentage of acid-resistant strains is higher in clinical isolates as compared to environmental isolates (Dykes and Moorhead, 2000). As it is formally feasible that the extent of acid stress encountered in the environment is not as severe as the one encountered within the host, one possible explanation for the difference in the frequency of acid resistance between the clinical and environmental isolates could be the absolute necessity for acid resistance in isolates capable of infecting humans (Dykes and Moorhead, 2000). Study of the acid stress response of *L. monocytogenes* has focused on mechanisms employed by this pathogen to withstand acid stress in the foods it contaminates, as well as mechanisms related to the ability of the pathogen to survive and escape the acid stress inflicted during passage through the gastric barrier and from the acidic environment of the phagosome during the intracellular period of the pathogen life cycle.

Surprisingly, although *L. monocytogenes* is a Gram-positive pathogen, many of the characterized mechanisms enabling it to withstand acid stress show similarity to those discussed previously herein as being employed by Gram-negative bacteria. Additionally, the phenomenon whereby acid stress induces cross-protection against other forms of stress is also common to *Listeria* spp. (Ryan *et al.*, 2008). Thus, cross-protection appears to be a general characteristic of the acid stress response across multiple microbial species. Moreover, it is worthwhile to note that the convergent evolution of similar mechanisms of acid resistance in diverse pathogen lineages is indicative of the critical nature, effectiveness and success of these pathways in combating acid stress. While an exhaustive review of all of the specific acid-resistance system in *Listeria* spp. is not possible here, we will discuss the role of heat shock proteins (HSPs), regulatory proteins, the F_0F_1-ATPase system, membrane restructuring, amino acid decarboxylase and deaminase systems, and the way these pathways are interlinked to provide acid protection for this organism. For an extensive review of Listeria's acid stress response, see Ryan *et al.* (2008).

The repair of acid-induced damage to macromolecules is a key component in the ability of *L. monocytogenes* to withstand acid stress. Since acidic conditions can damage both DNA and proteins, it is perhaps not surprising that dedicated repair systems for both types of macromolecule have been found

to be involved in conferring acid resistance. For instance, the *uvrA* gene, which is part of the 'SOS response' pathway that aids in the repair of damaged DNA, was shown to be important for recovery of *L. monocytogenes* after exposure of the bacterium to acidified media (Kim *et al.*, 2006b). In terms of protein repair, increased expression of chaperonin proteins such as DnaK (Hanawa *et al.*, 1999, 2002) and the GroEL-GroES system (Gahan *et al.*, 2001) have been noted during acid stress, suggesting that they are important during the process of acid adaptation. Furthermore, the serine protease *htrA* was shown to be involved in the response to acid exposure; growth of the *htrA* mutant was impaired under these conditions in comparison to its parental wild-type strain (Stack *et al.*, 2005). While the reason for this is not immediately obvious, it has been speculated that since proteases are involved in the salvage of nutrients from 'damaged' proteins, the upregulation of this class of HSPs should be expected (Stack *et al.*, 2005). Taken together, the identification of genes involved in DNA and protein repair as playing a role in the survival of *L. monocytogenes* in acidic conditions indicates the importance of these classes of gene products to repair damage induced by exposure to low pH.

The involvement of HSPs and other genes in the acid stress response is regulated on several levels, and the signals for their activation vary. For example, the involvement of the *htrA* gene product in acid response is regulated in *L. monocytogenes* by the TCST LisRK system (Stack *et al.*, 2005), which is a direct acid sensor. In addition, LisRK regulates expression of the gene that encodes the acid-induced HtrA-like serine protease (Stack *et al.*, 2005). Thus, in *L. monocytogenes* the LisRK TCST system is necessary to mount the full response to acid stress (Cotter *et al.*, 1999). Furthermore, *L. monocytogenes* also uses alternative sigma factors to enable transcriptional changes to various stress responses. In Gram-negative bacteria this is often accomplished by RpoS. However, in the Gram-positive *L. monocytogenes* the alternative sigma factor, SigB, is responsible for altered expression of genes in response to acid stress (Sue *et al.*, 2004). Accordingly, mutation of *sigB* results in decreased resistance to low pH (Ferreira *et al.*, 2001, 2003). Taken together, it is clear that *Listeria* spp. transcriptional regulators control and orchestrate the response to acid stress.

Like the Gram-negative pathogens that have been discussed throughout this chapter, *L. monocytogenes* has also developed mechanisms to deal with the accumulation of excess cytoplasmic protons, which is inherent to acid stress. Currently, two major types of systems are known to deal with the presence of excess protons. The first of these is the F_0F_1-ATPase, which appears to be essential in *L. monocytogenes* (Ryan *et al.*, 2008) as well as many other bacteria (Koebmann *et al.*, 2000). This enzyme takes an active role in pH homeostasis under acid shock conditions to enable fast recovery of cytoplasmic pH. To this end, the F_0F_1-ATPase forms a channel through the bacterial membrane and acts as an ATP synthase when protons move through the channel into the cell and as an ATP hydrolase when protons are extruded from the cell. pH-mediated induction of the system's components has been shown by 2-D PAGE, and the importance for surviving acid stress shown by partial mutagenesis (Phan-Thanh and Mahouin, 1999; Cotter *et al.*, 2000). The need of this system for ATP hydrolysis to remove protons from the cell creates an energetic burden. However, this burden is relieved partially by the amino acid de-carboxylase and deaminase systems that serve as the second major type of systems that deal with removing excess protons from the cell.

In a manner remarkably similar to the GAD system of *E. coli*, which is discussed in detail elsewhere in this chapter, *L. monocytogenes* also uses the glutamate decarboxylase system to maintain internal pH; this system is essential for *L. monocytogenes* to survive low pH (Cotter *et al.*, 2001). However, the complexity of the system in *L. monocytogenes* is higher than in *E. coli*; *L. monocytogenes* encodes three GAD homologues (*gadD1*, *gadD2* and *gadD3*) and two antiporter homologues (*gadT1* and *gadT2*) (Cotter *et al.*, 2005). The role of this system

was identified by examination of the induction of the various genes (Cotter *et al.*, 2001) and identification of the growth phase dependence and differential induction of the various components during the adaptation phase of the ATR (Wemekamp-Kamphuis *et al.*, 2004). The second, less studied system, which uses amino acids as a substrate and increases intracellular pH, is the arginine deaminase system. Once again, this system is used by Gram-negative bacteria as well as by several Gram-positive bacteria (Dong *et al.*, 2002; Spano *et al.*, 2004) and relies on the transport of the amino acid arginine into the cell and concomitant exchange of an ornithine molecule through an antiport mechanism. Within the cell, the system's enzyme catabolyses conversion of arginine into ornithine and produces ammonia (NH_3), which in the acidic environment of the cell consumes a proton to produce an ammonium ion (NH_4^+) and ATP. Mutation of components of the system revealed that the resulting mutant was impaired in its acid tolerance (Ryan *et al.*, 2009). Thus, the general theme of proton elimination from the cell's interior through utilization of amino acids has been adopted by markedly different pathogens that occupied the same environment and faced similar challenges.

Taken together, the ability of *L. monocytogenes* to survive sudden acid stress as well as to mount an ATR makes this hardy bacterium a formidable challenge to the food industry. The various systems at play must be considered in order to eliminate possible contamination and the downstream associated threat. Clearly, an understanding of the ability of this bacterium to 'prepare' for severe conditions by mounting an ATR must be considered during food processing. This is exemplified by increased acid resistance, which has been shown to be induced following mild acid exposure during poultry decontamination (Alonso-Hernando *et al.*, 2009). Overall, the complex relationship between the different systems and the tight regulation associated with each of them teaches us about the importance of dealing with acid stress in this bacterium's life cycle.

7.6 Through the Environment – Acid Stress and Combined Stressors

In order to complete the cycle of infection, the bacterial pathogens of the GI tract have to pass from one host to the next. For some, the infection route is unclear. For example, the only known reservoir for *H. pylori* is humans and non-human primates, and the means of transmission between hosts is predicted to be oral–oral or faecal–oral. However, virtually nothing is known about the infectious dose or the specifics of the process (Schwarz *et al.*, 2008). For other pathogens discussed herein, the infection route is better defined (Leclerc *et al.*, 2002). Regardless, outside the human host, the bacteria need to withstand changes in temperature, nutrient availability and other environmental factors that impose stress. In addition, many of these pathogens, especially those with a transient stage through the food supply, face acid stress (Ryan *et al.*, 2008). This acid stress can be traced to the actual food these microorganisms contaminate; apple cider is inherently acidic. Alternatively, it can be the result of materials used to treat the surface of the food and packaging where the bacterial contamination usually occurs, or due to processes during the manufacturing of the final product.

One should also remember that though most of these pathogens can infect and colonize the accidental humans they encounter, most are probably better adapted to thrive within an altogether different ecological niche. Because of this, some of their acid adaptation systems might actually have developed in order to deal with environmental stresses other than those found in the human host (Fig. 7.3).

For example, for *Salmonella* spp. it is clear that one of the natural reservoirs is avian. In chickens and other species of fowl, before entering the extreme pH of the stomach, bacteria pass through the ingluvies (crop), where the pH is between 4 and 5 (Rychlik and Barrow, 2005). This intermediate pH stress is similar to the *in vitro* laboratory pH where the ATR is induced (Foster and Moreno, 1999). Thus, understanding that the ATR can occur in this natural environmental reservoir helps

to provide a better understanding of the 'logical progression' of acid adaptation mechanisms from the bacterial environmental perspective to the human pathogenesis viewpoint.

7.7 Discussion and Future Leads

Despite the numerous studies that have shed light on some of the mechanisms employed by pathogenic bacteria to survive in the environment, as well as to breach the acid barrier created by the stomach and other segments of the GI tract, there are still clearly many open venues for research into this topic. While there are clearly too many of these to discuss, we propose the following three examples as prime areas of interest. First, an increased interest in the role of small molecules, either small non-coding (snc) RNA or small proteins, in the regulation and adaptation processes of bacteria has led researchers to examine the role of these molecules in acid stress survival. For example, in *E. coli* the snc-RNA, GadY, which is located in the intergenic region between *gadX* and *gadW*, has been shown to regulate the acid response (Opdyke *et al.*, 2004). Furthermore, other snc-RNAs have been identified and shown to be important in the stress response; the snc-RNA, GcvB, has been shown to be involved in acid stress by affecting RpoS expression (Jin *et al.*, 2009). Future studies aimed at the discovery of the role of other such molecules in acid stress response and development of greater understanding of their importance in the regulatory processes governing the stress response of bacterial cells seems to be an obvious target to pursue.

In addition to recent technological advances that are beginning to make it possible to address the function of small molecules in the acid stress response, these technologies are also beginning to ease the task of addressing the 'real-life' challenges facing an enteric pathogen. Notably, these pathogens do not encounter a single stress factor at a time. Rather, a real-life scenario would be one where the bacteria move from one multi-stress environment to the next. This

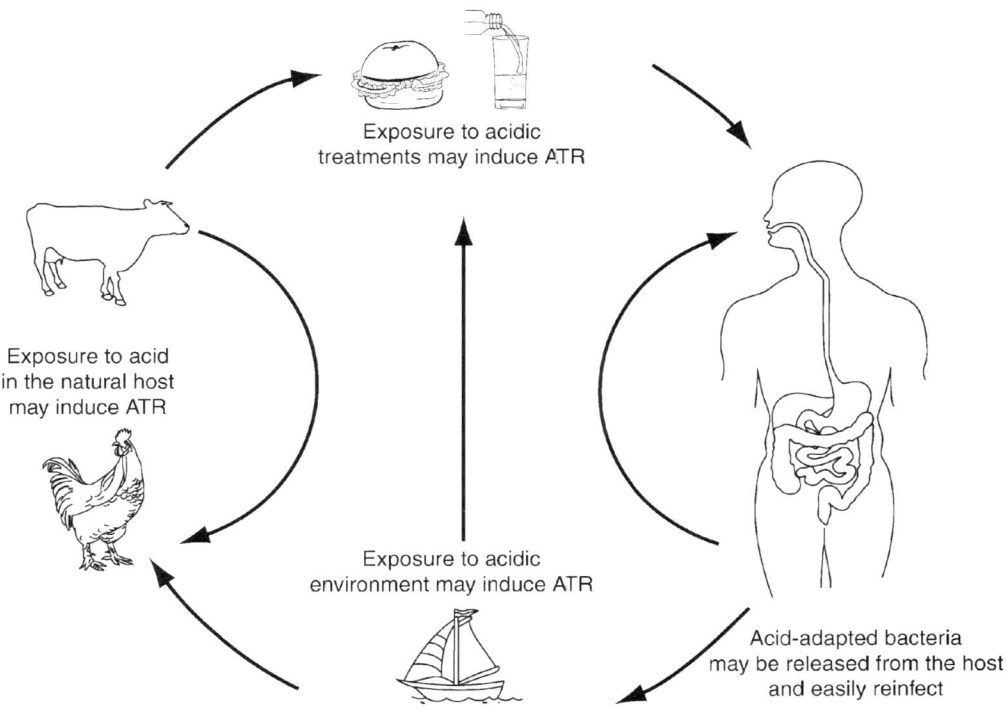

Fig. 7.3. The complex life cycle of enteric pathogens. A generic life cycle of a prototypical enteric pathogen is shown. The human host is often infected via the consumption of contaminated food or water. After reaching the preferred site of colonization, the pathogen multiplies in the GI tract; the numbers of produced progeny differ dramatically for each bacterial species. Subsequently, the bacteria are shed from the human host during defecation/diarrhoeal illness. The shed bacteria can then be ingested by a new host that is in proximity to the shedding event, or they can remain in the environment in a VBNC state or as free-living planktonic cells. These environmental bacteria can reinfect a human host by association with food products and/or drinking water. In addition, other, non-human hosts may serve as an intermediate host for these bacteria; these hosts often show no disease symptoms. The bacteria in the non-human host can be shed to infect other animals or to contaminate the environment. Additionally, since humans consume many of the animal reservoirs, food products can be contaminated and the life cycle is reinitiated. Intrinsic acid resistance and induced acid tolerance have the potential to impact each and every stage of the life cycle.

shift likely triggers a response much more complex than the one following a single stressor in the laboratory, and thus a major question is how this response differs in the face of multiple stressors. This is beginning to be addressed in *E. coli* O157:H7, which can survive in various foods and has been implicated in several outbreaks through consumption of contaminated apple juice.

For this pathogen, the global transcription patterns, before and after exposure to apple juice (pH = 3.5), were compared. Genes involved in the acid, osmotic, oxidative and envelope stress responses showed marked changes in their transcription patterns. In particular, pH and osmotic stress response genes, including *asr*, *osmC*, *osmB* and *osmY*, were induced significantly. Remarkably, 104 of the 331 genes induced in this model are O157:H7-specific (Bergholz *et al.*, 2009), which may indicate this emerging pathogen's specific evolution that has rendered it better suited for this specific niche.

Finally, in addition to the bacterial pathogens discussed herein, the ever-changing world of microorganisms is sure to challenge humans with new and emerging pathogens. An interesting example of a scenario where a newly identified pathogen of the GI tract shows acid stress machinery similar to the ones discussed above can be found with *Streptococcus mutans*. This bacterium, which has an oral–oral transmission route, occupies the mouth at the uppermost section of the GI tract. It has not been included in our discussion since it does not need to experience the acid barrier of the stomach or the volatile fatty acids of the lower GI tract in order to infect a new host. However, it is of interest to note that it shares many of the acid stress survival mechanisms described above (Matsui and Cvitkovitch, 2010). As exemplified by *E. coli*, the most probable source of emerging pathogens of the GI tract is the vast community of commensals colonizing the GI tract of humans and the GI tract of animals that come in close contact with humans. These potential pathogens already have the advantage of being well equipped to withstand acid stress found within their host. Indeed, among the emerging pathogens that cause diarrhoea, one can find bacteria like *C. difficile*, *B. fragilis* and other such commensals. Each of these has proven to be involved in human disease (for a recent review of emerging pathogens of the gastrointestinal tract, see Schlenker and Surawicz, 2009) and has the potential to become more than an anecdotal case report. Thus, clearly, there is a need to understand if acid-resistance mechanisms employed by these potential emerging pathogens are comparable to the well-studied systems of current pathogens.

Clearly, the digestive tract is a hostile place where acid lurks in the form of inorganic acid (HCl) in the stomach, short-chain fatty acids in the intestine and colon, and acidified phagolysosomes in macrophages. Thus, in addition to the other hurdles in front of them,

GI pathogens have learned to resist and survive in this treacherous environment. Overall, the extent and sophistication of the acid survival mechanisms used by bacterial pathogens of the digestive tract is amazing. Our awe at the diverse mechanisms employed by a single species can only be matched by the amazing fact that similarities between acid-resistance systems have evolved across diverse bacterial lineages. The current scientific era, where whole genomes and transcriptomes can be analysed with ease, and where we can follow the reaction to the environment on the single bacterium level, undoubtedly will lead to new discoveries and a better understanding of the survival mechanisms employed by pathogens as they face acid stress.

References

Anon. (1994) Schistosomes, liver flukes and *Helicobacter pylori*. IARC Working Group on the Evaluation of Carcinogenic Risks to Humans. Lyon, 7–14 June 1994. *IARC Monographs and Evaluation of Carcinogenic Risks to Humans* 61, 1–241.

Anon. (2004) Day care-related outbreaks of rhamnose-negative *Shigella sonnei* – six states, June 2001–March 2003. *MMWR Morbidity and Mortality Weekly Report* 53(3), 60–63.

Anon. (2005) The Nobel Prize in Physiology or Medicine 2005. Press release (http://nobelprize.org/nobel_prizes/medicine/laureates/2005/press.html, accessed 20 April 2010).

Anon. (2008a) Preliminary FoodNet data on the incidence of infection with pathogens transmitted commonly through food – 10 states, 2007. *MMWR Morbidity and Mortality Weekly Report* 57(14), 366–370.

Anon. (2008b) Salmonellosis, National Center for Zoonotic, Vector-Borne, and Enteric Diseases (http://www.cdc.gov/nczved/divisions/dfbmd/diseases/salmonellosis, accessed 22 April 2010).

Anon. (2009a) Cholera: global surveillance summary, 2008. *Weekly Epidemiological Records* 84(31), 309–324.

Anon. (2009b) Preliminary FoodNet Data on the incidence of infection with pathogens transmitted commonly through food – 10 states, 2008. *MMWR Morbidity and Mortality Weekly Report* 58(13), 333–337.

Adams, P., Fowler, R., Kinsella, N., Howell, G., Farris, M., Coote, P. and O'Connor, C.D. (2001) Proteomic detection of PhoPQ- and acid-mediated repression of *Salmonella* motility. *Proteomics* 1(4), 597–607.

Alexopoulos, E., Kanjee, U., Snider, J., Houry, W.A. and Pai, E.F. (2008) Crystallization and preliminary X-ray analysis of the inducible lysine decarboxylase from *Escherichia coli*. *Acta Crystallography Section F: Structural Biology and Crystallography Communications* 64(Pt 8), 700–706.

Alonso-Hernando, A., Alonso-Calleja, C. and Capita, R. (2009) Comparative analysis of acid resistance in *Listeria monocytogenes* and *Salmonella enterica* strains before and after exposure to poultry decontaminants. Role of the glutamate decarboxylase (GAD) system. *Food Microbiology* 26(8), 905–909.

Alpuche Aranda, C.M., Swanson, J.A., Loomis, W.P. and Miller, S.I. (1992) *Salmonella typhimurium* activates virulence gene transcription within acidified macrophage phagosomes. *Proceedings of the National Academy of Sciences of the United States of America* 89(21), 10079–10083.

Altier, C. (2005) Genetic and environmental control of *Salmonella* invasion. *Journal of Microbiology* 43, 85–92.

Alvarez-Ordonez, A., Fernandez, A., Bernardo, A. and Lopez, M. (2010) Arginine and lysine decarboxylases and the acid tolerance response of *Salmonella Typhimurium*. *International Journal of Food Microbiology* 136(3), 278–282.

Andrell, J., Hicks, M.G., Palmer, T., Carpenter, E.P., Iwata, S. and Maher, M.J. (2009) Crystal structure of the acid-induced arginine decarboxylase from *Escherichia coli*: reversible decamer assembly controls enzyme activity. *Biochemistry* 48(18), 3915–3927.

Arnold, C.N., McElhanon, J., Lee, A., Leonhart, R. and Siegele, D.A. (2001) Global analysis of *Escherichia coli* gene expression during the acetate-induced acid tolerance response. *Journal of Bacteriology* 183(7), 2178–2186.

Asaka, M., Kato, M. and Graham, D.Y. (2010) Prevention of gastric cancer by *Helicobacter pylori* eradication. *International Medicine* 49(7), 633–636.

Audia, J.P., Webb, C.C. and Foster, J.W. (2001) Breaking through the acid barrier: an orchestrated response to proton stress by enteric bacteria. *International Journal of Medical Microbiology* 291(2), 97–106.

Auger, E.A. and Bennett, G.N. (1989) Regulation of lysine decarboxylase activity in *Escherichia coli* K-12. *Archives in Microbiology* 151(5), 466–468.

Ayazi, S., Leers, J.M., Oezcelik, A., Abate, E., Peyre, C.G., Hagen, J.A., DeMeester, S.R., Banki, F., Lipham, J.C., DeMeester, T.R. and Crookes, P.F. (2009) Measurement of gastric pH in ambulatory esophageal pH monitoring. *Surgical Endoscopy* 23(9), 1968–1973.

Bang, I.S., Kim, B.H., Foster, J.W. and Park, Y.K. (2000) OmpR regulates the stationary-phase acid tolerance response of *Salmonella enterica* serovar Typhimurium. *Journal of Bacteriology* 182(8), 2245–2252.

Bang, I.S., Audia, J.P., Park, Y.K. and Foster, J.W. (2002) Autoinduction of the OmpR response regulator by acid shock and control of the *Salmonella enterica* acid tolerance response. *Molecular Microbiology* 44(5), 1235–1250.

Basso, D. and Plebani, M. (2004) *H. pylori* infection: bacterial virulence factors and cytokine gene polymorphisms as determinants of infection outcome. *Critical Reviews in Clinical Laboratory Science* 41(3), 313–337.

Bearson, B.L., Wilson, L. and Foster, J.W. (1998) A low pH-inducible, PhoPQ-dependent acid tolerance response protects *Salmonella typhimurium* against inorganic acid stress. *Journal of Bacteriology* 180(9), 2409–2417.

Bearson, S., Bearson, B. and Foster, J.W. (1997) Acid stress responses in enterobacteria. *FEMS Microbiology Letters* 147(2), 173–180.

Bearson, S.M., Benjamin, W.H. Jr, Swords, W.E. and Foster, J.W. (1996) Acid shock induction of RpoS is mediated by the mouse virulence gene *mviA* of *Salmonella typhimurium*. *Journal of Bacteriology* 178(9), 2572–2579.

Benjamin, M.M. and Datta, A.R. (1995) Acid tolerance of enterohemorrhagic *Escherichia coli*. *Applied and Environmental Microbiology* 61(4), 1669–1672.

Bergholz, T.M., Vanaja, S.K. and Whittam, T.S. (2009) Gene expression induced in *Escherichia coli* O157:H7 upon exposure to model apple juice. *Applied and Environmental Microbiology* 75(11), 3542–3553.

Bhagwat, A.A. and Bhagwat, M. (2008) Methods and tools for comparative genomics of foodborne pathogens. *Foodborne Pathogenic Diseases* 5(4), 487–497.

Bijlsma, J.J., Lie, A.L., Nootenboom, I.C., Vandenbroucke-Grauls, C.M. and Kusters, J.G. (2000) Identification of loci essential for the growth of *Helicobacter pylori* under acidic conditions. *Journal of Infectious Diseases* 182(5), 1566–1569.

Bijlsma, J.J., Waidner, B., Vliet, A.H., Hughes, N.J., Hag S., Bereswill, S., Kelly, D.J., Vandenbroucke-Grauls, C.M., Kist, M. and Kusters, J.G. (2002) The *Helicobacter pylori* homologue of the ferric uptake regulator is involved in acid resistance. *Infection and Immunity* 70(2), 606–611.

Bik, E.M., Eckburg, P.B., Gill, S.R., Nelson, K.E., Purdom, E.A., Francois, F., Perez-Perez, G., Blaser, M.J. and Relman, D.A. (2006) Molecular analysis of the bacterial microbiota in the human stomach. *Proceedings of the National Academy of Sciences of the United States of America* 103(3), 732–737.

Bina, J., Zhu, J., Dziejman, M., Faruque, S., Calderwood, S. and Mekalanos, J. (2003) ToxR regulon of *Vibrio* cholerae and its expression in *Vibrio*s shed by cholera patients. *Proceedings of the National Academy of Sciences of the United States of America* 100(5), 2801–2806.

Booth, I.R. (2002) Stress and the single cell: intrapopulation diversity is a mechanism to ensure survival upon exposure to stress. *International Journal of Food Microbiology* 78(1–2), 19–30.

Booth, I.R., Cash, P. and O'Byrne, C. (2002) Sensing and adapting to acid stress. *Antonie Van Leeuwenhoek* 81(1–4), 33–42.

Bourret, T.J., Porwollik, S., McClelland, M., Zhao, R., Greco, T., Ischiropoulos, H. and Vazquez-Torres, A. (2008) Nitric oxide antagonizes the acid tolerance response that protects *Salmonella* against innate gastric defenses. *PLoS.One* 3(3), e1833.

Boyle, E.C., Bishop, J.L., Grassl, G.A. and Finlay, B.B. (2007) *Salmonella*: from pathogenesis to therapeutics. *Journal of Bacteriology* 189(5), 1489–1495.

Brown, J.L., Ross, T., McMeekin, T.A. and Nichols, P.D. (1997) Acid habituation of *Escherichia coli* and the potential role of cyclopropane fatty acids in low pH tolerance. *International Journal of Food Microbiology* 37(2–3), 163–173.

Brown, M.R. and Kornberg, A. (2004) Inorganic polyphosphate in the origin and survival of species. *Proceedings of the National Academy of Sciences of the United States of America* 101(46), 16085–16087.

Brown, M.R. and Kornberg, A. (2008) The long and short of it – polyphosphate, PPK and bacterial survival. *Trends in Biochemical Sciences* 33(6), 284–290.

Butler, S.M., Nelson, E.J., Chowdhury, N., Faruque, S.M., Calderwood, S.B. and Camilli, A. (2006) Cholera stool bacteria repress chemotaxis to increase infectivity. *Molecular Microbiology* 60(2), 417–426.

Callaway, T.R., Carr, M.A., Edrington, T.S., Anderson, R.C. and Nisbet, D.J. (2009) Diet, *Escherichia coli* O157:H7, and cattle: a review after 10 years. *Current Issues in Molecular Biology* 11(2), 67–79.

Carpenter, B.M., Whitmire, J.M. and Merrell, D.S. (2009) This is not your mother's repressor: the complex role of Fur in pathogenesis. *Infection and Immunity* 77(7), 2590–2601.

Chang, Y.Y. and Cronan, J.E. Jr (1999) Membrane cyclopropane fatty acid content is a major factor in acid resistance of *Escherichia coli*. *Molecular Microbiology* 33(2), 249–259.

Chaveerach, P., ter Huurne, A.A., Lipman, L.J. and van Knapen, F. (2003) Survival and resuscitation of ten strains of *Campylobacter jejuni* and *Campylobacter coli* under acid conditions. *Applied and Environmental Microbiology* 69(1), 711–714.

Chiang, S.L. and Mekalanos, J.J. (1998) Use of signature-tagged transposon mutagenesis to identify *Vibrio cholerae* genes critical for colonization. *Molecular Microbiology* 27(4), 797–805.

Conte, J.E. Jr (2002) *Manual of Antibiotics and Infectious Diseases: Treatment and Prevention*, 9th edn. Lippincott Williams and Wilkins, Philadelphia, Pennsylvania.

Contreras, M., Thiberge, J.M., Mandrand-Berthelot, M.A. and Labigne, A. (2003) Characterization of the roles of NikR, a nickel-responsive pleiotropic autoregulator of *Helicobacter pylori*. *Molecular Microbiology* 49(4), 947–963.

Cotter, P.D., Emerson, N., Gahan, C.G. and Hill, C. (1999) Identification and disruption of *lisRK*, a genetic locus encoding a two-component signal transduction system involved in stress tolerance and virulence in *Listeria monocytogenes*. *Journal of Bacteriology* 181(21), 6840–6843.

Cotter, P.D., Gahan, C.G. and Hill, C. (2000) Analysis of the role of the *Listeria monocytogenes* F0F1-ATPase operon in the acid tolerance response. *International Journal of Food Microbiology* 60(2–3), 137–146.

Cotter, P.D., Gahan, C.G. and Hill, C. (2001) A glutamate decarboxylase system protects *Listeria monocytogenes* in gastric fluid. *Molecular Microbiology* 40(2), 465–475.

Cotter, P.D., Ryan, S., Gahan, C.G. and Hill, C. (2005) Presence of GadD1 glutamate decarboxylase in selected *Listeria monocytogenes* strains is associated with an ability to grow at low pH. *Applied and Environmental Microbiology* 71(6), 2832–2839.

Coynault, C., Robbe-Saule, V. and Norel, F. (1996) Virulence and vaccine potential of *Salmonella typhimurium* mutants deficient in the expression of the RpoS (sigma S) regulon. *Molecular Microbiology* 22(1), 149–160.

Danielli, A., Romagnoli, S., Roncarati, D., Costantino, L., Delany, I. and Scarlato, V. (2009) Growth phase and metal-dependent transcriptional regulation of the *fecA* genes in *Helicobacter pylori*. *Journal of Bacteriology* 191(11), 3717–3725.

Darwin, K.H. and Miller, V.L. (1999) Molecular basis of the interaction of *Salmonella* with the intestinal mucosa. *Clinical Microbiological Reviews* 12(3), 405–428.

Delany, I., Ieva, R., Soragni, A., Hilleringmann, M., Rappuoli, R. and Scarlato, V. (2005) *In vitro* analysis of protein-operator interactions of the NikR and *fur* metal-responsive regulators of coregulated genes in *Helicobacter pylori*. *Journal of Bacteriology* 187(22), 7703–7715.

Dimaline, R. and Varro, A. (2007) Attack and defence in the gastric epithelium – a delicate balance. *Experimental Physiology* 92(4), 591–601.

Dinsmore, J.E., Jackson, R.J. and Smith, S.D. (1997) The protective role of gastric acidity in neonatal bacterial translocation. *Journal of Pediatric Surgery* 32(7), 1014–1016.

Dong, T. and Schellhorn, H.E. (2010) Role of RpoS in virulence of pathogens. *Infection and Immunity* 78(3), 887–897.

Dong, Y., Chen, Y.Y., Snyder, J.A. and Burne, R.A. (2002) Isolation and molecular analysis of the gene cluster for the arginine deiminase system from *Streptococcus gordonii* DL1. *Applied and Environmental Microbiology* 68(11), 5549–5553.

Drevets, D.A. and Bronze, M.S. (2008) *Listeria monocytogenes*: epidemiology, human disease, and mechanisms of brain invasion. *FEMS Immunology and Medical Microbiology* 53(2), 151–165.

Dunn, B.E. and Phadnis, S.H. (1998) Structure, function and localization of *Helicobacter pylori* urease. *Yale Journal of Biological Medicine* 71(2), 63–73.

Dykes, G.A. and Moorhead, S.M. (2000) Survival of osmotic and acid stress by *Listeria monocytogenes* strains of clinical or meat origin. *International Journal of Food Microbiology* 56 (2–3), 161–166.

Fang, F.C., Libby, S.J., Buchmeier, N.A., Loewen, P.C., Switala, J., Harwood, J. and Guiney, D.G. (1992) The alternative sigma factor *katF* (*rpoS*) regulates *Salmonella* virulence. *Proceedings of the National Academy of Sciences of the United States of America* 89(24), 11978–11982.

Fang, Y., Jayaram, H., Shane, T., Kolmakova-Partensky, L., Wu, F., Williams, C., Xiong, Y. and Miller, C. (2009) Structure of a prokaryotic virtual proton pump at 3.2 A resolution. *Nature* 460(7258), 1040–1043.

Ferreira, A., O'Byrne, C.P. and Boor, K.J. (2001) Role of sigma(B) in heat, ethanol, acid, and oxidative stress resistance and during carbon starvation in *Listeria monocytogenes*. *Applied and Environmental Microbiology* 67(10), 4454–4457.

Ferreira, A., Sue, D., O'Byrne, C.P. and Boor, K.J. (2003) Role of *Listeria monocytogenes* sigma(B) in survival of lethal acidic conditions and in the acquired acid tolerance response. *Applied and Environmental Microbiology* 69(5), 2692–2698.

Foster, J.W. (1991) *Salmonella* acid shock proteins are required for the adaptive acid tolerance response. *Journal of Bacteriology* 173(21), 6896–6902.

Foster, J.W. (1993) The acid tolerance response of *Salmonella typhimurium* involves transient synthesis of key acid shock proteins. *Journal of Bacteriology* 175(7), 1981–1987.

Foster, J.W. (2004) *Escherichia coli* acid resistance: tales of an amateur acidophile. *Nature Reviews Microbiology* 2(11), 898–907.

Foster, J.W. and Hall, H.K. (1990) Adaptive acidification tolerance response of *Salmonella typhimurium*. *Journal of Bacteriology* 172(2), 771–778.

Foster, J.W. and Hall, H.K. (1991) Inducible pH homeostasis and the acid tolerance response of *Salmonella typhimurium*. *Journal of Bacteriology* 173(16), 5129–5135.

Foster, J.W. and Hall, H.K. (1992) Effect of *Salmonella typhimurium* ferric uptake regulator (*fur*) mutations on iron- and pH-regulated protein synthesis. *Journal of Bacteriology* 174(13), 4317–4323.

Foster, J.W. and Moreno, M. (1999) Inducible acid tolerance mechanisms in enteric bacteria. *Novartis Foundation Symposium* 221, 55–69.

Foster, J.W., Park, Y.K., Bang, I.S., Karem, K., Betts, H., Hall, H.K. and Shaw, E. (1994) Regulatory circuits involved with pH-regulated gene expression in *Salmonella typhimurium*. *Microbiology* 140(Pt 2), 341–352.

Gahan, C.G., O'Mahony, J. and Hill, C. (2001) Characterization of the *groESL* operon in *Listeria monocytogenes*: utilization of two reporter systems (*gfp* and *hly*) for evaluating *in vivo* expression. *Infection and Immunity* 69(6), 3924–3932.

Gale, E.F. and Epps, H.M. (1942) The effect of the pH of the medium during growth on the enzymic activities of bacteria (*Escherichia coli* and *Micrococcus lysodeikticus*) and the biological significance of the changes produced. *Biochemical Journal* 36(7–9), 600–618.

Gancz, H., Censini, S. and Merrell, D.S. (2006) Iron and pH homeostasis intersect at the level of Fur regulation in the gastric pathogen *Helicobacter pylori*. *Infection and Immunity* 74(1), 602–614.

Gandhi, M. and Chikindas, M.L. (2007) *Listeria*: a foodborne pathogen that knows how to survive. *International Journal of Food Microbiology* 113(1), 1–15.

Gangaiah, D., Kassem, I.I., Liu, Z. and Rajashekara, G. (2009) Importance of polyphosphate kinase 1 for *Campylobacter jejuni* viable-but-nonculturable cell formation, natural transformation, and antimicrobial resistance. *Applied and Environmental Microbiology* 75(24), 7838–7849.

Gao, X., Lu, F., Zhou, L., Dang, S., Sun, L., Li, X., Wang, J. and Shi, Y. (2009) Structure and mechanism of an amino acid antiporter. *Science* 324(5934), 1565–1568.

Gao, X., Zhou, L., Jiao, X., Lu, F., Yan, C., Zeng, X., Wang, J. and Shi, Y. (2010) Mechanism of substrate recognition and transport by an amino acid antiporter. *Nature* 463(7282), 828–832.

Giannella, R.A., Broitman, S.A. and Zamcheck, N. (1972) Gastric acid barrier to ingested microorganisms in man: studies *in vivo* and *in vitro*. *Gut* 13(4), 251–256.

Go, M.F. (2002) Review article: natural history and epidemiology of *Helicobacter pylori* infection. *Alimentary and Pharmacology Theory* 16(Suppl 1), 3–15.

Gold H.S. and Eisenstein B.I. (2000) Introduction to bacterial diseases. In: Mandell, G.L., Bennett, J.E. and Dolin, R. (eds) *Mandell, Douglas, and Bennett's Principles and Practice of Infectious Diseases*, 5th edn. Churchill Livingstone, Philadelphia, Pennsylvania, pp. 2065–2069.

Goodson, M. and Rowbury, R.J. (1989) Habituation to normally lethal acidity by prior growth of *Escherichia coli* at a sub-lethal acid pH value. *Letters in Applied Microbiology* 8(2), 77–79.

Goodson, M. and Rowbury, R.J. (1991) RecA-independent resistance to irradiation with u.v. light in acid-habituated *Escherichia coli*. *Journal of Applied Bacteriology* 70(2), 177–180.

Gorden, J. and Small, P.L. (1993) Acid resistance in enteric bacteria. *Infection and Immunity* 61(1), 364–367.

Greenacre, E.J. and Brocklehurst, T.F. (2006) The acetic acid tolerance response induces cross-protection to salt stress in *Salmonella typhimurium*. *International Journal of Food Microbiology* 112(1), 62–65.

Groisman, E.A. (2001) The pleiotropic two-component regulatory system PhoP-PhoQ. *Journal of Bacteriology* 183(6), 1835–1842.

Guina, T., Yi, E.C., Wang, H., Hackett, M. and Miller, S.I. (2000) A PhoP-regulated outer membrane protease of *Salmonella enterica* serovar Typhimurium promotes resistance to alpha-helical antimicrobial peptides. *Journal of Bacteriology* 182(14), 4077–4086.

Gupta, A., Polyak, C.S., Bishop, R.D., Sobel, J. and Mintz, E.D. (2004) Laboratory-confirmed shigellosis in the United States, 1989–2002: epidemiologic trends and patterns. *Clinical and Infectious Diseases* 38(10), 1372–1377.

Gyles, C.L. (2007) Shiga toxin-producing *Escherichia coli*: an overview. *Journal of Animal Sciences* 85(Suppl 13), E45–E62.

Hall, H.K. and Foster, J.W. (1996) The role of *fur* in the acid tolerance response of *Salmonella typhimurium* is physiologically and genetically separable from its role in iron acquisition. *Journal of Bacteriology* 178(19), 5683–5691.

Hanawa, T., Fukuda, M., Kawakami, H., Hirano, H., Kamiya, S. and Yamamoto, T. (1999) The *Listeria monocytogenes* DnaK chaperone is required for stress tolerance and efficient phagocytosis with macrophages. *Cell Stress and Chaperones* 4(2), 118–128.

Hanawa, T., Yamanishi, S., Murayama, S., Yamamoto, T. and Kamiya, S. (2002) Participation of DnaK in expression of genes involved in virulence of *Listeria monocytogenes*. *FEMS Microbiology Letters* 214(1), 69–75.

Hersh, B.M., Farooq, F.T., Barstad, D.N., Blankenhorn, D.L. and Slonczewski, J.L. (1996) A glutamate-dependent acid resistance gene in *Escherichia coli*. *Journal of Bacteriology* 178(13), 3978–3981.

House, B., Kus, J.V., Prayitno, N., Mair, R., Que, L., Chingcuanco, F., Gannon, V., Cvitkovitch, D.G. and Barnett, F.D. (2009) Acid-stress-induced changes in enterohaemorrhagic *Escherichia coli* O157: H7 virulence. *Microbiology* 155(Pt 9), 2907–2918.

Huang, C.H., Lee, I.L., Yeh, I.J., Liao, J.H., Ni, C.L., Wu, S.H. and Chiou, S.H. (2010) Upregulation of a non-heme iron-containing ferritin with dual ferroxidase and DNA-binding activities in *Helicobacter pylori* under acid stress. *Journal of Biochemistry* 147(4), 535–543.

Huang, D.B., Mohanty, A., DuPont, H.L., Okhuysen, P.C. and Chiang, T. (2006) A review of an emerging enteric pathogen: enteroaggregative *Escherichia coli*. *Journal of Medical Microbiology* 55(Pt 10), 1303–1311.

Iyer, R., Iverson, T.M., Accardi, A. and Miller, C. (2002) A biological role for prokaryotic ClC chloride channels. *Nature* 419(6908), 715–718.

Iyer, R., Williams, C. and Miller, C. (2003) Arginine-agmatine antiporter in extreme acid resistance in *Escherichia coli*. *Journal of Bacteriology* 185(22), 6556–6561.

Jackson, D.N., Davis, B., Tirado, S.M., Duggal, M., van Frankenhuyzen, J.K., Deaville, D., Wijesinghe, M.A., Tessaro, M. and Trevors, J.T. (2009) Survival mechanisms and culturability of

Campylobacter jejuni under stress conditions. *Antonie Van Leeuwenhoek* 96(4), 377–394.

Jamieson, D.J., Theiler, R.N. and Rasmussen, S.A. (2006) Emerging infections and pregnancy. *Emerging and Infectious Diseases* 12(11), 1638–1643.

Janssen, R., Krogfelt, K.A., Cawthraw, S.A., Pelt, W. van, Wagenaar, J.A. and Owen, R.J. (2008) Host–pathogen interactions in *Campylobacter* infections: the host perspective. *Clinical Microbiology Reviews* 21(3), 505–518.

Jeong, K.C., Hung, K.F., Baumler, D.J., Byrd, J.J. and Kaspar, C.W. (2008) Acid stress damage of DNA is prevented by Dps binding in *Escherichia coli* O157:H7. *BMC Microbiology* 8, 181.

Jin, Y., Watt, R.M., Danchin, A. and Huang, J.D. (2009) Small non-coding RNA GcvB is a novel regulator of acid resistance in *Escherichia coli*. *BMC Genomics* 10, 165.

Joerger, R.D., Sartori, C.A. and Kniel, K.E. (2009) Comparison of genetic and physiological properties of *Salmonella enterica* isolates from chickens reveals one major difference between serovar Kentucky and other serovars: response to acid. *Foodborne Pathogenic Diseases* 6(4), 503–512.

Kang, I.H., Kim, J.S., Kim, E.J. and Lee, J.K. (2007) Cadaverine protects *Vibrio vulnificus* from superoxide stress. *Journal of Microbiology and Biotechnology* 17(1), 176–179.

Kannan, G., Wilks, J.C., Fitzgerald, D.M., Jones, B.D., BonDurant, S.S. and Slonczewski, J.L. (2008) Rapid acid treatment of *Escherichia coli*: transcriptomic response and recovery. *BMC Microbiology* 8, 37.

Kaper, J.B., Morris, J.G. Jr and Levine, M.M. (1995) Cholera. *Clinical Microbiology Reviews* 8(1), 48–86.

Kasper, G.F. (1988) Natural sources and microbiological characteristics of *Campylobacter pylori*. *Scandanavian Journal of Gastroenterology, Supplement* 142, 14–15.

Kaur, P., Chakraborti, A. and Asea, A. (2010) Enteroaggregative *Escherichia coli*: an emerging enteric food borne pathogen. *Interdisciplinary Perspectives of Infectious Diseases* 2010, 254159.

Kim, B.H., Kim, S., Kim, H.G., Lee, J., Lee, I.S. and Park, Y.K. (2005) The formation of cyclopropane fatty acids in *Salmonella enterica* serovar Typhimurium. *Microbiology* 151(Pt 1), 209–218.

Kim, J.S., Choi, S.H. and Lee, J.K. (2006a) Lysine decarboxylase expression by *Vibrio vulnificus* is induced by SoxR in response to superoxide stress. *Journal of Bacteriology* 188(24), 8586–8592.

Kim, S.H., Gorski, L., Reynolds, J., Orozco, E., Fielding, S., Park, Y.H. and Borucki, M.K.

(2006b) Role of *uvrA* in the growth and survival of *Listeria monocytogenes* under UV radiation and acid and bile stress. *Journal Food Protection* 69(12), 3031–3036.

Kirkpatrick, C., Maurer, L.M., Oyelakin, N.E., Yoncheva, Y.N., Maurer, R. and Slonczewski, J.L. (2001) Acetate and formate stress: opposite responses in the proteome of *Escherichia coli*. *Journal of Bacteriology* 183(21), 6466–6477.

Koebmann, B.J., Nilsson, D., Kuipers, O.P. and Jensen, P.R. (2000) The membrane-bound $H^{(+)}$-ATPase complex is essential for growth of *Lactococcus lactis*. *Journal of Bacteriology* 182(17), 4738–4743.

Koelz, H.R. (1992) Gastric acid in vertebrates. *Scandanavian Journal of Gastroenterology, Supplement* 193, 2–6.

Koga, T., Sakamoto, F., Yamoto, A. and Takumi, K. (1999) Acid adaptation induces cross-protection against some environmental stresses in *Vibrio parahaemolyticus*. *Journal General Applied Microbiology* 45(4), 155–161.

Krukonis, E.S. and Dirita, V.J. (2003) From motility to virulence: sensing and responding to environmental signals in *Vibrio cholerae*. *Current Opinion of Microbiology* 6(2), 186–190.

Kuper, C. and Jung, K. (2005) CadC-mediated activation of the *cadBA* promoter in *Escherichia coli*. *Journal of Molecular Microbiology and Biotechnology* 10(1), 26–39.

Kwon, Y.M. and Ricke, S.C. (1998) Induction of acid resistance of *Salmonella typhimurium* by exposure to short-chain fatty acids. *Applied and Environmental Microbiology* 64(9), 3458–3463.

Leclerc, H., Schwartzbrod, L. and Dei-Cas, E. (2002) Microbial agents associated with waterborne diseases. *Critical Reviews in Microbiology* 28(4), 371–409.

Lee, I.S., Slonczewski, J.L. and Foster, J.W. (1994) A low-pH-inducible, stationary-phase acid tolerance response in *Salmonella typhimurium*. *Journal of Bacteriology* 176(5), 1422–1426.

Leyer, G.J. and Johnson, E.A. (1993) Acid adaptation induces cross-protection against environmental stresses in *Salmonella typhimurium*. *Applied and Environmental Microbiology* 59(6), 1842–1847.

Lin, J., Lee, I.S., Frey, J., Slonczewski, J.L. and Foster, J.W. (1995) Comparative analysis of extreme acid survival in *Salmonella typhimurium*, *Shigella flexneri*, and *Escherichia coli*. *Journal of Bacteriology* 177(14), 4097–4104.

Lin, J., Smith, M.P., Chapin, K.C., Baik, H.S., Bennett, G.N. and Foster, J.W. (1996) Mechanisms of acid resistance in enterohemorrhagic *Escherichia coli*. *Applied and Environmental Microbiology* 62(9), 3094–3100.

Loh, J.T., Gupta, S.S., Friedman, D.B., Krezel, A.M. and Cover, T.L. (2010) Analysis of protein expression regulated by the *Helicobacter pylori* ArsRS two-component signal transduction system. *Journal of Bacteriology* 192(8), 2034–2043.

Ma, Y., Hanning, I. and Slavik, M. (2009) Stress-induced adaptive tolerance response and virulence gene expression in *Campylobacter jejuni*. *Journal of Food Safety* 29, 126–143.

McGee, D.J. and Mobley, H.L. (1999) Mechanisms of *Helicobacter pylori* infection: bacterial factors. *Current Topics in Microbiology and Immunology* 241, 155–180.

Madoff L.C. and Kasper, D.L. (2004) Introduction to infectious disease: host–pathogen interactions. In: Kasper, D.L., Fauci, A.S., Longo, D.L., Braunwald, E., Hauser, S.L. and Jameson, J.L. (eds) *Harrison's Principles of Internal Medicine*, 16th edn. McGraw-Hill, New York, pp. 695–699.

Malki, A., Le, H.T., Milles, S., Kern, R., Caldas, T., Abdallah, J. and Richarme, G. (2008) Solubilization of protein aggregates by the acid stress chaperones HdeA and HdeB. *Journal of Biological Chemistry* 283, 13679–13687.

Manson, J.M., Rauch, M. and Gilmore, M.S. (2008) The commensal microbiology of the gastrointestinal tract. *Advances in Experimental Medical Biology* 635, 15–28.

Marshall, B.J. and Warren, J.R. (1984) Unidentified curved bacilli in the stomach of patients with gastritis and peptic ulceration. *Lancet* 1(8390), 1311–1315.

Matsui, R. and Cvitkovitch, D. (2010) Acid tolerance mechanisms utilized by Streptococcus mutans. *Future Microbiology* 5, 403–417.

Meneghini, R. (1997) Iron homeostasis, oxidative stress, and DNA damage. *Free Radical Biology and Medicine* 23, 783–792.

Merrell, D.S. and Camilli, A. (1999) The *cadA* gene of *Vibrio cholerae* is induced during infection and plays a role in acid tolerance. *Molecular Microbiology* 34(4), 836–849.

Merrell, D.S. and Camilli, A. (2000) Regulation of *Vibrio cholerae* genes required for acid tolerance by a member of the 'ToxR-like' family of transcriptional regulators. *Journal of Bacteriology* 182(19), 5342–5350.

Merrell, D.S., Bailey, C., Kaper, J.B. and Camilli, A. (2001) The ToxR-mediated organic acid tolerance response of *Vibrio cholerae* requires OmpU. *Journal of Bacteriology* 183, 2746–2754.

Merrell, D.S., Butler, S.M., Qadri, F., Dolganov, N.A., Alam, A., Cohen, M.B., Calderwood, S.B., Schoolnik, G.K. and Camilli, A. (2002a) Host-induced epidemic spread of the cholera bacterium. *Nature* 417, 642–645.

Merrell, D.S., Hava, D.L. and Camilli, A. (2002b) Identification of novel factors involved in colonization and acid tolerance of *Vibrio cholerae*. *Molecular Microbiology* 43(6), 1471–1491.

Miles, S., Piazuelo, M.B., Semino-Mora, C., Washington, M.K., Dubois, A., Peek, R.M. Jr, Correa, P. and Merrell, D.S. (2010) Detailed *in vivo* analysis of the role of *Helicobacter pylori* Fur in colonization and disease. *Infection and Immunity* 78(7), 3073–3082.

Miller, S.I., Kukral, A.M. and Mekalanos, J.J. (1989) A two-component regulatory system (*phoP phoQ*) controls *Salmonella typhimurium* virulence. *Proceedings of the National Academy of Sciences of the United States of America* 86(13), 5054–5058.

Miller, V.L., Taylor, R.K. and Mekalanos, J.J. (1987) Cholera toxin transcriptional activator *toxR* is a transmembrane DNA binding protein. *Cell* 48(2), 271–279.

Missiakas, D. and Raina, S. (1998) The extracytoplasmic function sigma factors: role and regulation. *Molecular Microbiology* 28(6), 1059–1066.

Muller, C., Bang, I.S., Velayudhan, J., Karlinsey, J., Papenfort, K., Vogel, J. and Fang, F.C. (2009) Acid stress activation of the sigma(E) stress response in *Salmonella enterica* serovar Typhimurium. *Molecular Microbiology* 71(5), 1228–1238.

Neely, M.N., Dell, C.L. and Olson, E.R. (1994) Roles of LysP and CadC in mediating the lysine requirement for acid induction of the *Escherichia coli* cad operon. *Journal of Bacteriology* 176(11), 3278–3285.

Oberto, J., Nabti, S., Jooste, V., Mignot, H. and Rouviere-Yaniv, J. (2009) The HU regulon is composed of genes responding to anaerobiosis, acid stress, high osmolarity and SOS induction. *PLoS One* 4(2), e4367.

Oliver, J.D. (2009) Recent findings on the viable but non-culturable state in pathogenic bacteria. *FEMS Microbiology Reviews* 34, 415–425.

O'Neal, C.R., Gabriel, W.M., Turk, A.K., Libby, S.J., Fang, F.C. and Spector, M.P. (1994) RpoS is necessary for both the positive and negative regulation of starvation survival genes during phosphate, carbon, and nitrogen starvation in *Salmonella typhimurium*. *Journal of Bacteriology* 176(15), 4610–4616.

Opdyke, J.A., Kang, J.G. and Storz, G. (2004) GadY, a small-RNA regulator of acid response genes in *Escherichia coli*. *Journal of Bacteriology* 186(20), 6698–6705.

Parkinson, J.S. (1993) Signal transduction schemes of bacteria. *Cell* 73(5), 857–871.

Paul, B. and Hirshfield, I. (2003) The effect of acid treatment on survival and protein expression of a laboratory K-12 strain *Escherichia coli*. *Research in Microbiology* 154(2), 115–121.

Peterson, K.M. and Mekalanos, J.J. (1988) Characterization of the *Vibrio cholerae* ToxR regulon: identification of novel genes involved in intestinal colonization. *Infection and Immunity* 56(11), 2822–2829.

Pflock, M., Finsterer, N., Joseph, B., Mollenkopf, H., Meyer, T.F. and Beier, D. (2006) Characterization of the ArsRS regulon of *Helicobacter pylori*, involved in acid adaptation. *Journal of Bacteriology* 188(10), 3449–3462.

Phan-Thanh, L. and Mahouin, F. (1999) A proteomic approach to study the acid response in *Listeria monocytogenes*. *Electrophoresis* 20(11), 2214–2224.

Pohl, D., Fox, M., Fried, M., Goke, B., Prinz, C., Monnikes, H., Rogler, G., Dauer, M., Keller, J., Lippl, F., Schiefke, I., Seidler, U. and Allescher, H.D. (2008) Do we need gastric acid? *Digestion* 77(3–4), 184–197.

Price-Carter, M., Fazzio, T.G., Vallbona, E.I. and Roth, J.R. (2005) Polyphosphate kinase protects *Salmonella enterica* from weak organic acid stress. *Journal of Bacteriology* 187(9), 3088–3099.

Prost, L.R. and Miller, S.I. (2008) The *Salmonellae* PhoQ sensor: mechanisms of detection of phagosome signals. *Cellular Microbiology* 10(3), 576–582.

Qin, J., Li, R., Raes, J., Arumugam, M., Burgdorf, K.S., Manichanh, C., Nielsen, T., Pons, N., Levenez, F., Yamada, T., Mende, D.R., Li, J., Xu, J., Li, S., Li, D., Cao, J., Wang, B., Liang, H., Zheng, H., Xie, Y., Tap, J., Lepage, P., Bertalan, M., Batto, J.M., Hansen, T., Le, P.D., Linneberg, A., Nielsen, H.B., Pelletier, E., Renault, P., Sicheritz-Ponten, T., Turner, K., Zhu, H., Yu, C., Li, S., Jian, M., Zhou, Y., Li, Y., Zhang, X., Li, S., Qin, N., Yang, H., Wang, J., Brunak, S., Dore, J., Guarner, F., Kristiansen, K., Pedersen, O., Parkhill, J., Weissenbach, J., Bork, P., Ehrlich, S.D. and Wang, J. (2010) A human gut microbial gene catalogue established by metagenomic sequencing. *Nature* 464(7285), 59–65.

Raja, N., Goodson, M., Chui, W.C., Smith, D.G. and Rowbury, R.J. (1991) Habituation to acid in *Escherichia coli*: conditions for habituation and its effects on plasmid transfer. *Journal of Applied Bacteriology* 70(1), 59–65.

Rangel, J.M., Sparling, P.H., Crowe, C., Griffin, P.M. and Swerdlow, D.L. (2005) Epidemiology of *Escherichia coli* O157:H7 outbreaks, United States, 1982–2002. *Emerging and Infectious Diseases* 11(4), 603–609.

Rao, A., Jump, R.L., Pultz, N.J., Pultz, M.J. and Donskey, C.J. (2006) *In vitro* killing of nosocomial pathogens by acid and acidified nitrite. *Antimicrobial Agents and Chemotherapy* 50(11), 3901–3904.

Rathman, M., Sjaastad, M.D. and Falkow, S. (1996) Acidification of phagosomes containing *Salmonella typhimurium* in murine macrophages. *Infection and Immunity* 64(7), 2765–2773.

Reid, A.N., Pandey, R., Palyada, K., Naikare, H. and Stintzi, A. (2008a) Identification of *Campylobacter jejuni* genes involved in the response to acidic pH and stomach transit. *Applied and Environmental Microbiology* 74(5), 1583–1597.

Reid, A.N., Pandey, R., Palyada, K., Whitworth, L., Doukhanine, E. and Stintzi, A. (2008b) Identification of *Campylobacter jejuni* genes contributing to acid adaptation by transcriptional profiling and genome-wide mutagenesis. *Applied and Environmental Microbiology* 74(5), 1598–1612.

Richard, H.T. and Foster, J.W. (2003) Acid resistance in *Escherichia coli*. *Advances in Applied Microbiology* 52, 167–186.

Ricke, S.C. (2003) Perspectives on the use of organic acids and short chain fatty acids as antimicrobials. *Poultry Science* 82(4), 632–639.

Rodriguez-Romo, L. and Yousef, A. (2005) Cross-protective effects of bacterial stress. In: Griffiths, M. (ed.) *Understanding Pathogen Behaviour: Virulence, Stress Response, and Resistance*. CPC Press, Woodhead Publishing in Food Science and Technology, New York, pp. 128–151.

Rowley, G., Stevenson, A., Kormanec, J. and Roberts, M. (2005) Effect of inactivation of *degS* on *Salmonella enterica* serovar Typhimurium *in vitro* and *in vivo*. *Infection and Immunity* 73(1), 459–463.

Russell, R.G., Blaser, M.J., Sarmiento, J.I. and Fox, J. (1989) Experimental *Campylobacter jejuni* infection in Macaca nemestrina. *Infection and Immunity* 57(5), 1438–1444.

Ryan, S., Hill, C. and Gahan, C.G. (2008) Acid stress responses in *Listeria monocytogenes*. *Advances in Applied Microbiology* 65, 67–91.

Ryan, S., Begley, M., Gahan, C.G. and Hill, C. (2009) Molecular characterization of the arginine deiminase system in *Listeria monocytogenes*: regulation and role in acid tolerance. *Environmental Microbiology* 11(2), 432–445.

Rychlik, I. and Barrow, P.A. (2005) *Salmonella*

stress management and its relevance to behaviour during intestinal colonisation and infection. *FEMS Microbiology Reviews* 29(5), 1021–1040.

Sachs, G., Weeks, D.L., Wen, Y., Marcus, E.A., Scott, D.R. and Melchers, K. (2005) Acid acclimation by *Helicobacter pylori*. *Physiology (Bethesda)* 20, 429–438.

Sahay, P., West, A.P., Birkenhead, D. and Hawkey, P.M. (1995) *Campylobacter jejuni* in the stomach. *Journal of Medical Microbiology* 43(1), 75–77.

Sansonetti, P. (2001) Phagocytosis of bacterial pathogens: implications in the host response. *Seminars in Immunology* 13(6), 381–390.

Sarker, S.A. and Gyr, K. (1992) Non-immunological defence mechanisms of the gut. *Gut* 33(7), 987–993.

Scarlato, V., Delany, I., Spohn, G. and Beier, D. (2001) Regulation of transcription in *Helicobacter pylori*: simple systems or complex circuits? *International Journal of Medical Microbiology* 291(2), 107–117.

Schlenker, C. and Surawicz, C.M. (2009) Emerging infections of the gastrointestinal tract. *Best Practice in Research of Clinical Gastroenterolgy* 23(1), 89–99.

Schubert, M.L. (2008) Gastric secretion. *Current Opinion in Gastroenterology* 24(6), 659–664.

Schwarz, S., Morelli, G., Kusecek, B., Manica, A., Balloux, F., Owen, R.J., Graham, D.Y., van der Merwe, S., Achtman, M. and Suerbaum, S. (2008) Horizontal versus familial transmission of *Helicobacter pylori*. *PLoS Pathogens* 4(10), e1000180.

Shahinian, M.L., Passaro, D.J., Swerdlow, D.L., Mintz, E.D., Rodriguez, M. and Parsonnel, J. (2000) *Helicobacter pylori* and epidemic *Vibrio cholerae* O1 infection in Peru. *Lancet* 355(9201), 377–378.

Shao, C., Zhang, Q., Tang, W., Qu, W., Zhou, Y., Sun, Y., Yu, H. and Jia, J. (2008) The changes of proteomes components of *Helicobacter pylori* in response to acid stress without urea. *Journal of Microbiology* 46(3), 331–337.

Shimada, T. and Arakawa, E. (2000) Current status of *Vibrio parahaemolyticus* food poisoning. *Journal of Antibacterial and Antifungal Agents* 28, 157–167.

Sivula, C.P., Bogomolnaya, L.M. and Andrews-Polymenis, H.L. (2008) A comparison of cecal colonization of *Salmonella enterica* serotype *Typhimurium* in white leghorn chicks and *Salmonella*-resistant mice. *BMC Microbiology* 8, 182.

Skouloubris, S., Labigne, A. and de Reuse, H. (1997) Identification and characterization of an aliphatic amidase in *Helicobacter pylori*. *Molecular Microbiology* 25(5), 989–998.

Skouloubris, S., Labigne, A. and de Reuse, H. (2001) The AmiE aliphatic amidase and AmiF formamidase of *Helicobacter pylori*: natural evolution of two enzyme paralogues. *Molecular Microbiology* 40(3), 596–609.

Small, P., Blankenhorn, D., Welty, D., Zinser, E. and Slonczewski, J.L. (1994) Acid and base resistance in *Escherichia coli* and *Shigella flexneri*: role of *rpoS* and growth pH. *Journal of Bacteriology* 176(6), 1729–1737.

Smith, J.L. (2003) The role of gastric acid in preventing foodborne disease and how bacteria overcome acid conditions. *Journal of Food Protection* 66(7), 1292–1303.

Soncini, F.C. and Groisman, E.A. (1996) Two-component regulatory systems can interact to process multiple environmental signals. *Journal of Bacteriology* 178(23), 6796–6801.

Spano, G., Chieppa, G., Beneduce, L. and Massa, S. (2004) Expression analysis of putative *arcA*, *arcB* and *arcC* genes partially cloned from *Lactobacillus plantarum* isolated from wine. *Journal of Applied Microbiology* 96(1), 185–193.

Stack, H.M., Sleator, R.D., Bowers, M., Hill, C. and Gahan, C.G. (2005) Role for HtrA in stress induction and virulence potential in *Listeria monocytogenes*. *Applied and Environmental Microbiology* 71(8), 4241–4247.

Stancik, L.M., Stancik, D.M., Schmidt, B., Barnhart, D.M., Yoncheva, Y.N. and Slonczewski, J.L. (2002) pH-dependent expression of periplasmic proteins and amino acid catabolism in *Escherichia coli*. *Journal of Bacteriology* 184(15), 4246–4258.

Stecher, B. and Hardt, W.D. (2008) The role of microbiota in infectious disease. *Trends in Microbiology* 16(3), 107–114.

Stock, A.M., Robinson, V.L. and Goudreau, P.N. (2000) Two-component signal transduction. *Annual Review in Biochemistry* 69, 183–215.

Sue, D., Fink, D., Wiedmann, M. and Boor, K.J. (2004) sigmaB-dependent gene induction and expression in *Listeria monocytogenes* during osmotic and acid stress conditions simulating the intestinal environment. *Microbiology* 150(Pt 11), 3843–3855.

Tanaka, Y., Kimura, B., Takahashi, H., Watanabe, T., Obata, H., Kai, A., Morozumi, S. and Fujii, T. (2008) Lysine decarboxylase of *Vibrio parahaemolyticus*: kinetics of transcription and role in acid resistance. *Journal of Applied Microbiology* 104(5), 1283–1293.

Taylor, R.K., Hall, M.N., Enquist, L. and Silhavy, T.J.

(1981) Identification of OmpR: a positive regulatory protein controlling expression of the major outer membrane matrix porin proteins of *Escherichia coli* K-12. *Journal of Bacteriology* 147(1), 255–258.

Tenaillon, O., Skurnik, D., Picard, B. and Denamur, E. (2010) The population genetics of commensal *Escherichia coli*. *Nature Reviews Microbiology* 8(3), 207–217.

Tennant, S.M., Hartland, E.L., Phumoonna, T., Lyras, D., Rood, J.I., Robins-Browne, R.M. and van Driel, I.R. (2008) Influence of gastric acid on susceptibility to infection with ingested bacterial pathogens. *Infection and Immunity* 76(2), 639–645.

Testerman, T.L., Vazquez-Torres, A., Xu, Y., Jones-Carson, J., Libby, S.J. and Fang, F.C. (2002) The alternative sigma factor sigmaE controls antioxidant defences required for *Salmonella* virulence and stationary-phase survival. *Molecular Microbiology* 43(3), 771–782.

Toledo, H., Valenzuela, M., Rivas, A. and Jerez, C.A. (2002) Acid stress response in *Helicobacter pylori*. *FEMS Microbiology Letters* 213(1), 67–72.

Torres, A.G. (2009) The *cad* locus of Enterobacteriaceae: more than just lysine decarboxylation. *Anaerobe* 15(1–2), 1–6.

Tucker, D.L., Tucker, N. and Conway, T. (2002) Gene expression profiling of the pH response in *Escherichia coli*. *Journal of Bacteriology* 184(23), 6551–6558.

Vanaja, S.K., Springman, A.C., Besser, T.E., Whittam, T.S. and Manning, S.D. (2010) Differential expression of virulence and stress fitness genes between *Escherichia coli* O157:H7 strains with clinical or bovine-biased genotypes. *Applied and Environmental Microbiology* 76(1), 60–68.

Wang, G., Hong, Y., Olczak, A., Maier, S.E. and Maier, R.J. (2006) Dual roles of *Helicobacter pylori* NapA in inducing and combating oxidative stress. *Infection and Immunity* 74(12), 6839–6846.

Warren, J.R. and Marshall, B.J. (1983) Unidentified curved bacilli on gastric epithelium in active chronic gastritis. *Lancet* 1(8336), 1273–1275.

Waterman, S.R. and Small, P.L. (1998) Acid-sensitive enteric pathogens are protected from killing under extremely acidic conditions of pH 2.5 when they are inoculated onto certain solid food sources. *Applied and Environmental Microbiology* 64(10), 3882–3886.

Wemekamp-Kamphuis, H.H., Wouters, J.A., de Leeuw, P.P., Hain, T., Chakraborty, T. and Abee, T. (2004) Identification of sigma factor sigma

B-controlled genes and their impact on acid stress, high hydrostatic pressure, and freeze survival in *Listeria monocytogenes* EGD-e. *Applied and Environmental Microbiology* 70(6), 3457–3466.

Wen, Y., Feng, J., Scott, D.R., Marcus, E.A. and Sachs, G. (2007) The HP0165-HP0166 two-component system (ArsRS) regulates acid-induced expression of HP1186 alpha-carbonic anhydrase in *Helicobacter pylori* by activating the pH-dependent promoter. *Journal of Bacteriology* 189(6), 2426–2434.

Whitmire, J.M., Gancz, H. and Merrell, D.S. (2007) Balancing the double-edged sword: metal ion homeostasis and the ulcer bug. *Current Medical Chemistry* 14(4), 469–478.

WHO (2000) The increasing incidence of human Campylobacteriosis. Report and Procceedings of a WHO Consultation of Experts (*http://whqlibdoc.who.int/hq/2001/WHO_CDS_CSR_APH_2001.7.pdf*, accessed 13 April 2010).

WHO (2008) The global burden of disease: 2004 update. World Health Organization, Geneva (*http://www.who.int/healthinfo/global_burden_disease/2004_report_update/en/index.html*, accessed 13 April, 2010).

WHO (2009a) Diarrhoea: why children are still dying and what can be done. World Health Organization, Geneva (*http://www.who.int/child_adolescent_health/documents/9789241598415/en/index.html*, accessed 13 April 2010).

WHO (2009b) Water sanitation and health (WSH). World Health Organization (*http://www.who.int/water_sanitation_health/diseases/Campylobacteriosis/en/*, accessed 13 April 2010).

Williams, C. and McColl, K.E. (2006) Review article: proton pump inhibitors and bacterial overgrowth. *Alimentary and Pharmacology Theory* 23(1), 3–10.

Windle, H.J., Kelleher, D. and Crabtree, J.E. (2007) Childhood *Helicobacter pylori* infection and growth impairment in developing countries: a vicious cycle? *Pediatrics* 119(3), e754–e759.

Winfield, M.D. and Groisman, E.A. (2003) Role of non-host environments in the lifestyles of *Salmonella* and *Escherichia coli*. *Applied and Environmental Microbiology* 69(7), 3687–3694.

Xu, H., Lee, H.Y. and Ahn, J. (2008) Cross-protective effect of acid-adapted *Salmonella enterica* on resistance to lethal acid and cold stress conditions. *Letters in Applied Microbiology* 47(4), 290–297.

Yang, J., Nie, H., Chen, L., Zhang, X., Yang, F., Xu,

X., Zhu, Y., Yu, J. and Jin, Q. (2007) Revisiting the molecular evolutionary history of *Shigella* spp. *Journal of Molecular Evolution* 64(1), 71–79.

Young, K.T., Davis, L.M. and Dirita, V.J. (2007) *Campylobacter jejuni*: molecular biology and pathogenesis. *Nature Reviews Microbiology* 5(9), 665–679.

8 Urease and the Bacterial Acid Stress Response

Peter T. Chivers

8.1 Introduction

Bacteria exhibit a range of optimal intracellular pH levels (Booth, 1985). Acid stress occurs when protons accumulate within the cell and lower the intracellular pH below optimal values. Acidic pH affects protein stability and activity as well as pH-dependent processes, such as proton-linked reaction equilibria and membrane potential (Booth, 1985). Loss of the proton motive force affects the ability of cells to synthesize ATP via the F_0/F_1-ATPase (White, 2007). Acidification can be generated by various mechanisms. The production of hydrochloric acid (HCl) by the human stomach is a well-known exogenously generated stress for several Gram-negative bacteria that either reside in or passage the stomach. Acid stress can also be self-generated via glycolytic end products (e.g. lactate) in Gram-positive bacteria (Madigan *et al.*, 2003), such as the *Streptococci*. There are some exceptions in adapting to low pH, for example, acetic acid bacteria such as *Acetobacter aceti* which have a physiology to function at low intracellular pH (Menzel and Gottschalk, 1985), but most non-acidophilic bacteria have mechanisms to buffer intra-cellular pH and allow survival or even growth under acid stress. These are two distinct situations, as survival requires restoring the *intracellular* pH to a normal level, while growth necessitates an additional elevation of *periplasmic* pH to restore the proton motive force necessary for full metabolic activity (ATP synthesis). Various mechanisms of pH homeostasis are well studied and discussed in detail elsewhere (see Gancz and Merrell, Chapter 7, this volume; Booth, 1985; Foster, 1999). These strategies include proton efflux and the generation of buffering molecules that consume protons. In this chapter the specific role of bacterial urease in the acid stress response will be discussed.

8.2 Urease and Acid Stress – An Overview

Enzymes can help to buffer acid stress by catalysing reactions either that consume protons or that generate reaction products that are in equilibrium with protons. Enzymes typically do not affect protonation equilibria themselves because this process is rapid in an aqueous environment. This chapter focuses on bacterial urease, a nickel-dependent enzyme that hydrolyses urea to produce ammonia (e.g. $NH_3 + H^+ \Leftrightarrow NH_4$). Many bacteria encode the genes for urease assembly and activity; in many cases the urease activity is used solely as a source of nitrogen. Various aspects of urease biochemistry and its role in cell function have been covered in great detail in previous reviews (Mobley and Hausinger, 1989; Mobley *et al.*, 1995b; Burne and Chen, 2000; Ciurli, 2007). This chapter will focus on urease activity in bacteria, where it has an experimentally demonstrated role in response to acid stress.

Differentiating the use of urease for acid resistance versus nitrogen utilization can be tested simply, using acid growth conditions in the presence or absence of urea with another nitrogen source present. If the addition of urea does not enhance acid resistance, then urease likely does not contribute to acid resistance. When genetically tractable, deletion of the genes for urease catalytic subunits will also provide an indication of their requirement. The transcriptional regulation of urease gene expression can also provide significant information regarding the biological role of ureolysis in a particular bacterial species. Different mechanisms of urease regulation will be a major focus of a later section of this chapter.

A subset of ureolytic bacteria colonizes different acidic environments within the human host (stomach, mouth, macrophage, bladder). Acidification in these environments occurs by different mechanisms. Bacteria from these different growth niches show variations in the catalytic properties of the urease enzyme, as well as in the mechanisms of transcriptional, post-transcriptional and post-translational regulation. The most prominent example for the role of urease in acid survival is for *Helicobacter pylori*, a Gram-negative pathogen of the human stomach. Consequently, the discussion will emphasize properties of the *H. pylori* urease and how its activity is controlled to permit growth in the acidic milieu of the stomach. Ureases from other bacteria, including both Gram-negative (*Yersinia*) and Gram-positive (*Streptococcus*

and *Mycobacterium*) bacteria, have also been shown to be important for survival under acid stress conditions in different growth environments. Differences in the catalytic properties of these enzymes, as well as the regulation of their expression and activity, provide an instructive basis for comparing and contrasting the ways in which urea hydrolysis contributes to the acid stress response.

8.2.1 The enzymatic properties of microbial ureases

Urease (urea aminohydrolase; EC 3.5.1.5) catalyses the hydrolysis of urea, a non-redox reaction, to generate ammonia (NH_3) and carbamate. Non-catalysed hydrolysis of carbamate generates a second equivalent of ammonia and one molecule of carbon dioxide (CO_2). Both NH_3 and CO_2 are in equilibrium with protons. The zinc-dependent carbonic anhydrase increases the rate of equilibration of CO_2 with bicarbonate ($CO_2 + H_2O \Leftrightarrow HCO_3^- + H^+ \Leftrightarrow H_2CO_3$). In contrast to the reactions catalysed by most of the known nickel-enzymes (Maroney, 1999), the urease reaction is a non-redox process and the Ni(II) ions are neither reduced nor oxidized. In principle, nickel-enzymes that use redox chemistry to couple proton reduction to product formation could buffer acid stress (Table 8.1). However, these reactions require the presence of substrates that are not always abundant or that present their own stress. For example, the H_2-evolving Ni-Fe hydrogenase-3 isozyme of *Escherichia coli* has been shown recently to facilitate survival under anaerobic acid stress conditions (Noguchi *et al.*, 2010). This activity requires the presence of reducing equivalents (e^-) and

Table 8.1. Nickel-dependent enzymes involved in microbial stress responses.

Enzyme	Reaction catalysed	Physiological role
Urease	$(NH_2)_2CO + H_2O \Leftrightarrow 2NH_3 + CO_2$	*N*-metabolism; acid stress
Ni-Fe hydrogenase	$2H^+ + 2e^- \Leftrightarrow H_2$	H_2 or formate oxidation
Glyoxalase	Methylglyoxal + GSH $\Leftrightarrow$ *S*-lactoyl glutathione	Methylglyoxal detoxification
Ni-superoxide dismutase	$2O_2^{\bullet-} + 2H^+ \Leftrightarrow O_2 + H_2O_2$	Superoxide detoxification

the absence of O_2. Similarly, NiSOD (superoxide dismutase), which is found in a limited number of bacterial species, could combat acid stress but requires the presence of the highly reactive superoxide molecule, which generates its own form of stress (see Imlay, 2008; Imlay and Hassett, Chapter 1, this volume).

The isolation of urease from plants was first described in 1926 in a classic study by Sumner (Sumner, 1926). However, it was several decades before the essential metal was identified as nickel (Dixon *et al.*, 1975). Urease was the first example of a nickel-dependent enzyme and thus indicated that nickel was a biologically required metal rather than a toxic element. Ureases are hetero-oligomeric proteins, typically composed of three different polypeptides (α, β, γ) that form higher order structures ($\alpha\beta\gamma)_3$. One exception to this arrangement is the *H. pylori* urease, wherein the genes encoding β and γ subunits are fused, resulting in a single polypeptide. Additionally, the *H. pylori* $\alpha\beta$ heterodimer forms a dodecamer structure (Ha *et al.*, 2001; Pinkse *et al.*, 2003).

Urea is hydrolysed slowly in water ($t_{1/2}$ ~3.6 y), with no change in rate from pH 2–12 (Zerner, 1991). The addition of nucleophiles, such as Tris buffer, will alter that rate. Urease catalysis proceeds with a half-life of approximately 1 ms. This represents a rate enhancement of ~10^{14}-fold, although from a strict mechanistic perspective this value is incorrect because the water versus urease catalysed hydrolysis proceeds through different chemical mechanisms (Stewart and Benkovic, 1995). Nevertheless, the rate enhancement enables a response to acid stress on a biologically relevant timescale. The observed rate of urea hydrolysis in the cell will depend on the amount of active urease, the Michaelis constant (K_m) for urea and the urea concentration. Thus, the urease expression level and its intrinsic catalytic properties will contribute significantly to the rate of ammonia production under a given growth condition. For example, a limiting concentration of urease with a low K_m will be saturated easily with substrate and thus be suited poorly to the high flux that would be required for survival under prolonged acid stress (Fersht, 1998).

The active site residues required for urea hydrolysis come from the α-subunit. These residues are required to coordinate the two nickel ions or assist in substrate binding during the catalytic cycle. The nickel ions coordinate a water molecule and activate it for urea hydrolysis. Active urease requires assembly proteins that deliver nickel to the active site of the assembly catalytic subunits (Kuchar and Hausinger, 2004). This process will be discussed later in the chapter in the context of regulation of urease activity.

The bimetallic urease active site has been observed in structurally non-homologous proteins. The general prevalence of this metal coordination suggests that it is not specific to either nickel ions or urease. Rather, it is generally well suited to metal ion-based catalysis in a variety of contexts. The structural and mechanistic basis for metal ion selectivity in these different enzymes has not been elucidated. Interestingly, recent reports indicate the discovery of a non-nickel urease (Stoof *et al.*, 2008), which will be discussed later in this chapter .

Ureases exhibit a pH optimum for their activity (Table 8.2). Because urea binding is constant as a function of pH, as observed for

Table 8.2. Microbial ureases and their catalytic properties.

Organism	K_m (mM)	pH optimum[a]	Acid/N-metabolism
H. pylori	0.17	7.5	Yes/unknown
K. pneumoniae	2.8	7.75	No/yes
S. salivarius	nd[b]	7.5	Yes/yes
M. tuberculosis	0.3	7	Yes/yes
Y. enterolitica	1.7	5.5	Yes/yes

Note: [a]The buffer in which urease activity is measured can affect this determination due to pH-dependent inhibition of the urease by some buffer salts, i.e. phosphate at acidic pH (Mobley and Hausinger, 1989); [b]not determined.

the enzymes from *Klebsiella aerogenes* and *Proteus mirabilis* (Todd and Hausinger, 1987; Breitenbach and Hausinger, 1988), the pH dependence of catalysis reflects the sensitivity of k_{cat}, and thus k_{cat}/K_m to protons. The pH-dependent rate profile is indicative of the presence of general acid/base catalytic groups in the active site. Urease can hydrolyse other amide-containing molecules to generate ammonia, but the catalytic parameters for these reactions are much less impressive compared to urea as a substrate.

There are acid stable ureases that exhibit a shift in pH optimum (Young *et al.*, 1996). The structural basis for this shift in catalytic activity is not understood, as the sequences of the proteins are generally similar. However, it seems likely that the electrostatic properties of the active site are altered in some way to lower the pK_a of the general acid and base catalytic groups.

8.2.2 Urea production in mammals

Urea is produced in large quantities by mammals to excrete excess nitrogen. The five-reaction enzymatic cycle that produces urea was discovered in 1932 (Krebs and Henseleit, 1932; Shambaugh, 1977). The two nitrogen atoms in urea come from ammonia (derived from glutamate) and the amino group of aspartate, with CO_2 providing the central carbon. The initial reactions occur in the liver mitochondrion and the final three steps take place in the cytosol. After production in the liver, urea circulates as a component of plasma (<10 mM). Saliva is a source of urea for oral and gastrointestinal bacteria (Stewart and Smith, 2005) in a variety of mammals, particularly ruminants. Urea is thought to be freely diffusible across the cell membrane, although evidence of urea channels has been mentioned in the literature (Sachs *et al.*, 2005). Urea is concentrated in the kidney for excretion in the urine and reaches much higher levels (0.5 M) (Griffith *et al.*, 1976). Ureolytic bacteria, which are associated with urinary tract infections (e.g. *Proteus mirabilis* and *Ureaplasma urealyticum*), cause an elevation of the mildly acidic pH that can result in precipitation of phosphate salts with

either manganese (struvite) or calcium (carbonate apatite) (Griffith *et al.*, 1976; Lerner *et al.*, 1989). This pathology is known as urolithiasis.

Other higher organisms (fish and birds) do not use urea to eliminate excess nitrogen, although urea can be used to cycle nitrogen and may be present in these species. Thus, bacterial ureases will show restricted roles in acid stress responses depending on the metabolic cycle used to eliminate excess nitrogen of the host organism. The high concentration of urea in the urine of mammalian and other terrestrial vertebrates means it will also be available to soil-dwelling ureolytic microbes that do not associate with ureogenic animals.

8.3 *Helicobacter pylori* Urease and Growth in the Gastric Environment

The best-known example of the role of urease in an acid stress response comes from the identification of the flagellated spiral Gram-negative bacterium, *H. pylori*, as a resident of the human stomach. The discovery of *H. pylori* and related *Helicobacter* species, their host range and other important physiological characteristics of this genus are described in excellent detail elsewhere (Mobley *et al.*, 2001). There are two main classifications of this genus based on growth habit (Solnick and Vandamme, 2001). The gastric *Helicobacter* colonize the stomach of a variety of mammals, and likely many more diverse higher organisms. The enterohepatic *Helicobacter* are found primarily in the intestinal tract and hepatobiliary system, both less acidic environments. These different growth niches are reflected in the presence of urease in the two groups. Gastric *Helicobacter* are uniformly urease positive, while the enterohepatic *Helicobacter* show a variable presence of urease. It is also notable that the presence of urease in enterohepatic *H. hepaticus* is not sufficient for growth at low pH, but aids in withstanding more moderate acid shock (Belzer *et al.*, 2005).

Growth under acidic conditions results in the upregulation of numerous *H. pylori* genes (McGowan *et al.*, 2003; Merrell *et al.*, 2003; Wen *et al.*, 2003; Bury-Mone *et al.*, 2004; Pflock *et al.*, 2006; Scott *et al.*, 2007), indicating that urease is not the sole determinant of growth in acid or the gastric milieu. These genes include those that express proteins with functions in chemotaxis and motility, cell envelope, energy metabolism and general stress responses. These activities will not be discussed further in this chapter. The various acid-induced transcriptional and physiological changes in *H. pylori* have been the subject of regular updates (Sachs *et al.*, 2000, 2003, 2005, 2006; Scott *et al.*, 2002; Stingl and de Reuse, 2005).

8.3.1 The importance of regulating urease activity

Urease activity must be regulated because the unchecked production of NH_3 at neutral pH will lead to alkalinization of the cell and subsequent death. It is also reactive and, in higher organisms, can have other effects on numerous metabolic processes (Cagnon and Braissant, 2007). Under nitrogen-limiting conditions, NH_3/NH_4^+ can be used for anabolic pathways. Indeed, in bacteria that do not use urease for an acid stress response, regulation of urease gene expression is under the control of the NtrC response regulator (Collins *et al.*, 1993). It has been observed that radiolabelled urea can be used by gastro-intestinal bacteria to generate amino acids that are absorbed by the intestine (Metges, 2000). Other ureases are inducible; for example, the UreR activator of *P. mirabilis* is an AraC-type regulator that upregulates urease gene expression in response to urea availability (Nicholson *et al.*, 1993; D'Orazio *et al.*, 1996). UreR is essential for *P. mirabilis* virulence in the bladder (Dattelbaum *et al.*, 2003). In many other cases, urease expression is constitutive and other means are used to regulate its activity. *H. pylori* urease provides the most striking example of post-transcriptional regulation of activity, as up to 10–15% of the total protein is composed of the UreAB subunits (Bauerfeind *et al.*, 1997). Much of this urease lacks nickel and is not active at neutral pH, otherwise the cytoplasm would rapidly become alkaline in the presence of urea.

8.3.2 The role of extracellular urease in *H. pylori*

The means by which urease contributes to acid resistance in *H. pylori* has been the subject of intense debate in the past. The initial controversy centred on the cell localization of the urease that contributed to acid acclimation. *H. pylori* ureases, as well as other microbially expressed ureases, are known to be found extracellularly and urea hydrolysis can be detected. Additionally, the urease catalytic subunits are immunogenic and *H. pylori*-positive individuals have sera with reactivity towards urease catalytic subunits (Xiang *et al.*, 1995; Kimmel *et al.*, 2000). It was proposed, based on additional experiments, that extracellular urease was responsible for acid resistance (Krishnamurthy *et al.*, 1998). However, there were several concerns regarding the contribution of extracellular urease to acid resistance. First, there was no good mechanism to explain the transport of *H. pylori* urease to the cell surface. The large size of the urease dodecamer would require a novel transport mechanism. In the absence of such a transport mechanism it was proposed that a subpopulation of cells exhibited an altruistic behaviour in which cell lysis elicited extracellular urease (Phadnis *et al.*, 1996). Second, the pH optimum for *H. pylori* urease catalysis was inconsistent with prolonged stability under extracellular acid stress conditions. It was also proposed that the unique quaternary structure of the *H. pylori* urease would contribute to acid stability (Ha *et al.*, 2001). Another problem with the extracellular urease model was that it did not provide a direct way to buffer the cytoplasm, and a mechanism for NH_3 import would be required. Several careful experiments demonstrated that intracellular urease was required for acid tolerance (Rektorschek *et al.*, 2000; Stingl *et al.*, 2001; Scott *et al.*, 2002). Most notably, the inner membrane urea channel (UreI) was shown to be essential for survival in acid and gastric colonization (Skouloubris *et al.*, 1998; Scott *et al.*, 2000; Weeks *et al.*, 2000). These data do not provide any indication of a biological role for extracellular urease in acid resistance, so its functional role remains ambiguous and difficult to test.

8.3.3 UreI

Urea must enter the cell in order for any intracellular urease activity to be useful in combating acid stress. For simple diffusion, this process will depend on the extracellular urea concentration. At normal physiological concentrations (<10 mM) this process is likely to be insufficient for buffering intracellular acidity. Several pathogenic bacteria that use urease for acid resistance express proteins required for urea transport (Weeks *et al.*, 2000; Sebbane *et al.*, 2002). Such transporters are not typically found in microbes that use urea solely as a nitrogen source. The pH-gated UreI urea channel has been studied extensively in *H. pylori*. UreI is an inner membrane protein with six-transmembrane helices. Deletion of *ureI* results in a loss of acidic tolerance and the ability of *H. pylori* to colonize the stomach. Experiments have shown that UreI is a pH-gated channel that allows urea entry only under acidic conditions, which leads to increased urease expression. These observations provide additional support for the role of intracellular rather than extracellular urease for acid tolerance. The pH gating relies on a single histidine residue that is predicted to be located in a periplasmic loop region of the protein (Bury-Mone *et al.*, 2001; Weeks and Sachs, 2001). The imidazole group of histidine residues has a pK_a value of between 6 and 7, depending on their local environment. Thus, His residues are well suited to a titratable acidic pH-gating mechanism. In this case, deprotonation of the His residue inactivates the channel by stabilizing a closed conformation. Interestingly, this His residue is not conserved in the UreI channel from *S. salivarius* (discussed below), so in that case the channel is active under all conditions (Weeks *et al.*, 2004).

Recently, a role for UreI in the diffusion of urease end products (NH_3, NH_4^+ and CO_2) has been suggested based on experimental studies (Scott *et al.*, 2010). These data suggest a critical role for UreI in allowing rapid equilibration of both substrate and products across the membrane. Once sufficient NH_4^+ accumulates in the periplasm, the pH will rise and the UreI channel will close and urease activity will diminish.

8.3.4 Transcriptional regulation of urease expression

Urease is expressed constitutively at neutral pH in *H. pylori*. There are very few examples of transcriptional regulation in *H. pylori*. None the less, there is evidence for upregulation of *ureAB* gene expression under acid conditions. This regulation is complex and depends on several proteins that respond to different stimuli.

NikR and Fur

The first protein found to regulate *ureAB* transcription was NikR. This protein was discovered first in *E. coli* (De Pina *et al.*, 1999; Chivers and Sauer, 2000; Rowe *et al.*, 2005), where it was a nickel-responsive repressor of genes encoding a specific nickel transporter. *H. pylori* NikR also functions as a repressor but, unlike *E. coli*, it regulates the expression of numerous genes (Contreras *et al.*, 2003). It likely acts to activate *ureAB* expression because it binds upstream of the RNA polymerase binding site in P_{ureAB} (van Vliet *et al.*, 2001, 2002). It is also likely that NikR serves to upregulate *ureAB* expression in response to increased intracellular nickel as a means of ensuring there is sufficient binding capacity for the available metal.

The iron-responsive Fur protein is also involved in the acid stress response in *H. pylori* (Ernst *et al.*, 2005a; Gancz *et al.*, 2006), although it does not regulate urease expression directly. However, there is evidence for regulation of *nikR* expression (Bury-Mone *et al.*, 2004), and vice versa. Thus, Fur levels will affect *ureAB* expression indirectly via NikR. Another example of the overlap in the regulatory network is observed in the pH-dependent expression of two amidases involved in the *H. pylori* acid stress response (van Vliet *et al.*, 2004). Fur regulates the amidases directly, while NikR elicits an indirect effect on their expression via regulation of Fur.

ArsRS

The major *H. pylori* transcriptional response to acid stress is mediated by the ArsRS two-component system (Pflock *et al.*, 2004). ArsR is a classical response regulator that is phosphorylated by the ArsS sensor protein. There are several ArsR targets within the cell, including *ureAB* and urease assembly proteins (Pflock *et al.*, 2006; Loh *et al.*, 2010). Mutational analysis has shown that one His residue as well as other titratable side chains contribute to the ability of ArsS to sense pH (Muller *et al.*, 2009).

8.3.5 Post-transcriptional regulation of urease expression

H. pylori also exhibits post-transcriptional regulation of urease activity (Akada *et al.*, 2000). The urease catalytic subunits (UreAB) and urease assembly proteins (UreIEFGH; described below) are synthesized from different transcripts. Both mRNAs exhibit pH-dependent stability, with increased half-life under acidic conditions. The transcript encoding the assembly proteins displays greater sensitivity to higher pH. The increased stability of the second transcript at acidic pH allows for more synthesis of the assembly proteins, leading to greater urease activation. The mechanism of this pH dependence is unknown. However, a recent study of the *H. pylori* transcriptome may shed further light on this mechanism of regulation (Sharma *et al.*, 2010).

8.3.6 Post-translational control of urease assembly

The assembly of the active urease enzyme has attracted much interest as a model for understanding metalloenzyme assembly (Kuchar and Hausinger, 2004), as with the Ni-Fe hydrogenase. Studies of nickel insertion in urease have been carried out primarily on the *K. aerogenes* enzyme. More recent studies have been extended to the *H. pylori* enzyme, and some key differences are apparent. The nickel requirement defines an early step in urease assembly, after synthesis of the structural and accessory polypeptides. In the absence of nickel, the enzyme is not active

and this affords an opportunity for post-translational control of urease assembly.

The specific roles of the various accessory proteins in urease assembly are still being dissected. To date, most of the interest has focused on the UreE protein because of its nickel-binding properties compared to the other urease assembly proteins. The UreG protein is a GTP-binding protein, with weak GTPase activity in the isolated protein in the absence of its partner proteins. The UreF and UreD (called UreH in *H. pylori*) proteins have less well-defined roles. Intriguingly, the UreE protein has weak affinity (mM) for nickel by itself, but the bound nickel ion becomes resistant to strong chelators when it is in the presence of the UreFGH proteins (Soriano *et al.*, 2000). This result suggests the formation of a stable complex that is competent for nickel insertion, but that requires hydrolysis of the UreG-bound GTP to overcome the energetic barrier for nickel insertion.

The assembly mechanism is more complex in *H. pylori* due to the participation of two proteins, HypA and B, normally involved in Ni-Fe hydrogenase assembly (Olson *et al.*, 2001; Mehta *et al.*, 2003a,b). Ni-Fe hydrogenases also have a complex active site, more so than urease, and associated assembly proteins are known (Blokesch *et al.*, 2002; Li and Zamble, 2009; Kaluarachchi *et al.*, 2010). The Ni-Fe hydrogenase of *H. pylori* plays an essential role in pathogenesis, most likely via energy metabolism (Olson and Maier, 2002; Maier, 2005). The molecular basis for this shared assembly cohort is not yet known. *H. pylori* HypA is analogous to UreE and possesses a weak nickel-binding site. *H. pylori* HypB is a GTPase, similar to UreG, and may possess nickel-binding activity (Dias *et al.*, 2008). A recent proteomics study also demonstrated a multi-protein complex required for assembly (Stingl *et al.*, 2008). *H. pylori* HypA exhibits a pH- and nickel-dependent conformational change that may influence its participation in one or more of the nickel-enzyme assembly pathways (Herbst *et al.*, 2010).

Recently, it has been shown that the urease assembly complex is recruited to the inner face of the cytoplasmic membrane in response to acid pH (Scott *et al.*, 2010), where it interacts with UreI. This process depends on the sensor kinase, HP0244. The molecular mechanism by which the localization occurs has not been elucidated. The localization of urease assembly near the UreI channel provides an efficient way both to regulate its activity and to provide rapid access to urea and a means of ready exit for NH_3/NH_4^+ and CO_2.

The localization of the assembly complex near the site of entry of nickel ions into the cell may also impact access to newly imported nickel ions. Gene deletions of assembly complex participants result in NikR activation and decreased nickel activation (Benanti and Chivers, 2009). This observation suggests intense competition for free nickel ions in the cell. The mechanistic basis of this competition is not clear, and may not necessarily be based on relative affinities of the different protein for available nickel ions. Localization of nickel-enzyme assembly near a nickel transporter could serve to increase the local nickel concentration and prevent NikR from easy access.

8.3.7 Mechanisms of nickel uptake in *H. pylori*

A central requirement for urease activity is the availability of nickel ions. Transition metals are transported into bacterial cells by specific proteins or protein complexes (Ma *et al.*, 2009). Several structural classes of nickel transporter have been identified and all can be found in bacteria capable of expressing urease, although individual species generally have only one or two of the structural classes (Zhang *et al.*, 2009). Nickel uptake is either ATP dependent (primary transport, such as *E. coli* NikABCDE) or relies on a chemical potential and co-transport via symport or antiport, or protons or cations (secondary transport, such as *H. pylori* NixA). There is a crucial distinction between these mechanisms, as the collapse of the proton motive force will affect secondary transporters immediately, whereas an ATP-dependent transporter can use chemical energy to import nickel ions, at least initially, before ATP pools are depleted. Specific transporters typically display affin-

ities for nickel that are consistent with its availability in most growth environments (K_m $\leq$ 10 nM). Metal transporter expression is usually regulated by nickel availability to prevent excess accumulation of metals. Nickel transporter expression can also be coupled with conditions that favour nickel-enzyme expression (Rowe *et al.*, 2005).

The full complement of proteins required for nickel acquisition in *H. pylori* has not been elucidated. The NixA protein, a secondary transporter, was identified via genetic complementation in *E. coli* (Mobley *et al.*, 1995a). Deletion of *nixA* does not eliminate completely the ability of *H. pylori* to assemble active urease, nor is the transporter essential for host colonization (Nolan *et al.*, 2002). These observations suggest the presence of an additional import mechanism. Studies of nickel-dependent gene regulation by NikR identified a gene encoding an outer membrane receptor (HP1512) as the likely candidate for involvement in nickel uptake (Davis *et al.*, 2006; Ernst *et al.*, 2006). Experimental characterization of HP1512 has revealed it is involved in nickel uptake, in a manner analogous of siderophore receptors (Schauer *et al.*, 2007). This suggests the import of a nickel-ligand complex, the identity of which remains unknown. Whether it is already present in the environment or is synthesized by *H. pylori* in response to nickel deficiency is not yet known. If FrpB4 imports a nickel complex, it must be dissociated before it can be imported by NixA in the absence of another transporter. Candidates for ABC-type transporters have been studied (Hendricks and Mobley, 1997; Davis and Mobley, 2005), but eliminating their function does not account for the nickel uptake seen in the absence of NixA. It is possible that there are growth conditions that influence expression of the unidentified nickel transporter, thus hampering its discovery.

8.4 Urease and Oral Health – *Streptococcus salivarius*

Streptococci are Gram-positive bacteria that generate acid metabolic end products, such as lactic acid. Several different species of

Streptococci are known to be part of the oral microflora. One of these species, *S. salivarius*, is ureolytic (Sissons *et al.*, 1988). Several studies of *S. salivarius* urease and its regulation implicate it in maintaining normal dental microflora and preventing the chronic acidification that leads to dental caries (Burne and Marquis, 2000). *S. salivarius* urease has a neutral pH optimum (Chen *et al.*, 1996), indicating that the intracellular pH is maintained in a more neutral range. The K_m of this urease is roughly 3 mM, consistent with the salivary urea concentration (3–10 mM). The presence of urease is essential for the urea-dependent acid stress response, but it is also required when urea provides the sole source of nitrogen (Chen *et al.*, 2000).

The gene cluster encoding the urease operon has some notable features. All of the genes required for urea uptake (*ureI*), urease assembly and catalysis (*ureABCEFGD*) and nickel uptake (*ureMQO*) are expressed from a single promoter (Chen *et al.*, 1998; Chen and Burne, 2003). Thus, urea uptake (UreI), urease synthesis (UreABCEFGD) and nickel transport (UreMQO) are all regulated in the same manner. There is no evidence for nickel-dependent regulation of the nickel transport, as has been observed or predicted in many other nickel-utilizing bacteria (De Pina *et al.*, 1999; Chivers and Sauer, 2000; Ernst *et al.*, 2005b; Ahn *et al.*, 2006; Rodionov *et al.*, 2006). Instead, the regulation of expression is dominated by acidic pH (Sissons *et al.*, 1990; Chen *et al.*, 1998; Li *et al.*, 2000). The acid-responsive gene expression depends on two upstream *cis*-elements in the region 5′ to *ureI* (Chen *et al.*, 2002). This regulation has been speculated to be part of a general acid stress response. There is also evidence that carbohydrate levels can induce urease expression further, but only under acidic conditions (Chen and Burne, 1996). The mechanistic basis for carbohydrate regulation has not been identified, but this is consistent with the organism tuning its pH-homeostatic machinery to the potential for generating further acidity. As noted above, the UreI of *S. salivarius* is not pH gated like that of *H. pylori* (Weeks *et al.*, 2004), indicating that control of urea import rests solely on transcriptional regulation of *ureI* expression.

S. salivarius urease activity is likely beneficial for oral health by reducing acidity and allowing the growth of beneficial acid-sensitive bacteria. In an insightful study, the urease gene cluster of *S. salivarius* was expressed heterologously in *S. mutans* (Clancy and Burne, 1997; Clancy *et al.*, 2000). *S. mutans* is a non-ureolytic species that is tolerant of acid growth conditions and is associated with the change of microflora associated with the formation of dental caries. When expressing *S. salivarius* urease, *S. mutans*-infected germ-free rats showed markedly improved dental health compared to the parent strain under normally cariogenic dietary conditions (Clancy *et al.*, 2000).

8.5 Urease and the *Yersinia* sp.

Yersiniae are pathogenic bacteria that encounter acid stress during passage through the stomach. The urease of *Yersinia* is distinct because it has an acidic pH optimum (Table 8.1). This was determined by examining the urease activity in permeabilized cells across a wide pH range (1.5–7). The optimum activity was seen at pH 5.5 (Young *et al.*, 1996), leading to the hypothesis that the regulation of *Yersinia* urease activity was controlled at least partially by activity. Thus, as the cytoplasm becomes more acidic, the urease activity increases and buffers the decrease in pH. The acidic shift in the maximum activity of *Yersinia* and related ureases is likely due to sequence-dependent modulation of the pKa values of the catalytic residues, which are identical to those found in the neutral optimum activity ureases. The *Yersinia* urease can function at neutral pH, as demonstrated by studies of a wild-type and urease mutant strain grown on urea as a sole nitrogen source (Young *et al.*, 1996). Only the wild-type strain was able to grow.

Interestingly, *Yersinia* urease activity in whole cells does not result in the alkalinization of the growth medium, suggesting that the buffering effect is constrained to the cytoplasm. This behaviour differs from *H. pylori* and *S. salivarius* that contain the UreI channel and neutralize their surrounding growth environment. *Yersinia* express a urea channel that is structurally distinct from UreI (Sebbane *et al.*, 2002) and is not pH gated. Thus, the properties of *Yersinia* urease, along with its mode of urea uptake, are consistent with the use of urease to survive but not grow in acid conditions.

Recent data from *Yersinia pseudotuberculosis* indicate that OmpR acts as a transcriptional activator of expression in response to acid stress (Hu *et al.*, 2009). Cells lacking *ompR* were unable to survive an acidic challenge in the presence of urea but were unaffected in growth at neutral pH. The regulation appears to be direct as the OmpR protein was shown to bind directly to fragments corresponding to predicted promoter regions.

8.6 Mycobacterium and Urease

Many *Mycobacterium* species are capable of expressing a urease. Interestingly, this urease probably plays a role in pathogenesis, wherein it causes alkalinization of the acidic phagocyte, preventing fusion with the lysosome (Gordon *et al.*, 1980). The kinetic properties of the enzyme are similar to the *H. pylori* enzyme with a neutral pH optimum and a low K_m (Clemens *et al.*, 1995), which is consistent with the lesser abundance of urea in the phagocyte. Evidence for the role of *Mycobacterium* urease in acid stress response versus nitrogen metabolism came from studies in which the enzyme was attached to beads and produced the same phenotypic effect as infection with cells (Sendide *et al.*, 2004). Thus, the activity of this urease interferes with pH-dependent macrophage signalling and allows bacterial propagation.

8.7 *Helicobacter mustelae* Urease – A Non-nickel-requiring Urease

An interesting variation on the metal dependence of urease activity has emerged from studies of *Helicobacter* species that colonize carnivores, as opposed to omnivores like humans. Nickel is obtained primarily from plants, roots and nuts. All of these are components of a diet unlike that of carnivores known to be colonized by *Helicobacter* sp.

Intriguingly, these species contain a second urease (UreA2B2). This urease shows nickel-dependent repression and iron-dependent activation of expression (Stoof *et al.*, 2008). Additionally, the *ureA2B2* gene cluster lacks the accessory proteins required for nickel site assembly. Deletion of the normal urease assembly proteins has no effect on UreA2B2 activity. Furthermore, the urease activity of strains expressing only UreA2B2 is sensitive to lysis, suggesting an active site that is labile to oxygen. These observations have led to the intriguing hypothesis that urease activity is coupled tightly to the prevailing dietary availability of metal ions. A similar Ni/Fe relationship has been observed for SOD activity in *Streptomyces* (Kim *et al.*, 1996, 1998a,b; Ahn *et al.*, 2006).

8.8 Summary

Urease has a well-established role in bacterial responses to acidic extracellular pH. The biochemical basis for this activity is simple – hydrolysis of urea leads to the production of two NH_3 molecules and one CO_2 molecule. Both of these entities can absorb protons and buffer the local pH. The regulation of urease activity is complex and reflects the growth niche of the bacterium. Uncovering the mechanisms of regulation has been illuminating from the perspectives of molecular biology and cell physiology. The recent discovery of a non-nickel-utilizing urease adds a further level of interest to the studies of this venerable enzyme.

References

Ahn, B.E., Cha, J., Lee, E.J., Han, A.R., Thompson, C.J. and Roe, J.H. (2006) Nur, a nickel-responsive regulator of the Fur family, regulates superoxide dismutases and nickel transport in *Streptomyces coelicolor*. *Molecular Microbiology* 59, 1848–1858.

Akada, J.K., Shirai, M., Takeuchi, H., Tsuda, M. and Nakazawa, T. (2000) Identification of the urease operon in *Helicobacter pylori* and its control by mRNA decay in response to pH. *Molecular Microbiology* 36, 1071–1084.

Bauerfeind, P., Garner, R., Dunn, B.E. and Mobley, H.L. (1997) Synthesis and activity of *Helicobacter pylori* urease and catalase at low pH. *Gut* 40, 25–30.

Belzer, C., Stoof, J., Beckwith, C.S., Kuipers, E.J., Kusters J.G. and van Vliet, A.H. (2005) Differential regulation of urease activity in *Helicobacter hepaticus* and *Helicobacter pylori*. *Microbiology* 151, 3989–3995.

Benanti, E.L. and Chivers, P.T. (2009) An intact urease assembly pathway is required to compete with NikR for nickel ions in *Helicobacter pylori*. *Journal of Bacteriology* 189, 2405–2408.

Blokesch, M., Paschos, A., Theodoratou, E., Bauer, A., Hube, M., Huth, S. and Böck, A. (2002) Metal insertion into NiFe-hydrogenases. *Biochemical Society Transactions* 30, 674–680.

Booth, I.R. (1985) Regulation of cytoplasmic pH in bacteria. *Microbiology Reviews* 49, 359–378.

Breitenbach, J.M. and Hausinger, R.P. (1988) *Proteus mirabilis* urease. Partial purification and inhibition by boric acid and boronic acids. *Biochemistry Journal* 250, 917–920.

Burne, R.A. and Chen, Y.Y. (2000) Bacterial ureases in infectious diseases. *Microbes and Infection* 2, 533–542.

Burne, R.A. and Marquis, R.E. (2000) Alkali production by oral bacteria and protection against dental caries. *FEMS Microbiology Letters* 193, 1–6.

Bury-Mone, S., Skouloubris, S., Labigne, A. and de Reuse, H. (2001) The *Helicobacter pylori* UreI protein: role in adaptation to acidity and identification of residues essential for its activity and for acid activation. *Molecular Microbiology* 42, 1021–1034.

Bury-Mone, S., Thiberge, J.M., Contreras, M., Maitournam, A., Labigne, A. and de Reuse, H. (2004) Responsiveness to acidity via metal ion regulators mediates virulence in the gastric pathogen *Helicobacter pylori*. *Molecular Microbiology* 53, 623–638.

Cagnon, L. and Braissant, O. (2007) Hyper-ammonemia-induced toxicity for the developing central nervous system. *Brain Research Reviews* 56, 183–197.

Chen, Y.Y. and Burne, R.A. (1996) Analysis of *Streptococcus salivarius* urease expression using continuous chemostat culture. *FEMS Microbiology Letters* 135, 223–229.

Chen, Y.Y. and Burne, R.A. (2003) Identification and characterization of the nickel uptake system for urease biogenesis in *Streptococcus salivarius* 57.I. *Journal of Bacteriology* 185, 6773–6779.

Chen, Y.Y., Clancy, K.A. and Burne, R.A. (1996) *Streptococcus salivarius* urease: genetic and biochemical characterization and expression in a dental plaque streptococcus. *Infection and Immunity* 64, 585–592.

Chen, Y.Y., Weaver, C.A., Mendelsohn, D.R. and Burne, R.A. (1998) Transcriptional regulation of the *Streptococcus salivarius* 57.I urease operon. *Journal of Bacteriology* 180, 5769–5775.

Chen, Y.Y., Weaver, C.A. and Burne, R.A. (2000) Dual functions of *Streptococcus salivarius* urease. *Journal of Bacteriology* 182, 4667–4669.

Chen, Y.Y., Betzenhauser, M.J. and Burne, R.A. (2002) Cis-acting elements that regulate the low-pH-inducible urease operon of *Streptococcus salivarius*. *Microbiology* 148, 3599–3608.

Chivers, P.T. and Sauer, R.T. (2000) Regulation of high affinity nickel uptake in bacteria. Ni^{2+}-dependent interaction of NikR with wild-type and mutant operator sites. *Journal of Biological Chemistry* 275, 19735–19741.

Ciurli, S. (2007) Urease: recent insights on the role of nickel. In: Sigel, A., Sigel, H. and Sigel, R.K.O. (eds) *Metal Ions in the Life Sciences: Nickel and its Surprising Impact in Nature*, Vol 2. John Wiley and Sons Ltd, Chichester, UK, pp. 214–278.

Clancy, A. and Burne, R.A. (1997) Construction and characterization of a recombinant ureolytic *Streptococcus mutans* and its use to demonstrate the relationship of urease activity to pH modulating capacity. *FEMS Microbiology Letters* 151, 205–211.

Clancy, K.A., Pearson, S., Bowen, W.H. and Burne, R.A. (2000) Characterization of recombinant, ureolytic *Streptococcus mutans* demonstrates an inverse relationship between dental plaque ureolytic capacity and cariogenicity. *Infection and Immunity* 68, 2621–2629.

Clemens, D.L., Lee, B.Y. and Horwitz, M.A. (1995) Purification, characterization, and genetic analysis of *Mycobacterium tuberculosis* urease, a potentially critical determinant of host–pathogen interaction. *Journal of Bacteriology* 177, 5644–5652.

Collins, C.M., Gutman, D.M. and Laman, H. (1993) Identification of a nitrogen-regulated promoter controlling expression of *Klebsiella pneumoniae* urease genes. *Molecular Microbiology* 8, 187–198.

Contreras, M., Thiberge, J.M., Mandrand-Berthelot, M.A. and Labigne, A. (2003) Characterization of the roles of NikR, a nickel-responsive pleiotropic autoregulator of *Helicobacter pylori*. *Molecular Microbiology* 49, 947–963.

D'Orazio, S.E., Thomas, V. and Collins, C.M. (1996) Activation of transcription at divergent urea-dependent promoters by the urease gene regulator UreR. *Molecular Microbiology* 21, 643–655.

Dattelbaum, J.D., Lockatell, C.V., Johnson, D.E. and Mobley, H.L. (2003) UreR, the transcriptional activator of the *Proteus mirabilis* urease gene cluster, is required for urease activity and virulence in experimental urinary tract infections. *Infection and Immunity* 71, 1026–1030.

Davis, G.S. and Mobley, H.L. (2005) Contribution of DppA to urease activity in *Helicobacter pylori* 26695. *Helicobacter* 10, 416–423.

Davis, G.S., Flannery, E.L. and Mobley, H.L. (2006) *Helicobacter pylori* HP512 is a nickel-responsive NikR-regulated outer membrane protein. *Infection and Immunity* 74, 6811–6820.

De Pina, K., Desjardin, V., Mandrand-Berthelot, M.A., Giordano, G. and Wu, L.F. (1999) Isolation and characterization of the *nikR* gene encoding a nickel-responsive regulator in *Escherichia coli*. *Journal of Bacteriology* 181, 670–674.

Dias, A.V., Mulvihill, C.M., Leach, M.R., Pickering, I.J., George, G.N. and Zamble, D.B. (2008) Structural and biological analysis of the metal sites of *Escherichia coli* hydrogenase accessory protein HypB. *Biochemistry* 47, 11981–11991.

Dixon, N.E., Gazzola, T.C., Bakeley, R.L. and Zerner, B. (1975) Letter: Jack bean urease (EC 3.5.1.5). A metalloenzyme. A simple biological role for nickel? *Journal of the American Chemical Society* 97, 4131–4133.

Ernst, F.D., Bereswill, S., Waidner, B., Stoof, J., Mader, U., Kusters, J.G., Kuipers, E.J., Kist, M., van Vliet, A.H. and Homuth, G. (2005a) Transcriptional profiling of *Helicobacter pylori* Fur- and iron-regulated gene expression. *Microbiology* 151, 533–546.

Ernst, F.D., Kuipers, E.J., Heijens, A., Sarwari, R., Stoof, J., Penn, C.W., Kusters, J.G. and van Vliet, A.H. (2005b) The nickel-responsive regulator *nikR* controls activation and repression of gene transcription in *Helicobacter pylori*. *Infection and Immunity* 73, 7252–7258.

Ernst, F.D., Stoof, J., Horrevoets, W.M., Kuipers, E.J., Kusters, J.G. and van Vliet, A.H. (2006) NikR mediates nickel-responsive transcriptional repression of the *Helicobacter pylori* outer membrane proteins FecA3 (HP1400) and FrpB4 (HP1512). *Infection and Immunity* 74, 6821–6828.

Fersht, A.R. (1998) *Structure and Mechanism in Protein Science: A Guide to Enzyme Catalysis and Protein Folding*. W.H. Freeman and Company, New York.

Foster, J.W. (1999) When protons attack: microbial strategies of acid adaptation. *Current Opinion in Microbiology* 2, 170–174.

Gancz, H., Censini, S. and Merrell, D.S. (2006) Iron and pH homeostasis intersect at the level of fur regulation in the gastric pathogen *Helicobacter pylori*. *Infection and Immunity* 74, 602–614.

Gordon, A.H., Hart, P.D. and Young, M.R. (1980) Ammonia inhibits phagosome-lysosome fusion in macrophages. *Nature* 286, 79–80.

Griffith, D.P., Musher, D.M. and Itin, C. (1976) Urease. The primary cause of infection-induced urinary stones. *Investigations in Urology* 13, 346–350.

Ha, N.C., Oh, S.T., Sung, J.Y., Cha, K.A., Lee, M.H. and Oh, B.H. (2001) Supramolecular assembly and acid resistance of *Helicobacter pylori* urease. *Nature Structural Biology* 8, 505–509.

Hendricks, J.K. and Mobley, H.L. (1997) *Helicobacter pylori* ABC transporter: effect of allelic exchange mutagenesis on urease activity. *Journal of Bacteriology* 179, 5892–5902.

Herbst, R.W., Perovic, I., Martin-Diaconescu, V., O'Brien, K., Chivers, P.T., Pochapsky, S., Pochapsky, T. and Maroney, M.J. (2010) The communication between the zinc and nickel sites in dimeric HypA: metal recognition and pH sensing. *Journal of the American Chemical Society* 132, 10338–10351.

Hu, Y., Lu, P., Wang, Y., Ding, L., Atkinson, S. and Chen, S. (2009) OmpR positively regulates urease expression to enhance acid survival of *Yersinia pseudotuberculosis*. *Microbiology* 155, 2522–2531.

Imlay, J.A. (2008) Cellular defenses against superoxide and hydrogen peroxide. *Annual Reviews in Biochemistry* 77, 755–776.

Kaluarachchi, H., Chan Chung, K.C. and Zamble, D.B. (2010) Microbial nickel proteins. *Natural Products Reports* 27, 681–694.

Kim, E.J., Chung, H.J., Suh, B., Hah, Y.C. and Roe, J.H. (1998a) Expression and regulation of the *sodF* gene encoding iron- and zinc-containing superoxide dismutase in *Streptomyces coelicolor* muller. *Journal of Bacteriology* 180, 2014–2020.

Kim, E.J., Chung, H.J., Suh, B., Hah, Y.C. and Roe, J.H. (1998b) Transcriptional and post-transcriptional regulation by nickel of *sodN* gene encoding nickel-containing superoxide dismutase from *Streptomyces coelicolor* muller. *Molecular Microbiology* 27, 187–195.

Kim, F.J., Kim, H.P., Hah, Y.C. and Roe, J.H. (1996) Differential expression of superoxide dismutases containing Ni and Fe/Zn in *Streptomyces coelicolor*. *European Journal of Biochemistry* 241, 178–185.

Kimmel, B., Bosserhoff, A., Frank, R., Gross, R., Goebel, W. and Beier, D. (2000) Identification of immunodominant antigens from *Helicobacter pylori* and evaluation of their reactivities with sera from patients with different gastroduodenal pathologies. *Infection and Immunity* 68, 915–920.

Krebs, H.A. and Henseleit, K. (1932) Untersuchungen über die Harnstoffbildung im Tierkörper. *Hoppe Seyler's Zeitschrift für Physiologische Chemie* 210, 33–36.

Krishnamurthy, P., Parlow, M., Zitzer, J.B., Vakil, N.B., Mobley, H.L., Levy, M., Phadnis, S.H. and Dunn, B.E. (1998) *Helicobacter pylori* containing only cytoplasmic urease is susceptible to acid. *Infection and Immunity* 66, 5060–5066.

Kuchar, J. and Hausinger, R.P. (2004) Biosynthesis of metal sites. *Chemical Reviews* 104, 509–525.

Lerner, S.P., Gleeson, M.J. and Griffith, D.P. (1989) Infection stones. *Journal of Urology* 141, 753–758.

Li, Y. and Zamble, D.B. (2009) Nickel homeostasis and nickel regulation: an overview. *Chemical Reviews* 109, 4617–4643.

Li, Y.H., Chen, Y.Y. and Burne, R.A. (2000) Regulation of urease gene expression by *Streptococcus salivarius* growing in biofilms. *Environmental Microbiology* 2, 169–177.

Loh, J.T., Gupta, S.S., Friedman, D.B., Krezel, A.M. and Cover, T.L. (2010) Analysis of protein expression regulated by the *Helicobacter pylori* ArsRS two-component signal transduction system. *Journal of Bacteriology* 192, 2034–2043.

Ma, Z., Jacobsen, F.E. and Giedroc, D.P. (2009) Coordination chemistry of bacterial metal transport and sensing. *Chemical Reviews* 109, 4644–4681.

McGowan, C.C., Necheva, A.S., Forsyth, M.H., Cover, T.L. and Blaser, M.J. (2003) Promoter analysis of *Helicobacter pylori* genes with enhanced expression at low pH. *Molecular Microbiology* 48, 1225–1239.

Madigan, M.T., Martinko, J.M. and Parker, J. (2003) *Brock Biology of Microorganisms*. Prentice Hall, Upper Saddle River, New Jersey.

Maier, R.J. (2005) Use of molecular hydrogen as an energy substrate by human pathogenic bacteria. *Biochemical Society Transactions* 33, 83–85.

Maroney, M.J. (1999) Structure/function relationships in nickel metallobiochemistry. *Current Opinion Chemical Biology* 3, 188–199.

Mehta, N., Benoit, S. and Maier, R.J. (2003a) Roles of conserved nucleotide-binding domains in accessory proteins, HypB and UreG, in the maturation of nickel-enzymes required for efficient *Helicobacter pylori* colonization. *Microbial Pathogensis* 35, 229–234.

Mehta, N., Olson, J.W. and Maier, R.J. (2003b) Characterization of *Helicobacter pylori* nickel metabolism accessory proteins needed for maturation of both urease and hydrogenase. *Journal of Bacteriology* 185, 726–734.

Menzel, U. and Gottschalk, G. (1985) The internal-pH of *Acetobacterium wieringae* and *Acetobacter aceti* during growth and production of acetic acid. *Archives of Microbiology* 143, 47–51.

Merrell, D.S., Goodrich, M.L., Otto, G., Tompkins, L.S. and Falkow, S. (2003) pH-regulated gene expression of the gastric pathogen *Helicobacter pylori*. *Infection and Immunity* 71, 3529–3539.

Metges, C.C. (2000) Contribution of microbial amino acids to amino acid homeostasis of the host. *Journal of Nutrition* 130, 1857S–1864S.

Mobley, H.L. and Hausinger, R.P. (1989) Microbial ureases: significance, regulation, and molecular characterization. *Microbiology Reviews* 53, 85–108.

Mobley, H.L., Garner, R.M. and Bauerfeind, P. (1995a) *Helicobacter pylori* nickel-transport gene NixA: synthesis of catalytically active urease in *Escherichia coli* independent of growth conditions. *Molecular Microbiology* 16, 97–109.

Mobley, H.L., Island, M.D. and Hausinger, R.P. (1995b) Molecular biology of microbial ureases. *Microbiology Reviews* 59, 451–480.

Mobley, H.L., Mendz, G.L. and Hazell, S.L. (2001) *Helicobacter pylori – Physiology and Genetics*. ASM Press, Washington, DC.

Muller, S., Gotz, M. and Beier, D. (2009) Histidine residue 94 is involved in pH sensing by histidine kinase ArsS of *Helicobacter pylori*. *PLoS One* 4, e6930.

Nicholson, E.B., Concaugh, E.A., Foxall, P.A., Island, M.D. and Mobley, H.L. (1993) *Proteus mirabilis* urease: transcriptional regulation by UreR. *Journal of Bacteriology* 175, 465473.

Noguchi, K., Riggins, D.P., Eldahan, K.C., Kitko, R.D. and Slonczewski, J.L. (2010) Hydrogenase-3 contributes to anaerobic acid resistance of *Escherichia coli*. *PLoS One* 5, e10132.

Nolan, K.J., McGee, D.J., Mitchell, H.M., Kolesnikow, T., Harro, J.M., O'Rourke, J., Wilson, J.E., Danon, S.J., Moss, N.D., Mobley, H.L. and Lee, A. (2002) *In vivo* behavior of a *Helicobacter pylori* SS1 *nixA* mutant with reduced urease activity. *Infection and Immunity* 70, 685–691.

Olson, J.W. and Maier, R.J. (2002) Molecular hydrogen as an energy source for *Helicobacter pylori*. *Science* 298, 1788–1790.

Olson, J.W., Mehta, N.S. and Maier, R.J. (2001) Requirement of nickel metabolism proteins

HypA and HypB for full activity of both hydrogenase and urease in *Helicobacter pylori*. *Molecular Microbiology* 39, 176–182.

Pflock, M., Dietz, P., Schar, J. and Beier, D. (2004) Genetic evidence for histidine kinase HP165 being an acid sensor of *Helicobacter pylori*. *FEMS Microbiology Letters* 234, 51–61.

Pflock, M., Finsterer, N., Joseph, B., Mollenkopf, H., Meyer, T.F. and Beier, D. (2006) Characterization of the ArsRS regulon of *Helicobacter pylori*, involved in acid adaptation. *Journal of Bacteriology* 188, 3449–3462.

Phadnis, S.H., Parlow, M.H., Levy, M., Ilver, D., Caulkins, C.M., Connors, J.B. and Dunn, B.E. (1996) Surface localization of *Helicobacter pylori* urease and a heat shock protein homolog requires bacterial autolysis. *Infection and Immunity* 64, 905–912.

Pinkse, M.W., Maier, C.S., Kim, J.I., Oh, B.H. and Heck, A.J. (2003) Macromolecular assembly of *Helicobacter pylori* urease investigated by mass spectrometry. *Journal of Mass Spectrometry* 38, 315–320.

Rektorschek, M., Buhmann, A., Weeks, D., Schwan, D., Bensch, K.W., Eskandari, S., Scott, D., Sachs, G. and Melchers, K. (2000) Acid resistance of *Helicobacter pylori* depends on the UreI membrane protein and an inner membrane proton barrier. *Molecular Microbiology* 36, 141–152.

Rodionov, D.A., Hebbeln, P., Gelfand, M.S. and Eitinger, T. (2006) Comparative and functional genomic analysis of prokaryotic nickel and cobalt uptake transporters: evidence for a novel group of ATP-binding cassette transporters. *Journal of Bacteriology* 188, 317–327.

Rowe, J.L., Starnes, G.L. and Chivers, P.T. (2005) Complex transcriptional control links *nikABCDE*-dependent nickel transport with hydrogenase expression in *Escherichia coli*. *Journal of Bacteriology* 187, 6317–6323.

Sachs, G., Scott, D., Weeks, D. and Melchers, K. (2000) Gastric habitation by *Helicobacter pylori*: insights into acid adaptation. *Trends in Pharmacology Science* 21, 413–416.

Sachs, G., Weeks, D.L., Melchers, K. and Scott, D.R. (2003) The gastric biology of *Helicobacter pylori*. *Annual Reviews in Physiology* 65, 349–369.

Sachs, G., Weeks, D.L., Wen, Y., Marcus, E.A., Scott, D.R. and Melchers, K. (2005) Acid acclimation by *Helicobacter pylori*. *Physiology (Bethesda)* 20, 429–438.

Sachs, G., Kraut, J.A., Wen, Y., Feng, J. and Scott, D.R. (2006) Urea transport in bacteria: acid acclimation by gastric *Helicobacter* spp. *Journal of Membrane Biology* 212, 71–82.

Schauer, K., Gouget, B., Carriere, M., Labigne, A. and de Reuse, H. (2007) Novel nickel transport mechanism across the bacterial outer membrane energized by the Tonb/Exbb/Exbd machinery. *Molecular Microbiology* 63, 1054–1068.

Scott, D.R., Marcus, E.A., Weeks, D.L., Lee, A., Melchers K., and Sachs, G. (2000) Expression of the *Helicobacter pylori ureI* gene is required for acidic pH activation of cytoplasmic urease. *Infection and Immunity* 68, 470–477.

Scott, D.R., Marcus, E.A., Weeks, D.L., and Sachs, G. (2002) Mechanisms of acid resistance due to the urease system of *Helicobacter pylori*. *Gastroenterology* 123, 187–195.

Scott, D.R., Marcus, E.A., Wen, Y., Oh, J. and Sachs, G. (2007) Gene expression *in vivo* shows that *Helicobacter pylori* colonizes an acidic niche on the gastric surface. *Proceedings of the National Academy of Sciences of the United States of America* 104, 7235–7240.

Scott, D.R., Marcus, E.A., Wen, Y., Singh, S., Feng, J. and Sachs, G. (2010) Cytoplasmic histidine kinase (HP0244)-regulated assembly of urease with UreI, a channel for urea and its metabolites, CO_2, NH_3, and $NH_4(+)$, is necessary for acid survival of *Helicobacter pylori*. *Journal of Bacteriology* 192, 94–103.

Sebbane, F., Bury-Mone, S., Cailliau, K., Browaeys-Poly, E., de Reuse, H. and Simonet, M. (2002) The *Yersinia pseudotuberculosis* Yut protein, a new type of urea transporter homologous to eukaryotic channels and functionally interchangeable *in vitro* with the *Helicobacter pylori* UreI protein. *Molecular Microbiology* 45, 1165–1174.

Sendide, K., Deghmane, A.E., Reyrat, J.M., Talal, A. and Hmama, Z. (2004) *Mycobacterium bovis* BCG urease attenuates major histocompatibility complex class II trafficking to the macrophage cell surface. *Infection and Immunity* 72, 4200–4209.

Shambaugh, G. E. 3rd (1977) Urea biosynthesis I. The urea cycle and relationships to the citric acid cycle. *American Journal of Clinical Nutrition* 30, 2083–2087.

Sharma, C.M., Hoffmann, S., Darfeuille, F., Reignier, J., Findeiss, S., Sittka, A., Chabas, S., Reiche, K., Hackermuller, J., Reinhardt, R., Stadler, P.F. and Vogel, J. (2010) The primary transcriptome of the major human pathogen *Helicobacter pylori*. *Nature* 464, 250–255.

Sissons, C.H., Hancock, E.M., Perinpanayagam, H.E. and Cutress, T.W. (1988) The bacteria responsible for ureolysis in artificial dental plaque. *Archives in Oral Biology* 33, 727–733.

Sissons, C.H., Perinpanayagam, H.E., Hancock, E.M. and Cutress, T.W. (1990) pH regulation of urease levels in *Streptococcus salivarius*. *Journal of Dental Research* 69, 1131–1137.

Skouloubris, S., Thiberge, J.M., Labigne, A. and de Reuse, H. (1998) The *Helicobacter pylori* UreI protein is not involved in urease activity but is essential for bacterial survival *in vivo*. *Infection and Immunity* 66, 4517–4521.

Solnick, J. and Vandamme, P. (2001) Taxonomy of the helicobacter genus. In: Mobley, H.L., Mendz, G.L. and Hazell, S.L. (eds) *Helicobacter pylori – Physiology and Genetics*. ASM Press, Washington, DC, pp. 39–52.

Soriano, A., Colpas, G.J. and Hausinger, R.P. (2000) UreE stimulation of GTP-dependent urease activation in the UreD-UreF-UreG-urease apoprotein complex. *Biochemistry* 39, 12435–12440.

Stewart, G.S. and Smith, C.P. (2005) Urea nitrogen salvage mechanisms and their relevance to ruminants, non-ruminants and man. *Nutrition Research Reviews* 18, 4962.

Stewart, J.D. and Benkovic, S.J. (1995) Transition-state stabilization as a measure of the efficiency of antibody catalysis. *Nature* 375, 388–391.

Stingl, K. and de Reuse, H. (2005) Staying alive overdosed: how does *Helicobacter pylori* control urease activity? *International Journal of Medical Microbiology* 295, 307–315.

Stingl, K., Uhlemann Em, E.M., Deckers-Hebestreit, G., Schmid, R., Bakker, E.P. and Altendorf, K. (2001) Prolonged survival and cytoplasmic pH homeostasis of *Helicobacter pylori* at pH 1. *Infection and Immunity* 69, 1178–1180.

Stingl, K., Schauer, K., Ecobichon, C., Labigne, A., Lenormand, P., Rousselle, J.C., Namane, A. and de Reuse, H. (2008) *In vivo* interactome of *Helicobacter pylori* urease revealed by tandem affinity purification. *Molecular Cellular Proteomics* 7, 2429–2441.

Stoof, J., Breijer, S., Pot, R.G., van der Neut, D., Kuipers, E.J., Kusters J.G. and van Vliet, A.H. (2008) Inverse nickel-responsive regulation of two urease enzymes in the gastric pathogen *Helicobacter mustelae*. *Environmental Microbiology* 10, 2586–2597.

Sumner, J. (1926) The isolation and crystallization of the enzyme urease. *Journal of Biological Chemistry* 69, 435–441.

Todd, M.J. and Hausinger, R.P. (1987) Purification and characterization of the nickel-containing multicomponent urease from *Klebsiella aerogenes*. *Journal of Biological Chemistry* 262, 5963–5667.

van Vliet, A.H., Kuipers, E.J., Waidner, B., Davies, B.J., de Vries, N., Penn, C.W., Vandenbroucke-Grauls, C.M., Kist, M., Bereswill, S. and Kusters, J.G. (2001) Nickel-responsive induction of urease expression in *Helicobacter pylori* is mediated at the transcriptional level. *Infection and Immunity* 69, 4891–4897.

van Vliet, A.H., Poppelaars, S.W., Davies, B.J., Stoof, J., Bereswill, S., Kist, M., Penn, C.W., Kuipers, E.J. and Kusters, J.G. (2002) NikR mediates nickel-responsive transcriptional induction of urease expression in *Helicobacter pylori*. *Infection and Immunity* 70, 2846–2852.

van Vliet, A.H., Kuipers, E.J., Stoof, J., Poppelaars, S.W. and Kusters, J.G. (2004) Acid-responsive gene induction of ammonia-producing enzymes in *Helicobacter pylori* is mediated via a metal-responsive repressor cascade. *Infection and Immunity* 72, 766–773.

Weeks, D.L. and Sachs, G. (2001) Sites of pH regulation of the urea channel of *Helicobacter pylori*. *Molecular Microbiology* 40, 1249–1259.

Weeks, D.L., Eskandari, S., Scott, D.R. and Sachs, G. (2000) A H$^+$-gated urea channel: the link between *Helicobacter pylori* urease and gastric colonization. *Science* 287, 482–485.

Weeks, D.L., Gushansky, G., Scott, D.R. and Sachs, G. (2004) Mechanism of proton gating of a urea channel. *Journal of Biological Chemistry* 279, 9944–9950.

Wen, Y., Marcus, E.A., Matrubutham, U., Gleeson, M.A., Scott, D.R. and Sachs, G. (2003) Acid-adaptive genes of *Helicobacter pylori*. *Infection and Immunity* 71, 5921–5939.

White, D. (2007) *The Physiology and Biochemistry of Prokaryotes*. Oxford University Press, New York.

Xiang, Z., Censini, S., Bayeli, P.F., Telford, J.L., Figura, N., Rappuoli, R. and Covacci, A. (1995) Analysis of expression of CagA and VacA virulence factors in 43 strains of *Helicobacter pylori* reveals that clinical isolates can be divided into two major types and that CagA is not necessary for expression of the vacuolating cytotoxin. *Infection and Immunity* 63, 94–98.

Young, G.M., Amid, D. and Miller, V.L. (1996) A bifunctional urease enhances survival of pathogenic *Yersinia enterocolitica* and *Morganella morganii* at low pH. *Journal of Bacteriology* 178, 6487–6495.

Zerner, B. (1991) Recent advances in the chemistry of an old enzyme, urease. *Bioorganic Chemistry* 19, 116–131.

Zhang, Y., Rodionov, D.A., Gelfand, M.S. and Gladyshev, V.N. (2009) Comparative genomic analyses of nickel, cobalt and vitamin B$_{12}$ utilization. *BMC Genomics* 10, 78–103.

Part 4
Nutrient Stress

9 Secretion Systems and Metabolism in the Pathogenic Yersiniae

Matthew S. Francis

9.1 Prologue

The genus *Yersinia* comprises 11 species, three of which have clear aetiology for causing human disease (*Y. pestis, Y. pseudotuberculosis* and *Y. enterocolitica*) (Smego *et al.*, 1999). The obligate pathogen *Y. pestis* is the most infamous of these, being the causal agent of plague, a bivalent disease that when left untreated is invariably fatal. The life cycle of *Y. pestis* is complex, being dependent on two diverse hosts – the invertebrate flea, *Xenopsylla cheopis*, and a mammalian host (usually wild rodents) (Perry and Fetherston, 1997; Smego *et al.*, 1999; Prentice and Rahalison, 2007; Stenseth *et al.*, 2008). Although capable of catastrophic consequences, plague in humans is accidental – a consequence of being infected with *Y. pestis* via the bite of an infected flea that has been forced from its normal rodent host. In brief, the initial stage of disease presents as swollen lymph nodes (buboes) and is termed bubonic plague, whereas the second stage is a more vigorous systemic infection that results in bacterial colonization of multiple tissue organs, including the lung. This form of disease is termed pneumonic plague; a highly contagious disease that enables bacteria to spread to new hosts rapidly and effectively via aerosol droplets. In light of this, global health organizations routinely list *Y. pestis* as a category A biowarfare agent. On the other hand, *Y. pseudo-tuberculosis* and *Y. enterocolitica* are essentially environmental bacteria that are capable of causing spasmodic enteric disease (known as yersiniosis) outbreaks linked to the ingestion of contaminated food or fluids (Koornhof *et al.*, 1999; Naktin and Beavis, 1999; Smego *et al.*, 1999). While these diseases cause gastrointestinal discomfort, they are usually self-limiting and rarely associated with systemic disease. In certain susceptible individuals, however, chronic reactive arthritic sequelae can be attributed to these bacteria (Vahamiko *et al.*, 2005).

On account of their ability to cause human disease, a prolific amount of information is available that describes these three human pathogens with respect to their ecology, epidemiology and the pathogenesis of disease. In contrast, very little information is available concerning the additional *Yersinia* species (*Y. frederiksenii, Y. intermedia, Y. kristensenii, Y. bercovieri, Y. mollaretii, Y. rohdei, Y. ruckeri, Y. aleksiciae, Y. mexicana* and *Y. aldovae*) (Sulakvelidze, 2000). However, they still might be clinically relevant given their propensity to harbour a moderate number of genes that encode for products known to be associated with pathogenicity by other non-*Yersinia* bacteria (Loftus *et al.*, 2002; Sulakvelidze, 2000). It is at least well established that *Y. ruckeri* is the causative agent of yersiniosis in infected salmonid fish, although the pathogenic mechanisms are comparatively poorly understood (Fernandez et al., 2007a; Tobback et al., 2007).

In the first part of this chapter, the presence of known and suggested protein secretion mechanisms in the Yersiniae is described. Where appropriate, these individual processes are discussed briefly in the context of their contribution to bacterial pathogenesis to help the reader gain an understanding of their physiological importance in the various unique environments of an infected host. As considerably more is known about the pathogenic mechanisms of human pathogenic Yersinia, these examples will dominate the discussion. Then, focus will turn to the consequences of adaptation of pathogenic Yersinia to their surrounding environment. Where possible, emphasis will be given to the crosstalk between metabolism and the temporal and spatial regulatory control of these important secretion systems. This connection ensures that Yersinia conserve their valuable energy reserves to maximize their survival in stressful environments and only synthesize energetically expensive virulence determinants, such as multi-component secretion systems, when they will have utmost benefit during host infections.

9.2 Protein Secretion Systems in the Yersiniae

Protein secretion by Gram-negative bacteria is a prerequisite for numerous physiological processes including adhesion, invasion, killing of potential competitors (toxicity), degradation/hydrolysis, movement, provision of nutrients, cell-to-cell communication, detoxification of the environment and cell wall biosynthesis and protein quality control. To perform this multiple tasking, bacteria employ several 'one-step' and 'two-step' secretion pathways to localize diverse protein substrates either into or across the cell envelope (consisting of the inner and outer membranes and the intervening periplasm) (Lee and Schneewind, 2001; Saier, 2006). One-step secretion refers to protein delivery from the bacterial cytoplasm to the exterior without any periplasmic intermediate, while two-step protein delivery implies an intermediate

periplasmic phase. A schematic illustration summarizing the various Yersinia protein secretion systems is provided in Fig. 9.1 and their possible secreted substrates are listed in Table 9.1.

In pathogenic bacteria, secretion systems not only have implications for understanding decisive mechanisms of pathogenesis, but also are the target of novel approaches for the treatment of infectious diseases including the development of secretion-blocking antibiotics and heterologous vaccine delivery vehicles (Hahn and von Specht, 2003; Aberg and Almqvist, 2007; Keyser et al., 2008; Panthel et al., 2008). This is pertinent given the widespread overuse of antibiotics and the rapid emergence of antibiotic-resistant pathogens, including those resistant to multiple drugs commonly used in the clinic (Levy and Marshall, 2004). This is true also of the pathogenic Yersiniae, with reports of emerging strains harbouring resistance to multiple antibiotics beginning to appear (Galimand et al., 2006).

9.2.1 Secretion across the inner (cytoplasmic) membrane

sec-dependent secretion

Sec-dependent export of proteins across the inner membrane is a universal and essential mechanism ubiquitous in all living organisms (Cao and Saier, 2003). The pathway has been studied most extensively in Escherichia coli (Driessen and Nouwen, 2008). A key component is the SecB molecular chaperone (Randall and Hardy, 2002). SecB maintains a newly synthesized pre-polypeptide in an unfolded state by binding to it as it exits from the ribosome. SecB also pilots the unfolded substrate to the membrane-spanning SecYEG translocon via a specific interaction with SecA, an ATPase that energizes the initial translocation of proteins through the translocation channel (Rusch and Kendall, 2007; Driessen and Nouwen, 2008). All sec-dependent substrates possess an N-terminal secretion signal that is cleaved upon transport through the inner membrane by specific signal peptidases (Paetzel et al., 2002).

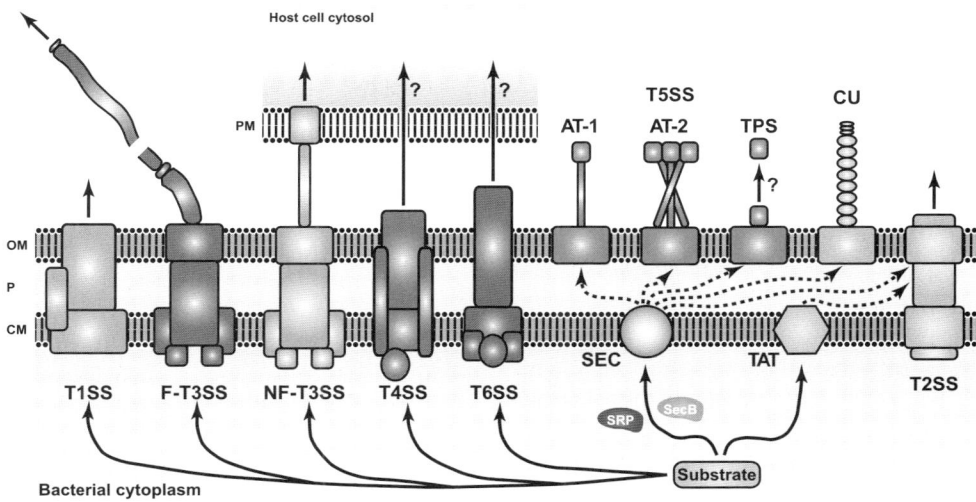

Fig. 9.1. Secretion systems of pathogenic Yersinia. *Yersinia* substrates secreted by the type I (T1SS), flagella (F-T3SS) and non-flagella type III (NF-T3SS), type IV (T4SS) and type VI (T6SS) secretion systems transverse the cytoplasmic membrane (CM) and outer membrane (OM) in one step. Some of these substrates are delivered directly across the plasma membrane (PM) into the cytosol of eukaryotic cells. Further substrate variants are secreted to the outside in two steps via the type II (T2SS), type V (T5SS) and chaperone-usher (CU) secretion systems. T5SSs can be divided into autotransporter 1 and 2 (AT-1 or AT-2, depending on substrate oligomerization status) and two-partner systems (TPS). All these two-step systems first require export of their cargo across the CM via the SecB or TAT (twin-arginine transport) pathways. The SRP (signal recognition particle) pathway is presumed to be responsible for the transport of integral inner membrane proteins. The arrows highlight the suggested substrate route taken during secretion. None of the secretion systems are drawn to scale and they are not intended to indicate their individual complexity. Type II, type IV and type VI secretion requires inherently complex systems, whereas the remaining systems (types I, V and CU) are comparatively simplistic (see text and Table 9.1 for details).

sec-dependent secretion in *Yersinia* has not been studied, but there seems little reason to doubt its essentiality to bacterial viability. For example, having the complete sequence of several *Yersinia* genomes enabled *in silico* mapping of the entire set of core components of the Sec translocon all of which were highly homologous to their counterparts in *E. coli* (Yen *et al.*, 2008).

Twin-arginine translocation (Tat)

Independent of the Sec pathway, the *tat*-dependent pathway is responsible for the secretion of fully folded proteins (Lee *et al.*, 2006; Sargent *et al.*, 2006). These proteins are synthesized as precursors with an N-terminal signal sequence containing characteristic 'twin-arginine' residues. The core translocon

apparatus is composed of the integral membrane proteins TatA, TatB and TatC. Pathogenic *Yersinia* are known to possess a functional Tat system (Lavander *et al.*, 2006; Shi *et al.*, 2007; Yen *et al.*, 2008). Multiple putative substrates have been suggested by *in silico* predictions of proteins containing an N-terminal sequence reminiscent of the twin-arginine signal sequences, but few of these have actually been verified by *in vivo* assays as *bona fide* substrates of the *Yersinia* Tat system. At least in *Y. pseudotuberculosis* some of these putative substrates could be important for pathogenesis, as a *tatC* knockout was attenuated in virulence during mouse infections (Lavander *et al.*, 2006). However, this attenuation was not apparent for an equivalent mutant of *Y. enterocolitica*, which remained fully virulent despite some obvious defects in certain metabolic pathways (Shi

Table 9.1. Known substrates of *Yersinia* protein secretion systems.[a]

Secretion system (species)	Substrate	Function
T1SS		
YrpDEF system	Yrp1	Metalloprotease
(*Y. ruckeri*)		
Unknown (?)	y3857	Metalloprotease (?)
(*Y. pestis*)		
T2SS		
Tts1 system	ChiY and EngY	Chitinase/chitin binding
(*Y. enterocolitica* only)	YE3650	Unknown (?)
TtS2 system (pathogenic	Unknown(?)	Unknown (?)
and non-pathogenic)		
Non-flagella T3SS		
Ysc-Yop system	YopE, YopH, YopT and YpkA	Antiphagocytosis
(all human pathogenic		
Yersinia)		
	YopJ	Immune suppression; apoptosis
	YopM	Unknown (?)
	Insecticidal-like toxin complex	Unknown (?)
	(YitABC and YipAB)	
Ysa-Ysp system	YspA, YspE, YspF, YspI, YspL, YspM	Unknown (?)
(*Y. enterocolitica* only)	and YspP	
	YspK	Immune suppression (?)
Flagella T3SS		
Polar flagellar system	FlgL, FlgK, FlgD and FliC	Flagella filament components
(enteropathogenic		(swimming)
Yersinia)		
	YplA (*Y. enterocolitica* only)	Phospholipase
	Additional Fops (Flagella outer	Unknown (?)
	proteins)	
T4SS		
tra cluster	Unknown (?)	Virulence
(*Y. ruckeri*)		
T5SS		
AT-1	YapC	Adhesin (?); autoaggregation (?);
(*Y. pestis* and		biofilm formation (?)
Y. pseudotuberculosis)	YapE	Adhesin; autoaggregation
	YapA, YapG, YapH, YapK and YapL	Unknown (?)
AT-2	YadA	Adhesin; autoaggregation; serum
(enteropathogenic		resistance
Yersinia)		
AT-2	YadB,YadC	Virulence (adhesin?)
(*Y. pestis* and *Y.*	YapF, M and YapN	Unknown (?)
pseudotuberculosis)		
Atypical AT (?)	Invasin	Adhesin
(enteropathogenic		
Yersinia)		
TPS (?)	RscBAC (Ytps1)	Adhesin (?)
(*Y. enterocolitica* and		
Y. pestis)		
TPS (?)	Ytps2 and Ytps4	Adhesin (?)
(*Y. pestis*)		
TPS (?)	YhlBA (Ytps3)	Haemolysin (?)
(*Y. ruckeri* and		
Y. pestis)		

Table 9.1. *Continued.*

Secretion system (species)	Substrate	Function
T6SS		
YPO0499-YPO0516 (*Y. pestis*)	Unknown (?)	Unknown (?)
CU		
PsaABC system (*Y. pestis* and *Y. pseudotuberculosis*)	pH 6 antigen	Antiphagocytosis; adherance
Caf system (*Y. pestis*)	F1 capsular antigen	Antiphagocytosis
Unknown (?) (*Y. pestis*)	y0561 and y0563	Biofilm formation (?)
MyfABC system (*Y. enterocolitica*)	Myf fibrillar antigen	Adhesin (?)

Notes: [a]Generally, only those secretion systems that have been verified *in vivo* are listed. However, several additional secretion systems, especially encoding T5SSs, T6SSs and CU systems, are predicted in the genome of pathogenic *Yersinia* on the basis of *in silico* mapping. (?), either unknown or not confirmed.

et al., 2007). These differences suggest that the Tat system could have customized functions in the various pathogenic *Yersinia*.

Signal recognition particle (SRP)

The SRP-dependent pathway is responsible for targeting proteins destined for integration into the inner membrane to the Sec translocon (Luirink *et al.*, 2005). A key component is SRP, which recognizes the hydrophobic character of the nascent polypeptide. SRP, comprised of a protein called Fth and a 4.5S RNA molecule, targets ribosome/nascent polypeptide chains to the SRP receptor, FtsY. In turn, FtsY directs the SRP-ribosome complex to the Sec translocon (Shan and Walter, 2005). SRP is therefore involved in co-translational protein targeting, making it distinct from the SecB targeting system, which occurs post-translationally. The relatively abundant YidC inner membrane protein is also implicated in the integration of proteins into the membrane as they are being exported through the Sec translocon (Kol *et al.*, 2008). Based on evolutionary conserved features of Fth, FtsY and YidC from *E. coli* with those from *Y. pestis*, it has been suggested that these similar functions are preserved in *Yersinia* (Yen *et al.*, 2008).

9.2.2 One-step protein secretion to the outer membrane or beyond

To remind the reader, one-step secretion is the process by which proteins are transported from the bacterial cytoplasm to the outer membrane or beyond without a periplasmic intermediate. This process can be performed by a number of dedicated secretion systems. The secreted substrate is usually not processed during transport, nor is transport through the inner membrane directly dependent on the Sec or Tat translocon. However, one should always keep in mind that these inner membrane translocons are still invariably needed to localize and/or assemble correctly the relevant components of each one-step secretion machine in the bacterial envelope, well before substrate secretion can occur.

Type I secretion (ABC transporter-dependent pathway)

The type I secretion system (T1SS) is composed of a three-component translocon; an outer membrane protein (OMP) and two cytoplasmic membrane proteins, an ATP-binding cassette (ABC) protein and a membrane fusion protein (MFP) (Delepelaire,

2004). Secreted substrates are generally closely related proteins belonging to the toxin family or hydrolytic enzyme family, which includes proteases, nucleases, glucanases and lipases. They are characterized by a non-cleaved C-terminal secretion signal located in the last 60 amino acids. Because of this C-terminal position, secretion occurs only after translation. It is possible that cytoplasmic chaperones are therefore needed to keep substrates unfolded prior to secretion. Individual C-terminal secretion signals are recognized specifically by cognate ABC proteins. This recognition triggers the assembly of a functional translocon together with the OMP and the MFP. This generates a hollow contiguous conduit to allow substrate passage through the entire cell envelope and release into the external environment. T1SSs are analogous to many physiologically important transport systems. Archetypal systems exist for import/export of sugar metabolites, antibiotics and metallic ions (Davidson *et al.*, 2008). In addition, *E. coli* TolC functions as the OMP component for the secretion of several type I substrates, as well as in drug and cation efflux. This is most likely true in other bacteria, for TolC family proteins are ubiquitous among Gram-negative bacteria (Koronakis *et al.*, 2004).

In pathogenic *Yersinia*, reports of ABC transporter-dependent protein secretion are scarce. Scrutiny of the *Y. pestis* genome has identified four ABC and MFP pairs; the former containing motifs indicative of being energized by ATP hydrolysis, suggesting that these units specifically transport proteins rather than nutrients or drugs (Yen *et al.*, 2008). Searching for glycine-rich repeats within *Y. pestis* open reading frames (ORFs) – which are used as general predictors of T1SS substrates – identified one ORF with potential to encode for a metalloprotease that may represent a likely T1SS substrate (Delepelaire, 2004). None the less, the only definitive example of a T1SS in *Yersinia* comes from work on the fish pathogen, *Y. ruckeri*. This organism uses a T1SS to secrete the Yrp1 metalloprotease, an essential determinant of virulence (Fernandez *et al.*, 2002).

Type III secretion in motility and intoxication of eukaryotic cells

The ability to evade innate immune responses of an animal, plant, fish or insect host or to establish a symbiotic relationship with the same is common to many bacteria. This ability seems most often mediated by a type III secretion system (T3SS). T3SSs allow bacteria to translocate (inject) effector proteins directly from the cytoplasm into target eukaryotic host cells (Francis *et al.*, 2004; Galan and Wolf-Watz, 2006). A T3SS resembles a hypodermic syringe composed of 20–25 proteins that span the bacterial envelope. These structures are reminiscent of the swimming organelle – flagella – reflecting that T3SSs used by bacteria in contact with eukaryotic cells most likely have evolved from the flagella biosynthesis machine (Saier, 2004). Two major protein classes are secreted by the non-flagella T3SS: the toxic effectors and the translocators (Mueller *et al.*, 2008; Francis, 2010). The latter form translocon pores in infected cell plasma membranes through which the effectors may gain entry into the cell cytosol. The effector arsenal translocated into host cells can vary widely among different bacteria, but all usually possess an enzymatic activity designed to subvert host cell signalling for the bacteria's benefit (Galan, 2009). Thus, the effector toolkit available to each bacterium has evolved over many years of living in close association with eukaryotic cells and subsequently has been tailored to fit their individual lifestyle. Most often, each secreted substrate requires a dedicated cytoplasmic chaperone for their pre-secretory stabilization and/or efficient secretion (Francis, 2010). In addition, a signal sequence for protein secretion resides in the N-terminus of the protein, but no obvious consensus sequence exists among the secreted substrates. There are conflicting details about the nature of this signal – it could be a combination of both mRNA and amino acid sequence (Lloyd *et al.*, 2001; Cornelis, 2003; Sorg *et al.*, 2005). There is also solid support for a chaperone-mediated secretion signal (Francis, 2010).

Yersinia spp. represent a model organism for the study of T3SSs. All three human pathogens possess a highly homologous Ysc

(*Yersinia* secretion)-Yop (*Yersinia* outer protein) T3SS that is encoded on a large virulence plasmid. This Ysc-Yop T3SS is absolutely essential for pathogenicity and consequently has been studied extensively (Cornelis *et al.*, 1998; Fallman and Gustavsson, 2005; Viboud and Bliska, 2005). A small but potent effector arsenal is targeted to the cytosol of eukaryotic cells by the Ysc-Yop T3SS. Internalized effectors modulate phagocytic, apoptotic and proinflammatory cellular signalling mechanisms to enable *Yersinia* to replicate primarily extracellularly (Viboud and Bliska, 2005; Matsumoto and Young, 2009). Curiously, the chromosome of each species also encodes the full repertoire of a second non-flagella T3SS. Significantly, these are not related to each other, suggesting independent acquisition (Troisfontaines and Cornelis, 2005; Matsumoto and Young, 2009). While this second system in *Y. enterocolitca*, termed the Ysa (*Yersinia* secretion apparatus)-Ysp (*Yersinia* secreted protein) T3SS, has been studied in some detail and is implicated in virulence (for example, Foultier *et al.*, 2002; Haller *et al.*, 2000; Matsumoto and Young, 2006; Venecia and Young, 2005), for the most part their individual functions remain obscure.

Additional substrates of *Yersinia* T3SSs might also be components of insecticidal toxin complexes (Gendlina *et al.*, 2007). In the entomopathogenic *Photorhabdus luminescens* bacterium such complexes are associated with genetic loci encoding three conserved toxin components – the *tcaAB/tcdA*-like A component, *tcaC/tcdB*-like B component and *tccC*-like C component – that are required for parasitism of the insect host (Waterfield *et al.*, 2001). Broadly found in several *Yersinia* species (Waterfield *et al.*, 2007; Fuchs *et al.*, 2008), it is not yet clear what role these 'toxins' play in the life cycle of *Yersinia* bacteria. Present understanding suggests they could be important for *Y. pestis* proliferation and transmission in the flea vector (Erickson *et al.*, 2007; Gendlina *et al.*, 2007) and for aiding environmental survival of other Yersiniae, perhaps by assisting in the avoidance of invertebrate predation or by establishing an invertebrate mutualistic association (Bresolin *et al.*, 2006; Erickson *et al.*, 2007; Pinheiro and Ellar, 2007; Fuchs *et al.*, 2008). It has even been

suggested that the *Yersinia* variants have been adapted specifically to promote survival in the mammalian host (Tennant *et al.*, 2005; Gendlina *et al.*, 2007; Waterfield *et al.*, 2007; Hares *et al.*, 2008). Whatever the role, their presence in the *Yersinia* genomes represents a truly fascinating example of bacterial evolutionary adaptation.

As implied earlier, the final stages in flagella assembly – notably the secretion of filament-associated proteins – requires an archetypal T3SS (Chevance and Hughes, 2008; Minamino *et al.*, 2008). Flagella-mediated *Yersinia* motility is required to facilitate target cell contact and/or cell invasion, thereby maximizing the pathogenic effects of the infecting bacterium (Young *et al.*, 2000; McNally *et al.*, 2007). Moreover, the flagella T3SS of *Yersinia* is also capable of secreting heterologous substrates, including substrates normally secreted by the non-flagella T3SSs (Young *et al.*, 1999; Young and Young, 2002b). Whether this secretion promiscuity actually contributes to *Yersinia* pathogenesis is not established. However, there is some precedent for this based on the observation that different *Yersinia* T3SSs can deliver non-cognate effectors into the cytosol of target eukaryotic cells (Young and Young, 2002a).

Type IV secretion

Plasmid transfer by means of conjugation is the archetypal type IV secretion system (T4SS). Type IV secretion has since been adapted by some bacteria to transfer proteins and/or DNA-protein complexes into a wide array of target cell types (Christie *et al.*, 2005; Backert and Meyer, 2006). The T4SS is composed of ~10–15 proteins that assemble into a machine spanning the entire bacterial envelope. Substrates destined for secretion do not share any obvious consensus secretion signal sequence motif. However, most substrates routinely possess a cluster of positively charged residues at their C-termini that could be a T4S signal. T4SSs bear no relation to T3SSs, most probably due to the former having evolved from an ancestral conjugation system while the latter having evolved from an ancient system conferring motility. Curiously though, these different pathways apparently achieve the same end result: effector translocation into eukaryotic cells. With this, it is not surprising that type III and IV protein secretion is rarely a feature of the same bacterium; the majority of pathogenic *Yersinia* apparently lack any T4SS (Yen *et al.*, 2008). Only a few strains of *Y. pseudotuberculosis* reportedly harbour a T4SS (Eppinger *et al.*, 2007), while a putative T4SS in *Y. ruckeri* has been implicated recently in virulence of this fish pathogen (Mendez *et al.*, 2009). To date, however, it is unknown what role these putative T4SSs may play in the infection cycle of these *Yersinia* strains and, in particular, what kind of substrates they may deliver into eukaryotic cells or their relationship with existing T3SSs in the same bacterium.

Type VI secretion

Another mechanism for the translocation of effector toxins into target eukaryotic cells can involve type VI secretion systems (T6SSs) (Filloux *et al.*, 2008; Pukatzki *et al.*, 2009). The mechanism of T6SSs is poorly understood due to their very recent discovery. The system may consist of up to 25 proteins; most of these are novel proteins unrelated to any protein belonging to other independent bacterial secretion systems. A limited number of secreted substrates have been reported, but it is not yet clear how these are targeted for secretion. In 2002 the first hint of novel putative protein secretion loci (now known to be a T6SS cluster) in *Yersinia* was reported (Folkesson *et al.*, 2002). Further *in silico* scrutiny of the sequenced *Yersinia* genomes has since confirmed that multiple T6SS gene clusters exist (Bingle *et al.*, 2008; Heermann and Fuchs, 2008; Yen *et al.*, 2008). With the identification of these T6SS loci, it is now necessary to define their secretion cargo and to determine their function. In a recent study, however, deletion of one of the T6SS loci from *Y. pestis* did not compromise this mutant's ability to colonize either the flea vector or mice during infections modelling bubonic or pneumonic plague (Robinson *et al.*, 2009). Thus, further research is needed in order to reveal the impact of type VI secretion on *Yersinia* infection biology or environmental survival.

9.2.3 Two-step protein secretion to the outer membrane or beyond

The process of two-step protein secretion is first dependent on the Sec or Tat translocon for transport of substrate across the inner membrane. This results in a periplasmic intermediate that is usually proteolytically processed into the mature substrate. Sometimes, however, further substrate processing can occur following secretion across the outer membrane. Transport across the outer membrane requires additional specialized secretion systems.

General secretory pathway (type II secretion) and related systems

Type II secretion systems (T2SSs) are known as the main terminal branch of the general secretory (or Sec-dependent) pathway (GSP) (Cianciotto, 2005; Johnson *et al.*, 2006). This system is capable of secreting a diverse array of fully folded protein substrates across the outer membrane; a process that involves an apparatus usually consisting of 12–16 different proteins. Homologues of some T2SS components also form the basis for type IV pilus biogenesis used by bacteria to adhere to eukaryotic cells and to perform flagella independent twitching motility (Nunn, 1999; Peabody *et al.*, 2003). The secretin – an integral outer membrane multimeric protein of about 12–15 units – is a key feature of T2SSs and type IV pili. It forms a large singular pore through which secreted proteins are transported across the outer membrane. Related proteins serve a similar role in T3SSs. A helical pilus composed of one major pilin molecule and several different minor pilins is another feature of T2SSs and type IV pili. A specific pre-pilin peptidase is responsible for cleaving off the N-terminal leader peptide of pre-pilins, followed by methylation of the processed pilins to allow assembly into pilus-like structures (Paetzel *et al.*, 2002). A high degree of secretion substrate specificity exists between the T2SSs. However, the molecular mechanism of this intrinsic specificity remains unknown. No primary sequence identity among the substrates is evident, and crystal structures of a few substrates have failed to reveal any common structural motif.

In *Y. enterocolitica* a T2SS gene cluster designated Tts1 (*Yersinia* type II secretion 1) identified solely in so-called 'high pathogenicity' isolates was necessary for full virulence in murine infections (Iwobi *et al.*, 2003). Identified secreted substrates possess chitin- and oligosaccharide-binding properties, suggesting this to be an important virulence strategy; although they may also contribute to survival in the environment (Shutinoski *et al.*, 2010). Additionally, a second T2SS gene cluster (Tts2) of no known function is widespread among pathogenic and non-pathogenic *Yersinia* (Iwobi *et al.*, 2003; Yen *et al.*, 2008; Shutinoski *et al.*, 2010). Furthermore, *Y. pestis* also possess another putative T2SS that may be the most minimalistic system reported so far – only four components (Yen *et al.*, 2008). Finally, a type IV pilus needed for virulence has been observed in a restricted set of clinical and veterinary isolates of *Y. pseudotuberculosis* (Collyn *et al.*, 2002). This *pil* locus is encoded on a pathogenicity island that is also present in *Y. enterocolitica* but not in *Y. pestis* (Collyn *et al.*, 2004; Yen *et al.*, 2008).

In recent years, a so-called Tad (tight adherence) macromolecular transport locus has also surfaced. This locus is a major subtype of the T2SS cluster encoding for the assembly of adhesive Flp (fimbrial low-molecular weight protein) pili (Tomich *et al.*, 2007). This pilus assembly system is widely distributed in prokaryotes, including pathogenic *Yersinia* (Planet *et al.*, 2003; Yen *et al.*, 2008). In several bacteria, Flp pili are implicated strongly in the colonization of, and biofilm formation on, a variety of surfaces (Tomich *et al.*, 2007). However, such a role is not obvious from the study of Flp pili function, suggesting an alternative function for this locus in the Yersiniae.

Type V secretion

Widely distributed among pathogenic bacteria, type V secretion systems (T5SSs) represent a distinct terminal branch of the GSP. T5SSs are the most simplistic secretion machines known, existing as an auto-transporter (AT) – which can be subdivided further into two subfamilies, the monomeric AT-1 and trimeric AT-2 – or as a two-partner system (TPS) (Cotter *et al.*, 2005; Dautin and Bernstein, 2007; Mazar and Cotter, 2007). Although this nomenclature implies at most two components are needed for secretion, in reality substrate secretion also depends on outer membrane assembly factors such as the Bam (Omp85) complex and periplasmic folding factors (for example, Purdy *et al.*, 2007; Bodelon *et al.*, 2009; Ieva and Bernstein, 2009; Ruiz-Perez *et al.*, 2009; Sauri *et al.*, 2009). Nevertheless, T5SSs provide a solution to the secretion of essentially large proteins with certain folding characteristics – many of which are implicated in bacterial pathogenesis (Newman and Stathopoulos, 2004). An AT-1 substrate has a modular domain structure composed of a N-terminal *sec* secretion signal, an internal passenger domain-containing effector function and a large C-terminal domain that forms an outer membrane-spanning β-barrel with a central hydrophilic core (Dautin and Bernstein, 2007). In AT-2, the substrates function as trimers and each subunit monomer contributes a much shorter C-terminal domain that oligomerizes in the outer membrane to form a trimeric trans-location channel (Cotter *et al.*, 2005). Additionally, in TPS the passenger domain protein (generically referred to as TpsA) also contains a 'TPS' motif for specific recognition by the cognate β-barrel-containing partner protein (TpsB) (Mazar and Cotter, 2007).

Arguably the best-characterized T5SS in pathogenic *Yersinia* is that of YadA, a prototypical member of AT-2 that forms trimeric structures in the outer membrane (Cotter *et al.*, 2005). YadA is a multifaceted adhesin involved in binding to both eukaryotic cells and the extracellular matrix, and is also required for autoaggregation and serum resistance (El Tahir and Skurnik, 2001). YadA therefore has a critical role in virulence, especially in the enteric pathogen *Y. enterocolitica*; *yadA* is a pseudogene in *Y. pestis*. Two homologues of YadA in *Y. pestis* and *Y. pseudotuberculosis*, designated YadB and YadC, are also proposed to act in concert to exert a virulence function important for the onset of bubonic plague, although it is unclear if these act as a bacterial surface adhesin

(Forman *et al.*, 2008). In addition, *in silico* analysis of the *Y. pestis* genome predicts three other distinct proteins, designated YapF, YapM and YapN, which also belong to this AT-2 family (Yen *et al.*, 2007). These have not been studied experimentally.

A further seven AT-1 proteins were predicted; YapA, YapC, YapE, YapG, YapH, YapK and YapL (Yen *et al.*, 2007). In a surrogate *E. coli* host YapC could mediate binding to eukaryotic cells and contribute to auto-aggregation and biofilm formation, although none of these facets were perturbed in *Y. pestis* lacking *yapC* (Felek *et al.*, 2008). On the other hand, YapE, which also mediates auto-agglutination and binding to eukaryotic cells, is required for efficient colonization of *Y. pestis* in a mouse model simulating bubonic plague (Lawrenz *et al.*, 2009). Significantly, alleles encoding all ten Yaps are also present in enteropathogenic *Y. pseudotuberculosis* (Yen *et al.*, 2007; Lawrenz *et al.*, 2009). In all likelihood, gaining an understanding of how these AT systems function will benefit current views on *Yersinia* biology.

The bacterial cell-surface associated protein, invasin, is another important cellular adhesin and virulence determinant of the enteric *Yersinia* (Isberg *et al.*, 2000; Grassl *et al.*, 2003). Although invasin secondary structure does not conform to other proteins typically belonging to the AT superfamily, its surface localization is dependent on an N-terminal β-barrel domain that is indicative of AT secretion (Newman and Stathopoulos, 2004; Yen *et al.*, 2007). Thus, invasin may represent another distinct class of AT proteins (*aka* AT-3).

The remaining T5SS family is the TPS comprising of two components. In *Y. pestis* four TPSs have been suggested based on *in silico* data and designated Ytps1–4 (Yen *et al.*, 2008). In each case, the two partner proteins (designated A and B) are encoded by genes within the same locus. Phylogenetic studies also indicated that Ytps1A, Ytps2A and Ytps4A might all function as surface adhesins, while Ytps3A was most likely a haemolysin (Yen *et al.*, 2008). In fact, the Ytps3 system could possibly be analogous to the *yhlBA* secretion system involved in promoting *Y. ruckeri* virulence, since both share a relatedness to the *Serratia*-type haemolysin (Fernandez *et al.*, 2007b; Yen *et al.*, 2008). In *Y. enterocolitica* the *rscBAC* locus encodes a putative TPS system analogous to the HMW1 and HMW2 adhesin systems of *Haemophilus influenzae* (Nelson *et al.*, 2001). While the function of the secreted RscA substrate is unknown, deletion of *rscA* alters the kinetics of *Y. enterocolitica* infection, enhancing the systemic dissemination of mutant bacteria (Nelson *et al.*, 2001). Based on shared homologies to the HMW1 system, the *rscBAC* locus might be equivalent to the Ytps1 system of *Y. pestis*. Additional putative TPS loci have been identified recently in *Y. enterocolitica* (Heermann and Fuchs, 2008) and, via the Kyoto Encyclopedia of Genes and Genomes (KEGG) bioinformatics resource webserver (Kanehisa and Goto, 2000), in *Y. pseudotuberculosis* (unpublished observations). However, they remain uncharacterized and their functional conservation within the different Yersiniae is not established.

Chaperone-usher secretion

The chaperone-usher (CU) pathway is highly conserved among Gram-negative bacteria. It enables the assembly of a diverse array of multi-subunit fibres on the bacterial surface (Zavialov *et al.*, 2007; Waksman and Hultgren, 2009). The most-studied example of this is the biogenesis of type I pili, such as the Pap and Fim systems of uropathogenic *E. coli*. These structures consist of a major pilus subunit protruding from the bacterial surface containing at the tip a specific adhesin(s) that mediates microbial attachment to host tissues and the formation of biofilms. Other 'poly-adhesin' structures can form a capsule circumventing the bacterial surface to facilitate the evasion of phagocytic host defences (Zavialov *et al.*, 2007). The two key components in their biogenesis are a periplasmic molecular 'chaperone' that receives the subunits as they are transported through the Sec-translocon and then delivers them to the integral outer membrane 'usher' protein, which serves as the platform on which the growing surface organelles are assembled.

In *Y. pestis* two 'classical' CU systems are very well characterized: the capsular F1 antigen (Caf1 system) involved in anti-phagocytic activities (Knight, 2007) and the pH 6 antigen (Psa system) involved in antiphagocytosis and adherence (Lindler and Tall, 1993; Yang et al., 1996). Related to these systems are the Myf fimbriae, a putative adhesive organelle of Y. enterocolitica (Iriarte et al., 1993). A further seven intact systems belonging to the classical CU family have been reported in Y. pestis (Felek et al., 2007; Runco et al., 2008; Yen et al., 2008). When these loci were expressed in E. coli, novel filamentous appendages could be visualized on the bacterial surface and, in some cases, this influenced the adhesiveness towards eukaryotic cells as well as biofilm formation (Felek et al., 2007). According to the KEGG bioinformatics resource webserver (Kanehisa and Goto, 2000), multiple loci are also annotated as CU systems in enteropathogenic Yersinia (unpublished observations). Eventual discovery of their functions potentially could enhance our understanding of mechanisms used by Yersinia to colonize surfaces, exist as multicellular communities or evade host immune defences.

9.3 Physiological Mechanisms Underlying Protein Secretion and Virulence Control

Pathogenic Yersinia can occupy very diverse infection niches ranging from invertebrate to vertebrate hosts and, armed with the ability to negate host immune responses, can permit both extracellular and intracellular prolifer-ation. With this competence comes the ability to sense and respond to a diverse array of environmental cues in order to reprogramme their gene expression profiles quickly to facilitate persistence and survival. The recent access to fully sequenced genomes of various Yersinia isolates has enabled utilization of genome-wide approaches such as tran-scriptomics and proteomics to gain some insight into Yersinia responsiveness to physio-chemical and nutritional stresses and the varied host environment (for example, Motin et al., 2004; Chromy et al., 2005; Lawson et al., 2006; Han et al., 2007; Rosso et al., 2008; Pieper et al., 2009). Access to the complete genome

sequence has also allowed in silico mapping of putative virulence determinants and regulatory pathways on a large scale (Marceau, 2005; Yen et al., 2008) that can be followed up with a genetic reductionist approach to pry apart individual roles of large protein families. This type of approach was used to examine the role of all two-component regulator systems (TCSs) existing in Y. pseudotuberculosis (Flamez et al., 2008) and Y. pestis (O'Loughlin et al., 2010). These collective approaches have enhanced dramatically the rate of novel discovery that can now be scrutinized further with powerful molecular approaches to define better the mechanistic regulatory interplay in minute detail. Reprogramming of bacterial gene expression can occur at multiple levels: transcription, mRNA stability, translation, protein activation and protein turnover, all of which can involve diverse physical and chemical inputs. Some examples of these are given in overview in the ensuing sections. Together, they demonstrate the ingenuity and adaptability of pathogenic Yersinia as they manage to control temporal and spatial production of their numerous virulence factors while occupying diverse environ-mental niches.

9.3.1 Physiological stress responses in the cytoplasm

The stationary phase (general stress) response

The alternate sigma factor, σ^S (also known as σ^{38} or RpoS), has a global role in regulating bacterial tolerance to stress – including starvation, acidic, heat and oxidative stress conditions (Hengge-Aronis, 2002; Dong and Schellhorn, 2010). At least in E. coli, multiple levels of transcriptional, translational and post-translational regulation apparently control RpoS output (Hengge-Aronis, 2002). Some of these factors are mentioned in this chapter for their role in controlling aspects of Yersinia virulence; these include the cAMP-CRP system, the small molecule (p)ppGpp and the ClpXP protease (see below). Despite these regulatory connections, and the likelihood that the *Yersinia* RpoS regulon contains many notable members – especially if observations in *E. coli* are any guide (Lacour and Landini, 2004; Patten *et al.*, 2004) – *Yersinia* lacking the *rpoS* allele are still capable of *in vivo* survival to the same degree as parental bacteria (Badger and Miller, 1995; Iriarte *et al.*, 1995). This contrasts with the dramatic attenuating effects of *rpoS* knockouts in several other bacterial pathogens (Dong and Schellhorn, 2010). In *Yersinia*, therefore, RpoS function probably overlaps with other undisclosed factors in the regulation of virulence gene expression during the stationary phase.

The heat shock response

To safeguard against the dangers of protein misfolding and aggregation in the cytoplasm, bacteria produce molecular chaperones and proteases that participate in essential protein folding and degradation control to ensure that proteins fold correctly and maintain functionality (Wickner *et al.*, 1999; Dougan *et al.*, 2002). During extreme physiological stress, these cytoplasmic chaperones and proteases assume even greater responsibility. This is reflected by a significant elevation in their production; a phenomenon known as the heat shock stress response that is controlled principally by another alternate sigma factor, σ^H (also known as σ^{32} or RpoH) (Wick and Egli, 2004; Guisbert *et al.*, 2008). Most of these σ^H-controlled factors utilize ATP hydrolysis to prevent protein aggregation and assist with protein folding, refolding or degradation in the bacterial cytoplasm (Gottesman, 1996; Wickner *et al.*, 1999; Dougan *et al.*, 2002). A microarray analysis of the heat shock regulon in *Y. pestis* has confirmed the elevated expression of genes encoding several putative cytoplasmic chaperones and/or proteases (Han *et al.*, 2005). These elevated levels also correspond to the thermoregulation of multiple virulence genes including the F1 capsule, integral outer membrane proteins including Ail, pH 6 fimbriae, the Ysc-Yop T3SS and a T6SS (Motin *et al.*, 2004; Chromy *et al.*, 2005; Han *et al.*, 2005), suggesting that regulated proteolysis might be an important modulator of *Yersinia*

virulence gene expression. Indeed, direct support for this comes from work with the Lon and Clp protease systems, which have been shown to affect expression levels of T3SS products (Jackson *et al.*, 2004; Falker *et al.*, 2006), the ail adhesin (Pederson *et al.*, 1997), invasin and motility (Badger *et al.*, 2000) and the global regulator of virulence gene expression, RovA (Herbst *et al.*, 2009). Thus, the physiological response to heat stress that is borne out through elevated levels of molecular chaperones and proteases has dramatic repercussions on regulated expression of multiple *Yersinia* virulence factors.

9.3.2 Extracytoplasmic stress responses

Cpx pathway and the sigmaE system

Quality control of protein folding in the periplasm and outer membrane also requires input from protein folding and degradation factors. Their roles become even more pronounced when bacteria experience so-called extracytoplasmic stresses that can compromise bacterial envelope integrity and protein folding in the periplasm. Such periplasmic-located quality control factors include the Skp chaperone, the Dsb disulfide bond catalyst and isomerization system, various peptidyl-prolyl cis/trans isomerases (PPIases) and the DegP/HtrA serine protease (Duguay and Silhavy, 2004; Wick and Egli, 2004; Ades, 2008). All function in an ATP-exclusion vacuum to overcome issues of protein folding and refolding in the periplasm. Their production is under the control of two principal extracytoplasmic stress (ECS) responsive pathways: an alternate sigma factor σ^E (otherwise known as RpoE) and the CpxA sensor kinase/CpxR response regulator TCS (Duguay and Silhavy, 2004; Wick and Egli, 2004; Ades, 2008) (Fig. 9.2).

Recent evidence suggests that these systems not only perform housekeeping functions involved in overseeing bacterial envelope biogenesis, but also are critical for assembly of surface-located virulence factors in bacterial pathogens (Raivio, 2005; Dorel *et al.*, 2006; Rowley *et al.*, 2006).

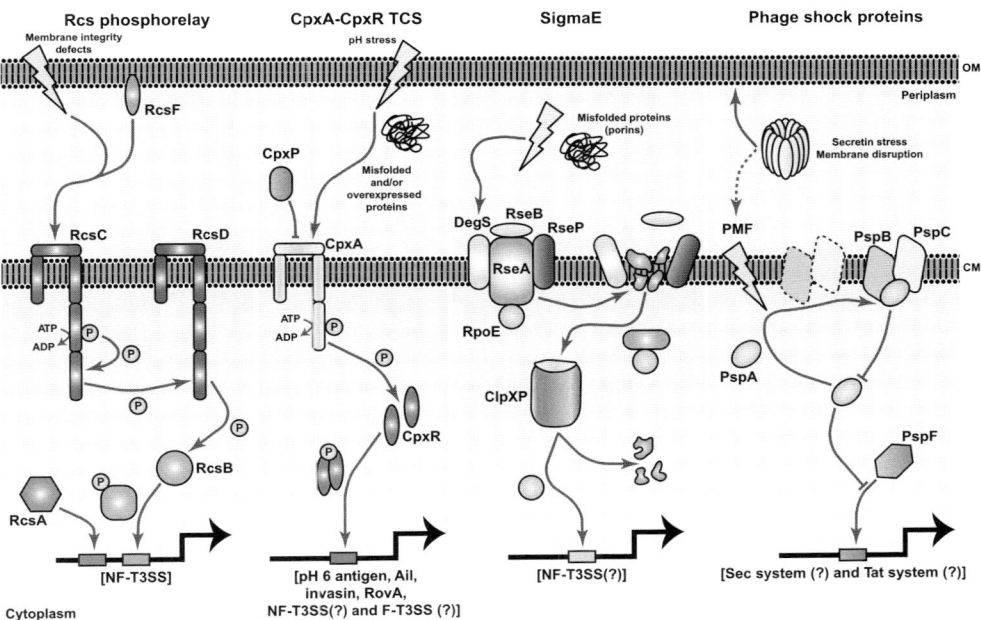

Fig. 9.2. Extracytoplasmic stress sensing in Yersinia. The Rcs phosphorelay system and the Cpx two-component system (TCS) both rely on sensor kinase (RscC and CpxA, respectively) autophosphorylation to initiate a complex (Rcs) or simple (Cpx) relay of phosphate (P) to the terminal phosphate acceptor – the response regulator RcsB and CpxR, respectively. The phosphorylated response regulators in turn effect changes in transcriptional output from responsive promoters. In the case of RcsB, this may sometimes require cooperation with RcsA. Activation of the Rcs phosphorelay occurs by RcsF-dependent and RcsF-independent extracytoplasmic stress (ECS) signals (lightning bolt). Normally repressed by the action of CpxP, Cpx pathway activation occurs in response to ECS that includes low pH and protein overexpression and/or their misfolding. Similarly, the stress of porin misfolding can initiate a proteolytic cascade involving the DegS, RseP and ClpXP proteases that cooperate to liberate free RpoE sigma factor into the cytoplasm. Cytoplasmic RpoE can recruit the RNA polymerase core enzyme (not shown) to establish transcription of a specific gene set. The phage shock proteins are induced by secretin complexes that are presumably localized incorrectly to the CM. This may dissipate the proton gradient associated with proton motive force (PMF) generation that is considered to induce a mechanism that culminates in PspA sequestration by the PspB and PspC proteins. In the absence of PspA inhibition, the cytoplasmic PspF transcription factor is free to initiate the phage shock response. Generally, knowledge of the function of these ECS response systems in Yersinia is limited. In a few cases, their involvement in the control of substrate secretion (indicated in parenthesis) is relatively well established. Secretion systems followed by a question mark are examples where the molecular mechanism governing the regulatory connection remains elusive.

With this in mind, we explored the contribution of the σ^E and CpxRA pathways in aspects of ECS sensing and Yersinia pathogenicity. Interestingly, transcription of adhesin genes encoding invasin, pH 6 antigen and Ail, along with the global regulator, RovA, are influenced negatively by the direct action of an activated Cpx system [Liu and Francis, unpublished data] (Carlsson et al., 2007b). Moreover, T3SS gene expression and/ or subsequent component assembly in the bacterial envelope is also regulated by the σ^E and Cpx pathways [Liu and Francis, unpublished data] (Carlsson et al., 2007a). However, it is not yet clear whether this involvement involves a direct interaction of σ^E or the phosphorylated CpxR response regulator with T3S gene promoters. It might also be that σ^E- and/or Cpx-dependent production of periplasmic protein folding and degradation factors function to ensure the correct assembly of T3SS components in the bacterial envelope. For example, the periplasmic disulfide oxidoreductase, DsbA, is necessary for flagella and non-flagella T3SS assembly [Carlsson and Francis, unpublished data] (Jackson and Plano, 1999) and for function of the Caf1M chaperone in F1 capsule assembly (Zav'yalov et al., 1997). In addition, production of the DegP protease of Yersinia is induced in vivo and is necessary for intracellular survival in macrophages and for full virulence in the mouse model of infection (Li et al., 1996; Yamamoto et al., 1996; Williams et al., 2000; Heusipp et al., 2004). This suggests that DegP function targets a Yersinia virulence mechanism(s) such as the Ysc-Yop type III secretion [Carlsson and Francis, unpublished data]. However, since loss of DegP protease activity leads to the accumulation of multiple uncharacterized proteins in Y. pestis whole cell lysates (Williams et al., 2000), various virulence mechanisms could be affected. Periplasmic PPIases also appear to play some role in the assembly of Yersinia virulence factors in the bacterial envelope [Obi and Francis, unpublished data]. Thus, the roles attributed to these factors in other bacterial pathogens (Raivio, 2005; Dorel et al., 2006; Rowley et al., 2006) also seem likely to extend to the pathogenic Yersiniae. Such observations reinforce the notion that virulence factor quality control is linked strongly to ECS sensing mechanisms in Yersinia and other related pathogens.

When discussing aspects of virulence factor assembly in the bacterial envelope, it is also pertinent to highlight how these processes are influenced by their immediate surroundings: more specifically, outer membrane composition and the lipopolysaccharide (LPS), lipoprotein and integral membrane protein content. For example, LPS is critical for the activity of omptins, such as the surface-located plasminogen (Pla) protease of Y. pestis (Kukkonen et al., 2004), and also affects the levels and/or function of Y. enterocolitica surface-located virulence factors such as the adhesins, YadA, Ail and invasin, as well as flagella-mediated type III secretion

(Bengoechea et al., 2004). In addition, it appears that the length of the Ysc-Yop T3SS needle extending from the Yersinia surface has also been measured precisely to project sufficiently beyond other surface appendages such as the YadA adhesin (Mota et al., 2005). Moreover, various lipoproteins are known to impact on Yersinia pathogenicity (Burghout et al., 2004; Sha et al., 2008; Tidhar et al., 2009). Comparative global gene expression profiles of a parent and its mutant lacking the Braun lipoprotein indicated that lipoproteins probably can influence Yersinia infections in multiple ways, as a primary virulence factor, a modulator of the host innate immune response or by impacting on the activity of other established Yersinia virulence factors. Thus, the interplay between intrinsic defensive properties of the bacterial envelope and the virulence factor armoury is a striking feature of pathogenic Yersinia – indeed, of all bacterial pathogens.

Phage shock proteins and maintenance of proton motive force

Another extracytoplasmic stress response pathway called the phage shock protein (PSP) response is needed for bacterial survival when the ion permeability barrier function of the cytoplasmic (inner) membrane has been compromised (Darwin, 2005) (Fig. 9.2). Since dissipation of the proton motive force (PMF) is a consequence of inner membrane disruption, the primary function of the PSPs may well be to restore the PMF in damaged membranes induced by stress (Kleerebezem et al., 1996; Jovanovic et al., 2006; Kobayashi et al., 2007). Efficient sec-dependent and tat-dependent secretion across the inner membrane also requires monitoring by the PSP system (Jones et al., 2003; DeLisa et al., 2004). As stipulated already in this chapter, these secretion systems are a prerequisite for two-step protein secretion across the outer membrane and also for initial component assembly in the bacterial envelope of one-step secretion systems. Moreover, a highly specific and foremost inducer of the PSP response is overexpression of the outer membrane secretin proteins, probably caused by their mislocalization to the inner membrane (Guilvout et al., 2006; Seo et al., 2007). This is significant because these outer membrane secretins are essential for processes like type II and type III secretion, as well as type IV pili assembly (Thanassi, 2002; Yen *et al.*, 2002).

Taken together, it seems reasonable therefore to assume that an intact PSP system may also have relevance to the function of protein secretion across the outer membrane. Indeed, an intact PSP system is required for *Y. enterocolitica* survival during active Ysc-Yop type III secretion when the YscC secretin is naturally overexpressed (Darwin and Miller, 2001; Green and Darwin, 2004). Strangely, however, the PSP system was not required for actual T3SS assembly and function (Darwin and Miller, 1999). In light of this, the recent observation that Ysc-Yop T3SS is reliant on PMF for function (Wilharm *et al.*, 2004) is significant given that it occurs apparently regardless of the sentinel role played by PSPs in maintenance of the proton gradient across the inner membrane. Given the pressing need to understand how T3SSs are energized (Wilharm *et al.*, 2007; Galan, 2008; Minamino *et al.*, 2008), these findings warrant more study to understand this apparent disparity better. It also seems worthwhile to investigate the role, if any, played by the PSP system in other secretin-dependent secretion pathways of *Yersinia*.

The Rcs phosphorelay

Bacterial sensing of the environment can be performed by prototypical histidine kinase and response regulator two-component pathways (such as the CpxA-CpxR system, and also see below) or by far more complex signal transduction phosphorelay systems such as the Rcs pathway, which is largely exclusive to the Enterobacteriaceae family (Majdalani and Gottesman, 2005; Huang *et al.*, 2006) (Fig. 9.2). RcsC, the sensor kinase, first transfers a phosphoryl group to the intermediate phosphotransfer protein, RcsD, and then on to RcsB, the response regulator. Phosphorylated RcsB is able to bind DNA, but this may sometimes require input from the auxiliary protein, RcsA. An outer membrane lipoprotein-like factor, RcsF, can also be involved in activating the Rcs

phosphorelay. In addition to RcsF-dependent signalling, many additional RcsF-independent inputs to RcsC are also known; most of these are expected to affect cell envelope integrity and/or composition (Majdalani and Gottesman, 2005; Huang *et al.*, 2006). Initially defined as a regulator of colonic acid (capsule) production in *E. coli* and the synthesis of type I capsular polysaccharides in other bacteria, the Rcs regulon includes many genes involved in a variety of processes ranging from cell division, flagella biosynthesis and motility, small regulatory RNA biosynthesis, biofilm formation and pathogenicity (Majdalani and Gottesman, 2005; Huang *et al.*, 2006).

Interestingly, *rcs* loci are present in all *Yersinia* sp., but *rcsD* is a natural pseudogene in *Y. pestis* (Hinchliffe *et al.*, 2008). This suggests that the Rcs phosphorelay system is maintained specifically in enteropathogenic *Yersinia* for a reason. Sure enough, the survival of enteropathogenic Yersiniae grown under a variety of environmental stresses (Hinchliffe *et al.*, 2008) or during the initial stages of gastrointestinal colonization in a murine model of infection (Venecia and Young, 2005) both require an intact Rcs phosphorelay pathway. Additionally, microarray data indicate that the Rcs regulon in *Y. pseudotuberculosis* is extensive, with many targets encoding for factors predicted either to influence the bacterial envelope or to serve as determinants for host and/or environmental survival (Hinchliffe *et al.*, 2008). In *Y. enterocolitica* one distinctive virulence factor under Rcs control is the chromosomal-encoded Ysa-Ysp T3SS (Venecia and Young, 2005). Similar to the contributions made by the *rcs* loci in *Yersinia* virulence, the *ysa-ysp* loci also play an important role during the early gastrointestinal stage of murine infections (Venecia and Young, 2005). This suggests that tight temporal and spatial control of *ysa-ysp* expression by the Rcs phosphorelay system is a critical strategy promoting initial enteric *Yersinia* colonization prior to systemic dissemination into deeper tissue. Future work directed at understanding the precise mechanisms of Rcs-mediated temporal control of other survival strategies is therefore destined to provide valuable insight into key aspects of *Yersinia* pathogenicity.

Additional two-component regulatory systems

Human pathogenic *Yersinia* occupy a wide array of ecological niches; *Y. pestis* has a strict life cycle alternating between flea vector and mammalian host, while the foodborne enteropathogens possess environmental niches in soil and water, along with spasmodic mammalian host infections. Yersiniae therefore encounter and survive in environments that are in a constant state of physico-chemical flux. Pathogenic *Yersinia* sense these environmental cues to coordinate temporal and spatial control of virulence gene expression, which is especially important if the genes are unlinked and therefore scattered around the entire genome. Most often, this regulatory task is fulfilled by an array of TCSs that essentially serve as the sensory organs of bacteria. Depending on the species, about 25–30 putative TCSs exist within the genome of pathogenic *Yersinia*. The importance of a few of these in regulation of virulence gene expression is illustrated by their requirement for intracellular survival and pathogenicity (Flamez *et al.*, 2008; O'Loughlin *et al.*, 2010). For example, by possibly responding to Mg^{2+} and Ca^{2+} concentrations, the role of the PhoQ sensor kinase-PhoP response regulator TCS in *Yersinia* virulence appears to stem from an involvement in promoting intracellular growth and survival (Oyston *et al.*, 2000; Grabenstein *et al.*, 2004; O'Loughlin *et al.*, 2010) and/or in negatively regulating biofilm formation (Sun *et al.*, 2009b). Another TCS composed of the EnvZ sensor kinase and the OmpR response regulator is also essential for *Yersinia* virulence. Probably, this is because it is required for *Yersinia* survival in high osmotic and acid conditions (Dorrell *et al.*, 1998; Brzostek *et al.*, 2003; Flamez *et al.*, 2008; Hu *et al.*, 2009a) and regulates genes encoding pathogenicity factors such as motility (Hu *et al.*, 2009b) and invasin (Brzostek *et al.*, 2007). In *Y. ruckeri*, pathogenesis requires a functional BarA-UvrYTCS for invasion of host epithelial cells and to resist oxidative stress (Dahiya and Stevenson, 2010). Hence, TCSs form important bridges between the external environment and survival and/or virulence gene regulation in pathogenic *Yersinia*.

9.3.3 RovA – a global regulator of virulence

RovA is a member of the MarR/SlyA family of transcriptional regulators. It was first identified in *Yersinia* because it was required for transcription of the invasin adhesin, a putative autotransporter (Revell and Miller, 2000; Nagel *et al.*, 2001). Whole-genome analyses have since indicated a global regulatory role for RovA in pathogenic *Yersinia* (Cathelyn *et al.*, 2006, 2007). Among the RovA regulon are additional virulence strategies including the prominent pH 6 antigen CU system and other uncharacterized potential secretion systems. RovA levels are therefore controlled tightly at multiple levels in accordance with the prevailing environmental growth conditions (Fig. 9.3).

In the first instance, an auto-amplification loop overcomes H-NS-mediated silencing to promote *rovA* transcription (Heroven *et al.*, 2004; Tran *et al.*, 2005; Lawrenz and Miller, 2007). Especially at lower temperature, RovA can outcompete H-NS successfully for similar binding sequences in the *rovA* promoter. RovA binding therefore activates *rovA* transcription. Transcription of *rovA* is refined further by additional inputs; positive regulation might also engage LeuO, a LysR-like regulator apparently involved in the stringent response (Lawrenz and Miller, 2007), while negative regulation involves another LysR-type protein, RovM (Heroven and Dersch, 2006). This negative regulatory loop probably senses nutritional status since its output is refined by the participation of a Csr-type carbon storage regulator system controlling *rovM* expression (Heroven *et al.*, 2008). Our own work has also demonstrated that an activated CpxA-CpxR pathway moves to repress *rovA* transcription directly during occasions of ECS [Liu, Obi and Francis, unpublished data] (Carlsson *et al.*, 2007b). In addition, RovA levels and activity are affected post-transcriptionally; temperature-dependent conformational changes render RovA less able to bind to target DNA sequences and to resist proteolysis by endogenous proteases (Herbst *et al.*, 2009). It is interesting that RovA levels are controlled tightly by multiple pathways that integrate diverse

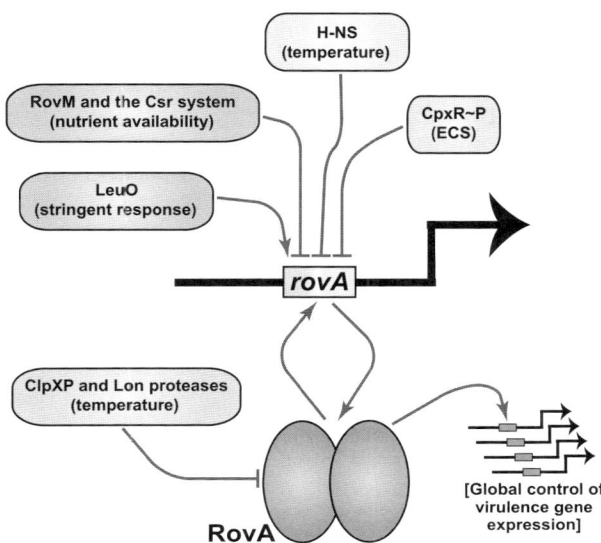

Fig. 9.3. Environmental control of the global virulence gene regulator RovA. RovA is controlled strictly at both the transcriptional and post-transcriptional levels in response to multiple environmental cues. These cues lead to an induction (arrow) or reduction (blunted line) of cytoplasmic RovA. In response to low temperature, *rovA* transcription undergoes auto-amplification in which RovA competes successfully with the H-NS repressor for binding to the *rovA* promoter. However, auto-amplification is also kept in check at elevated temperature by the thermoregulated proteolysis of RovA by ClpXP and Lon proteases. In response to nutritional stress (RovM) and the stringent response (LeuO), the levels of *rovA* transcription are further refined. Finally, CpxA-CpxR activation by ECS also functions to repress *rovA* transcription. RovA regulation by multiple pathways suggests that levels of this transcriptional regulator are controlled tightly to ensure that infecting bacteria restrict virulence factor production to those occasions where they will have maximal benefit.

environmental cues. The dynamic nature of this regulation is obviously required for RovA to impart significant global control on virulence gene transcription in *Yersinia* sp.

9.3.4 Nucleotide-based secondary messengers

Carbon catabolite repression – cAMP-CRP

When members of the Enterobacteriaceae sense carbon-limiting environments, crosstalk between the phosphoenolpyruvate-carbohydrate phosphotransferase system, adenylate cyclase (producer of the second messenger cyclic AMP – cAMP) and CRP (cAMP receptor protein) is often used to stimulate the production of alternate carbon-scavenging pathways (Deutscher, 2008; Gorke and Stulke, 2008) (Fig. 9.4). This enables bacteria to

sequester and utilize other carbon sources, but only when their preferred source is unavailable – usually glucose because it is readily accessible and promotes rapid growth.

Generally referred to as carbon catabolite repression, preferential glucose utilization therefore inhibits the expression of genes required for the metabolism of other secondary carbon nutrients (Deutscher, 2008; Gorke and Stulke, 2008). On the other hand, glucose limitation effectively stimulates the activation of adenylate cyclase to produce high levels of cAMP. In turn, cAMP is recognized by CRP, a global activator of transcription initiation (Lawson *et al.*, 2004; Won *et al.*, 2009). Crucially, CRP is active and able to target upstream DNA promoter sequences to stimulate gene transcription only when complexed with cAMP. Thus, such a global regulatory mechanism allows

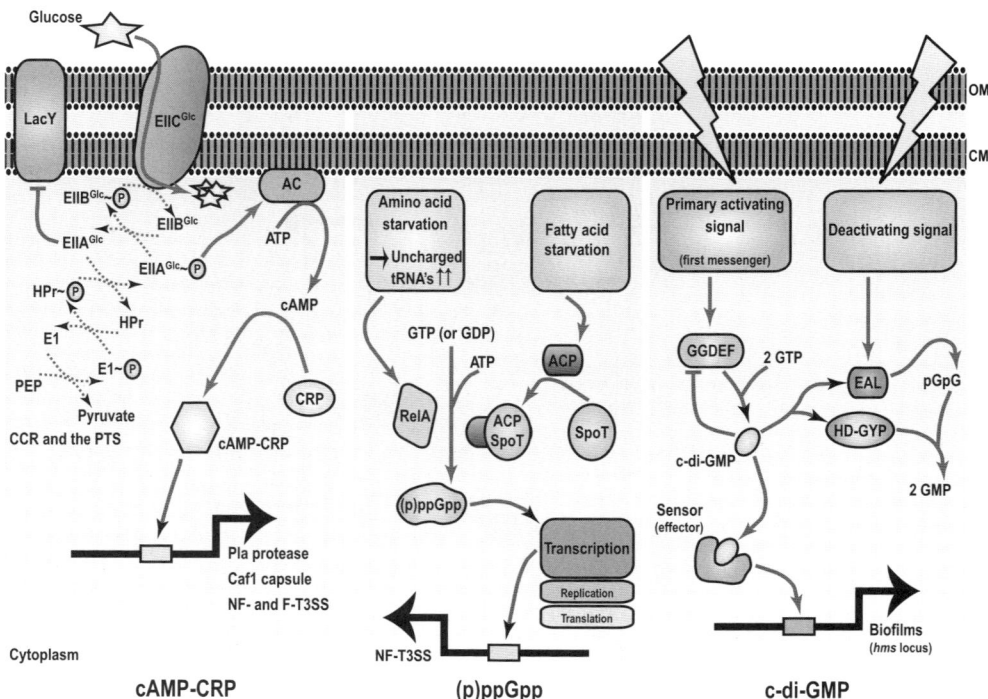

cAMP-CRP   **(p)ppGpp**   **c-di-GMP**

Fig. 9.4. Secretion control by nucleotide-based secondary messengers. Although not studied extensively, the major nucleotide-based second messengers identified in *Yersinia* are cAMP, (p)ppGpp and c-di-GMP. Our understanding of their function in *Yersinia* is therefore derived primarily from studies on the respective *E. coli* systems. The cAMP molecule is part of the carbon catabolite repression (CCR) pathway. The phosphorylation state of the EIIA domain of the glucose transporter is integral to this control, which is determined by the phosphoenolpyruvate (PEP)-carbohydrate phosphotransferase system (PTS). In the presence of glucose (Glc) as the favoured carbon source, non-phosphorylated EIIA^{Glc} accumulates and can inhibit transporters of non-preferred carbohydrates (such as the lactose transporter, LacY). In contrast, low glucose levels lead to phosphorylated EIIA^{Glc} (EIIA^{Glc}~P) accumulation that activates adenylate cyclase (AC) to stimulate production of cAMP, which is then bound by the cAMP receptor protein (CRP). cAMP-CRP complexes can bind and activate the promoters of other catabolic genes, along with various *Yersinia* secretion systems. The alarmone, (p)ppGpp, is synthesized by RelA in response to amino acid starvation in cooperation with a complex of acyl carrier protein (ACP) and SpoT that forms during depletion of fatty acids. By direct binding to the RNA polymerase, (p)ppGpp can effect transcription of a plethora of genes. It is also known to have both direct and indirect effects on translation and DNA replication. In *Yersinia* (p)ppGpp affects non-flagella (NF-)T3SSs, although how this occurs is currently unknown. Finally, primary environmental and/or cellular signals stimulate diguanylate cyclases characterized by GGDEF domains to generate c-di-GMP. Effector molecules, which can be either protein or RNA, engage c-di-GMP to affect some regulatory output. In *Yersinia* c-di-GMP production by the GGDEF containing protein HmsT induces Hms-mediated biofilm formation at ambient temperature. In this case, the sensor/effector molecule involved in this process is presumed to be HmsR, although this is not well established. When an upper threshold of c-di-GMP is reached, a process of feedback inhibition of GGDEF activity is invoked. Secondary deactivating signals can also initiate a c-di-GMP degradation pathway involving phosphodiesterase activity possessed by proteins containing either EAL or HD-GYP domains. HmsP possesses phosphodiesterase activity and, accordingly, regulates *Yersinia* biofilm formation negatively.

free-living bacteria to maximize their growth rate in an ultra-competitive and micro-biologically diverse environment.

It is not surprising therefore that such a regulatory mechanism has been adapted to fine-tune virulence gene expression specifically in several bacterial pathogens. The pathogenic Yersiniae are no exception, with a few studies implicating CRP in the regulation of virulence gene expression (Fig. 9.4). CRP was shown to affect the gene expression of as much as 6% of the *Y. pestis* genome, although the number of direct promoter targets was determined to be considerably fewer (Zhan *et al.*, 2008). CRP was necessary to activate positively the *pla* locus encoding for the surface-located Pla protease (Kim *et al.*, 2007; Zhan *et al.*, 2008). The Pla protease is essential for *Y. pestis* virulence through its role in enhancing bacterial spread by activating the plasminogen-plasmin proteolytic cascade to dissolve fibrin complexes and to cause tissue damage (Suomalainen *et al.*, 2007). The function of other secreted virulence factors of *Yersinia* affected by cAMP-CRP activity include the Ysc-Yop, Ysa-Ysp and flagella T3SSs, as well as the capsular F1 antigen CU system (Petersen and Young, 2002; Zhan *et al.*, 2008, 2009). Precisely how CRP and cAMP levels exert these regulatory effects on multi-component biological processes is currently unclear. However, expression from the *sycOypkAyopJ* locus, encoding for a T3SS chaperone (SycO) and two translocated effector substrates (YpkA and YopJ) of the Ysc-Yop T3SS, is repressed directly by cAMP-CRP binding to upstream regulatory sequences (Zhan *et al.*, 2009). These CRP targets are specialized *Yersinia* virulence factors encoded by genes located on extrachromosomal plasmids that have been acquired horizontally during evolution. This could occur only through the subsequent introduction of promoter elements responsive to cAMP-CRP. Such findings provide illustration of how the pressing need of *Yersinia* to survive in the environment and during host infections has conjured effective use of normal metabolic processes to achieve correct spatial and temporal control of virulence gene expression.

Stringent response and (p)ppGpp

The bacterial stringent response is defined broadly as a starvation-induced response that leads to accumulation of guanosine poly-phosphate molecules – represented collect-ively as (p)ppGpp (Potrykus and Cashel, 2008; Srivatsan and Wang, 2008) (Fig. 9.4). Being the focus of much attention over several decades, (p)ppGpp molecules are prominent second messengers that can initiate dramatic 'reprogramming' of multiple physiological processes within the bacterial cell in response to changing nutrient availability and environ-mental stress (Potrykus and Cashel, 2008; Srivatsan and Wang, 2008). With profound effects on replication, transcription initiation and translation, it is no wonder that these small nucleotide second messengers are also integral to the control of bacterial virulence gene expression (Braeken *et al.*, 2006). In brief, two enzymatic pathways are essentially responsible for (p)ppGpp synthesis, although these can deviate depending on the bacterial species (Potrykus and Cashel, 2008; Srivatsan and Wang, 2008) (Fig. 9.4). At least in the Enterobacteriaceae, the first pathway is mediated by RelA in response to amino acid starvation and the subsequent accumulation of uncharged tRNAs. The second pathway is dependent on an interaction between SpoT and acyl carrier protein caused by a depriv-ation of carbohydrates, fatty acids or inorganic ions. In *Y. pestis* both RelA and SpoT are essential for the accumulation of (p)ppGpp (Sun *et al.*, 2009a). The inability to synthesize (p)ppGpp in a *Y. pestis relA*, *spoT* double mutant induced autoaggregation at low temperature, reduced the capacity to synthesize and secrete Yops of the Ysc-Yop T3SS and attenuated virulence (Sun *et al.*, 2009a). In fact, such was the attenuation of this strain that it may have potential as a vaccine strain. The precise mechanism by which *Y. pestis* uses (p)ppGpp to regulate the synthesis and secretion of Yops remains to be studied in any detail.

Planktonic versus sessile Yersinia: the c-di-GMP signalling network in biofilm formation

Arguably, the most prominent second mes-senger is bis-(3′-5′)-cyclic dimeric guanosine monophosphate (c-di-GMP; cyclic diguany-late) (Fig. 9.4). In bacteria it regulates cell-surface associate traits that can be seen as alterations in secretion, cell adhesion and motility, leading to biofilm formation (Tamayo *et al.*, 2007; Hengge, 2009; Schirmer and Jenal, 2009). In very general terms, c-di-GMP regulates the transition between planktonic and sessile lifestyles through inverse repression of motility and pathogenicity factors and stimulation of biofilm formation. Ultimately, this increases the survivability of bacterial pathogens in extreme environments, creating persistent bacterial reservoirs that bode well for subsequent infections. c-di-GMP is synthesized from two GTP molecules by diguanylate cyclases containing GGDEF domains. Specific phosphodiesterases associ-ated with EAL or HD-GYP domains are responsible for c-di-GMP degradation (Tamayo *et al.*, 2007; Hengge, 2009; Schirmer and Jenal, 2009) (Fig. 9.4). Interestingly, GGDEF and EAL domains can occur in the same protein or be present in multiple proteins encoded by any given bacterial genome. This creates potential for a complex and puzzling network of signalling events to generate the appropriate effector output. For example, in *Y. enterocolitica* where the effect of c-di-GMP regulatory control has not been studied, as many as 22 GGDEF and/or EAL domain-containing proteins have been identi-fied from the sequenced genome (Heermann and Fuchs, 2008). Interestingly, in the obligate *Y. pestis* pathogen, the numbers are smaller, with only eight such proteins suggested from genome sequence scrutiny (Darby, 2008). Nevertheless, these observations point to the existence of complex c-di-GMP-mediated regulatory networks in control of virulence and biofilm formation in *Yersinia*.

Yersinia and biofilm formation has received most attention in the plague-causing bacterium. The ability of *Y. pestis* to form biofilms in the flea vector contributed to its evolution as a dangerous pathogen because this was an integral feature of bacterial transmission from vector to animal host (Darby, 2008; Hinnebusch and Erickson, 2008). The *hms* locus encoded on a chromosomal pathogenicity island is largely responsible for the biosynthesis and secretion of an exopolysaccharide material that is used to form a highly aggregative biofilm (Darby *et al.*, 2002; Jarrett *et al.*, 2004; Kirillina *et al.*, 2004). Bacteria produce this material only at low temperature (flea body temperature) and this is required for infection and blockage of the flea proventriculus (Hinnebusch *et al.*, 1996; Jarrett *et al.*, 2004). This promotes bacterial transmission because a feeding flea regurgitates an infected blood meal. Unlinked to the *hms* locus are two regulatory genes, *hmsT* and *hmsP*; *hmsT* mutants have diminished biofilm-forming capacity, con-trasting with *hmsP* mutants that display enhanced biofilm formation (Jones *et al.*, 1999; Kirillina *et al.*, 2004; Bobrov *et al.*, 2005). Significantly, HmsT possesses diguanylate cyclase activity and HmsP displays phospho-diesterase activity. Thus, *Y. pestis* biofilms are initiated through synthesis of c-di-GMP by HmsT and are repressed by the degradative activity of HmsP (Kirillina *et al.*, 2004; Bobrov *et al.*, 2005; Simm *et al.*, 2005). A complex of regulatory and structural Hms proteins form in the inner membrane and it is likely that c-di-GMP levels control extra polysaccharide production and biofilm formation (Bobrov *et al.*, 2008).

c-di-GMP signalling is probably not the sole control network of *Y. pestis* biofilm formation. In other bacteria, quorum sensing – a process used by bacteria to determine the population density of self or other organisms and to regulate gene expression accordingly – is often implicated in regulation of multi-cellular sessile behaviour (Irie and Parsek, 2008). Although roles for quorum sensing-dependent regulation of virulence gene expression in enteropathogenic *Yersinia* are becoming apparent (Atkinson *et al.*, 2006), it is not entirely clear if this mode of regulation is applicable to *Y. pestis* pathogenicity, vector transmission and biofilm development (Bobrov *et al.*, 2007).

9.3.5 The SsrA-SmpB (tmRNP) system

A fundamental capacity of all bacterial cells is the ability to target for proteolytic degradation non-functional proteins translated from stalled ribosomes located on damaged mRNA templates lacking appropriate stop codons. In a process termed *trans*-translation, this enables bacteria to perform protein quality control functions (folding and degradation) and to release and redirect stalled ribosomes for renewed *de novo* translation (Keiler, 2008; Richards *et al.*, 2008; Hayes and Keiler, 2010). The key component of this quality control system is the small stable transfer-messenger RNA (tmRNA; also known as SsrA or 10S RNA) that contains an ORF with coding capacity for an 11-amino acid peptide. In association with the small protein, SmpB, an SsrA-SmpB complex (also termed tmRNP for transfer-messenger ribonucleoprotein) is recruited by EF-Tu to a stalled ribosome on truncated mRNA. This induces release and decay of the existing mRNA template; a process designed to avert further ribosome stalling events. In turn, the ribosome switches to the tmRNA/SsrA template permitting translation of the ORF, appending the 11-amino acid product on to the C-terminus of the truncated nascent polypeptide. The tagged protein is then targeted for degradation by multiple cytoplasmic proteases concomitant with ribosomal subunit release in preparation for their recycling (Keiler, 2008; Richards *et al.*, 2008; Hayes and Keiler, 2010).

Defects in stress responsiveness, DNA replication, growth and cell division, antimicrobial sensitivity and pathogenicity have been observed in bacteria lacking SsrA or SmpB (Keiler, 2007). In bacteria this system therefore has the capacity to modulate distinct physiological processes at the level of gene expression. So it has been observed in *Y. pseudotuberculosis*; an *ssrA-smpB* mutant was more sensitive to hostile environments – including inside macrophages – and was attenuated in the murine model of infection (Okan *et al.*, 2006). This avirulence phenotype was linked to a drastic reduction in Ysc-Yop type III secretion and in flagella-mediated motility. The Ysc-Yop type III secretion deficiency apparently resulted from a specific loss of ribosome rescue function, whereas the motility defect was probably induced by loss of both ribosome rescue and protein tagging functions (Okan *et al.*, 2006). Moreover, Ysc-Yop T3SS was impaired at the level of *ysc-yop* transcription initiation due to reduced levels of the specific AraC-like transcriptional activator, LcrF (Okan *et al.*, 2006). (More information about the role of LcrF in transcriptional control of the Ysc-Yop T3SS can be seen later in this chapter.) Thus, the SsrA-SmpB system plays an important role in controlling T3SS gene expression in *Yersinia*, which is necessary to maintain bacterial infectivity and survival inside a host.

9.3.6 DNA adenine methylation and pathogenicity

A common post-replicative methylation event involves DNA adenine methyltransferase (Dam), a modifying enzyme that functions in bacteria without any cognate restriction endonuclease (Lobner-Olesen *et al.*, 2005; Wion and Casadesus, 2006). Dam is responsible for catalysing the transfer of a methyl group from *S*-adenosyl-L-methionine to the N^6 position of adenine in the sequence 5′-GATC-3′; an event having major implications for basic bacterial processes involving DNA, such as replication initiation, DNA segregation, transcription, transposition and mutation repair (Lobner-Olesen *et al.*, 2005; Wion and Casadesus, 2006). Furthermore, Dam-dependent methylation can even influence virulence gene expression in various bacterial pathogens, including the Yersiniae (Heusipp *et al.*, 2007; Marinus and Casadesus, 2009). When not lethal, *dam* mutants in some strains of *Y. pestis* and *Y. pseudotuberculosis* caused attenuation in a mouse model of infection (Robinson *et al.*, 2005; Taylor *et al.*, 2005). Virulence attenuation was also observed for *Y. pseudotuberculosis* Dam+ overexpressing strains (Julio *et al.*, 2001). These attenuated strains also provide effective protection against subsequent challenge by wild-type bacteria, potentially making them suitable as the basis of future vaccines (Julio *et al.*, 2001; Robinson *et al.*, 2005; Taylor *et al.*, 2005). The *Y. enterocolitica dam* gene is essential but its overexpression does modulate virulence factor production, leading to enhanced motility and invasion of eukaryotic cells, and also to an increase in LPS molecules lacking O-antigen side chains (Falker *et al.*, 2005, 2006, 2007). Modulating Dam activity in *Y. enterocolitica* and *Y. pseudotuberculosis* also deregulates Ysc-Yop type III secretion, causing constitutive Yops release even when grown in repressive conditions such as low temperature and/or high calcium ions (Julio *et al.*, 2001, 2002; Falker *et al.*, 2006). At least in *Y. enterocolitica* this effect was post-translational, with Dam overexpression causing an elevation of *clpXP* expression, prompting ClpXP-dependent degradation of the Ysc-Yop regulator, LcrG (Falker *et al.*, 2006). Interestingly, a second DNA adenine methyltransferase, denoted YamA, has been reported in *Y. pseudotuberculosis* that is necessary for full virulence in a murine intragastric model of infection (Pouillot *et al.*, 2007). However, the cause of this attenuation is unknown.

Adenine methylation is therefore widely implicated in control of *Yersinia* virulence gene expression. In most cases, however, the actual molecular basis for this regulation is not known. In addition to the intriguing post-translational effect of LcrG degradation by ClpXP (Falker *et al.*, 2006), another attractive hypothesis is that altered DNA methylation could modulate the affinity by which transcriptional repressors and/or activators are able to bind regulatory DNA elements upstream of various virulence genes (Wion and Casadesus, 2006; Heusipp *et al.*, 2007). This would affect the levels of virulence gene transcription and subsequent bacterial infectivity. At present, however, there is no experimental proof that such a regulatory mechanism occurs in reality. Until the precise mechanisms are characterized sufficiently, doubt will therefore remain over whether the effects of reducing or elevating Dam production on *Yersinia* virulence are direct, or whether they are an indirect result of disrupting some fundamental cellular process caused by modulating Dam activity.

9.3.7 Ysc-Yop type III secretion and the low calcium response – a genuine crosstalk with metabolism

Transcriptional regulation

Temperature is a key environmental cue influencing transcriptional control of multiple genes, including those encoding for the Ysc-Yop T3SS. For the most part, low temperature repression of *ysc-yop* and *lcr* (low calcium response) genes is mediated by YmoA, a small histone-like protein (Cornelis *et al.*, 1991) (Fig. 9.5).

When bound to intrinsically curved DNA within specific promoter elements, YmoA effects silencing of gene transcription. However, changes in DNA topology induced by elevated temperature reduced YmoA affinity to DNA. Also under these conditions, YmoA stability is diminished through concerted proteolysis by the ClpXP and Lon proteases (Jackson *et al.*, 2004). In response to this thermoregulated YmoA inactivity, *lcrF* is probably the first T3SS gene to be transcribed. It encodes for the AraC-like transcriptional activator, LcrF, needed for transcription of all other Ysc-Yop T3SS genes (Yother *et al.*, 1986; Cornelis *et al.*, 1989; Lambert de Rouvroit *et al.*, 1992). This family of transcriptional regulators is characterized by a C-terminal helix-turn-helix DNA-binding motif that allows specific LcrF binding to DNA sequences of *ysc-yop* gene promoters. Since *lcrF* transcription is also subject to positive-feedback auto-regulatory control augmenting LcrF production, levels of *ysc-yop* transcription are amplified rapidly. Thus, temperature, YmoA and LcrF control the levels of *lcrF* transcription, while LcrF in turn directs levels of *ysc-yop* transcription. Admittedly, this view of *ysc-yop* transcriptional control is overly simplistic. A putative hairpin structure in the *lcrF* mRNA transcript suggests that translation of LcrF is also thermoresponsive (Hoe and Goguen, 1993). Moreover, as discussed already in this chapter, the tmRNP system influences the level of *lcrF* transcription, with dramatic consequences (Okan *et al.*, 2006). Finally, a newly identified LysR-type transcriptional regulator, YtxR, also operates a dominant transcriptional off switch for *ysc-yop* expression (Axler-DiPerte *et al.*, 2009). YtxR is anticipated to compete with LcrF for binding to overlapping DNA sequences within the *ysc-yop* promoter regions. In reality, therefore, multiple environmental conditions underpin transcriptional control of the Ysc-Yop T3SS (Fig. 9.5). No doubt more contributing factors will be uncovered in time; for example, possibly after the regulatory

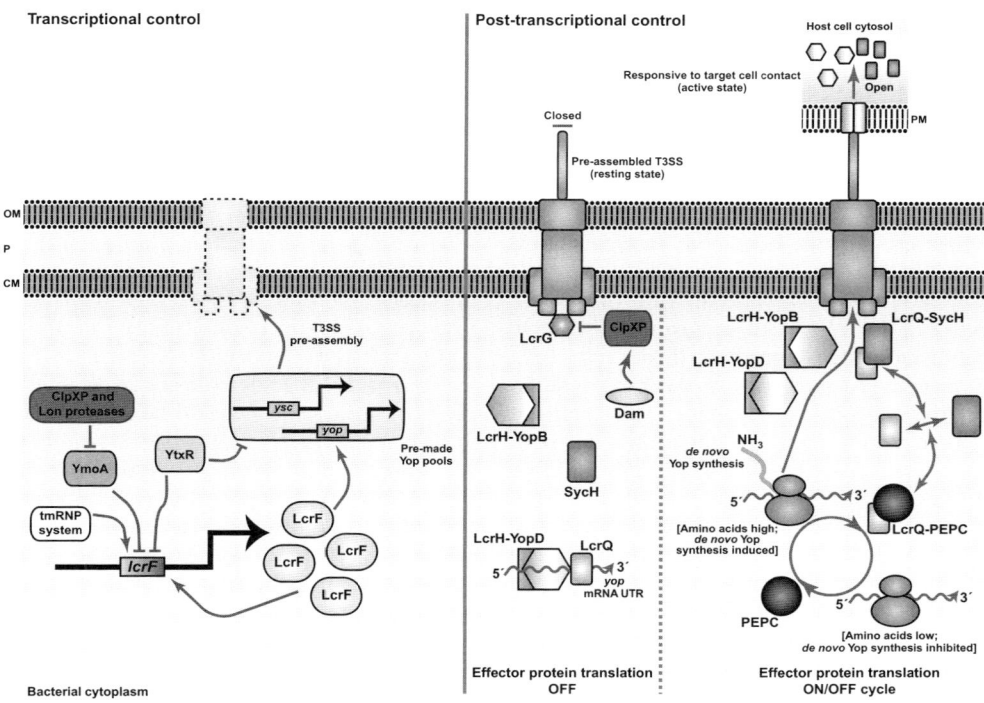

Fig. 9.5. Regulation of the Ysc-Yop T3SS. Transcriptional control centres on the activity of the LcrF transcriptional activator that is necessary for transcription of *ysc* and *yop* genes. Early transcription of *yop* genes probably generates stores of effectors in readiness for secretion immediately on target host cell contact. Early transcription of *ysc* genes drives the production of Ysc components that pre-assemble into the secretion apparatus. Levels of LcrF are controlled by temperature-dependent repression by the DNA-binding protein, YmoA. At elevated temperature, however, repression is relieved by reduced binding activity and by the protease-dependent degradation of YmoA. YtxR also turns off *lcrF* transcription by disrupting the auto-amplification loop associated with direct LcrF binding to the endogenous promoter. In a manner that is not understood, *lcrF* transcription also requires input from the transfer-messenger ribonucleoprotein (tmRNP) system. Post-transcriptional control of Ysc-Yop secretion is also multifactorial. In a resting state, the secretion apparatus is gated (closed) both internally (LcrG) and externally (LcrH). This causes a cytoplasmic build-up of the regulatory complex involving YopD, LcrH and LcrQ. When bound to the untranslated region (UTR) of *yop* mRNA, this complex probably inhibits ribosome access to the shine-dalgarno sequence to restrict effector translation. When activated in response to target cell contact, LcrG might be degraded by the ClpXP protease system in response to high DNA adenine methylation (Dam) activity. The secretion machine is opened to allow concerted release of repressive factors such as YopD and LcrQ. In turn, ribosomes gain access to *yop* mRNA to initiate *de novo* synthesis of Yops. During intense Yop synthesis, levels of available amino acids quickly become exhausted. Some of these are replenished by the activity of phosphoenol pyruvate carboxylase (PEPC), which can be inhibited by an interaction with LcrQ. Newly synthesized Yops are stabilized by customized T3S chaperones such as LcrH and SycH. After Yops secretion, these free T3SS chaperones can bind LcrQ, releasing PEPC to replenish amino acid pools once more, permitting another round of Yops production. As their stability will again require their customized T3SS chaperones, LcrQ will be freed to interact and inhibit PEPC for another time. In this way, successive rounds of Yops production are interspersed with periods of inaction that is concomitant with LcrQ-dependent control of PEPC activity and associated restoration of amino acid reserves. The molecular mechanism governing CpxA-CpxR, cAMP-CRP and (p)ppGpp regulation of Ysc-Yop T3SS is not yet elucidated and is therefore omitted from this schematic diagram.

mechanisms governing the CpxA-CpxR, cAMP-CRP- and (p)ppGpp-dependent effects on Ysc-Yop T3SS have been resolved.

Post-transcriptional regulation

Human pathogenic *Yersinia* spp. display a fascinating nutritional dependency for Ca^{2+} when growing at host body temperatures, such that in the absence of Ca^{2+} they soon stop replicating DNA and dividing (Brubaker, 2007). However, in these growth-restrictive conditions (high temperature and low Ca^{2+}), a pronounced induction of the Ysc-Yop T3SS occurs (Brubaker, 2007). This quirky phenomenon is known as the low calcium response (LCR) (Goguen *et al.*, 1984). While the molecular basis for the LCR is still mysterious, it could involve defects in PMF generation (as ATP pools are low) and an inability to reduce toxic levels of Na^+ from the cytoplasm (Brubaker, 2005, 2007). *Yersinia* are toxic to Na^+ only during active Ysc-Yop type III secretion, since a Ca^{2+} surplus even at high temperature simultaneously represses Ysc-Yop activity and relieves Na^+ sensitivity (Fowler *et al.*, 2009). It is not known why *Yersinia* is so sensitive to elevated Na^+ ions, but it must be due to the inhibition of some essential cellular process needed for growth (Brubaker, 2007; Fowler *et al.*, 2009). It follows that excessive Na^+ build-up inside *Yersinia* is most probably linked to the specific action of one or more Ysc-Yop components.

Although this notion remains speculative, specific regulatory components of the Ysc-Yop T3SS that can somehow sense Ca^{2+} levels would be logical candidates for influencing the LCR. The regulatory molecule, LcrQ, is a good case in point; especially given its exclusivity to the *Yersinia* Ysc-Yop system – no homologue exists in any other bacteria. LcrQ has the ability to interact with various T3SS chaperones, which are normally involved in pre-secretory stabilization and/or efficient and specific secretion of cognate substrates (Francis, 2010). An interaction of LcrQ with the T3SS chaperone, LcrH, and its cognate substrate, YopD, mediates the post-transcriptional repression of Ysc-Yop type III secretion at a time when the actual T3SS is closed (Williams and Straley, 1998; Cambronne

et al., 2000; Francis *et al.*, 2001; Anderson *et al.*, 2002; Cambronne and Schneewind, 2002) (Fig. 9.5). The tripartite complex is thought to bind to the 5′-untranslated region of *yop* mRNA and deny access of ribosomes to the mRNA template (Anderson *et al.*, 2002; Cambronne and Schneewind, 2002). On induction through Ca^{2+} depletion and elevated temperature (in the laboratory) or through target cell contact (during an animal infection), the repressive complex is destabilized because an open T3SS actively promotes LcrQ and YopD secretion to deplete their levels inside the cytoplasm. Presumably, ribosomes can then regain access to each mRNA transcript, permitting translation and an upsurge of substrate synthesis and secretion (Pettersson *et al.*, 1996; Wulff-Strobel *et al.*, 2002; Cambronne *et al.*, 2004) (Fig. 9.5). Thus, the ability of LcrQ to engage T3SS chaperones is important for the post-transcriptional control of Ysc-Yop type III secretion.

However, LcrQ also associates with phosphoenol pyruvate carboxylase (PEPC) involved in the biosynthesis of oxaloacetate-derived amino acids (Schmid *et al.*, 2009) (Fig. 9.5). The consequence of this interaction is to inhibit PEPC activity, effectively reducing the levels of these oxaloacetate amino acid derivatives. A model is proposed whereby, following active type III synthesis and secretion of pre-made pools of substrate, freed T3SS chaperone is able to sequester LcrQ from PEPC. In turn, liberated PEPC replenishes amino acid pools to initiate a subsequent round of T3SS substrate synthesis. Solubility and stability of these newly synthesized Yop substrates necessitates recruitment of their cognate T3SS chaperones. This releases LcrQ again to seek out and neutralize PEPC activity to reduce unnecessary synthesis of amino acids (Schmid *et al.*, 2009). Thus, the dual ability of LcrQ to bind both T3SS chaperones and PEPC might ensure amino acids are available when the synthesis of T3SS substrates are most needed – during target cell contact in an infected host (Schmid *et al.*, 2009) (Fig. 9.5). In essence, type III synthesis and secretion depletes *Yersinia* of oxaloacetate-derived amino acids, which in all likelihood would restrict bacterial growth (i.e. akin to the LCR). Since amino acid levels flux between high and low depending on T3SS activity and the PEPC-LcrQ interaction, this suggests that LcrQ function could be a bridge between *Yersinia* T3SS, central metabolism and the LCR.

9.4 Conclusions

Historically, studies of *Yersinia* pathogenicity have focused overwhelmingly on the Ysc-Yop T3SS encoded by a common virulence plasmid present among all three species pathogenic to humans. Since its inception in the 1980s, work on this secretion system has provided a wealth of information concerning bacteria–host cell interactions; to the point of contributing to the emergence and recognition of a new research field – cellular microbiology. Even now, the field of *Yersinia* infection biology continues to trailblaze new discoveries. The various *Yersinia* genome-sequencing projects have allowed hitherto unprecedented insight into the evolution of a bacterial pathogen; all the while exposing researchers to many new and fascinating virulence and/or survival factors that further expand (and amaze) current perceptions of bacterial–host interactions. With this comes the need to place these discoveries into a physiological context. A continuing research focus is therefore to understand better how this diverse group of *Yersinia* pathogens taps into environmental and nutritional cues to control their customized armoury of virulence/survival factors spatially and temporally. Like previous achievements in the field, new developments promise to benefit our knowledge on diverse regulatory control mechanisms. No doubt an appreciation of these regulatory connections will heighten selection of the most appropriate targets for the design and development of eagerly anticipated anti-*Yersinia* therapeutic drugs and vaccines.

Acknowledgements

Thanks are extended to the numerous local and international *Yersinia* researchers – their experimental and intellectual contributions have provided a continued impetus for greater understanding of *Yersinia* biology. Apologies are extended to those colleagues whose work was not cited in this article due to space limitations. Work in the Francis laboratory is performed within the framework of the Umeå Centre for Microbial Research (UCMR) Linnaeus Programme. Past and present grant support from the Swedish Research Council, Carl Tryggers Foundation for Scientific Research, Umeå University Basic Science-Oriented Biotechnology Research Fund, Umeå University Foundation for Medical Research, Swedish Cystic Fibrosis Association Research Fund and Swedish Medical Association is gratefully acknowledged.

References

Aberg, V. and Almqvist, F. (2007) Pilicides-small molecules targeting bacterial virulence. *Organic and Biomolecular Chemistry* 5, 1827–1834.

Ades, S.E. (2008) Regulation by destruction: design of the sigmaE envelope stress response. *Current Opinion in Microbiology* 11, 535–540.

Anderson, D.M., Ramamurthi, K.S., Tam, C. and Schneewind, O. (2002) YopD and LcrH regulate expression of *Yersinia enterocolitica* YopQ by a posttranscriptional mechanism and bind to *yopQ* RNA. *Journal of Bacteriology* 184, 1287–1295.

Atkinson, S., Sockett, R.E., Camara, M. and Williams, P. (2006) Quorum sensing and the lifestyle of *Yersinia*. *Current Issues in Molecular Biology* 8, 1–10.

Axler-DiPerte, G.L., Hinchliffe, S.J., Wren B.W. and Darwin, A.J. (2009) YtxR acts as an overriding transcriptional off switch for the *Yersinia enterocolitica* Ysc-Yop type 3 secretion system. *Journal of Bacteriology* 191, 514–524.

Backert, S. and Meyer, T.F. (2006) Type IV secretion systems and their effectors in bacterial pathogenesis. *Current Opinion in Microbiology* 9, 207–217.

Badger, J.L. and Miller, V.L. (1995) Role of RpoS in survival of *Yersinia enterocolitica* to a variety of environmental stresses. *Journal of Bacteriology* 177, 5370–5373.

Badger, J.L., Young, B.M., Darwin, A.J. and Miller, V.L. (2000) *Yersinia enterocolitica* ClpB affects levels of invasin and motility. *Journal of Bacteriology* 182, 5563–5571.

Bengoechea, J.A., Najdenski, H. and Skurnik, M.

(2004) Lipopolysaccharide O antigen status of *Yersinia enterocolitica* O:8 is essential for virulence and absence of O antigen affects the expression of other *Yersinia* virulence factors. *Molecular Microbiology* 52, 451–469.

Bingle, L.E., Bailey, C.M. and Pallen, M.J. (2008) Type VI secretion: a beginner's guide. *Current Opinion in Microbiology* 11, 3–8.

Bobrov, A.G., Kirillina, O. and Perry, R.D. (2005) The phosphodiesterase activity of the HmsP EAL domain is required for negative regulation of biofilm formation in *Yersinia pestis*. *FEMS Microbiology Letters* 247, 123–130.

Bobrov, A.G., Bearden, S.W., Fetherston, J.D., Khweek, A.A., Parrish, K.D. and Perry, R.D. (2007) Functional quorum sensing systems affect biofilm formation and protein expression in *Yersinia pestis*. *Advances in Experimental Medical Biology* 603, 178–191.

Bobrov, A.G., Kirillina, O., Forman, S., Mack, D. and Perry, R.D. (2008) Insights into *Yersinia pestis* biofilm development: topology and co-interaction of Hms inner membrane proteins involved in exopolysaccharide production. Environmental Microbiology, 10: 1419–1432.

Bodelon, G., Marin, E. and Fernandez, L.A. (2009) Role of periplasmic chaperones and BamA (YaeT/Omp85) in folding and secretion of intimin from enteropathogenic *Escherichia coli* strains. *Journal of Bacteriology* 191, 5169–5179.

Braeken, K., Moris, M., Daniels, R., Vanderleyden, J. and Michiels, J. (2006) New horizons for (p) ppGpp in bacterial and plant physiology. *Trends in Microbiology* 14, 45–54.

Bresolin, G., Morgan, J.A., Ilgen, D., Scherer, S. and Fuchs, T.M. (2006) Low temperature-induced insecticidal activity of *Yersinia enterocolitica*. *Molecular Microbiology* 59, 503–512.

Brubaker, R.R. (2005) Influence of Na(+), dicarboxylic amino acids, and pH in modulating the low-calcium response of *Yersinia pestis*. *Infection and Immunity* 73, 4743–4752.

Brubaker, R.R. (2007) Intermediary metabolism, Na+, the low calcium-response, and acute disease. *Advances in Experimental Medical Biology* 603, 116–129.

Brzostek, K., Raczkowska, A. and Zasada, A. (2003) The osmotic regulator OmpR is involved in the response of *Yersinia enterocolitica* O:9 to environmental stresses and survival within macrophages. *FEMS Microbiology Letters* 228, 265–271.

Brzostek, K., Brzostkowska, M., Bukowska, I., Karwicka, E. and Raczkowska, A. (2007) OmpR negatively regulates expression of invasin in *Yersinia enterocolitica*. *Microbiology* 153, 2416–2425.

Burghout, P., Beckers, F., de Wit, E., van Boxtel, R., Cornelis, G.R., Tommassen, J. and Koster, M. (2004) Role of the pilot protein YscW in the biogenesis of the YscC secretin in *Yersinia enterocolitica*. *Journal of Bacteriology* 186, 5366–5375.

Cambronne, E.D. and Schneewind, O. (2002) *Yersinia enterocolitica* type III secretion: *yscM1* and *yscM2* regulate *yop* gene expression by a posttranscriptional mechanism that targets the 5′ untranslated region of yop mRNA. *Journal of Bacteriology* 184, 5880–5893.

Cambronne, E.D., Cheng, L.W. and Schneewind, O. (2000) LcrQ/YscM1, regulators of the *Yersinia yop* virulon, are injected into host cells by a chaperone-dependent mechanism. *Molecular Microbiology* 37, 263–273.

Cambronne, E.D., Sorg, J.A. and Schneewind, O. (2004) Binding of SycH chaperone to YscM1 and YscM2 activates effector *yop* expression in *Yersinia enterocolitica*. *Journal of Bacteriology* 186, 829–841.

Cao, T.B. and Saier, Jr, M.H. (2003) The general protein secretory pathway: phylogenetic analyses leading to evolutionary conclusions. *Biochimica Biophysica Acta* 1609, 115–125.

Carlsson, K.E., Liu, J., Edqvist, P.J. and Francis, M.S. (2007a) Extracytoplasmic-stress-responsive pathways modulate type III secretion in *Yersinia pseudotuberculosis*. *Infection and Immunity* 75, 3913–3924.

Carlsson, K.E., Liu, J., Edqvist, P.J. and Francis, M.S. (2007b) Influence of the Cpx extracytoplasmic-stress-responsive pathway on *Yersinia* sp. – eukaryotic cell contact. *Infection and Immunity* 75, 4386–4399.

Cathelyn, J.S., Crosby, S.D., Lathem, W.W., Goldman, W.E. and Miller, V.L. (2006) RovA, a global regulator of *Yersinia pestis*, specifically required for bubonic plague. *Proceedings of the National Academy of Sciences of the United States of America* 103, 13514–13519.

Cathelyn, J.S., Ellison, D.W., Hinchliffe, S.J., Wren, B.W. and Miller, V.L. (2007) The RovA regulons of *Yersinia enterocolitica* and *Yersinia pestis* are distinct: evidence that many RovA-regulated genes were acquired more recently than the core genome. *Molecular Microbiology* 66, 189–205.

Chevance, F.F. and Hughes, K.T. (2008) Coordinating assembly of a bacterial macromolecular machine. *Nature Reviews Microbiology* 6, 455–465.

Christie, P.J., Atmakuri, K., Krishnamoorthy, V., Jakubowski, S. and Cascales, E. (2005) Biogenesis, architecture, and function of bacterial type IV secretion systems. *Annual Reviews in Microbiology* 59, 451–485.

Chromy, B.A., Choi, M.W., Murphy, G.A., Gonzales, A.D., Corzett, C.H., Chang, B.C., Fitch, J.P. and McCutchen-Maloney, S.L. (2005) Proteomic characterization of *Yersinia pestis* virulence. *Journal of Bacteriology* 187, 8172–8180.

Cianciotto, N.P. (2005) Type II secretion: a protein secretion system for all seasons. *Trends in Microbiology* 13, 581–588.

Collyn, F., Lety, M.A., Nair, S., Escuyer, V., Ben Younes, A., Simonet, M. and Marceau, M. (2002) *Yersinia pseudotuberculosis* harbors a type IV pilus gene cluster that contributes to pathogenicity. *Infection and Immunity* 70, 6196–6205.

Collyn, F., Billault, A., Mullet, C., Simonet, M. and Marceau, M. (2004) YAPI, a new *Yersinia pseudotuberculosis* pathogenicity island. *Infection and Immunity* 72, 4784–4790.

Cornelis, G.R. (2003) How Yops find their way out of *Yersinia. Molecular Microbiology* 50, 1091–1094.

Cornelis, G., Sluiters, C., de Rouvroit, C.L. and Michiels, T. (1989) Homology between *virF*, the transcriptional activator of the *Yersinia* virulence regulon, and AraC, the *Escherichia coli* arabinose operon regulator. *Journal of Bacteriology* 171, 254–262.

Cornelis, G.R., Sluiters, C., Delor, I., Geib, D., Kaniga, K., Lambert de Rouvroit, C., Sory, M.P., Vanooteghem, J.C. and Michiels, T. (1991) *ymoA*, a *Yersinia enterocolitica* chromosomal gene modulating the expression of virulence functions. *Molecular Microbiology* 5, 1023–1234.

Cornelis, G.R., Boland, A., Boyd, A.P., Geuijen, C., Iriarte, M., Neyt, C., Sory, M.P. and Stainier, I. (1998) The virulence plasmid of *Yersinia*, an antihost genome. *Microbiology and Molecular Biology Reviews* 62, 1315–1352.

Cotter, S.E., Surana, N.K. and St Geme, 3rd, J.W. (2005) Trimeric autotransporters: a distinct subfamily of autotransporter proteins. *Trends in Microbiology* 13, 199–205.

Dahiya, I. and Stevenson, R.M. (2010) The UvrY response regulator of the BarA-UvrY two-component system contributes to Yersinia ruckeri infection of rainbow trout. *Archives of Microbiology* 192, 541–547.

Darby, C. (2008) Uniquely insidious: *Yersinia pestis* biofilms. *Trends in Microbiology* 16, 158–164.

Darby, C., Hsu, J.W., Ghori, N. and Falkow, S. (2002).*Caenorhabditis elegans*: plague bacteria biofilm blocks food intake. *Nature* 417, 243–244.

Darwin, A.J. (2005) The phage-shock-protein response. *Molecular Microbiology* 57, 621–628.

Darwin, A.J. and Miller, V.L. (1999) Identification of *Yersinia enterocolitica* genes affecting survival in an animal host using signature-tagged transposon mutagenesis. *Molecular Microbiology* 32, 51–62.

Darwin, A.J. and Miller, V.L. (2001) The *psp* locus of *Yersinia enterocolitica* is required for virulence and for growth *in vitro* when the Ysc type III secretion system is produced. *Molecular Microbiology* 39, 429–444.

Dautin, N. and Bernstein, H.D. (2007) Protein secretion in Gram-negative bacteria via the autotransporter pathway. *Annual Reviews in Microbiology* 61, 89–112.

Davidson, A.L., Dassa, E., Orelle, C. and Chen, J. (2008) Structure, function, and evolution of bacterial ATP-binding cassette systems. *Microbiology and Molecular Biology Reviews* 72, 317–364.

Delepelaire, P. (2004) Type I secretion in Gram-negative bacteria. *Biochimica Biophysica Acta* 1694, 149–161.

DeLisa, M.P., Lee, P., Palmer, T. and Georgiou, G. (2004) Phage shock protein PspA of *Escherichia coli* relieves saturation of protein export via the Tat pathway. *Journal of Bacteriology* 186, 366–373.

Deutscher, J. (2008) The mechanisms of carbon catabolite repression in bacteria. *Current Opinion in Microbiology* 11, 87–93.

Dong, T. and Schellhorn, H.E. (2010) Role of RpoS in virulence of pathogens. *Infection and Immunity* 78, 887–897.

Dorel, C., Lejeune, P. and Rodrigue, A. (2006) The Cpx system of *Escherichia coli,* a strategic signaling pathway for confronting adverse conditions and for settling biofilm communities? *Research in Microbiology* 157, 306–314.

Dorrell, N., Li, S.R., Everest, P.H., Dougan, G. and Wren, B.W. (1998) Construction and characterisation of a *Yersinia enterocolitica* O:8 *ompR* mutant. *FEMS Microbiology Letters* 165, 145–151.

Dougan, D.A., Mogk, A. and Bukau, B. (2002) Protein folding and degradation in bacteria: to degrade or not to degrade? That is the question. *Cell Molecular Life Sciences* 59, 1607–1616.

Driessen, A.J. and Nouwen, N. (2008) Protein translocation across the bacterial cytoplasmic membrane. *Annual Reviews in Biochemistry* 77, 643–667.

Duguay, A.R. and Silhavy, T.J. (2004) Quality control in the bacterial periplasm. *Biochimica Biophysica Acta* 1694, 121–134.

El Tahir, Y. and Skurnik, M. (2001) YadA, the multifaceted *Yersinia* adhesin. *International Journal of Medical Microbiology* 291, 209–218.

Eppinger, M., Rosovitz, M.J., Fricke, W.F., Rasko, D.A., Kokorina, G., Fayolle, C., Lindler, L.E., Carniel, E. and Ravel, J. (2007) The complete genome sequence of *Yersinia pseudotuberculosis* IP31758, the causative agent of Far East scarlet-like fever. *PLoS Genetics* 3, e142.

Erickson, D.L., Waterfield, N.R., Vadyvaloo, V., Long, D., Fischer, E.R., Ffrench-Constant, R. and Hinnebusch, B.J. (2007) Acute oral toxicity of *Yersinia pseudotuberculosis* to fleas: implications for the evolution of vector-borne transmission of plague. *Cellular Microbiology* 9, 2658–2666.

Falker, S., Schmidt, M.A. and Heusipp, G. (2005) DNA methylation in *Yersinia enterocolitica*: role of the DNA adenine methyltransferase in mismatch repair and regulation of virulence factors. *Microbiology* 151, 2291–2299.

Falker, S., Schmidt, M.A. and Heusipp, G. (2006) Altered Ca(2+) regulation of Yop secretion in *Yersinia enterocolitica* after DNA adenine methyltransferase overproduction is mediated by Clp-dependent degradation of LcrG. *Journal of Bacteriology* 188, 7072–7081.

Falker, S., Schilling, J., Schmidt, M.A. and Heusipp, G. (2007) Overproduction of DNA adenine methyltransferase alters motility, invasion, and the lipopolysaccharide O-antigen composition of *Yersinia enterocolitica*. *Infection and Immunity* 75, 4990–4997.

Fallman, M. and Gustavsson, A. (2005) Cellular mechanisms of bacterial internalization counteracted by *Yersinia*. *International Reviews in Cytology* 246, 135–188.

Felek, S., Runco, L.M., Thanassi, D.G. and Krukonis, E.S. (2007) Characterization of six novel chaperone/usher systems in *Yersinia pestis*. *Advances in Experimental Medical Biology* 603, 97–105.

Felek, S., Lawrenz, M.B. and Krukonis, E.S. (2008) The *Yersinia pestis* autotransporter YapC mediates host cell binding, autoaggregation and biofilm formation. *Microbiology* 154, 1802–1812.

Fernandez, L., Secades, P., Lopez, J.R., Marquez, I. and Guijarro, J.A. (2002) Isolation and analysis of a protease gene with an ABC transport system in the fish pathogen *Yersinia ruckeri*: insertional mutagenesis and involvement in virulence. *Microbiology* 148, 2233–2243.

Fernandez, L., Mendez, J. and Guijarro, J.A. (2007a) Molecular virulence mechanisms of the fish pathogen *Yersinia ruckeri*. *Veterinary Microbiology* 125, 1–10.

Fernandez, L., Prieto, M. and Guijarro, J.A. (2007b) The iron- and temperature-regulated haemolysin YhlA is a virulence factor of *Yersinia ruckeri*. *Microbiology* 153, 483–489.

Filloux, A., Hachani, A. and Bleves, S. (2008) The bacterial type VI secretion machine: yet another player for protein transport across membranes. *Microbiology* 154, 1570–1583.

Flamez, C., Ricard, I., Arafah, S., Simonet, M. and Marceau, M. (2008) Phenotypic analysis of *Yersinia pseudotuberculosis* 32777 response regulator mutants: new insights into two-component system regulon plasticity in bacteria. *International Journal of Medical Microbiology* 298, 193–207.

Folkesson, A., Lofdahl, S. and Normark, S. (2002) The *Salmonella enterica* subspecies I specific centisome 7 genomic island encodes novel protein families present in bacteria living in close contact with eukaryotic cells. *Research in Microbiology* 153, 537–545.

Forman, S., Wulff, C.R., Myers-Morales, T., Cowan, C., Perry, R.D. and Straley, S.C. (2008) *yadBC* of *Yersinia pestis*, a new virulence determinant for bubonic plague. *Infection and Immunity* 76, 578–587.

Foultier, B., Troisfontaines, P., Muller, S., Opperdoes, F.R. and Cornelis, G.R. (2002) Characterization of the ysa pathogenicity locus in the chromosome of *Yersinia enterocolitica* and phylogeny analysis of type III secretion systems. *Journal of Molecular Evolution* 55, 37–51.

Fowler, J.M., Wulff, C.R., Straley, S.C. and Brubaker, R.R. (2009) Growth of calcium-blind mutants of *Yersinia pestis* at 37 degrees C in permissive Ca^{2+}-deficient environments. *Microbiology* 155, 2509–2521.

Francis, M.S. (2010) Type III secretion chaperones: a molecular toolkit for all occasions. In: Durante, P. and Colucci, L. (eds) *Handbook of Molecular Chaperones: Roles, Structures and Mechanisms*, Chapter 2. Nova Science Publishers, Hauppauge, New York, Inc, pp. 79–147.

Francis, M.S., Lloyd, S.A. and Wolf-Watz, H. (2001) The type III secretion chaperone LcrH co-operates with YopD to establish a negative, regulatory loop for control of Yop synthesis in *Yersinia pseudotuberculosis*. *Molecular Microbiology* 42, 1075–1093.

Francis, M.S., Schesser, K., Forsberg, Å. and Wolf-Watz, H. (2004) Type III secretion systems in animal- and plant-interacting bacteria. In: Cossart, P., Boquet, P., Normark, S. and Rappuoli, R. (eds) *Cellular Microbiology*. American Society for Microbiology Press, 2nd edn. Washington, DC, pp. 361–392.

Fuchs, T.M., Bresolin, G., Marcinowski, L., Schachtner, J. and Scherer, S. (2008) Insecticidal genes of *Yersinia* spp.: taxonomical distribution, contribution to toxicity towards *Manduca sexta* and *Galleria mellonella*, and evolution. *BMC Microbiology* 8, 214.

Galan, J.E. (2008) Energizing type III secretion machines: what is the fuel? *Nature Structural Molecular Biology* 15, 127–128.

Galan, J.E. (2009) Common themes in the design and function of bacterial effectors. *Cell Host Microbe* 5, 571–579.

Galan, J.E. and Wolf-Watz, H. (2006) Protein delivery into eukaryotic cells by type III secretion machines. *Nature* 444, 567–573.

Galimand, M., Carniel, E. and Courvalin, P. (2006) Resistance of *Yersinia pestis* to antimicrobial agents. *Antimicrobial Agents and Chemotherapy* 50, 3233–3236.

Gendlina, I., Held, K.G., Bartra, S.S., Gallis, B.M., Doneanu, C.E., Goodlett, D.R., Plano, G.V. and Collins, C.M. (2007) Identification and type III-dependent secretion of the *Yersinia pestis* insecticidal-like proteins. *Molecular Microbiology* 64, 1214–1227.

Goguen, J.D., Yother, J. and Straley, S.C. (1984) Genetic analysis of the low calcium response in *Yersinia pestis* mud1(Ap lac) insertion mutants. *Journal of Bacteriology* 160, 842–848.

Gorke, B. and Stulke, J. (2008) Carbon catabolite repression in bacteria: many ways to make the most out of nutrients. *Nature Reviews Microbiology* 6, 613–624.

Gottesman, S. (1996) Proteases and their targets in *Escherichia coli*. *Annual Reviews Genetics* 30, 465–506.

Grabenstein, J.P., Marceau, M., Pujol, C., Simonet, M. and Bliska, J.B. (2004) The response regulator PhoP of *Yersinia pseudotuberculosis* is important for replication in macrophages and for virulence. *Infection and Immunity* 72, 4973–4984.

Grassl, G.A., Bohn, E., Muller, Y., Buhler, O.T. and Autenrieth, I.B. (2003) Interaction of *Yersinia enterocolitica* with epithelial cells: invasin beyond invasion. *International Journal of Medical Microbiology* 293, 41–54.

Green, R.C. and Darwin, A.J. (2004) PspG, a new member of the *Yersinia enterocolitica* phage shock protein regulon. *Journal of Bacteriology* 186, 4910–4920.

Guilvout, I., Chami, M., Engel, A., Pugsley, A.P. and Bayan, N. (2006) Bacterial outer membrane secretin PulD assembles and inserts into the inner membrane in the absence of its pilotin. *EMBO Journal* 25, 5241–5249.

Guisbert, E., Yura, T., Rhodius, V.A. and Gross, C.A. (2008) Convergence of molecular, modeling, and systems approaches for an understanding of the *Escherichia coli* heat shock response. *Microbiology and Molecular Biology Reviews* 72, 545–554.

Hahn, H.P. and von Specht, B.U. (2003) Secretory delivery of recombinant proteins in attenuated Salmonella strains: potential and limitations of Type I protein transporters. *FEMS Immunology and Medical Microbiology* 37, 87–98.

Haller, J.C., Carlson, S., Pederson, K.J. and Pierson, D.E. (2000) A chromosomally encoded type III secretion pathway in *Yersinia enterocolitica* is important in virulence. *Molecular Microbiology* 36, 1436–1446.

Han, Y., Zhou, D., Pang, X., Zhang, L., Song, Y., Tong, Z., Bao, J., Dai, E., Wang, J., Guo, Z., Zhai, J., Du, Z., Wang, X., Huang, P. and Yang, R. (2005) DNA microarray analysis of the heat- and cold-shock stimulons in *Yersinia pestis*. *Microbes Infection* 7, 335–348.

Han, Y., Qiu, J., Guo, Z., Gao, H., Song, Y., Zhou, D. and Yang, R. (2007) Comparative transcriptomics in *Yersinia pestis*: a global view of environmental modulation of gene expression. *BMC Microbiology* 7, 96.

Hares, M.C., Hinchliffe, S.J., Strong, P.C., Eleftherianos, I., Dowling, A.J., ffrench-Constant, R.H. and Waterfield, N. (2008) The *Yersinia pseudotuberculosis* and *Yersinia pestis* toxin complex is active against cultured mammalian cells. *Microbiology* 154, 3503–3517.

Hayes, C.S. and Keiler, K.C. (2010) Beyond ribosome rescue: tmRNA and co-translational processes. *FEBS Letters* 584, 413–419.

Heermann, R. and Fuchs, T.M. (2008) Comparative analysis of the *Photorhabdus luminescens* and the *Yersinia enterocolitica* genomes: uncovering candidate genes involved in insect pathogenicity. *BMC Genomics* 9, 40.

Hengge, R. (2009) Principles of c-di-GMP signalling in bacteria. *Nature Reviews Microbiology* 7, 263–273.

Hengge-Aronis, R. (2002) Signal transduction and regulatory mechanisms involved in control of the sigma(S) (RpoS) subunit of RNA polymerase. *Microbiology and Molecular Biology Reviews* 66, 373–395.

Herbst, K., Bujara, M., Heroven, A.K., Opitz, W., Weichert, M., Zimmermann, A. and Dersch, P. (2009) Intrinsic thermal sensing controls proteolysis of Yersinia virulence regulator RovA. *PLoS Pathogens* 5, e1000435.

Heroven, A.K. and Dersch, P. (2006) RovM, a novel LysR-type regulator of the virulence activator gene *rovA*, controls cell invasion, virulence and motility of *Yersinia pseudotuberculosis*. *Molecular Microbiology* 62, 1469–1483.

Heroven, A.K., Nagel, G., Tran, H.J., Parr, S. and Dersch, P. (2004) RovA is autoregulated and antagonizes H-NS-mediated silencing of invasin and *rovA* expression in *Yersinia pseudotuberculosis. Molecular Microbiology* 53, 871–888.

Heroven, A.K., Bohme, K., Rohde, M. and Dersch, P. (2008) A Csr-type regulatory system, including small non-coding RNAs, regulates the global virulence regulator RovA of *Yersinia pseudotuberculosis* through RovM. *Molecular Microbiology* 68, 1179–1195.

Heusipp, G., Nelson, K.M., Schmidt, M.A. and Miller, V.L. (2004) Regulation of *htrA* expression in *Yersinia enterocolitica. FEMS Microbiology Letters* 231, 227–235.

Heusipp, G., Falker, S. and Schmidt, M.A. (2007) DNA adenine methylation and bacterial pathogenesis. *International Journal of Medical Microbiology* 297, 1–7.

Hinchliffe, S.J., Howard, S.L., Huang, Y.H., Clarke, D.J. and Wren, B.W. (2008) The importance of the Rcs phosphorelay in the survival and pathogenesis of the enteropathogenic yersiniae. *Microbiology* 154, 1117–1131.

Hinnebusch, B.J. and Erickson, D.L. (2008) *Yersinia pestis* biofilm in the flea vector and its role in the transmission of plague. *Current Topics in Microbiology and Immunology* 322, 229–248.

Hinnebusch, B.J., Perry, R.D. and Schwan, T.G. (1996) Role of the *Yersinia pestis* hemin storage (*hms*) locus in the transmission of plague by fleas. *Science* 273, 367–370.

Hoe, N.P. and Goguen, J.D. (1993) Temperature sensing in *Yersinia pestis*: translation of the LcrF activator protein is thermally regulated. *Journal of Bacteriology* 175, 7901–7909.

Hu, Y., Lu, P., Wang, Y., Ding, L., Atkinson, S. and Chen, S. (2009a) OmpR positively regulates urease expression to enhance acid survival of *Yersinia pseudotuberculosis. Microbiology* 155, 2522–2531.

Hu, Y., Wang, Y., Ding, L., Lu, P., Atkinson, S. and Chen, S. (2009b) Positive regulation of *flhDC* expression by OmpR in *Yersinia pseudotuberculosis. Microbiology* 155, 3622–3631.

Huang, Y.H., Ferrieres, L. and Clarke, D.J. (2006) The role of the Rcs phosphorelay in Enterobacteriaceae. *Research in Microbiology* 157, 206–212.

Ieva, R. and Bernstein, H.D. (2009) Interaction of an autotransporter passenger domain with BamA during its translocation across the bacterial outer membrane. *Proceedings of the National Academy of Sciences of the United States of America* 106, 19120–19125.

Iriarte, M., Vanooteghem, J.C., Delor, I., Diaz, R., Knutton, S. and Cornelis, G.R. (1993) The Myf fibrillae of *Yersinia enterocolitica. Molecular Microbiology* 9, 507–520.

Iriarte, M., Stainier, I. and Cornelis, G.R. (1995) The

rpoS gene from *Yersinia enterocolitica* and its influence on expression of virulence factors. *Infection and Immunity* 63, 1840–1847.

Irie, Y. and Parsek, M.R. (2008) Quorum sensing and microbial biofilms. *Current Topics in Microbiology and Immunology* 322, 67–84.

Isberg, R.R., Hamburger, Z. and Dersch, P. (2000) Signaling and invasin-promoted uptake via integrin receptors. *Microbes and Infection* 2, 793–801.

Iwobi, A., Heesemann, J., Garcia, E., Igwe, E., Noelting, C. and Rakin, A. (2003) Novel virulence-associated type II secretion system unique to high-pathogenicity *Yersinia enterocolitica. Infection and Immunity* 71, 1872–1879.

Jackson, M.W. and Plano, G.V. (1999) DsbA is required for stable expression of outer membrane protein YscC and for efficient Yop secretion in *Yersinia pestis. Journal of Bacteriology* 181, 5126–5130.

Jackson, M.W., Silva-Herzog, E. and Plano, G.V. (2004) The ATP-dependent ClpXP and Lon proteases regulate expression of the *Yersinia pestis* type III secretion system via regulated proteolysis of YmoA, a small histone-like protein. *Molecular Microbiology* 54, 1364–1378.

Jarrett, C.O., Deak, E., Isherwood, K.E., Oyston, P.C., Fischer, E.R., Whitney, A.R., Kobayashi, S.D., DeLeo, F.R. and Hinnebusch, B.J. (2004) Transmission of *Yersinia pestis* from an infectious biofilm in the flea vector. *Journal of Infectious Diseases* 190, 783–792.

Johnson, T.L., Abendroth, J., Hol, W.G. and Sandkvist, M. (2006) Type II secretion: from structure to function. *FEMS Microbiology Letters* 255, 175–186.

Jones, H.A., Lillard, Jr, J.W. and Perry, R.D. (1999) HmsT, a protein essential for expression of the haemin storage (Hms+) phenotype of *Yersinia pestis. Microbiology* 145(Pt 8), 2117–2128.

Jones, S.E., Lloyd, L.J., Tan, K.K. and Buck, M. (2003) Secretion defects that activate the phage shock response of *Escherichia coli. Journal of Bacteriology* 185, 6707–6711.

Jovanovic, G., Lloyd, L.J., Stumpf, M.P., Mayhew, A.J. and Buck, M. (2006) Induction and function of the phage shock protein extracytoplasmic stress response in *Escherichia coli. Journal of Biological Chemistry* 281, 21147–21161.

Julio, S.M., Heithoff, D.M., Provenzano, D., Klose, K.E., Sinsheimer, R.L., Low, D.A. and Mahan, M.J. (2001) DNA adenine methylase is essential for viability and plays a role in the pathogenesis of *Yersinia pseudotuberculosis* and *Vibrio cholerae. Infection and Immunity* 69, 7610–7615.

Julio, S.M., Heithoff, D.M., Sinsheimer, R.L., Low,

D.A. and Mahan, M.J. (2002) DNA adenine methylase overproduction in *Yersinia pseudotuberculosis* alters YopE expression and secretion and host immune responses to infection. *Infection and Immunity* 70, 1006–1009.

Kanehisa, M. and Goto, S. (2000) KEGG: Kyoto encyclopedia of genes and genomes. *Nucleic Acids Research* 28, 27–30.

Keiler, K.C. (2007) Physiology of tmRNA: what gets tagged and why? *Current Opinion in Microbiology* 10, 169–175.

Keiler, K.C. (2008) Biology of trans-translation. *Annual Reviews in Microbiology* 62, 133–151.

Keyser, P., Elofsson, M., Rosell, S. and Wolf-Watz, H. (2008) Virulence blockers as alternatives to antibiotics: type III secretion inhibitors against Gram-negative bacteria. *Journal of International Medicine* 264, 17–29.

Kim, T.J., Chauhan, S., Motin, V.L., Goh, E.B., Igo, M.M. and Young, G.M. (2007) Direct transcriptional control of the plasminogen activator gene of *Yersinia pestis* by the cyclic AMP receptor protein. *Journal of Bacteriology* 189, 8890–8900.

Kirillina, O., Fetherston, J.D., Bobrov, A.G., Abney, J. and Perry, R.D. (2004) HmsP, a putative phosphodiesterase, and HmsT, a putative diguanylate cyclase, control Hms-dependent biofilm formation in *Yersinia pestis*. *Molecular Microbiology* 54, 75–88.

Kleerebezem, M., Crielaard, W. and Tommassen, J. (1996) Involvement of stress protein PspA (phage shock protein A) of *Escherichia coli* in maintenance of the protonmotive force under stress conditions. *EMBO Journal* 15, 162–171.

Knight, S.D. (2007) Structure and assembly of *Yersinia pestis* F1 antigen. *Advances in Experimental Medical Biology* 603, 74–87.

Kobayashi, R., Suzuki, T. and Yoshida, M. (2007) *Escherichia coli* phage-shock protein A (PspA) binds to membrane phospholipids and repairs proton leakage of the damaged membranes. *Molecular Microbiology* 66, 100–109.

Kol, S., Nouwen, N. and Driessen, A.J. (2008) Mechanisms of YidC-mediated insertion and assembly of multimeric membrane protein complexes. *Journal of Biological Chemistry* 283, 31269–31273.

Koornhof, H.J., Smego, Jr, R.A. and Nicol, M. (1999) Yersiniosis. II: the pathogenesis of *Yersinia* infections. *European Journal of Clinical Microbiology and Infectious Diseases* 18, 87–112.

Koronakis, V., Eswaran, J. and Hughes, C. (2004) Structure and function of TolC: the bacterial exit duct for proteins and drugs. *Annual Reviews in Biochemistry* 73, 467–489.

Kukkonen, M., Suomalainen, M., Kyllonen, P., Lahteenmaki, K., Lang, H., Virkola, R., Helander, I.M., Holst, O. and Korhonen, T.K. (2004) Lack of O-antigen is essential for plasminogen activation by *Yersinia pestis* and Salmonella enterica. *Molecular Microbiology* 51, 215–225.

Lacour, S. and Landini, P. (2004) SigmaS-dependent gene expression at the onset of stationary phase in *Escherichia coli*: function of sigmaS-dependent genes and identification of their promoter sequences. *Journal of Bacteriology* 186, 7186–7195.

Lambert de Rouvroit, C., Sluiters, C. and Cornelis, G.R. (1992) Role of the transcriptional activator, VirF, and temperature in the expression of the pYV plasmid genes of *Yersinia enterocolitica*. *Molecular Microbiology* 6, 395–409.

Lavander, M., Ericsson, S.K., Broms, J.E. and Forsberg, A. (2006) The twin arginine translocation system is essential for virulence of *Yersinia pseudotuberculosis*. *Infection and Immunity* 74, 1768–1776.

Lawrenz, M.B. and Miller, V.L. (2007) Comparative analysis of the regulation of *rovA* from the pathogenic yersiniae. *Journal of Bacteriology* 189, 5963–5975.

Lawrenz, M.B., Lenz, J.D. and Miller, V.L. (2009) A novel autotransporter adhesin is required for efficient colonization during bubonic plague. *Infection and Immunity* 77, 317–326.

Lawson, C.L., Swigon, D., Murakami, K.S., Darst, S.A., Berman, H.M. and Ebright, R.H. (2004) Catabolite activator protein: DNA binding and transcription activation. *Current Opinion in Structural Biology* 14, 10–20.

Lawson, J.N., Lyons, C.R. and Johnston, S.A. (2006) Expression profiling of *Yersinia pestis* during mouse pulmonary infection. *DNA Cell Biology* 25, 608–616.

Lee, F.A., Tullman-Ercek, D. and Georgiou, G. (2006) The bacterial twin-arginine translocation pathway. *Annual Reviews in Microbiology* 60, 373–395.

Lee, V.T. and Schneewind, O. (2001) Protein secretion and the pathogenesis of bacterial infections. *Genes and Development* 15, 1725–1752.

Levy, S.B. and Marshall, B. (2004) Antibacterial resistance worldwide: causes, challenges and responses. *Nature Medicine* 10, S122–129.

Li, S.R. Dorrell, N., Everest, P.H., Dougan, G. and Wren, B.W. (1996) Construction and characterization of a *Yersinia enterocolitica* O:8 high-temperature requirement (*htrA*) isogenic mutant. *Infection and Immunity* 64, 2088–2094.

Lindler, L.E. and Tall, B.D. (1993) *Yersinia pestis* pH 6 antigen forms fimbriae and is induced by intracellular association with macrophages. *Molecular Microbiology* 8, 311–324.

Lloyd, S.A., Forsberg, Å., Wolf-Watz, H. and Francis, M.S. (2001) Targeting exported substrates to the Yersinia TTSS: different functions for different signals? *Trends in Microbiology* 9, 367–371.

Lobner-Olesen, A., Skovgaard, O. and Marinus, M.G. (2005) Dam methylation: coordinating cellular processes. *Current Opinion in Microbiology* 8, 154–160.

Loftus, C.G., Harewood, G.C., Cockerill, 3rd, F.R. and Murray, J.A. (2002) Clinical features of patients with novel Yersinia species. *Digestive Diseases and Sciences* 47, 2805–2810.

Luirink, J., von Heijne, G., Houben, E. and de Gier, J.W. (2005) Biogenesis of inner membrane proteins in *Escherichia coli*. *Annual Reviews in Microbiology* 59, 329–355.

McNally, A., La Ragione, R.M., Best, A., Manning, G. and Newell, D.G. (2007) An aflagellate mutant *Yersinia enterocolitica* biotype 1A strain displays altered invasion of epithelial cells, persistence in macrophages, and cytokine secretion profiles *in vitro*. *Microbiology* 153, 1339–1349.

Majdalani, N. and Gottesman, S. (2005) The Rcs phosphorelay: a complex signal transduction system. *Annual Reviews in Microbiology* 59, 379–405.

Marceau, M. (2005) Transcriptional regulation in Yersinia: an update. *Current Issues in Molecular Biology* 7, 151–177.

Marinus, M.G. and Casadesus, J. (2009) Roles of DNA adenine methylation in host–pathogen interactions: mismatch repair, transcriptional regulation, and more. *FEMS Microbiology Reviews* 33, 488–503.

Matsumoto, H. and Young, G.M. (2006) Proteomic and functional analysis of the suite of Ysp proteins exported by the Ysa type III secretion system of *Yersinia enterocolitica* Biovar 1B. *Molecular Microbiology* 59, 689–706.

Matsumoto, H. and Young, G.M. (2009) Translocated effectors of *Yersinia*. *Current Opinion in Microbiology* 12, 94–100.

Mazar, J. and Cotter, P.A. (2007) New insight into the molecular mechanisms of two-partner secretion. *Trends in Microbiology* 15, 508–515.

Mendez, J., Fernandez, L., Menendez, A., Reimundo, P., Perez-Pascual, D., Navais, R. and Guijarro, J.A. (2009) A chromosomally located *traHIJKCLMN* operon encoding a putative type IV secretion system is involved in the virulence of *Yersinia ruckeri*. *Applied and Environmental Microbiology* 75, 937–945.

Minamino, T., Imada, K. and Namba, K. (2008) Mechanisms of type III protein export for bacterial flagellar assembly. *Molecular Biosystems* 4, 1105–1115.

Mota, L.J., Journet, L., Sorg, I., Agrain, C. and Cornelis, G.R. (2005) Bacterial injectisomes: needle length does matter. *Science* 307, 1278.

Motin, V.L., Georgescu, A.M., Fitch, J.P., Gu, P.P., Nelson, D.O., Mabery, S.L., Garnham, J.B., Sokhansanj, B.A., Ott, L.L., Coleman, M.A., Elliott, J.M., Kegelmeyer, L.M., Wyrobek, A.J., Slezak, T.R., Brubaker, R.R. and Garcia, E. (2004) Temporal global changes in gene expression during temperature transition in *Yersinia pestis*. *Journal of Bacteriology* 186, 6298–6305.

Mueller, C.A., Broz, P. and Cornelis, G.R. (2008) The type III secretion system tip complex and translocon. *Molecular Microbiology* 68, 1085–1095.

Nagel, G., Lahrz, A. and Dersch, P. (2001) Environmental control of invasin expression in *Yersinia pseudotuberculosis* is mediated by regulation of RovA, a transcriptional activator of the SlyA/Hor family. *Molecular Microbiology* 41, 1249–1269.

Naktin, J. and Beavis, K.G. (1999) *Yersinia enterocolitica* and *Yersinia pseudotuberculosis*. *Clinical Laboratory Medicine* 19, 523–536, vi.

Nelson, K.M., Young, G.M. and Miller, V.L. (2001) Identification of a locus involved in systemic dissemination of *Yersinia enterocolitica*. *Infection and Immunity* 69, 6201–6208.

Newman, C.L. and Stathopoulos, C. (2004) Autotransporter and two-partner secretion: delivery of large-size virulence factors by Gram-negative bacterial pathogens. *Critical Reviews in Microbiology* 30, 275–286.

Nunn, D. (1999) Bacterial type II protein export and pilus biogenesis: more than just homologies? *Trends in Cellular Biology* 9, 402–408.

O'Loughlin, J.L., Spinner, J.L., Minnich, S.A. and Kobayashi, S.D. (2010) *Yersinia pestis* two-component gene regulatory systems promote survival in human neutrophils. *Infection and Immunity* 78, 773–782.

Okan, N.A., Bliska, J.B. and Karzai, A.W. (2006) A role for the SmpB-SsrA system in *Yersinia pseudotuberculosis* pathogenesis. *PLoS Pathogens* 2, e6.

Oyston, P.C., Dorrell, N., Williams, K., Li, S.R., Green, M., Titball, R.W. and Wren, B.W. (2000) The response regulator PhoP is important for survival under conditions of macrophage-induced stress and virulence in *Yersinia pestis*. *Infection and Immunity* 68, 3419–3425.

Paetzel, M., Karla, A., Strynadka, N.C. and Dalbey, R.E. (2002) Signal peptidases. *Chemistry Reviews* 102, 4549–4580.

Panthel, K., Meinel, K.M., Sevil Domenech, V.E., Trulzsch, K. and Russmann, H. (2008) Salmonella type III-mediated heterologous antigen delivery: a versatile oral vaccination strategy to induce cellular immunity against infectious agents and tumors. *International Journal of Medical Microbiology* 298, 99–103.

Patten, C.L., Kirchhof, M.G., Schertzberg, M.R., Morton, R.A. and Schellhorn, H.E. (2004) Microarray analysis of RpoS-mediated gene expression in *Escherichia coli* K-12. *Molecular Genetics and Genomics* 272, 580–591.

Peabody, C.R., Chung, Y.J., Yen, M.R., Vidal-Ingigliardi, D., Pugsley, A.P. and Saier, Jr, M.H. (2003) Type II protein secretion and its relationship to bacterial type IV pili and archaeal flagella. *Microbiology* 149, 3051–3072.

Pederson, K.J., Carlson, S. and Pierson, D.E. (1997) The ClpP protein, a subunit of the Clp protease, modulates ail gene expression in *Yersinia enterocolitica. Molecular Microbiology* 26, 99–107.

Perry, R.D. and Fetherston, J.D. (1997) *Yersinia pestis* – etiologic agent of plague. *Clinical Microbiology Reviews* 10, 35–66.

Petersen, S. and Young, G.M. (2002) Essential role for cyclic AMP and its receptor protein in *Yersinia enterocolitica* virulence. *Infection and Immunity* 70, 3665–3672.

Pettersson, J., Nordfelth, R., Dubinina, E., Bergman, T., Gustafsson, M., Magnusson, K.E. and Wolf-Watz, H. (1996) Modulation of virulence factor expression by pathogen target cell contact. *Science* 273, 1231–1233.

Pieper, R., Huang, S.T., Robinson, J.M., Clark, D.J., Alami, H., Parmar, P.P., Perry, R.D., Fleischmann, R.D. and Peterson, S.N. (2009) Temperature and growth phase influence the outer-membrane proteome and the expression of a type VI secretion system in *Yersinia pestis. Microbiology* 155, 498–512.

Pinheiro, V.B. and Ellar, D.J. (2007) Expression and insecticidal activity of *Yersinia pseudotuberculosis* and *Photorhabdus luminescens* toxin complex proteins. *Cellular Microbiology* 9, 2372–2380.

Planet, P.J., Kachlany, S.C., Fine, D.H., DeSalle, R. and Figurski, D.H. (2003) The widespread colonization island of *Actinobacillus actinomycetemcomitans. Nature Genetics* 34, 193–198.

Potrykus, K. and Cashel, M. (2008) (p)ppGpp: still magical? *Annual Reviews in Microbiology* 62, 35–51.

Pouillot, F., Fayolle, C. and Carniel, E. (2007) A putative DNA adenine methyltransferase is involved in *Yersinia pseudotuberculosis* pathogenicity. *Microbiology* 153, 2426–2434.

Prentice, M.B. and Rahalison, L. (2007) Plague. *Lancet* 369, 1196–1207.

Pukatzki, S., McAuley, S.B. and Miyata, S.T. (2009) The type VI secretion system: translocation of effectors and effector-domains. *Current Opinion in Microbiology* 12, 11–17.

Purdy, G.E., Fisher, C.R. and Payne, S.M. (2007) IcsA surface presentation in *Shigella flexneri* requires the periplasmic chaperones DegP, Skp, and SurA. *Journal of Bacteriology* 189, 5566–5573.

Raivio, T.L. (2005) Envelope stress responses and Gram-negative bacterial pathogenesis. *Molecular Microbiology* 56, 1119–1128.

Randall, L.L. and Hardy, S.J. (2002) SecB, one small chaperone in the complex milieu of the cell. *Cell Molecualr Life Sciences* 59, 1617–1623.

Revell, P.A. and Miller, V.L. (2000) A chromosomally encoded regulator is required for expression of the *Yersinia enterocolitica inv* gene and for virulence. *Molecular Microbiology* 35, 677–685.

Richards, J., Sundermeier, T., Svetlanov, A. and Karzai, A.W. (2008) Quality control of bacterial mRNA decoding and decay. *Biochimica Biophysica Acta* 1779, 574–582.

Robinson, J.B., Telepnev, M.V., Zudina, I.V., Bouyer, D., Montenieri, J.A., Bearden, S.W., Gage, K.L., Agar, S.L., Foltz, S.M., Chauhan, S., Chopra, A.K. and Motin, V.L. (2009) Evaluation of a *Yersinia pestis* mutant impaired in a thermoregulated type VI-like secretion system in flea, macrophage and murine models. *Microbial Pathogenesis* 47, 243–251.

Robinson, V.L., Oyston, P.C. and Titball, R.W. (2005) A *dam* mutant of *Yersinia pestis* is attenuated and induces protection against plague. *FEMS Microbiology Letters* 252, 251–256.

Rosso, M.L., Chauvaux, S., Dessein, R., Laurans, C., Frangeul, L., Lacroix, C., Schiavo, A., Dillies, M.A., Foulon, J., Coppee, J.Y., Medigue, C., Carniel, E., Simonet, M. and Marceau, M. (2008) Growth of *Yersinia pseudotuberculosis* in human plasma: impacts on virulence and metabolic gene expression. *BMC Microbiology* 8, 211.

Rowley, G., Spector, M., Kormanec, J. and Roberts, M. (2006) Pushing the envelope: extracytoplasmic stress responses in bacterial pathogens. *Nature Reviews Microbiology* 4, 383–394.

Ruiz-Perez, F., Henderson, I.R., Leyton, D.L., Rossiter, A.E., Zhang, Y. and Nataro, J.P. (2009) Roles of periplasmic chaperone proteins in the biogenesis of serine protease autotransporters of Enterobacteriaceae. *Journal of Bacteriology* 191, 6571–6583.

Runco, L.M., Myrczek, S., Bliska, J.B. and Thanassi, D.G. (2008) Biogenesis of the fraction 1 capsule and analysis of the ultrastructure of *Yersinia pestis*. *Journal of Bacteriology* 190, 3381–3385.

Rusch, S.L. and Kendall, D.A. (2007) Oligomeric states of the SecA and SecYEG core components of the bacterial Sec translocon. *Biochimica Biophysica Acta* 1768, 5–12.

Saier, M.H. Jr (2004) Evolution of bacterial type III protein secretion systems. *Trends in Microbiology* 12, 113–115.

Saier, M.H. Jr (2006) Protein secretion and membrane insertion systems in gram-negative bacteria. *Journal of Membrane Biology* 214, 75–90.

Sargent, F., Berks, B.C. and Palmer, T. (2006) Pathfinders and trailblazers: a prokaryotic targeting system for transport of folded proteins. *FEMS Microbiology Letters* 254, 198–207.

Sauri, A., Soprova, Z., Wickstrom, D., de Gier, J.W., Van der Schors, R.C., Smit, A.B., Jong, W.S. and Luirink, J. (2009) The Bam (Omp85) complex is involved in secretion of the auto-transporter haemoglobin protease. *Microbiology* 155, 3982–3991.

Schirmer, T. and Jenal, U. (2009) Structural and mechanistic determinants of c-di-GMP signalling. *Nature Reviews Microbiology* 7, 724–735.

Schmid, A., Neumayer, W., Trulzsch, K., Israel, L., Imhof, A., Roessle, M., Sauer, G., Richter, S., Lauw, S., Eylert, E., Eisenreich, W., Heesemann, J. and Wilharm, G. (2009) Cross-talk between type three secretion system and metabolism in *Yersinia*. *Journal of Biological Chemistry* 284, 12165–12177.

Seo, J., Savitzky, D.C., Ford, E. and Darwin, A.J. (2007) Global analysis of tolerance to secretin-induced stress in *Yersinia enterocolitica* suggests that the phage-shock-protein system may be a remarkably self-contained stress response. *Molecular Microbiology* 65, 714–727.

Sha, J., Agar, S.L., Baze, W.B., Olano, J.P., Fadl, A.A., Erova, T.E., Wang, S., Foltz, S.M., Suarez, G., Motin, V.L., Chauhan, S., Klimpel, G.R., Peterson, J.W. and Chopra, A.K. (2008) Braun lipoprotein (Lpp) contributes to virulence of yersiniae: potential role of Lpp in inducing bubonic and pneumonic plague. *Infection and Immunity* 76, 1390–1409.

Shan, S.O. and Walter, P. (2005) Co-translational protein targeting by the signal recognition particle. *FEBS Letters* 579, 921–926.

Shi, Z.Y., Wang, H., Gu, L., Cui, Z.G., Wu, L.F., Kan, B., Pang, B., Wang, X., Xu, J.G. and Jing, H.Q. (2007) Pleiotropic effect of *tatC* mutation on metabolism of pathogen *Yersinia enterocolitica*. *Biomedical Environmental Science* 20, 445–449.

Shutinoski, B., Schmidt, M.A. and Heusipp, G. (2010) Transcriptional regulation of the Yts1 type-II secretion systems of *Yersinia enterocolitica* and identification of secretion substrates. *Molecular Microbiology* 75, 676–691.

Simm, R., Fetherston, J.D., Kader, A., Romling, U. and Perry, R.D. (2005) Phenotypic convergence mediated by GGDEF-domain-containing proteins. *Journal of Bacteriology* 187, 6816–6823.

Smego, R.A., Frean, J. and Koornhof, H.J. (1999) Yersiniosis I: microbiological and clinicoepidemiological aspects of plague and non-plague *Yersinia* infections. *European Journal of Clinical Microbiology and Infectious Diseases* 18, 1–15.

Sorg, J.A., Miller, N.C. and Schneewind, O. (2005) Substrate recognition of type III secretion machines – testing the RNA signal hypothesis. *Cellular Microbiology* 7, 1217–1225.

Srivatsan, A. and Wang, J.D. (2008) Control of bacterial transcription, translation and replication by (p)ppGpp. *Current Opinion in Microbiology* 11, 100–105.

Stenseth, N.C., Atshabar, B.B., Begon, M., Belmain, S.R., Bertherat, E., Carniel, E., Gage, K.L., Leirs, H. and Rahalison, L. (2008) Plague: past, present, and future. *PLoS Medicine* 5, e3.

Sulakvelidze, A. (2000) Yersiniae other than *Y. enterocolitica*, *Y. pseudotuberculosis*, and *Y. pestis*: the ignored species. *Microbes and Infection* 2, 497–513.

Sun, W., Roland, K.L., Branger, C.G., Kuang, X. and Curtiss, 3rd, R. (2009a) The role of *relA* and *spoT* in *Yersinia pestis* KIM5 pathogenicity. *PLoS One* 4, e6720.

Sun, Y.C., Koumoutsi, A. and Darby, C. (2009b) The response regulator PhoP negatively regulates *Yersinia pseudotuberculosis* and *Yersinia pestis* biofilms. *FEMS Microbiology Letters* 290, 85–90.

Suomalainen, M., Haiko, J., Ramu, P., Lobo, L., Kukkonen, M., Westerlund-Wikstrom, B., Virkola, R., Lahteenmaki, K. and Korhonen, T.K. (2007) Using every trick in the book: the Pla surface protease of *Yersinia pestis*. *Advances in Experimental and Medical Biology* 603, 268–278.

Tamayo, R., Pratt, J.T. and Camilli, A. (2007) Roles of cyclic diguanylate in the regulation of bacterial pathogenesis. *Annual Reviews in Microbiology* 61, 131–48.

Taylor, V.L., Titball, R.W. and Oyston, P.C. (2005) Oral immunization with a dam mutant of *Yersinia pseudotuberculosis* protects against plague. *Microbiology* 151, 1919–1926.

Tennant, S.M., Skinner, N.A., Joe, A. and Robins-Browne, R.M. (2005) Homologues of insecticidal toxin complex genes in *Yersinia enterocolitica* biotype 1A and their contribution to virulence. *Infection and Immunity* 73, 6860–6867.

Thanassi, D.G. (2002) Ushers and secretins: channels for the secretion of folded proteins across the bacterial outer membrane. *Journal of Molecular Microbiology Biotechnology* 4, 11–20.

Tidhar, A., Flashner, Y., Cohen, S., Levi, Y., Zauberman, A., Gur, D., Aftalion, M., Elhanany, E., Zvi, A., Shafferman, A. and Mamroud, E. (2009) The NlpD lipoprotein is a novel *Yersinia pestis* virulence factor essential for the development of plague. *PLoS One* 4, e7023.

Tobback, E., Decostere, A., Hermans, K., Haesebrouck, F. and Chiers, K. (2007) *Yersinia ruckeri* infections in salmonid fish. *Journal Fish Diseases* 30, 257–268.

Tomich, M., Planet, P.J. and Figurski, D.H. (2007) The *tad* locus: postcards from the widespread colonization island. *Nature Reviews Microbiology* 5, 363–375.

Tran, H.J., Heroven, A.K., Winkler, L., Spreter, T., Beatrix, B. and Dersch, P. (2005) Analysis of RovA, a transcriptional regulator of *Yersinia pseudotuberculosis* virulence that acts through antirepression and direct transcriptional activation. *Journal of Biological Chemistry* 280, 42423–42432.

Troisfontaines, P. and Cornelis, G.R. (2005) Type III secretion: more systems than you think. *Physiology (Bethesda)* 20, 326–339.

Vahamiko, S., Penttinen, M.A. and Granfors, K. (2005) Aetiology and pathogenesis of reactive arthritis: role of non-antigen-presenting effects of HLA-B27. *Arthritis Research and Therapy* 7, 136–141.

Venecia, K. and Young, G.M. (2005) Environmental regulation and virulence attributes of the Ysa type III secretion system of *Yersinia enterocolitica* biovar 1B. *Infection and Immunity* 73, 5961–5977.

Viboud, G.I. and Bliska, J.B. (2005) *Yersinia* outer proteins: role in modulation of host cell signaling responses and pathogenesis. *Annual Reviews in Microbiology* 59, 69–89.

Waksman, G. and Hultgren, S.J. (2009) Structural biology of the chaperone-usher pathway of pilus biogenesis. *Nature Reviews Microbiology* 7, 765–774.

Waterfield, N.R., Bowen, D.J., Fetherston, J.D., Perry, R.D. and ffrench-Constant, R.H. (2001) The *tc* genes of *Photorhabdus*: a growing family. *Trends in Microbiology* 9, 185–191.

Waterfield, N., Hares, M., Hinchliffe, S., Wren, B. and ffrench-Constant, R. (2007) The insect toxin complex of *Yersinia*. *Advances Experimental Medicine and Biology* 603, 247–257.

Wick, L.M. and Egli, T. (2004) Molecular components of physiological stress responses in *Escherichia coli*. *Advances in Biochemical Engineering and Biotechnology* 89, 1–45.

Wickner, S., Maurizi, M.R. and Gottesman, S. (1999) Posttranslational quality control: folding, refolding, and degrading proteins. *Science* 286, 1888–1893.

Wilharm, G., Lehmann, V., Krauss, K., Lehnert, B., Richter, S., Ruckdeschel, K., Heesemann, J. and Trulzsch, K. (2004) *Yersinia enterocolitica* type III secretion depends on the proton motive force but not on the flagellar motor components MotA and MotB. *Infection and Immunity* 72, 4004–4009.

Wilharm, G., Dittmann, S., Schmid, A. and Heesemann, J. (2007) On the role of specific chaperones, the specific ATPase, and the proton motive force in type III secretion. *International Journal of Medical Microbiology* 297, 27–36.

Williams, A.W. and Straley, S.C. (1998) YopD of *Yersinia pestis* plays a role in negative regulation of the low-calcium response in addition to its role in translocation of Yops. *Journal of Bacteriology* 180, 350–358.

Williams, K., Oyston, P.C., Dorrell, N., Li, S., Titball, R.W. and Wren, B.W. (2000) Investigation into the role of the serine protease HtrA in *Yersinia pestis* pathogenesis. *FEMS Microbiology Letters* 186, 281–286.

Wion, D. and Casadesus, J. (2006) N6-methyl-adenine: an epigenetic signal for DNA-protein interactions. *Nature Reviews Microbiology* 4, 183–192.

Won, H.S., Lee, Y.S., Lee, S.H. and Lee, B.J. (2009) Structural overview on the allosteric activation of cyclic AMP receptor protein. *Biochimica Biophysica Acta* 1794, 1299–1308.

Wulff-Strobel, C.R., Williams, A.W. and Straley, S.C. (2002) LcrQ and SycH function together at the Ysc type III secretion system in *Yersinia pestis* to impose a hierarchy of secretion. *Molecular Microbiology* 43, 411–423.

Yamamoto, T., Hanawa, T., Ogata, S. and Kamiya, S. (1996) Identification and characterization of the *Yersinia enterocolitica gsrA* gene, which protectively responds to intracellular stress induced by macrophage phagocytosis and to extracellular environmental stress. *Infection and Immunity* 64, 2980–2987.

Yang, Y., Merriam, J.J., Mueller, J.P. and Isberg, R.R. (1996) The *psa* locus is responsible for thermoinducible binding of *Yersinia pseudotuberculosis* to cultured cells. *Infection and Immunity* 64, 2483–2489.

Yen, M.R., Peabody, C.R., Partovi, S.M., Zhai, Y., Tseng, Y.H. and Saier, M.H. (2002) Protein-translocating outer membrane porins of

Gram-negative bacteria. *Biochimica Biophysica Acta* 1562, 6–31.

Yen, Y.T., Karkal, A., Bhattacharya, M., Fernandez, R.C. and Stathopoulos, C. (2007) Identification and characterization of autotransporter proteins of *Yersinia pestis* KIM. *Molecular Membrane Biology* 24, 28–40.

Yen, Y.T., Bhattacharya, M. and Stathopoulos, C. (2008) Genome-wide *in silico* mapping of the secretome in pathogenic *Yersinia pestis* KIM. *FEMS Microbiology Letters* 279, 56–63.

Yother, J., Chamness, T.W. and Goguen, J.D. (1986) Temperature-controlled plasmid regulon associated with low calcium response in *Yersinia pestis*. *Journal of Bacteriology* 165, 443–447.

Young, B.M. and Young, G.M. (2002a) Evidence for targeting of Yop effectors by the chromosomally encoded Ysa type III secretion system of *Yersinia enterocolitica*. *Journal of Bacteriology* 184, 5563–5571.

Young, B.M. and Young, G.M. (2002b) YplA is exported by the Ysc, Ysa, and flagellar type III secretion systems of *Yersinia enterocolitica*. *Journal of Bacteriology* 184, 1324–1334.

Young, G.M., Schmiel, D.H. and Miller, V.L. (1999) A new pathway for the secretion of virulence factors by bacteria: the flagellar export apparatus functions as a protein-secretion system. *Proceedings of the National Academy of Sciences of the United States of America* 96, 6456–6461.

Young, G.M., Badger, J.L. and Miller, V.L. (2000) Motility is required to initiate host cell invasion by *Yersinia enterocolitica*. *Infection and Immunity* 68, 4323–4326.

Zavialov, A., Zav'yalova, G., Korpela, T. and Zav'yalov, V. (2007) FGL chaperone-assembled fimbrial polyadhesins: anti-immune armament of Gram-negative bacterial pathogens. *FEMS Microbiology Review* 31, 478–514.

Zav'yalov, V.P., Chernovskaya, T.V., Chapman, D.A., Karlyshev, A.V., MacIntyre, S., Zavialov, A.V., Vasiliev, A.M., Denesyuk, A.I., Zav'yalova, G.A., Dudich, I.V., Korpela, T. and Abramov, V.M. (1997) Influence of the conserved disulphide bond, exposed to the putative binding pocket, on the structure and function of the immunoglobulin-like molecular chaperone Caf1M of *Yersinia pestis*. *Biochemical Journal* 324(Pt 2), 571–578.

Zhan, L., Han, Y., Yang, L., Geng, J., Li, Y., Gao, H., Guo, Z., Fan, W., Li, G., Zhang, L., Qin, C., Zhou, D. and Yang, R. (2008) The cyclic AMP receptor protein, CRP, is required for both virulence and expression of the minimal CRP regulon in *Yersinia pestis* biovar microtus. *Infection and Immunity* 76, 5028–5037.

Zhan, L., Yang, L., Zhou, L., Li, Y., Gao, H., Guo, Z., Zhang, L., Qin, C., Zhou, D. and Yang, R. (2009) Direct and negative regulation of the *sycO-ypkA-ypoJ* operon by cyclic AMP receptor protein (CRP) in *Yersinia pestis*. *BMC Microbiology* 9, 178.

10 Response of *Neisseria gonorrhoeae* to Oxygen Limitation and Excess

Jeffery A. Cole

10.1 Introduction to the Response of Pathogenic *Neisseria* to Oxygen Starvation

Two Neisserial species, *Neisseria meningitidis* and *N. gonorrhoeae*, are obligate human pathogens. Much of their genomes are almost identical, yet they inhabit different environmental niches. Meningococci can be found in the nasopharynx of up to a third of the human population, where they survive essentially as commensal bacteria. However, they can also cross the blood–brain barrier and cause the frequently fatal disease, meningitis. In contrast, gonococci are found in microaerobic biofilms on mucosal surfaces of genital organs, usually surrounded by anaerobes such as lactobacilli (Zheng *et al.*, 1994; St Amant *et al.*, 2002). They also can cross the epithelial cell layer and they provoke a localized inflammatory response by polymorphonuclear leukocytes that expose the bacteria to an oxidative burst of hydrogen peroxide and superoxide. For both of these bacteria, their commensal and pathogenic lifestyles require the ability to cope with oxygen-sufficient environments. They differ in that, as a commensal bacterium, the meningococcus is found in a predominantly oxygen-sufficient environment, whereas the gonococcus has adapted to cope better with oxygen-limited survival (Snyder *et al.*, 2005). To state that some of the genes that differ between the two species reflect their different lifestyles might seem so obvious as to be trite: if so, why should the gonococcus, which predominantly inhabits a microaerobic environment, be an even more prolific source of cytochromes and maintain a higher respiratory capacity than the meningococcus, which inhabits more aerobic environments? One might guess why both organisms are profligate in their expression of defence mechanisms against reactive oxygen species and reactive nitrogen species generated as part of the defence mechanisms of their human hosts, but what is the advantage to bacteria of duplicating such mechanisms?

This chapter will focus on how gonococci adapt to starvation of oxygen. How the transcription of key genes is regulated will be reviewed and attempts will be made to rationalize why overlapping biochemical pathways have been maintained by bacteria with such small genomes (typically about 2.2 or 2.3 Mb). This is a challenge because, despite the presence in the human body of closely related non-pathogenic commensal bacteria such as *N. lactamica* or *N. cinerea*, very little is known about their physiology and biochemistry other than can be deduced from bioinformatics analysis of their genomes. While this is a powerful approach, it provides only insights and hypotheses, not experimental proof of what differentiates a pathogen from a commensal organism. Much of the literature on pathogenic *Neisseria* focuses on pathogenicity determinants and surprisingly little is known about their basic physiology and biochemistry. However, a good starting point is the pioneering work of Morse and his colleagues on the growth and physiology of these bacteria, especially how they generate energy for growth or accumulate iron required for survival in the human body (Winter and Morse, 1975; Morse, 1978; Morse *et al.*, 1979).

10.2 Oxygen-sufficient and Oxygen-limited Growth of Gonococci and Meningococci

Pathogenic bacteria enjoy the luxury of living in an environment with central heating and three meals a day. While the meningococcus is moderately robust, gonococci are difficult to handle in the laboratory because cultures die rapidly, even when left on the bench for short periods at room temperature. Both organisms require rich media for growth and grow well only when supplied with a limited range of sources of carbon and energy. Both glucose and especially lactate are abundant *in vivo* at locations colonized by gonococci, the latter being produced by lactobacilli that dominate biofilms of the lower gastrointestinal tract (Zheng *et al.*, 1994; St Amant *et al.*, 2002). This provides an obvious rationale as to why they have evolved to use lactate as their preferred growth substrate and why glucose also supports abundant growth (Morse and Bartenstein, 1974; Britigan *et al.*, 1988). All of the members of the genus *Neisseria* contain a conventional cytochrome bc_1 complex typical of the mammalian electron transfer chain, but only a single type of cytochrome oxidase, cytochrome cbb_3, which is essential for survival. In the few bacteria in which this cytochrome oxidase has been characterized there are alternative pathways for oxygen reduction, and the cytochrome cbb_3 pathway appears to fulfil a specialized role. For example, in Rhizobia this cytochrome oxidase maintains the very low ambient concentrations of oxygen that are essential for nitrogen fixation in the root nodules of leguminous plants (Preisig *et al.*, 1993, 1996). Is it true that the Neisserial

cytochrome oxidase also has an extremely high affinity for oxygen? This has never been shown experimentally. If so, one can speculate that this is an obvious requirement for gonococci in their natural environment, but why should the meningococcus need to maintain a very low concentration of oxygen?

Both gonococci and meningococci were considered initially to be obligate aerobes. However, both can adapt to oxygen-limited growth by exploiting a truncated denitrification pathway in which nitrite, but not nitrate, is reduced via nitric oxide (NO) to nitrous oxide (Clark *et al.*, 1987, 1988; Householder *et al.*, 2000; Rock *et al.*, 2005). The nitrite reductase that catalyses the first step in denitrification is a member of the NirK family of copper-containing nitrite reductases (Boulanger and Murphy, 2002). It has been designated AniA (the major <u>a</u>naerobically <u>i</u>nduced protein). AniA is an outer membrane lipoprotein (Cannon, 1989; Hoehn and Clark, 1992), but how electrons are transferred across the periplasm to AniA has been only partially revealed. Reduction of the product of nitrite reduction, NO, is catalysed by a single subunit quinol dehydrogenase, NorB. In both of these denitrification reactions, electrons are transferred to the electron acceptor via the quinone pool, so we assume, without experimental proof, that energy is conserved as electrons, which are passed via flavoprotein dehydrogenases from physiological substrates such as NADH and succinate to ubiquinone.

In many types of bacteria, genes for anaerobic respiratory pathways are regulated coordinately to ensure that toxic intermediates do not accumulate. What is beyond doubt is that NO is a toxic intermediate of the Neisserial denitrification pathway (Householder *et al.*, 2000; Anjum *et al.*, 2002; Overton *et al.*, 2006), so superficially we might expect expression of *aniA* and *norB* to be regulated coordinately. However, to cope with NO generated as part of the oxidative burst of the human host, it could be fatal if nitrite was the signal to which these genes responded because there would be minimal protection against host-generated NO in the absence of nitrite. Conversely, what would be the physiological advantage of maintaining genes for a supplementary energy yielding pathway such as nitrite reduction that could not be used in the presence of its substrate unless the toxic intermediate, NO, was available first? This introduction therefore highlights just a few of the challenges faced by pathogens that must also cope with the stress imposed by varying concentrations of oxygen. It also raises some of the many questions currently unanswered.

10.2.1 The multiplicity of *c*-type cytochromes in pathogenic *Neisseria*

The introduction above states that pathogenic *Neisseria* have only two electron transfer pathways terminating in the reduction of oxygen to water and nitrite via NO to nitrous oxide. Nevertheless, gonococci are a surprisingly prolific source of *c*-type cytochromes (Strange and Judd, 1994). Bacterial *c*-type cytochromes are synthesized as pre-apoproteins that are secreted across the cytoplasmic membrane via the Sec

pathway. The leader sequence is then cleaved and haem is attached covalently to two cysteine residues organized as a C-X-X-C-H motif, where X is any amino acid. Although there are multiple pathways for the post-translational assembly of *c*-type cytochromes (see, for example, Fulop *et al.*, 2009), many bacteria use one of two alternative cytochrome *c* maturation pathways: in *Neisseria*, system II is used (Kranz *et al.*, 1998; Allen *et al.*, 2003). Furthermore, as some of the Neisserial *c*-type cytochromes are lipoproteins, their leader peptides are cleaved by signal peptidase 2 rather than the more commonly used signal peptidase 1 (Turner *et al.*, 2003). Cytochrome *c* synthesis and assembly is therefore a major biosynthetic pathway that places heavy demands on nutrient resources. Nevertheless, genes encoding seven *c*-type cytochromes can be recognized in the genome sequence of meningococci, with an additional cytochrome, a cytochrome *c* peroxidase (Ccp), present in gonococci (Fig. 10.1: Turner *et al.*, 2003; Li *et al.*, 2010).

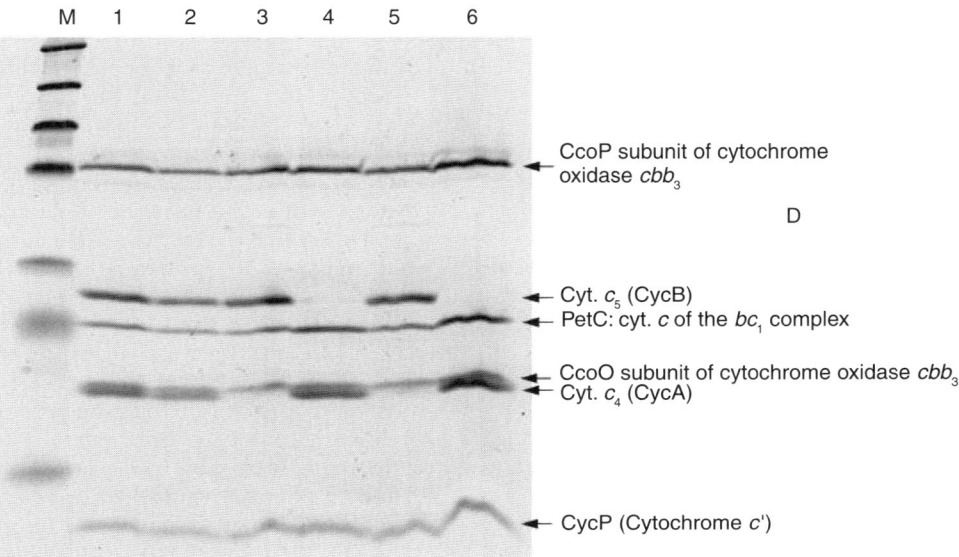

Fig. 10.1. SDS-PAGE gel of the *cccA*, *cycA* and *cycB* mutants stained for covalently-bound haem. *N. gonorrhoeae* strains F62 (lane 1), JCGC851 (Δc_2, lane 2), JCGC800 (Δc_4, lane 3), JCGC850 (Δc_5, lane 4), JCGC853 ($\Delta c_2\Delta c_4$, lane 5) and JCGC852 ($\Delta c_2\Delta c_5$, lane 6) were grown on plates at 37°C in a candle jar overnight. Samples were collected, lysed and separated by SDS-PAGE. The gel was then haem-stained to detect *c*-type cytochromes. Adapted from Li *et al.* (2010).

This suggests that the maintenance of multiple electron transfer pathways is essential for gonococcal survival in the human body. Two of these cytochromes, CcoP and CcoO, are components of the terminal cytochrome oxidase, cytochrome cbb_3. Genes for five other Neisserial c-type cytochromes include $cycP$ encoding cytochrome c', which has been implicated in protection against NO toxicity (Turner et $al.$, 2005); $petC$, encoding the cytochrome c_1 component of the cytochrome bc_1 complex; and three genes, $cccA$, $cycA$ and $cycB$, encoding cytochromes that have been designated cytochromes c_2, c_4 and c_5. By staining proteins separated by SDS PAGE for covalently bound haem, we and others have demonstrated the accumulation of six c-type cytochromes during aerobic growth of gonococci, including both cytochromes c_4 and c_5 (Fig. 10.1: Strange and Judd, 1994; Turner et $al.$, 2003, 2005; Li et $al.$, 2010). Ccp is expressed at a low level only during oxygen-limited growth (Lissenden et $al.$, 2000; Turner et $al.$, 2003; Whitehead et $al.$, 2007). Cytochrome c_2 has not been visualized in either gonococci or meningococci (Deeudom et $al.$, 2006; Sevastsyanovich et $al.$, 2009; Li et $al.$, 2010): consequently, its role, if any, in electron transfer pathways remains to be defined.

10.2.2 The gonococcal aerobic and anaerobic respiratory chains

Downs and Jones (1975) were among the earliest groups to propose that in $Azotobacter$ $vinelandii$ cytochromes c_4 and c_5 might provide alternative but parallel pathways for electron transfer from the cytochrome bc_1 complex to the terminal oxidase (reviewed by Haddock and Jones, 1977). Single mutations resulting in loss of one or other of these cytochromes had little effect on oxygen-sufficient growth or on rates of respiration in the presence of physiological substrates (Ng et $al.$, 1995; Rey and Maier, 1997). The explanation came much later with the realization that there were multiple cytochrome oxidases in $A.$ $vinelandii$, of which only one was the homologue of the $Bradyrhizobium$ $japonicum$ high-affinity oxidase that originally was designated FixNOQP

because of its role in promoting nitrogen fixation, but is referred to now as cytochrome oxidase cbb_3 (Kolonay et $al.$, 1994; D'Mello et $al.$, 1997).

Publication of the complete genome sequences of gonococci and meningococci revealed that, in contrast to $A.$ $vinelandii$, there was only a single cytochrome oxidase, cytochrome oxidase cbb_3. We therefore proposed that, as in $Azotobacter$, cytochromes c_4 and c_5 might provide alternative but parallel pathways for electron transfer from the cytochrome bc_1 complex to this terminal oxidase. Recently, this proposal has been shown to be correct (Li et $al.$, 2010). Gonococcal mutants lacking one of these cytochromes were constructed easily but, as in $Azotobacter$, respiration rates of both strains were only slightly lower than those of the parent. However, it was impossible to construct a double mutant defective in both cytochromes. When an isopropyl-thio-ß-D-galactoside (IPTG)-inducible, ectopically expressed copy of one of the cytochrome genes was also integrated into the chromosome, the double mutant could then be constructed, but growth and survival were totally dependent on the presence of IPTG (Li et $al.$, 2010). These results indicate that retention of one or other pathway is essential for gonococcal survival and growth. They also establish that there is no third major pathway that conserves sufficient energy for growth – even during oxygen-limited growth in the presence of abundant nitrite. Figure 10.2 is a cartoon of the current understanding of how proton motive force is generated and how electrons are transferred via the energy-conserving NADH dehydrogenase, Nuo, and ubiquinone to the cytochrome bc_1 complex. While it is clear that both cytochromes c_4 and c_5 provide the proposed alternative electron transfer pathways to cytochrome oxidase, it is not clear whether they are functionally equivalent or whether they interact preferentially with one or other of the CcoP or CcoQ subunits of cytochrome oxidase.

When starved of oxygen, gonococci increase rates of synthesis of a small group of proteins, the most prominent of which originally was designated AniA (Clarke et $al.$, 1987). The AniA protein is undetectable in

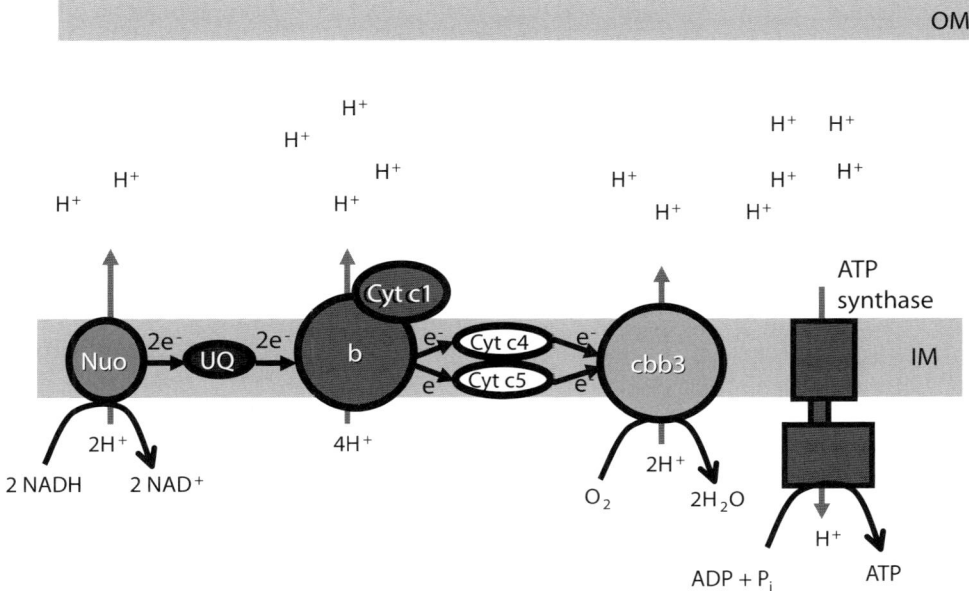

Fig. 10.2. Current understanding of the pathway of electron transfer from the physiological substrate, NADH, to oxygen. While there is strong evidence that cytochromes c_4 and c_5 provide the indicated alternative pathways of electron transfer between the cytochrome bc_1 complex and cytochrome oxidase cbb_3, details of interactions between the various cytochromes are unknown. The indicated sites of energy conservation are based on experimental data from bacteria other than Neisseriaceae. IM: cytoplasmic membrane; OM: outer membrane.

oxygen-sufficient cultures, so the presence of antibodies to AniA in sera from gonorrhoea patients is clear proof that gonococci *in vivo* are exposed to oxygen-limited conditions (Clark *et al.*, 1988). Although AniA is a member of the copper-containing NirK family of nitrite reductases, its amino acid sequence differs substantially from NirK proteins from other denitrifiers such as *Paracoccus denitrificans* and *Pseudomonas stutzeri*. It is also atypical of other NirK nitrite reductases in that it is a lipoprotein attached to the outer membrane (Hoehn and Clark, 1992). There is currently only limited information about how electrons are transferred across the periplasm to the outer membrane. Mutants defective in the *cccA*, *cycA* and *cycB* genes are all partially defective in nitrite reduction (Hopper, Li and Cole, unpublished observations), but clearly there are multiple pathways of electron transfer from the cytoplasmic membrane to AniA. Particularly interesting is the demon-

stration of a direct role of the CcoP component of the cytochrome oxidase itself in nitrite reduction. Unlike the di-haem CcoP subunits of the cytochrome cbb_3 oxidase from most other types of bacteria (including the meningococcus), the gonococcal CcoP has a large, C-terminal extension that includes a third haem group. Site-directed mutagenesis was used to show that this third haem group was essential for maximum rates of nitrite reduction, but played no detectable role in aerobic respiration (Hopper *et al.*, 2009). This can explain why mutants defective in cytochromes c_4 and c_5 are partially defective in nitrite reduction, but leaves open any possible role for cytochrome c_2.

The product of nitrite reduction is NO, which in turn is reduced to nitrous oxide by a single subunit NO reductase, NorB. Most of the previously characterized NO reductases contain two subunits, NorB and a *c*-type cytochrome, NorC. The N-terminal extension of gonococcal NorB lacks a haem-binding motif of the second subunit, NorC, so no *c*-type cytochrome is involved in NO reduction. As in *Nitrobacter*, the gonococcal NorB accepts electrons directly from the quinol pool without passing through the cytochrome bc_1 complex. Gonococci lack all except one of the genes for molybdopterin biosynthesis, and also structural genes for nitrate reduction. Not only are they unable to reduce nitrate to nitrite, but they also fail to respond to the presence of nitrate in the environment. Although genes for nitrous oxide reduction to dinitrogen are present on the chromosome, the enzyme is non-functional due to three frame-shift mutations that result in premature translation termination of NosR, NosZ and NosD. Thus, the gonococcal denitrification pathway is truncated to the reduction of nitrite to nitrous oxide.

A major area of uncertainty is how electrons are transferred between the cytoplasmic electron transfer components and outer membrane lipoproteins such as the nitrite reductase, NirK (AniA). This topic will be considered in a later section in the context of the roles of outer membrane lipoproteins in defence against reactive oxygen species, reactive nitrogen species and electron transfer to nitrite.

10.3 Stresses Generated During Aerobic Growth

Reactive oxygen species are generated during aerobic growth of many organisms. One source of hydrogen peroxide is the chemical interaction of molecular oxygen with reduced flavoproteins (Imlay and Fridovich, 1991; González-Flecha and Demple, 1995; Storz and Imlay, 1999). In contrast, a major source of superoxide is believed to be the chemical oxidation by molecular oxygen of the semiquinone form of ubiquinone at the Q_0 site of the cytochrome bc_1 complex (Fig. 10.3).

N. gonorrhoeae was described originally as an obligately aerobic bacterium that was moderately fastidious, requiring a rich culture medium and a microaerobic environment in which to grow (Reyn, 1974; Winter and Morse, 1975). Despite its microaerobic lifestyle *in vivo*, some widely used strains such as F62 can be grown in the laboratory in fully aerated batch cultures. Although rates of oxidation of physiological substrates such as glucose or lactate are almost the same as for the parent

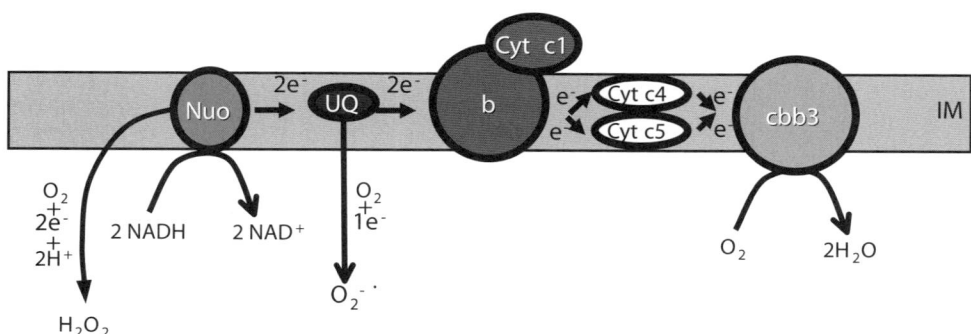

Fig. 10.3. Proposed sites for the generation of the reactive oxygen species hydrogen peroxide and superoxide as side products of aerobic respiration.

strain, single mutants defective in cytochromes c_4 or c_5 are much more sensitive than strain F62 to high levels of aeration (Li *et al.*, 2010). The artificial electron donor, ascorbate-reduced tetramethyl-*p*-phenylene diamine, TMPD, donates electrons directly into the terminal components of the electron transfer chain and can be used to estimate the capacity of electron transfer between the cytochrome bc_1 complex and the cytochrome oxidase cbb_3. Rates of oxidation of ascorbate-reduced TMPD are significantly lower in the cytochrome c_4 or c_5 mutants than in the parental strain (Li *et al.*, 2010). The obvious interpretation of these results is that the electron-transfer capacity between physiological substrates and the cytochrome bc_1 complex in the parent strain is less than that of the terminal components between the cytochrome bc_1 complex and cytochrome oxidase; however, the opposite is true for the *cycA* and *cycB* mutants (Li *et al.*, 2010). Loss of one or other pathway to cytochrome oxidase would have two detrimental effects, both leading to the production of increased quantities of reactive oxygen species. First, the ambient concentrations of oxygen have been shown to be higher in cultures of the mutants than of the parent (Li *et al.*, 2010). Secondly, the steady-state concentrations of reduced forms of flavoproteins and the quinone pool will also be higher in mutants defective in electron transfer to the terminal cytochrome oxidase. This 'double whammy' is sufficient to explain why the mutants are more sensitive to oxygen toxicity than the parent strain. It illustrates a fundamental principle that underlies the mechanisms that regulate levels of gene expression: not only must primary processes such as rates of electron transfer to oxygen, nitrite or NO be regulated, but also mechanisms that deal with secondary consequences of these processes must be integrated into the primary metabolic response. This leads to another, currently unanswered question: do the electron transfer pathways that detoxify hydrogen peroxide, NO and other secondary reactive oxygen and reactive nitrogen species also contribute significantly to energy conservation?

Despite the rational explanation for the sensitivity of mutants defective in electron transfer components to oxygen, this sensitivity is surprising in view of reports that the products of many genes appear to protect pathogenic *Neisseria* against reactive oxygen species during infection (reviewed by Seib *et al.*, 2006). Some of the gonococcal defence mechanisms, which are listed in Table 10.1, are designed primarily to protect the bacteria from reactive oxygen species generated by the host or against hydrogen peroxide generated by Lactobacilli that co-inhabit their niche; others provide protection against by-products of their own metabolism. Bacteria with small genomes cannot afford the luxury of maintaining redundancy, so there must be selective pressure to maintain the multiplicity of mechanisms that provide protection against reactive oxygen species. Occam's razor leads to the prediction that each mechanism protects against a different threat, so at this point it might be helpful to consider the merits and limitations of the methods available to sort out the physiological roles of each mechanism. The next three sections will therefore first introduce the major transcription factors that regulate the switch from aerobic to oxygen-limited growth. Then, some of the limitations of a systems biology approach to studies of gene regulation will be considered before details of how gonococci deal with reactive oxygen and reactive nitrogen species are presented. The key point is that none of these physiological responses can be considered in isolation.

10.4 Regulation of the Switch from Aerobic to Oxygen-limited Growth

The best-characterized mechanisms by which bacteria regulate gene expression are repression by proteins that prevent high-level expression and activation by transcription factors that upregulate expression. Both mechanisms regulate the rate of transcription initiation; both often involve proteins binding upstream or in the immediate vicinity of the transcription start site. Adaptation of *E. coli* to oxygen starvation involves the interplay between many transcription factors that are absent from pathogenic *Neisseria*, but four of them are also present in the gonococcus and

Table 10.1. Proteins that protect gonococci against oxidative stress.

Gene	Regulated by[a]	Cellular location	Proposed function
ccp	FNR; NarQP(+); PerR(+)	Outer membrane	Reduces H_2O_2 to H_2O
gor	OxyR (+)	Cytoplasm	Glutathione oxidoreductase
grx2	H_2O_2 (+)	Cytoplasm	?Glutaredoxin-like protein: resistance to H_2O_2
grx3	H_2O_2 (+)	Cytoplasm	?Glutaredoxin-like protein: resistance to H_2O_2
katA	OxyR (−)	Cytoplasm	Dismutates H_2O_2 to O_2 and H_2O
laz	Unknown	Outer membrane	?Reduction of peroxides
mntABC	PerR (−)	Cytoplasmic membrane	Mn^{2+} uptake, which protects against oxidative stress
msrAB	Ecf + H_2O_2 (+)	Outer membrane	Methionine sulfoxide reductase
priA	Unknown	Cytoplasm	DNA helicase: repair of oxidatively damaged DNA
prx	OxyR (−)	Cytoplasm	Resistance to H_2O_2
recN/recJ	Fur (+); H_2O_2 (+)	Cytoplasm	?Repair of oxidatively damaged DNA
sco	Unknown	Cytoplasm	Thiol:disulfide oxidoreductase; peroxiredoxin
sodB	Fur (+)	Cytoplasm	Possibly provides weak protection against superoxide
trx1	H_2O_2 (+)	Cytoplasm	?Resistance to H_2O_2
trxB	Nm1R	Cytoplasm	Putative thioredoxin reductase
tlpA	Unknown	Cytoplasm	?Thioredoxin-like protein: resistance to H_2O_2

Note: [a] (+), transcription activation; (−), transcription repressed; Ecf, extracytoplasmic function sigma factor. Much of this table was compiled from information in Seib *et al.* (2006, 2007), which reference original sources of the information listed.

meningococcus: FNR, NarP, NsrR and Fur. Also involved is an ArsR-like protein and three more regulators, OxyR, PerR and NmlR (for a description of NmlR, see Kidd, Chapter 5, this volume), which have been implicated in providing inducible defences against reactive oxygen species (Table 10.2).

In both the Enterobacteriaceae and Neisseriaceae the presence of oxygen is sensed by the [4Fe-4S] iron-sulfur cluster of dimeric FNR, so designated because it was identified originally as the regulator of fumarate and nitrate reduction in *E. coli*. It is the reduced form of FNR that is active as a transcription factor. When anaerobic bacteria are exposed to oxygen, the [4Fe-4S] iron-sulfur cluster changes sequentially to a [2Fe-2S] cluster and then disintegrates to yield an iron-free monomeric polypeptide that is unable to bind to DNA. FNR is synthesized constitutively, with only small changes in abundance between aerobic and anaerobic cultures (Khoroshilova *et al.*, 1997). However,

the [4Fe-4S] centre can be reconstituted in the absence of oxygen, enabling pre-formed FNR to be recycled. Two gonococcal genes encoding proteins of known function are among the few transcripts that are activated by FNR (Overton *et al.*, 2006). These are the nitrite reductase gene, *aniA*, and *ccp* that encodes a cytochrome *c* peroxidase. The latter observation highlights yet another puzzle: why should a protein implicated in recovery from oxidative stress be synthesized only during anaerobic growth – and then only at a very low level?

In *E. coli*, nitrate, and to a much lesser extent nitrite, is sensed by two, two-component regulatory systems, NarX-NarL and NarQ-NarP (Chiang *et al.*, 1992; Rabin and Stewart, 1993; Wang *et al.*, 1999). The NarX environmental sensor interacts preferentially with NarL rather than with NarP; however, the sensor, NarQ, has little preference for one response regulator relative to the other (Noriega *et al.*, 2010). Both response regulators,

Table 10.2. Transcription factors that regulate genes involved or implicated in adaptation to the availability of oxygen, reactive oxygen species, or reactive nitrogen species.

Gene	Signalling ligand	Examples of genes regulated[a]
Fnr	Inactivated by O_2 – also NO	*aniA*(+); *ccp*(+); also a small RNA
narQ narP	Unknown: not nitrate	*aniA*(+); *ccp*(−)
nsrR	Nitric oxide	*aniA*(−); *norB*(−); *dnrN*(−)
Fur	Repression activated by Fe^{2+}	*aniA*(−); *sodB*(−); *recN*(−); *norB*(**)
oxyR	Hydrogen peroxide	*katA*(+); *gor*(−); *prx*(−)
perR	Unknown: possibly a metal ion, Mn^{2+}?	*mntABC*(−)
nmlR	Unknown?	*adhC*(+); *trxB*(+); *copA*(+)
arsR	Unknown: possibly a metal ion?	*norB*(−)

Note: [a](+), transcription activated; (−), repression by the transcription factor; **see text for explanation.

NarL and NarP, bind to inverted heptamer DNA target sites arranged as inverted repeats separated by two base pairs (Tyson *et al.*, 1994; Darwin *et al.*, 1997). The three major differences between the two systems are that NarL, but not NarP, is able to regulate transcription from single binding sites; the phosphatase activity of NarX, but not NarQ, is activated strongly in the presence of nitrite; and NarQ has a higher affinity for nitrate than NarX. These properties make the NarQ-NarP system more relevant to the response of Enterobacteriaceae to the µM concentrations of nitrate and nitrite found in the bodies of warm-blooded animals. Conversely, the NarX-NarL system might confer a selective advantage during life outside the human body; for example, in a wastewater treatment works where higher concentrations of nitrate are available. Analysis of transcripts repressed by phosphorylated NarP also suggests that the NarQ-NarP system might be specialized to regulate the transitions between fermentation and respiration supported by limited availability of nitrate or nitrite (Constantinidou *et al.*, 2006).

A two-component regulatory system, designated NarQ-NarP, also activates transcription of Neisserial *aniA*, but attempts to define the ligand to which NarQ responds have been unsuccessful (Lissenden *et al.*, 2000; Whitehead *et al.*, 2007). When expressed in *E. coli*, the gonococcal NarQ appeared to be ligand-insensitive, locked on in its active state, suggesting that gonococci were unable to respond to nitrate or nitrite. It is possible that Neisserial NarQ responds to a ligand that is either always available in the rich media used for laboratory experiments, or it

responds *in vivo* to a ligand generated by the human host (Clark *et al.*, 2010).

NsrR is a repressor of the Rtf2 family of transcription factors (Rodionov *et al.*, 2005). Like FNR, it binds iron that serves as an environmental sensor, but the active site is an iron-sulfur cluster that binds NO with very high affinity (Tucker *et al.*, 2008; Yuki *et al.*, 2008; Isabella *et al.*, 2009). The iron atoms of NsrR are nitrosylated by NO, rendering the protein unable to bind to its DNA target site. In the absence of NO, the Neisserial NsrR binds as a repressor to the intergenic region between *aniA* and the divergently transcribed *norB* gene, which encodes the NO reductase, NorB (Overton *et al.*, 2006; Rock *et al.*, 2007; Isabella *et al.*, 2008). The physiological significance of NsrR regulation is twofold. First, it enables high-level synthesis of NorB, even in the presence of oxygen, when FNR is inactive, providing protection against NO generated by the nitrosative burst of the host. Secondly, it makes *Neisseria* vulnerable to NO toxicity should the situation arise that there is a burst of NO production from nitrite during sudden oxygen starvation, a scenario that has been confirmed experimentally by exposing a gonococcal *nsrR* mutant to a high concentration of nitrite (Overton *et al.*, 2008). Gonococci protect themselves from this potential self-inflicted death with different mechanisms. First, NsrR is a repressor of *aniA*, so expression of *aniA* and *norB* are coordinated by NsrR as soon as NO accumulates sufficiently to inactivate NsrR. Secondly, the constitutively synthesized outer membrane lipoprotein, CycP (cytochrome c'), buffers accumulation of free NO (Cross *et al.*, 2001; Turner *et al.*, 2005).

Finally, the product of the *dnrN* gene, known variously as DnrN, YtfE and RIC (for repair of iron centres) provides an effective repair mechanism that recycles iron and iron-sulfur proteins damaged by nitrosative stress (Justino *et al.*, 2006, 2007; Overton *et al.*, 2006, 2008). The role of the NmlR system also seems to protect the bacteria from NO or other reactive nitrogen and oxygen species (see Kidd, Chapter 5, this volume).

The final, less dramatic level of transcription control is mediated by a combination of two proteins, the Fur protein, which represses synthesis of genes, including *aniA*, in response to excess iron, and a second repressor, ArsR. From studies of the meningococcus, it was suggested originally that Fur activated expression of *norB* directly (Delaney *et al.*, 2004). Subsequent studies with the very similar *norB* promoter in the gonococcus revealed a partial overlap of the Fur-binding site with a binding site for ArsR. In the absence of Fur or under conditions of iron deficiency, ArsR binds to the *norB* regulatory region well upstream of the promoter and represses *norB* expression about fourfold. ArsR might therefore be an important regulator *in vivo*, where iron deficiency (and hence inactivation of Fur) is likely to occur. Under conditions of iron sufficiency, Fur competes with ArsR and prevents repression (Isabella *et al.*, 2008). This indirect mechanism of transcription activation is reminiscent of the role of NarL at the *E. coli nirB* promoter, which is to prevent repression by displacing the integration host factor (IHF) from an upstream site (Wu *et al.*, 1998; Browning *et al.*, 2000). Clarke *et al.* (2010) have discussed in greater detail possible mechanisms of transcription activation at the *aniA* and *norB* promoters. They also discuss at length the likely significance of the ability to adapt to anaerobic or oxygen-limited growth to pathogenicity.

The rationale for this regulation of *norB* expression by Fur is assumed to be that excess iron is toxic because of its ability to catalyse Fenton reactions, resulting in production of highly reactive oxygen species. This then introduces yet another level of complexity in understanding how bacteria in general respond to starvation of oxygen. In a potential low redox environment, more ferrous iron will be available, necessitating a safety valve to prevent self-destruction promoted by excess iron. This highlights a link between the generation of reactive nitrogen species and reactive oxygen species that can occur simultaneously if oxygen and nitrite are both available in limited supply. This point will be considered in further detail later in this chapter.

10.5 Limitations and Advantages of a Combined Physiological and Genetic Approach to Functional Analysis of Stress Responses

In the decade 2000–2009, 'systems biology' was the buzz phrase to guide our research. This embraces the concept that in order to understand the physiology of an organism, the function of a reaction in that organism, or even human health itself, every component of the organism must be determined, from the genome, transcriptome, proteome and metabolome to the kinetic constraints of each biochemical reaction that determine each factor. In practice, we are still limited by the methods available to study each of the above, and the mathematical capacity to deal with the outputs. Microbial physiologists use all of these tools. However, even studies of an organism as limited in its ability to respond to environmental changes as the gonococcus emphasize the limitations of the methods currently available. First, a few of the most commonly used methods available will be introduced; then, examples will be provided to explain the relevance of each point to our current knowledge of the gonococcal response to changes in oxygen availability.

One commonly used method to study how bacteria adapt from oxygen-sufficient to oxygen-limited growth is to generate mutants defective in a component proposed to be important for adaptation. As this approach requires the ability to identify a phenotype that depends on that mutation, it is useful only if the response is not masked by overlapping activities that can replace the mutated gene functionally under the conditions of the laboratory tests. Note the circular argument that is developing.

A more recent approach is to determine the complete transcriptome under specified conditions. An extremely powerful method is comparative transcriptomics, in which mRNA levels in a mutant or a parent strain are compared to determine how gene expression is altered by the mutation; or to compare the transcriptomes under two or more different growth conditions as a lead to which gene product might become more or less important as environmental conditions change. However, some gene products are so important for survival that they are expressed constitutively, so comparative regulation teaches us nothing about their function. Indeed, their importance can be overlooked in the race to find genes that respond to environmental signals via transcription factors.

The limitation of both the genetic and gene regulation approaches can be avoided in part by a combination of physiological and bioinformatics approaches. Often overlooked is the importance of the cellular location of a particular component. An enzyme located in the cytoplasm might be extremely active, but be in the wrong cellular compartment to protect against the prevalent threat to survival. Indeed, examples will be cited in which a correctly located but less active or less abundant component is the key to survival in a hostile environment. Metabolomic and biochemical approaches are mutually informative; the former cannot provide rate data without the latter, but kinetic parameters are usually obtained from studies with proteins isolated from the cell in which they function. We assay what we can detect, but we do not always ask the correct question that leads to understanding their physiological role. Specific examples relevant to adaptation to different levels of oxygen availability will now be presented.

10.5.1 The defence against hydrogen peroxide provided by the anaerobically induced cytochrome *c* peroxidase, Ccp

As a first example, the gonococcal *ccp* gene encodes a cytochrome *c* peroxidase that catalyses the reduction of hydrogen peroxide (and possibly other peroxides as well) to water (Turner *et al.*, 2003; Seib *et al.*, 2006). This clearly implicates Ccp as part of a protection mechanism against one of the two major reactive oxygen species generated during aerobic growth. Surprisingly, however, no Ccp is detectable in gonococci during aerobic growth because *ccp* expression is induced by FNR only when oxygen is severely limited. Furthermore, even in cultures incubated in an anaerobic jar, the *ccp* mutant is just as resistant to hydrogen peroxide as the parent strain. The explanation is simple: any hydrogen peroxide that crosses the periplasm and cytoplasmic membrane will be decomposed immediately by the very active catalase located in the cytoplasm. Only when the *kat* gene has first been inactivated will a phenotype due to a secondary *ccp* mutation become detectable (Turner *et al.*, 2003). But why should a defence mechanism against a reactive oxygen compound be induced when it is not required? Two mutually compatible explanations can be offered immediately. First, gonococci typically colonize biofilms in an anaerobic environment dominated by anaerobes, including Lactobacilli that generate hydrogen peroxide. This peroxide clearly enters gonococci through the outer membrane, which is therefore the critical location for a first line of defence. This explains how FNR-dependent expression of *ccp* becomes a rational evolutionary strategy.

A second, equally plausible explanation why Ccp accumulates anaerobically is that for gonococci to cross the epithelial cell layer, they must pass through an anaerobic environment before emerging into the bloodstream, where they will be attacked immediately by the oxidative burst of host phagocytes. Pre-induction of Ccp would ensure that they were ready to defend themselves against this oxidative burst. Other fascinating examples of FNR-dependent gene expression are coming to light; for example, in *Shigella* virulence. *Shigella flexneri* is a major source of dysentery. Its ability to enter epithelial cells depends on the delivery of the invasion plasmid antigen through a type 3 secretion system. Initially, needle components are assembled by *Shigella* in the oxygen-limited gastrointestinal tract. When the needle has reached a predefined length, it contacts the well-aerated layer surrounding the mucosa. Oxygen diffusing from the capillary network at the tips of the villi inactivates FNR, allowing antigen secretion through needles, followed by bacterial entry. As in the gonococcus, FNR controls antigen secretion while the bacteria are anaerobic, but activates the switch to attack mode in response to oxygen availability (Marteyn *et al.*, 2010).

10.5.2 Parallel pathways for electron transfer to cytochrome oxidase

It was noted above that those single mutants defective in *cycA* (encoding cytochrome c_4) or *cycB* (encoding cytochrome c_5) reduced oxygen using physiological substrates at rates comparable to the parent strain. Although they provide the only routes of electron transfer to cytochrome oxidase, the capacity of each branch is sufficient to support almost the maximal rate of oxygen reduction. However, three different assays revealed much clearer phenotypes. First, both of the single mutants reduce oxygen at about 50% of the rate of the parent in the presence of the artificial electron donor, ascorbate-reduced TMPD. Secondly, both mutants are inhibited strongly by excess aeration, but grow at the same rate as the parent when the oxygen supply is limited. Finally, the single mutants reduce nitrite more slowly than the parent. This example is instructive, for several reasons. The first point is obvious: the phenotype detected depends critically on the assay used – or put another way, on the question asked. Secondly, mutations in single genes failed to reveal the common physiological role of these cytochromes. Finally, they show that the excess capacity provided by duplicating a pathway is not redundant but physiologically important to provide protection against reactive oxygen species that would be generated when only one of the two pathways is functioning. As noted elsewhere (Li *et al.*, 2010), the high respiratory capacity of gonococci prevents the accumulation of oxygen, which would

generate toxic oxygen species such as hydrogen peroxide, superoxide and hydroxyl radicals. We therefore propose that excess capacity for oxygen reduction provides defence against reactive oxygen species generated as by-products of their own metabolism. Independent physiological roles for each of these cytochromes might be revealed when their possible roles in nitrite reduction are understood better (see below). However, it is already known that the third haem group of the C-terminal extension of the CcoP subunit of the cytochrome oxidase donates electrons directly or indirectly to AniA, so effects of *cycA* or *cycB* mutations on rates of nitrite reduction are predicted (Hopper *et al.*, 2009).

10.5.3 Importance of other outer membrane lipoproteins in provision of defence against reactive oxygen and reactive nitrogen species

At least three outer membrane lipoproteins other than Ccp have been implicated in protection against potentially toxic metabolites. These are the single gonococcal nitrite reductase, AniA; the NO-binding protein, CycP; and the blue copper protein, Laz (Trees and Spinola, 1990). While their location suggests that their primary roles are to protect the cytoplasm and cytoplasmic membrane from attack from toxic species generated by other bacteria or by the host, they are all redox proteins that potentially could be part of alternative energy-generating electron transfer pathways. Their extracytoplasmic location precludes their direct involvement in proton pumping, but they can still accept electrons from ubiquinol, regenerating ubiquinone for further reduction by energy-conserving dehydrogenases such as a succinate dehydrogenase or the NADH dehydrogenase, Nuo. As implied above, a major gap in current knowledge is the ability to estimate electron fluxes through these outer membrane lipoproteins compared with through the respiratory chain to cytochrome oxidase cbb_3.

10.5.4 Pathways for electron transfer to outer membrane lipoproteins

A further key unanswered question is how electrons are transferred from physiological substrates such as NADH and succinate to the four outer membrane proteins with potential redox activity. Hopper *et al.* (2009) demonstrated recently that the C-terminal extension of CcoP included a third haem group that transferred electrons to AniA (Fig. 10.4). However, it is unknown whether CcoP contacts AniA directly, or whether another protein mediates this electron transfer.

In other bacteria such as *Alcaligenes faecalis*, *P. pantotrophus* and some pseudomonads, the blue copper protein azurin transfers electrons directly to NirK (Kukutani *et al.*, 1981; Zumft, 1997), while in other bacteria *c*-type cytochromes have been implicated in this process. The gonococcal genome includes genes for Laz, a pseudo-azurin related to other members of the azurin family (Trees and Spinola, 1990; Hoehn and Clark, 1992), and a small *c*-type cytochrome that we have designated cytochrome c_2. However, a major role for Laz in nitrite reduction was discounted for several reasons. First, a *laz* mutant reduced nitrite at a rate comparable to that of the parent strain (Cannon, 1989). Secondly, a *laz* mutant was far more sensitive to inhibition by hydrogen peroxide, implicating Laz in protection against reactive oxygen species rather than in nitrite reduction (Seib *et al.*, 2004; Wu *et al.*,

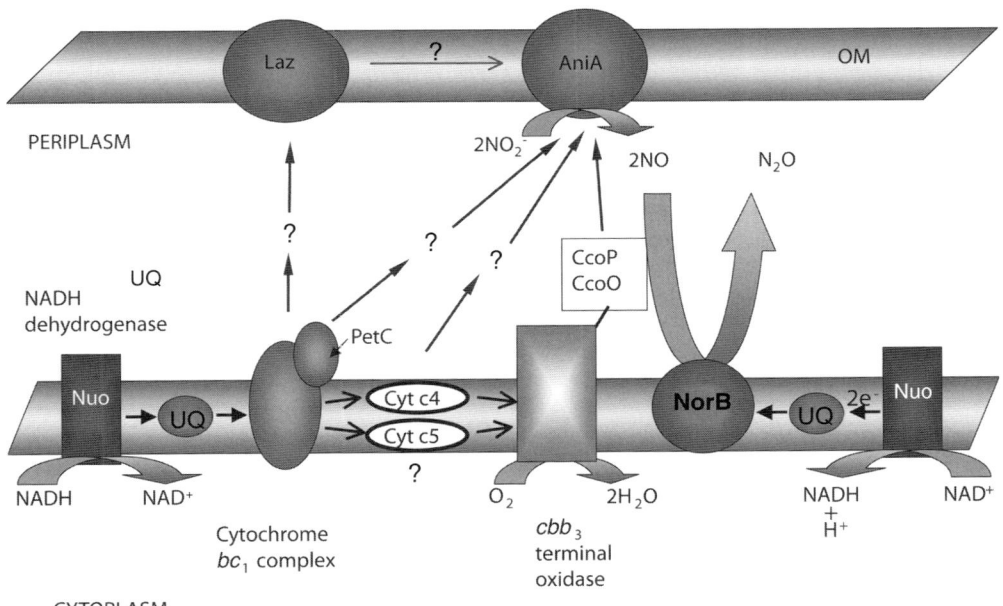

Fig. 10.4. Known and possible alternative pathways for electron transfer from the cytochrome bc_1 complex and the CcoP subunit of cytochrome oxidase to NO, nitrite and to outer membrane lipoproteins, CycP and Laz. Solid arrows indicate electron transfer pathways for which there is direct experimental evidence. Question marks indicate that current evidence is either equivocal or unavailable. They leave open the question whether additional proteins, for example the small cytochrome c_2 (product of the *cccA* gene), might also be involved. Also unknown is whether CcoP also transfers electrons to Ccp, CycP and Laz; whether Laz and cytochrome c_5 are both able to transfer electrons to any or all three of the other lipoproteins in addition to AniA.

2005). This second proposal is consistent with the report that pseudo-azurin is a direct electron donor to cytochrome c peroxidase, Ccp, in *P. pantotrophus* (Pauleta *et al.*, 2004). Note that none of these observations rule out a role for Laz in gonococcal nitrite reduction if it is accepted that, as in the electron transfer pathway to oxygen, there are multiple pathways for transferring electrons across the periplasm and that the maximum flux through more than one component is adequate to sustain normal rates of nitrite reduction. Figure 10.4 shows some of the many possible routes of electron transfer across the periplasm. Key questions that remain to be answered include how many points of electron exit there are from the electron transfer components in the cytoplasmic membrane; how many alternative intermediates transfer electrons to the four outer membrane lipoproteins; and how specific are these intermediates for the individual lipoproteins. Deeudom *et al.* (2008) have presented evidence that the meningococcal cytochrome c_5 is essential for nitrite reduction. If this is correct, gonococcal cytochrome c_5 and the CcoP subunit of cytochrome oxidase would provide the only two routes of electron transfer out of the cytoplasmic membrane, and a double mutant defective in both of these components will be totally defective in nitrite reduction. Moir (personal communication) has evidence that this is indeed correct.

10.6 Other Defence Mechanisms to Combat Oxidative Stress

Seib *et al.* (2006) have comprehensively reviewed mechanisms by which pathogenic *Neisseria* protect themselves against damage caused by reactive oxygen species. However, very little is known about the biochemical mechanism by which protection or damage repair is achieved. Despite the similarities between gonococci and meningococci, there are also striking differences. It has been mentioned already that meningococci lack the anaerobically inducible *ccp* gene encoding cytochrome c peroxidase. This possibly reflects the fact that meningococci, unlike gonococci, do not colonize an anaerobic environment. The two organisms also differ in their response to extracellular manganese (Seib *et al.*, 2004). Gonococci are inhibited by $[Mn^{2+}]$ above $100\,\mu M$, but lower concentrations provide protection against both superoxide and hydrogen peroxide, the latter independently of catalase activity. In contrast, meningococci are resistant to mM concentrations of Mn^{2+}, which apparently plays no role in protection against oxidative stress. Conversely, gonococci are an abundant source of catalase activity, possibly reflecting their exposure to Lactobacilli that generate hydrogen peroxide. A further striking difference is the relative lack of superoxide dismutase activity (Sod) in gonococci (Tseng *et al.*, 2001). Although there is a *sodB* gene present, it is expressed poorly unless induced by excess iron and by hydrogen peroxide. In other bacteria it is SodA that provides major protection against superoxide generated during aerobic growth, so this difference between gonococci and meningococci again reflects differences in their exposure to oxygen in their normal environmental niches.

Glutathione (GSH) is widely believed to provide the first line of protection against oxidative stress in the cytoplasm of many types of bacteria. It is present at a very high concentration, 5 mM, in *E. coli* and at an even higher concentration of above 15 mM in the gonococcus (Archibald and Duong, 1986). GSH becomes oxidized to form a disulfide when it repairs oxidative damage and is regenerated by the glutathione oxidoreductase, Gor. Other proteins of the thioredoxin and glutaredoxin families are thiol:disulfide oxidoreductases that reduce incorrect disulfide bonds in proteins which have been damaged oxidatively. Once toxic oxygen species are formed or accumulate in the cytoplasm, a wide range of molecules are prone to damage and so a range of repair mechanisms is required to reverse the damage. Four gonococcal proteins have been annotated as thioredoxins and a further three as glutaredoxins. At least two proteins, RecN and the DNA helicase, PriA, have been implicated in the repair of oxidatively damaged DNA (Stohl *et al.*, 2005).

Cysteine residues also provide the active sites of enzymes that reduce alkyl hydroperoxides. This function is achieved in the gonococcus by a peroxiredoxin, which is induced by hydrogen peroxide as part of the OxyR regulon. Seib *et al.* (2006) reported that a gonococcal *prx* mutant was surprisingly resistant to hydrogen peroxide, presumably because the catalase activity of the mutant was higher than that of the parent. The underlying regulatory mechanism remains to be defined.

Hydrogen peroxide also induces the synthesis of various proteins implicated in protection. They include catalase, Gor and Prx, which are components of the OxyR regulon, Fur and SodB, which is regulated by Fur in response to iron availability. The transcriptomic study of Stohl *et al.* (2005) identified many more proteins that were regulated differentially on exposure to hydrogen peroxide, but their precise roles and even whether their response was direct or indirect remained unclear. Some of the top candidates for providing protection against oxidative stress include a putative metalloprotease, NG1686, and a protein of unknown function, NG0554 (Stohl *et al.*, 2005).

Methionine residues are targets especially sensitive to oxidative damage, forming methionine sulfoxides. Gonococci synthesize a tri-functional methionine sulfoxide reductase, Msr, with an N-terminal disulfide oxidoreductase, a central MsrA-like domain and a C-terminal MsrB-like domain (Skaar *et al.*, 2002).

10.6.1 Regulation of genes providing defence against oxidative stress

PerR is related structurally to Fur and typically regulates the response of Gram-positive bacteria to peroxidative stress. The same function in many Gram-negative bacteria is fulfilled by OxyR. The gonococcus is a rare exception in that it maintains both proteins. The gonococcal OxyR is a repressor for the catalase gene, *katA*, but an activator of *gor* and *prx* (Stohl *et al.*, 2005). PerR regulates

another small group of genes, including repression of the *mntABC* operon required for manganese uptake, and hence for the Mn^{2+} protection against oxidative stress. The *adhC*, *trxB* and *copA* genes are activated by NmlR (see Kidd, Chapter 5, this volume).

10.6.2 Links between oxidative and nitrosative stress

In oxygen-limited environments pathogenic bacteria are exposed to four reactive nitrogen species, NO, peroxynitrite, dinitrogen trioxide and *S*-nitrosoglutathione (GSNO). The key source of nitrosative stress for the gonococcus is NO. This molecule arises from three sources: as a product of nitrite reduction at the cell surface by AniA; as part of the defence mechanism, the nitrosative burst, of host cells; and as a product of other bacteria in its environment. Note that although denitrifying bacteria are rarely found in the human gastrointestinal tract, NO is a product of nitrate and nitrite reduction by enteric bacteria that reduce the bulk of the nitrite to ammonia. Both peroxynitrite and dinitrogen trioxide are very toxic yet unstable species formed by the interaction of NO with superoxide or oxygen, respectively (Koppenol *et al.*, 1992; Weiss, 2006). As gonococci are found in environments where both oxygen and nitrite (and hence NO) are likely to be limiting, it is difficult to assess whether these species are major or minor threats to their survival. Finally, in the presence of metal ions, cytoplasmic NO can nitrosylate GSH to generate GSNO. As gonococci supplement oxygen-dependent growth by nitrite reduction to NO, there is a continuous threat of exposure to each of these reactive nitrogen species and direct links between adaptation to oxygen-limited growth and defence against reactive nitrogen species. Key players in protection against these threats include cytochrome *c'* (CycP), which binds NO generated by AniA at the outer membrane or generated by the nitrosative burst of the host, and DnrN, which repairs damage caused by NO that enters the cytoplasm. The roles of these proteins in providing protection against nitrosative stress will be reviewed briefly below. There is also a protein, Bcp, which is sufficiently similar to *E. coli* AhpC to be considered as a candidate for peroxynitrite reductase activity (Seib *et al.*, 2006).

10.6.3 Regulation of genes providing defence against nitrosative stress

The key regulator that responds to NO is the repressor protein, NsrR, which represses a small group of genes. As NsrR is a cytoplasmic DNA-binding protein, it provides a panic distress response only to NO that crosses the cytoplasmic membrane. Expression of three genes is induced strongly by NO: these are AniA, the nitrite reductase that will generate yet more NO; NorB, which reduces NO to nitrous oxide; and DnrN, which repairs damage caused by NO. Note that these three proteins are located in three separate cellular compartments, so each must function independently, presumably by interacting with other cellular components that are not synthesized as part of the NsrR regulon. The location of NorB in the cytoplasmic membrane provides a barrier that prevents exogenous NO from entering the cytoplasm. As inactivation of the NsrR repressor function is reversible and dependent on the nitrosylation of the iron centre of the protein, there must also be a mechanism for the removal of the NO group from the iron centre.

A second DNA-binding protein, NmlR, is a Neisserial MerR-like transcription factor. As an *nmlR* mutant is more sensitive to killing by GSNO than the parent strain, NmlR has been implicated as an activator of genes for defence against reactive nitrogen species (Kidd *et al.*, 2005; Kidd, Chapter 5, this volume).

DnrN

One of the few transcripts repressed most strongly by NsrR is *dnrN*, a homologue of the *E. coli ytfE* (Justino *et al.*, 2006, 2007). The name RIC, for repair of iron centres, has been proposed for these proteins (Overton *et al.*, 2008). The primary function of DnrN (RIC) is to remove the nitric oxide moiety from inactivated nitrosylated proteins such as NO-

damaged transcription factors Fnr, NsrR and Fur.

Although it is assumed that proteins like FNR and NsrR itself are regenerated by DnrN, as shown in Fig. 10.5, many questions surrounding the mechanism by which this is achieved remain to be answered.

For example, does DnrN regenerate the active form of the damaged protein directly, or initiate a chain of reactions that might include removal of the nitrosylated iron atom and then replace it? Also not clear is whether the NO is released unchanged, or whether it is transferred to an acceptor molecule, for example, glutathione. If the latter is correct, other open questions are whether and, if so, how GSH is regenerated from GSNO. One possibility is that there is a reductive pathway in which GSNO is reduced via GSNHOH to GSH and ammonia (via the AdhC driven pathway), but this is currently purely speculative. Other formal possibilities remain to be excluded. Whatever the answers to these questions, we already know that DnrN is critical for enabling gonococci to adapt to a sudden change in the availability of oxygen or nitrite (Overton *et al.*, 2008).

10.6.4 Dual function of another outer membrane protein, cytochrome *c′*

Although NorB largely prevents NO from reaching the cytoplasm and DnrN can restore function to damaged proteins, sudden exposure to NO before NsrR repression has been relieved can be lethal. A second outer membrane lipoprotein, cytochrome *c′*, provides a front line of defence against sudden attack, for example, by NO generated by macrophages as gonococci cross the epithelial cell layer. This cytochrome potentially also can protect gonococci against NO generated by nitrite reduction, acting as a buffer to decrease the concentration of free NO until NorB has been synthesized. Thus, cytochrome *c′* also fulfils two physiological roles as a first line of defence against external attack and to prevent self-induced destruction during nitrite reduction.

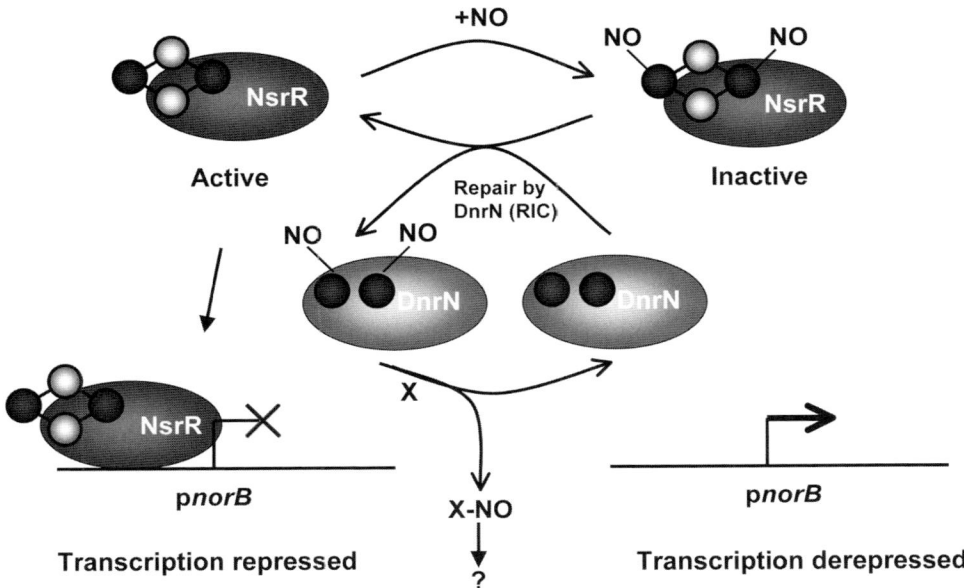

Fig. 10.5. Repair of iron-sulfur proteins damaged by nitrosative stress. While the role of DnrN (also known as RIC, for repair of iron centres, and YtfE) is well established, it is unknown whether DnrN is recycled, for example, by the pathway shown. If there is indeed a recycling pathway, this could be achieved either by spontaneous release of nitric oxide unchanged, or by its reduction via a hydroxamate to ammonia, with or without transfer to an acceptor molecule indicated in the figure as X. The possibility that X is a thiol such as glutathione is discussed in the text.

10.7 Take-home Messages and Their Implications for Pathogenicity

The first take-home message from this chapter is that pathogenic bacteria have not evolved to survive the artificial stresses imposed by life in conical flasks in orbital shakers. In laboratory experiments we provide unnatural excesses of some nutrients, and severe deprivation of others. In their natural environments bacteria are exposed to multiple substrate limitations that, in the context of pathogenic *Neisseria*, mean they will need to be able to cope simultaneously with limited supplies of different carbon sources, oxygen, hydrogen peroxide and other peroxides, ferrous iron, nitrite and NO. Transcription factors like FNR, NsrR, OxyR and Fur will be neither

completely inactive nor fully active. Most of the proteins mentioned in passing will therefore be present at submaximum levels *in vivo*, each contributing towards maintenance of homeostasis. The selective advantage conferred by gene regulation is the ability to adjust enzyme levels in response to gradual changes in the bacterial environment, not the transfer of a conical flask from a shaker to an anaerobic cabinet. While this point is again painfully obvious, it helps us understand why the iron-sulfur centre of the meningococcal FNR is more oxygen-resistant than its gonococcal counterpart (Rock *et al.*, 2007). It also rationalizes the constitutive synthesis of cytochrome *c'*, required to protect against the onslaught from outside the bacteria of nitrosative stress imposed by the always-aerobic host cells. It also explains why Ccp is maintained by gonococci but not by meningococci, and is consistent with the greater flexibility of the gonococcal than the meningococcal electron transfer pathways.

Gonococci are known to exploit host defence mechanisms for their survival; for example, the ability to sialylate their lipo-oligosaccharide using host-derived CMP-NANA (Nairn *et al.*, 1988; Parsons *et al.*, 1988). Reduction of hydrogen peroxide by Laz and NO by NorB might therefore be further examples of gonococcal exploitation of what otherwise might be lethal threats, providing classic examples of making a virtue of a necessity. The central roles of NsrR and CycP in maintaining low intracellular and extracellular concentrations of NO not only ensures that nitrosative damage to key regulators of oxygen metabolism like FNR and Fur can be repaired but also provides an auxiliary electron sink for generation of proton motive force when the supply of oxygen is limited. Furthermore, Clark *et al.* (2010) have proposed that the ability of many gonococcal strains to regulate ambient concentrations of NO has profound effects on the host pro-inflammatory and anti-inflammatory responses to infection. We have noted elsewhere that, as the gonococcus encounters oxygen limitation in the host, it becomes 'primed' for pathogenesis by inducing AniA expression in order to respire the μM concentrations of nitrite present in the host. Nitrite respiration would generate NO, deactivating NsrR and thus further inducing expression of AniA and allowing NorB to be produced. The gonococcus would now be 'primed' for exposure to and rapid detoxification of RNS generated by the host.

Acknowlegements

The author gratefully acknowledges Amanda Hopper and Drs Nick Tovell, Ying Li and Sudesh Mohan for help with the production of the figures for this chapter and for their many contributions to research in his laboratory.

References

Allen, J.W.A., Daltrop, O., Stevens, J.M. and Ferguson, S.J. (2003) C-type cytochromes: diverse structures and biogenesis systems pose evolutionary problems. *Philosophical Transactions of the Royal Society of London B Biological Sciences* 358, 255–266.

Anjum, M.F., Stevanin, T.M., Read, R.C. and Moir, J.W. (2002) Nitric oxide metabolism in Neisseria meningitidis. *Journal of Bacteriology* 184, 2987–2993.

Archibald, F.S. and Duong, M.N. (1986) Superoxide dismutase and oxygen toxicity defenses in the genus *Neisseria*. *Infection and Immunity* 51, 631–641.

Boulanger, M.J. and Murphy, M.E. (2002) Crystal structure of the soluble domain of the major anaerobically induced outer membrane protein (AniA) of pathogenic *Neisseria*: a new class of copper-containing nitrite reductases. *Journal of Molecular Biology* 315, 1111–1127.

Britigan, B.E., Klapper, D., Svendsen, T. and Cohen, M.S. (1988) Phagocyte-derived lactate stimulates oxygen consumption by *Neisseria gonorrhoeae*. An unrecognized aspect of the oxygen metabolism of phagocytosis. *Journal of Clinical Investigation* 81, 318–324.

Browning, D.F., Cole, J.A. and Busby, S.J. (2000) Suppression of FNR-dependent transcription activation at the *Escherichia coli nir* promoter by Fis, IHF and H-NS: modulation of transcription initiation by a complex nucleo-protein assembly. *Molecular Microbiology* 37, 12581269.

Cannon, J.G. (1989) Conserved lipoproteins of pathogenic *Neisseria* species bearing the H.8 epitope: lipid-modified azurin and H.8 epitope. *Clinical Microbiology Reviews* 2(suppl), S1–S4.

Chiang, R.C., Cavicchioli, R. and Gunsalus, R.P. (1992) Identification and characterization of NarQ, a second nitrate sensor for nitrate-dependent gene regulation in *Escherichia coli*. *Molecular Microbiology* 14, 1913–1923.

Clark, V.L., Campbell, L.A., Palermo, D.A., Evans, T.M. and Klimpel, K.W. (1987) Induction and repression of outer membrane proteins during anaerobic growth of *Neisseria gonorrhoeae*. *Infection and Immunity* 55, 1359–1364.

Clark, V.L., Knapp, J.S., Thompson, S. and Klimpel, K.W. (1988) Presence of antibodies to the major anaerobically induced gonococcal outer membrane protein in sera from patients with gonococcal infections. *Microbial Pathogenesis* 5, 381–390.

Clark, V.L., Isabella, V.M., Barth, K. and Overton, T.W. (2010) Regulation and function of the neisserial denitrification pathway: life with limited oxygen. In: Genco, C.A. and Wetzler, L. (eds) *Neisseria: Molecular Mechanisms and Pathogenesis.* Caister Academic Press, Norfolk, UK, pp. 19–39.

Constantinidou, C., Hobman, J.L., Griffiths, L., Patel, M.D., Penn, C.W., Cole, J.A. and Overton, T.W. (2006) A reassessment of the FNR regulon and transcriptomic analysis of the effects of nitrate, nitrite, NarXL, and NarQP as *Escherichia coli* adapts from aerobic to anaerobic growth. *Journal of Biological Chemistry* 281, 4802–4815.

Cross, R., Lloyd, D., Poole, R.K. and Moir, J.W. (2001) Enzymatic removal of nitric oxide catalyzed by cytochrome *c′* from *Rhodobacter capsulatus. Journal of Bacteriology* 183, 5000–5004.

Darwin, A.J., Tyson, K.L., Busby, S.J. and Stewart, V. (1997) Differential regulation by the homologous response regulators NarL and NarP of *Escherichia coli* K-12 depends on DNA binding site arrangement. *Molecular Microbiology* 25, 583–595.

Deeudom, M., Rock, J. and Moir, J. (2006) Organization of the respiratory chain of *Neisseria meningitidis. Biochemical Society Transactions* 34, 139–142.

Deeudom, M., Koomey, M. and Moir, J.W. (2008) Roles of *c*-type cytochromes in respiration in *Neisseria meningitidis. Microbiology* 154, 2857–2864.

Delaney, I., Rappuoli, R. and Scarlato, V. (2004) Fur functions as an activator and as a repressor of putative virulence genes in *Neisseiria meningitidis. Molecular Microbiology* 52, 1081–1090.

D'Mello, R., Purchase, D., Poole, R.K. and Hill, S. (1997) Expression and content of terminal oxidases in *Azotobacter vinelandii* grown with excess NH_4^+ are modulated by oxygen supply. *Microbiology* 143, 231–237.

Downs, A.J. and Jones, C.W. (1975) Respiration-linked proton translocation in *Azotobacter vinelandii. FEBS Letters* 60, 42–46.

Fulop, V., Sam, K.A., Ferguson, S.J., Ginger, M.L. and Allen, J.W. (2009) Structure of a trypanosomatid mitochondrial cytochrome *c* with heme attached via only one thioether bond and implications for the substrate recognition requirements of heme lyase. *FEBS Journal* 276, 2822–2832.

González-Flecha, B. and Demple, B. (1995) Metabolic sources of hydrogen peroxide in aerobically growing *Escherichia coli. Journal of Biological Chemistry* 270, 13681–13687.

Haddock, B.A. and Jones, C.W. (1977) Bacterial respiration. *Bacteriological Reviews* 41, 47–99.

Hoehn, G.T. and Clark, V.L. (1992) The major anaerobically induced outer membrane protein of *Neisseria gonorrhoeae*, Pan1, is a lipoprotein. *Infection and Immunity* 60, 4704–4708.

Hopper, A., Tovell, N. and Cole, J.A. (2009) A physiologically significant role in nitrite reduction of the CcoP subunit of the cytochrome oxidase. *FEMS Microbiology Letters* 301, 232–240.

Householder, T.C., Fozo, E.M., Cardinale, J.A. and Clark, V.L. (2000) Gonococcal nitric oxide reductase is encoded in a single gene, *norB*, which is required for anaerobic growth and is induced by nitric oxide. *Infection and Immunity* 68, 5241–5246.

Imlay, J. and Fridovich, I. (1991) Assay of metabolic superoxide production in *Escherichia coli. Journal of Biological Chemistry* 266, 6957–6965.

Isabella, V.M., Wright, L.F., Barth, K., Spence, J.M., Grogan, S., Genco, C.A. and Clarke, V.L. (2008) *cis*- and *trans*-acting elements involved in regulation of *norB* (*norZ*), the gene encoding nitric oxide reductase in *Neisseria gonorrhoeae. Microbiology* 154, 226–239.

Isabella, V.M., Lapek, J.D. Jr, Kennedy, E.M. and Clark, V. (2009) Functional analysis of NsrR, a nitric oxide-sensing Rrf2 repressor in *Neisseria gonorrhoeae. Molecular Microbiology* 71, 227–239.

Justino, M.C., Almeida, C.C., Goncalves, V.L., Teixeira, M. and Saraiva, L.M. (2006) *Escherichia coli* YtfE is a di-iron protein with an important function in assembly of iron sulphur clusters. *FEMS Microbiology Letters* 257, 278–284.

Justino, M.C., Almeida, C.C., Goncalves, V.L., Teixeira, M. and Saraiva, L.M. (2007) *Escherichia coli* di-iron YtfE protein is necessary for the repair of stress-damaged iron-sulphur clusters. *Journal of Biological Chemistry* 282, 10352–10359.

Khoroshilova, N., Popescu, C., Münck, E., Beinert, H. and Kiley, P.J. (1997) Iron-sulfur cluster disassembly in the FNR protein of *Escherichia coli* by O2: [4Fe-4S] to [2Fe-2S] conversion with loss of biological activity. *Proceedings of the National Academy of Sciences of the United States of America* 94, 6087–6092.

Kidd, S.P., Potter, A.J., Apicella, M.A., Jennings, M.P. and McEwan, A.G. (2005) NmlR of *Neisseria gonorrhoeae*: a novel redox responsive transcription factor from the MerR family. *Molecular Microbiology* 57, 1676–1689.

Kolonay, J.F. Jr, Moshiri, F., Gennis, R.B., Kaysser, T.M. and Maier, R.J. (1994) Purification and characterization of the cytochrome *bd* complex from *Azotobacter vinelandii*: comparison to the complex from *Escherichia coli. Journal of Bacteriology* 176, 4177–4181.

Koppenol, W.H., Moreno, J.J., Pryor, W.A., Ischiropoulos, H. and Beckman, J.S. (1992) Peroxynitrite, a cloaked oxidant formed by nitric oxide and superoxide. *Chemical Research in Toxicology* 5, 834–842.

Kranz, R., Lill, R., Goldman, B., Bonnard, G. and Merchant, S. (1998) Molecular mechanisms of cytochrome *c* biogenesis: three distinct systems. *Molecular Microbiology* 29, 383–396.

Kukutani, T., Watanabe, H., Arima, K. and Beppu, T. (1981) Purification and properties of a copper-containing nitrite reductase from a denitrifying bacterium *Alcaligenes faecalis* strain S-6. *Journal of Biochemistry (Tokyo)* 89, 453–461.

Li, Y., Hopper, A., Overton, T.W., Squire, D.J.P., Cole, J. and Tovell, N. (2010) Organisation of the electron transfer chain to oxygen in the obligate human pathogen, *Neisseria gonorrhoeae*: roles for cytochromes c_4 and c_5, but not cytochrome c_2, in oxygen reduction. *Journal of Bacteriology* 192, 2395–2406.

Lissenden, S., Mohan, S, Overton, T., Regan, T., Crooke, H., Cardinale, A., Householder, T.C., Adams, P., O'Conner, D., Clarke, V.L., Smith, H. and Cole, J.A. (2000) Identification of transcription activators that regulate gonococcal adaptation from aerobic to anaerobic or oxygen-limited growth. *Molecular Microbiology* 37, 839–855.

Marteyn, B., West, N., Browning, D., Cole, J., Shaw, J., Palm, F., Mourniers, J., Prévost, M.-C., Sansonetti, P. and Tang, C. (2010) Modulation of *Shigella* virulence in response to available oxygen *in vivo*. *Nature* 465, 355–358.

Morse, S.A. (1978) The biology of the gonococcus. *CRC Critical Reviews of Microbiology* 7, 93–189.

Morse, S.A. and Bartenstein, L. (1974) Factors affecting autolysis of *Neisseria gonorrhoeae*. *Proceedings of the Society for Experimental Biology and Medicine* 145, 1418–1421.

Morse, S.A., Cacciapuoti, A.F. and Lysko, P.G. (1979) Physiology of *Neissseria gonorrhoeae*. *Advances in Microbial Physiology* 20, 251–320.

Nairn, A.C., Cole, J.A., Patel, P.V., Parsons, N.J., Fox, J.E. and Smith, H. (1988) Cytidine 5'-monophospho-N-acetyl neuraminic acid or a related compound is the low M_r factor from human red blood cells which induces gonococcal resistance to killing by human serum. *Journal of General Microbiology* 134, 3295–3306.

Ng, T.C.N., Laheri, A.N. and Maier, R.J. (1995) Cloning, sequencing and mutagenesis of the cytochrome c_4 gene from *Azotobacter vinelandii*: characterisation of the mutant strain and a proposed new branch in the respiratory chain. *Biochimica et Biophysica Acta* 1230, 119–129.

Noriega, C.E., Lin, H.Y., Chen, L.L., Williams, S.B. and Stewart, V. (2010) Asymmetric cross-regulation between the nitrate-responsive NarX-NarL and NarQ-NarP two-component regulatory systems from *Escherichia coli*. *Molecular Microbiology* 75, 394–412.

Overton, T.W., Whitehead, R., Li, Y., Snyder, L.A., Saunders, N.J., Smith, H. and Cole, J.A. (2006) Coordinated regulation of the *Neisseria gonorrhoeae* truncated denitrification pathway by the nitric oxide-sensitive repressor, NsrR, and nitrite-insensitive NarQ-NarP. *Journal of Biological Chemistry* 281, 33115–33126.

Overton, T.W., Justino, M.C., Li, Y., Baptista, J.M., Melo, A.M., Cole, J.A. and Saraiva, L.M. (2008) Widespread distribution in pathogenic bacteria of di-iron proteins that repair oxidative and nitrosative damage to iron-sulphur centres. *Journal of Bacteriology* 190, 2004–2013.

Parsons, N.J., Patel, P.V., Tan, E.L., Andrade, J.R.C., Nairn, A.C., Goldner, M., Cole, J.A. and Smith, H. (1988) Cytidine 5'-monophospho-N-acetyl neuraminic acid and a low molecular weight factor from human blood cells induce lipopolysaccharide alteration in gonococci when conferring resistance to killing by human serum. *Microbial Pathogenesis* 5, 303–309.

Pauleta, S.R., Guerlesquin, F., Goodhew, C.F., Devreese, B., Van Beeumen, J., Pereira, A.S., Moura, I. and Pettigrew, G.W. (2004) *Paracoccus pantotrophus* pseudoazurin is an electron donor to cytochrome *c* peroxidase. *Biochemistry* 43, 11214–11225.

Preisig, O., Anthamatten, D. and Hennecke, H. (1993) Genes for a microaerobically induced complex in *Bradyrhizobium japonicum* are essential for a nitrogen-fixing symbiosis. *Proceedings of the National Academy of Sciences of the United States of America* 90, 3309–3313.

Preisig, O., Zufferey, R., Thöny-Meyer, L., Appleby, C.A. and Hennecke, H. (1996) A high-affinity cbb_3-type cytochrome oxidase terminates the symbiosis-specific respiratory chain of *Bradyrhizobium japonicum*. *Journal of Bacteriology* 178, 1532–1538.

Rabin, R.S. and Stewart, V. (1993) Dual response regulators (NarL and NarP) interact with dual sensors (NarX and NarQ) to control nitrate- and nitrite-regulated gene expression in *Escherichia coli* K-12. *Journal of Bacteriology* 175, 3259–3268.

Rey, L. and Maier, R.J. (1997) Cytochrome *c* terminal oxidase pathways of *Azotobacter vinelandii*: analysis of cytochrome c_4 and c_5 mutants and up-regulation of cytochrome *c*-dependent pathways with N_2 fixation. *Journal of Bacteriology* 179, 7191–7196.

Reyn, A. (1974) Gram negative cocci and coccibacilli. In: Buchanan, R.E. and Gibbons, N.E. (eds) *Bergey's Manual of Determinative Bacteriology.* Williams and Wilkin, Baltimore, Maryland, pp. 427–433.

Rock, J.D., Thomson, M.J., Read, R.C. and Moir, J.W. (2007) Regulation of denitrification genes in *Neisserial meningitidis* by nitric oxide and the repressor, NsrR. *Journal of Bacteriology* 189, 1138–1144.

Rock, M.D., Mahnane, M.R., Anjum, M.F., Shaw, J.G., Read, R.C. and Moir, J.W. (2005) The pathogen *Neisseria meningitidis* requires oxygen, but supplements growth by denitrification. Nitrite, nitric oxide and oxygen control respiratory flux at genetic and metabolic levels. *Molecular Microbiology* 58, 800–809.

Rodionov, D.A., Dubchak, I.L., Arkin, A.P., Alm, E.J. and Gelfand, M.S. (2005) Dissimilatory metabolism of nitrogen oxides in bacteria: comparative reconstruction of transcriptional networks. *Public Library of Science Computational Biology* e55.

St Amant, D.C., Valentin-Bon, T. and Jerse, A.E. (2002) Inhibition of *Neisseria gonorrhoeae* by *Lactobacillus* species that are commonly isolated from the female genital tract. *Infection and Immunity* 70, 7169–7171.

Seib, K.L., Tseng, H.J., McEwan, A.G., Apicella, M.A. and Jennings, M.P. (2004) Defenses against oxidative stress in *Neisseria gonorrhoeae* and *Neisseria meningitidis*: distinctive systems for distinctive lifestyles. *Journal of Infectious Diseases* 190, 136–147.

Seib, K.L., Wu, H.-J., Kidd, S.P., Apicella, M.A., Jennings, M.P. and McEwan, A.G. (2006) Defenses against oxidative stress in *Neisseria gonorrhoeae*: a system tailored for a challenging environment. *Microbiology and Molecular Biology Reviews* 70, 344–361.

Seib, K.L., Wu, H.J., Srikhanta, Y.N., Edwards, J.L., Falsetta, M.L., Hamilton, A.J., Maguire, T.L., Grimmond, S.M., Apicella, M.A., McEwan, A.G. and Jennings, M.P. (2007) Characterization of the OxyR regulon of *Neisseria gonorrhoeae*. *Molecular Microbiology* 63, 54–68.

Sevastsyanovich, Y., Alfasi, S., Overton, T., Hall, R., Jones, J., Hewitt, C. and Cole, J. (2009) Exploitation of GFP fusion proteins and stress avoidance as a generic strategy for the production of high quality recombinant proteins. *FEMS Microbiology Letters* 299, 86–94.

Skaar, E.P., Tobiasen, D.M., Quick, J., Judd, R.C., Weissbach, H., Etienne, F., Brot, N. and Seifert, H.S. (2002) The outer membrane localization of the *Neisseria gonorrhoeae* MsrA/B is involved in survival against reactive oxygen species.

Proceedings of the National Academy of Sciences of the United States of America 99, 10108–10113.

Snyder, L.A., Davies, J.K., Ryan, C.S. and Saunders, N.J. (2005) Comparative overview of the genomic and genetic differences between pathogenic Neisseria strains and species. *Plasmid* 54, 191–218.

Stohl, E.A., Criss, A. K. and Seifert, H.S. (2005) The transcriptome response of *Neisseria gonorrhoeae* to hydrogen peroxide reveals genes with previously uncharacterized roles in oxidative damage protection. *Molecular Microbiology* 58, 520–532.

Storz, G. and Imlay, J.A. (1999) Oxidative stress. *Current Opinion in Microbiology* 2, 188–194.

Strange, J.C. and Judd, R.C. (1994) Identification of heme containing proteins of *Neisseria gonorrhoeae*. In: Conde-Glez, C.J., Morse, S., Rice, P., Sparling, F. and Calderón, E. (eds) *Pathobiology and Immunology of Neisseriaceae.* Instituto Nacional de Salud Pública Press, Morelos, Mexico, pp. 417–423.

Trees, D.L. and Spinola, S.M. (1990) Localization of immune response to the lipid modified azurin of pathogenic Neisseria. *Journal of Infectious Diseases* 161, 336–339.

Tseng, H.J., Srikhanta, Y., McEwan, A.G. and Jennings, M.P. (2001) Accumulation of manganese in *Neisseria gonorrhoeae* correlates with resistance to oxidative killing by superoxide anion and is independent of superoxide dismutase activity. *Molecular Microbiology* 40, 1175–1186.

Tucker, N., Hicks, M.G., Clarke, T.A., Crack, J.C., Chandra, G., Le Brun, N.E., Dixon, R. and Hutchings, M.I. (2008) The transcriptional repressor protein NsrR senses nitric oxide directly via a [2Fe-2S] cluster. *Public Library of Science ONE,* 3, e3623.

Turner, S.M., Reid, E.G., Smith, H. and Cole, J.A. (2003) A novel cytochrome *c* peroxidase in *Neisseria gonorrhoeae*, a lipoprotein from a Gram-negative bacterium. *Biochemical Journal* 373, 865–873.

Turner, S.M., Moir, J.W., Griffiths, L., Overton, T.W., Smith, H. and Cole, J.A. (2005) Mutational and biochemical analysis of cytochrome *c′*, a nitric oxide-binding lipoprotein important for adaptation of *Neisseria gonorrhoeae* to oxygen-limited growth. *Biochemical Journal* 388, 545–553.

Tyson, K.L., Cole, J.A. and Busby, S.J. (1994) Nitrite and nitrate regulation at the promoters of two *Escherichia coli* operons encoding nitrite reductase: identification of common target heptamers for both NarP- and NarL-dependent regulation. *Molecular Microbiology* 13, 1045–1055.

Wang, H., Tseng, C.P. and Gunsalus, R.P. (1999) The *napF* and *narG* nitrate reductase operons in *Escherichia coli* are differentially expressed in response to sub-micromolar concentrations of nitrate but not nitrite. *Journal of Bacteriology* 181, 5303–5308.

Weiss, B. (2006) Evidence for mutagenesis by nitric oxide during nitrate metabolism in *Escherichia coli*. *Journal of Bacteriology* 188, 829–833.

Whitehead, R.N., Overton, T.W., Snyder, L.A.S., McGowan, S.J., Smith, H., Cole, J.A. and Saunders, N.J. (2007) The small FNR regulon of *Neisseria gonorrhoeae*: comparison with the larger *Escherichia coli* FNR regulon and interaction with the NarQ-NarP regulon. *BMC Genomics* 8, 35.

Winter, D.B. and Morse, S.S. (1975) Physiology and metabolism of pathogenic *Neisseria*: partial characterisation of the respiratory chain of *Neisseria gonorrhoeae*. *Journal of Bacteriology* 123, 631–636.

Wu, H.C., Tyson, K.L., Cole, J.A. and Busby, S.J.W. (1998) Regulation of the *E. coli nir* operon by two transcription factors: a new mechanism to account for co-dependence on two activators. *Molecular Microbiology* 27, 493–505.

Wu, H.J., Seib, K.L., Edwards, J.L., Apicella, M.A., McEwan, A.G. and Jennings, M.P. (2005) Azurin of pathogenic bacteria spp. is involved in defense against hydrogen peroxide and survival in cervical epithelial cells. *Infection and Immunity* 73, 8444–8448.

Yuki, E.T., Elbaz, M.A., Nakano, M.M. and Moënne-Loccoz, P. (2008) Transcription factor NsrR from *Bacillus subtilis* senses nitric oxide with a 4Fe-4S cluster. *Biochemistry* 47, 13084–13092.

Zheng, H.Y., Alcorn, T.M. and Cohen, M.S. (1994) Effects of H_2O_2-producing lactobacilli on *Neisseria gonorrhoeae* growth and catalase activity. *Journal of Infectious Diseases* 170, 1209–1215.

Zumft, W.G. (1997) Cell biology and molecular basis of denitrification. *Microbiology and Molecular Biology Reviews* 61, 533–616.

Part 5
Metal Ions and Pathogenic Bacteria

11 Copper and Zinc Stress in Bacteria

Selina R. Clayton, Karin Heurlier, Taku Oshima
and Jon L. Hobman*

11.1 Introduction

A number of the transition metals – Mn, Fe, Co, Ni, Cu, Zn, Mo and, to a lesser extent, V and W – are essential micronutrients required for diverse essential cellular functions in bacteria such as electron transfer, dioxygen utilization and osmotic balance; they can be essential components of metalloenzymes, enzyme cofactors, stabilizers of DNA and enzymes through electrostatic interactions, or can have structural roles in proteins and nucleic acids (Hobman, 2007). The essential transition metals can exert toxic effects on cells at higher concentrations, so bacteria need to maintain appropriate levels of these metals by balancing uptake, trafficking, allocation and metal-removal mechanisms.

This chapter will discuss the role that two essential but potentially toxic metals, copper (Cu) and zinc (Zn), play in bacterial nutrition, how bacteria handle these metals and the potential and known roles that these metals play in infection by pathogenic bacteria, or how Cu and Zn are used as antimicrobials.

11.2 Copper

Copper has long been known to humans, being one of the most easily extractable and workable metals. It is now known that as well as its industrial uses, copper is also an essential micronutrient for aerobic organisms (White *et al.*, 2009), but is toxic to both prokaryotes and eukaryotes in excess quantities. Copper is most likely to have become biologically important to organisms when the Earth's atmosphere became aerobic, with anaerobic organisms not demonstrating use of copper (Solioz *et al.*, 2010). The chemical properties of copper explain why it is both essential and toxic, particularly to aerobic biological systems. Copper is a d-block transition metal and exists in three forms: as metallic copper metal (0), cuprous copper (I) (Cu^{1+}) and cupric copper (II) (Cu^{2+}). There is also a fourth unstable state (Cu^{3+}) found in some crystalline copper compounds. Metallic copper readily oxidizes in air to copper (I) and then to copper (II). Copper (I) is a soft Lewis acid, which prefers coordination to S and N groups (soft Lewis bases), such as cysteine sulfhydryl (SH) groups or methionine thioether linkages, and/or histidine nitrogens in proteins, as well as to oxygen groups in other biological molecules. Copper (II) is a moderate strength Lewis acid preferring coordination to other ligands such as sulfate and nitrite. Many of the non-essential and toxic metals such as mercury and silver are thiophilic soft Lewis acids (Lippard and

* Corresponding author.

Berg, 1994; Reilly, 2004; Hobman, 2007). The preferred coordination of metals to biological molecules can also be viewed in terms of the stability of metal complexes. Divalent copper has the highest affinity of the divalent essential metals for binding to metalloproteins according to the Irving–Williams series of the order of stability of metal complexes: $Ca^{2+} < Mg^{2+} < Mn^{2+} < Fe^{2+} < Co^{2+} < Ni^{2+} < Cu^{2+} > Zn^{2+}$ (Fraústo da Silva and Williams, 2001; Waldron and Robinson, 2009). Copper has the highest affinity for proteins when compared to other essential metals; it coordinates directly to cysteine, methionine and histidine side chains in proteins, but its oxidation state determines its coordination environments so that copper (I) prefers tetrahedral coordination geometry, while copper (II) prefers a square planar coordination. Coordination to structurally important amino acids means that in addition to copper's role in redox cycling, it can also contribute to thermodynamic stability of protein structures (Osman and Cavet, 2008).

Copper (I) has a $3d^{10}$ closed orbital, while copper (II) has an incompletely filled $3d^9$ orbital. As copper can exhibit variable oxidation states, it can oxidize some substrates directly and also participate in the generation of reactive oxygen species (ROS), seen through Fenton-like reactions. This is due to the high redox potential of the copper (I)/(II) couple and ready cycling between them, meaning that copper can act as either an electron donor or acceptor, both in redox active enzymes or in the electron transport chain. There are over 30 known enzymes which require copper for functionality in higher organisms including cytochrome oxidase, superoxide dismutase (Cu, Zn-Sod) and amine oxidase, as well as electron transport proteins such as the azurins and plastocyanins (Solioz *et al.*, 2010). In bacteria there are fewer known copper-containing enzymes and these appear to be found exclusively outside the cytoplasm, being membrane-bound (Grampositive) or membrane-bound/periplasmic (Gram-negative) (Osman and Cavet, 2008; Solioz *et al.*, 2010).

11.2.1 Human uses of copper

Copper is widely distributed in the environment in various ores or as the copper metal, and is found at a mean concentration of 50 mg kg^{-1} in the Earth's crust, ranging from <1 mg kg^{-1} to 700 mg kg^{-1} in soils (Sparks, 2004). Copper can be released into the environment at high concentrations due to anthropogenic activities such as mining, smelting or other manufacturing processes. The technology to refine and work metallic copper has been used for as long as 5000 years, and was used to make bronze alloys for weapons and tools. Copper has been used medically for at least 4000 years: as an antiseptic and as a means to purify water. It has also been used for many centuries as a wound treatment, and is recognized as an antimicrobial and antiviral in medicine (Borkow and Gabbay, 2009). Copper compounds are also widely used as fungicides and wood preservatives in antifouling paints, and as molluscicides (Borkow and Gabbay, 2009). In agriculture copper has been used as an antimicrobial, algicide, pesticide and antifungal agent: copper sulfate solutions were used as an antifungal treatment of seed grains in the 18th century, and in the late 19th century Bordeaux mixture (copper sulfate and calcium hydroxide) and Burgundy mixture (copper sulfate and sodium carbonate) were widely used to control mildew on grape vines (Bremner, 1998). Copper salts are also commonly used as dietary supplements or are added to farm animal feed as both a growth promoter, enhancing the feed-to-body-weight conversion rate, and as an inhibitor of postweaning diarrhoea. As the levels of copper sulfate added to these feeds is far higher than the levels required to maintain nutritional health, there have been suggestions that the addition of copper (and zinc) may maintain the stability of intestinal flora or select against incoming pathogenic bacteria (Hasman *et al.*, 2006a). The European Union has banned the use of antibiotics as growth promoters in animal feeds, but still allows copper (as copper sulfate) at up to 170 ppm in pig feed, 25 ppm in chicken feed and 10 ppm in calf feed. The USA allows antibiotics to be added to animal feeds and does not restrict the levels of copper and zinc supplements (and also allows organic arsenic compounds such as Roxarsone to be added to chicken and turkey feeds), with typically between 125 and 250 ppm copper sulfate added to pig feed (Hasman *et al.*, 2006a; Sapotka *et al.*, 2007). Perhaps unsurprisingly, there have been several examples of increased copper resistance/copper tolerance in bacteria isolated from pigs fed on copper-supplemented feed (Williams *et al.*, 1993) or from agricultural crops treated with copper pesticides (Cooksey *et al.*, 1990).

11.2.2 Copper deficiency

Copper deficiency has important cellular consequences. In bacteria, insufficient copper in the growth medium results in reduced cellular growth (Kershaw *et al.*, 2005). In plants, response to copper availability and cellular copper homeostasis is controlled by uptake through roots, copper trafficking by copper chaperones and P-type ATPases, and by control of the levels of copper-containing proteins (Yamasaki *et al.*, 2008). In higher animals, dietary copper deficiency can cause anaemia and neurological defects, and increased susceptibility to infection. In sheep, swayback is a condition that is caused by copper deficiency in mid to late pregnancy ewes, which results in lambs that are characterized clinically by their lack of coordination and weakness in their hind limbs. In humans, copper deficiency is rare as copper is widely available in food and water, but has been seen in premature babies and malnourished children (Reilly, 2004). There is evidence that copper uptake in higher organisms can be interfered with by competing dietary metals such as zinc and iron. Zinc plays a significant role in influencing the binding of copper to metallothionein in pigs, sheep and calves, with zinc deficiency inhibiting copper binding by metallothionein (Bremner, 1998).

Copper homeostasis in humans is controlled at the cellular level in terms of allocation of copper to enzymes and regulation of free copper levels, but uptake into the body is through the gastrointestinal tract, where between 30 and 50% of ingested copper (Cu^{2+}) is absorbed in the small intestine and small amounts in the stomach. Copper uptake in the small intestine is controlled by the copper membrane transporter (CMT1). ATOX1 transports some copper in the cell to the trans-Golgi network, while metallothionein binds other copper ions. These molecules contain a number of cysteine residues capable of binding multiple zinc and copper ions. Their function remains undefined, but it has been suggested they provide a zinc store to aid in the homeostasis of these physiologically important metals. Metallothioneins are present intracellularly, primarily in the cytoplasm, and extracellularly in bodily fluids such as blood plasma and urine. As copper concentrations rise, ATP7a, copper transporting P-type ATPase, exports excess copper into the portal vein in the liver. Although liver cells also contain CMT1, ATOX1 and metallothionein, the ATP7b protein, another copper-transporting P-type ATPase, links copper to ceruloplasmin for release into the bloodstream, or for elimination in the bile. Menkes' disease is caused by a genetic defect in the ATP7a copper-transporting P-type ATPase, which is widely distributed in cells, with the exception of the liver. Defects in ATP7a can result indirectly in failure to uptake copper efficiently from food, and to defects in copper distribution throughout the body, which ultimately causes a copper deficiency in critical tissues such as the brain in sufferers. The symptoms of Menkes' disease include seizures, mental retardation and failure to thrive, and sufferers who are not injected with copper supplements have seriously shortened lives, many failing to survive beyond childhood.

11.2.3 Copper excess and toxicity

In plants copper toxicity results in chlorosis of the leaves and stunted growth, with only plant species that tolerate the higher levels of copper growing on the affected soil (Foye *et al.*, 1978). In farm animals copper toxicity has been known from both naturally occurring copper intoxication and incorrect usage of animal feeds containing copper supplements, but sensitivities to copper vary between animals. Pigs can tolerate large quantities of copper in their diet, cattle less, and sheep are very sensitive to copper toxicity (Bremner, 1998; Gaetke and Chow, 2003). Chronic copper toxicity primarily affects the liver, leading to cirrhosis and hepatic copper accumulation, with subsequent damage to kidneys, brain and other tissues. In higher animals copper poisoning through the ingestion of large quantities of copper in a single dose, as opposed to cumulative copper toxicity effects, leads to severe gastrointestinal pain and disturbances and damage to the gastrointestinal tract, liver and kidneys (Gaetke and Chow, 2003).

The human body contains ~100 mg of copper, which primarily is found bound tightly to metalloproteins such as ceruloplasmin in the blood, albumin, or metallothioneins. The normal dietary intake of copper in humans is 1–4 mg day^{-1} (Reilly, 2004). Acute copper toxicity causes nausea, vomiting, diarrhoea and jaundice, and an excess copper intake of 1000 times the normal dietary requirement can be lethal, but this appears to be a rare event. Chronic copper toxicity, characterized by an accumulation of copper in the liver, is of greater concern, particularly in neonates (Bremner, 1998).

Wilson's disease in humans is caused by a defect in the ATP7b protein, a copper-transporting P-type ATPase, similar to the Menkes' protein. This defect causes copper to accumulate in the body, particularly the liver. Symptoms include liver disease and neuropsychiatric problems, with elevated urinary copper levels and deposition of copper in the iris of the eye other indicators of Wilson's disease. Treatment is initially in the form of chelation therapy using penicillamine, but subsequently zinc may be added to the diet to increase metallothionein production, which binds copper in the gut.

11.3 Bacterial Copper Homeostasis and Resistance

11.3.1 Copper toxicity in bacteria

Copper exerts toxic effects within the cell if free copper levels become too high, with the level of free copper being controlled tightly in *Escherichia coli* and being estimated at less than one atom cell^{-1} (Changela *et al.*, 2003). The toxic effects of copper have been widely ascribed to the generation of hydroxyl radicals, hydrogen peroxide and superoxide, which damage proteins, lipids and DNA through oxidation and peroxidation reactions. These reactions can be demonstrated *in vitro* (reviewed in Gaetke and Chow, 2003; Solioz *et al.*, 2010), and excess copper does induce production of some oxidative stress transcriptional responses in *E. coli* (Kershaw *et al.*, 2005). Another mechanism of copper toxicity may be by depletion of the sulfhydryl pool, because of the strong affinity of copper for sulfur-containing amino acids. However, it has been argued that as free copper levels in the cell are very low, Fenton-like reactions may not be significant *in vivo* and many Gram-positive organisms are tolerant of hydrogen peroxide (Solioz *et al.*, 2010). Some recent work has contradicted the 'copper as a free radical generator' view for *E. coli*. In this work, copper-loaded *E. coli* was shown to be less sensitive to H_2O_2 than *E. coli* grown under limited copper conditions, and copper decreased the rate of H_2O_2-induced DNA damage. High intracellular copper also impaired iron-mediated H_2O_2 killing of cells (Macomber *et al.*, 2007; Macomber and Imlay, 2009). One novel mechanism of copper toxicity in bacteria has been reported to be damage to iron-sulfur (Fe-S) clusters in a number of enzymes, by displacing iron. In particular, increasing intracellular copper (I) levels in the cytoplasm damages isopropylmalate synthetase, responsible for branched-chain amino acid synthesis, and fumarase A (Chillappagari *et al.*, 2010b; Kim *et al.*, 2010).

11.3.2 Copper export in Gram-negative bacteria

Unlike other essential metals such as iron, zinc and magnesium, there appear to be no copper-specific uptake proteins in *E. coli*, which have one of the best-understood copper homeostasis systems. It has been argued that this is because copper is not required in the cytoplasm, but there have been several reports that there are copper uptake systems in other bacteria. In most bacteria without copper uptake systems, copper efflux is used primarily to control intracellular copper. There is also a role for chelation of copper by glutathione, and for repair of Fe-S clusters in proteins (Solioz *et al.*, 2010), as well as some form of copper chaperoning in the periplasm of Gram-negative bacteria. Efflux of copper from the cytoplasm in Gram-positive and -negative bacteria is via P1-type ATPases or root nodulation cell division (RND) family efflux proteins.

E. coli possesses two chromosomally encoded export systems which are used to maintain very low levels of free copper in the cytoplasm. The first is a cytoplasmic membrane-bound P-type ATPase, CopA, which exports copper ions to the periplasm. P-type ATPases are ubiquitously distributed membrane proteins involved in the transport of specific ions using energy from ATP. *copA* is part of the same regulon as *cueO*, which encodes a periplasmic multi-copper oxidase, and expression of both *copA* and *cueO* is regulated by CueR. The second copper export system is the *cus* system, which augments copper tolerance in *E. coli*, when levels of copper overwhelm the *cue* system during aerobic growth, but also plays a major role in copper homeostasis during anaerobic growth, possibly because CueO loses function (Outten *et al.*, 2001). The *cus* operon encodes a copper/silver efflux system, encoded on the *cusCFBA* operon, regulated by the two-component system, CusRS (Munson *et al.*, 2000; Franke *et al.*, 2001; Outten *et al.*, 2001). The *cus* system comprises the three components of a CBA transporter plus a small periplasmic protein, CusF, which plays an important but not essential role in copper resistance. This protein is a periplasmic metallochaperone,

may be a copper-sequestering protein and may have further ancillary roles (Bagai *et al.*, 2008; Kim *et al.*, 2010). CusC is 97% identical to IbeB from the neonatal meningitis strain, *E. coli* K1, involved in brain microvascular endothelial cell invasion (Huang *et al.*, 1999; Munson *et al.*, 2000). Recently, overexpression of an outer membrane lipoprotein, NlpE, was shown to increase expression of the multi-drug exporters, AcrD and MdtAB, which export copper in *E. coli* K-12 (Nishino *et al.*, 2010).

Salmonella enterica serovar Typhimurium uses a somewhat different copper homeostasis system compared to *E. coli* (Pontel and Soncini, 2009). While there are homologues of the *cueR*, *copA* and *cueO* genes from *E. coli* in *S.* Typhimurium, the *cus* system is absent, but an additional periplasmic protein, CueP, exists which, particularly under anaerobic conditions, contributes to copper resistance (Pontel and Soncini, 2009). There is an additional gene cluster in *S.* Typhimurium comprised of another CueR homologue (GolS), a P-type ATPase (GolT) and a chaperone (GolB, resembling CopZ and the yeast Atx1 copper chaperone), which have been reported to confer resistance to gold (Checa *et al.*, 2007). Recent work suggests that the *gol* system is a copper not gold efflux system (Osman *et al.*, 2010).

11.3.3 Regulation of copper homeostasis in *E. coli* and other Gram-negative bacteria

Copper efflux systems found in *E. coli* appear, in the absence of any identified regulated copper import system, to be the major means by which internal copper levels are controlled. Data also suggest that the envelope stress response is one of the other major cellular responses to the deleterious effects of excess copper (see Hobman *et al.*, 2007). The *cue* and *cus* systems are regulated by different regulatory circuits. The *cue* system (CopA and CueO) is regulated at the transcriptional level by CueR, a MerR family regulator, and the *cus* system is regulated by CusRS, a two-component regulator system (Fig. 11.1). While the *cue* system is induced under very low

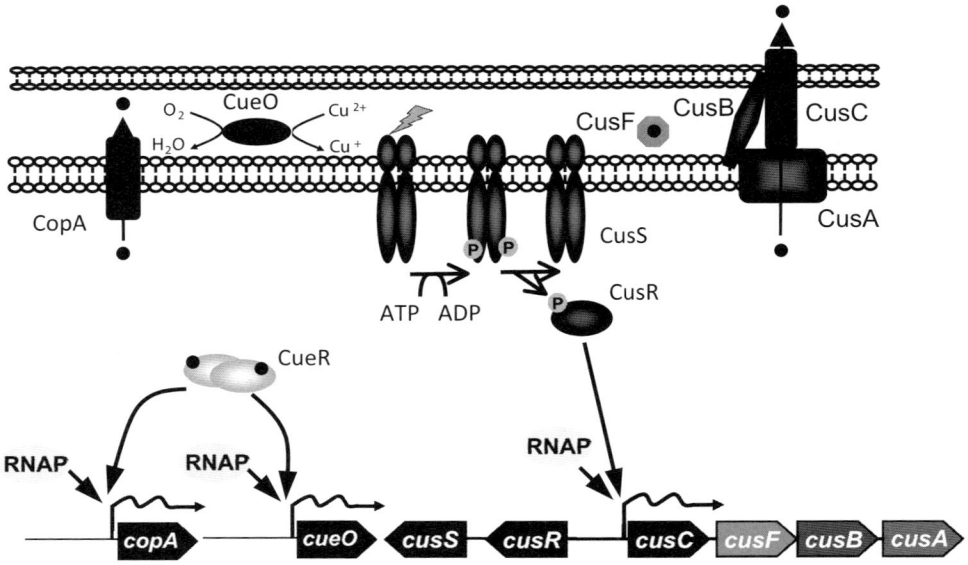

Fig. 11.1. Copper export in *Escherichia coli.* CueR regulates the expression of *copA* and *cueO* in response to cytoplasmic free copper. CopA is a P1-type ATPase and CueO is a multi-copper oxidase. Expression of the *cus* copper efflux system structural genes (*cusCFBA*) is regulated by the two-component sensor/regulator proteins, CusRS. CusS senses copper at the inner membrane. CusCBA encodes a tripartite efflux pump, and CusF is a copper chaperone.

external copper concentrations, the *cus* system has been reported to be induced under higher external levels of copper, and may be important under anaerobic conditions (Munson *et al.*, 2000).

The levels of free copper in the cytoplasm are believed to be very low in *E. coli* (Finney and O'Halloran, 2003) and, because of the very high affinity of CueR for copper, the CueR regulon appears in *E. coli* to be the major copper homeostasis system for 'normal' copper levels. CueR is well characterized (Outten *et al.*, 2000; Petersen and Moller, 2000; Stoyanov *et al.*, 2001; Changela *et al.*, 2003; Hobman *et al.*, 2005), with homologues found in a number of other bacteria, and regulation of copper efflux by a MerR family regulator appears to be a conserved regulatory circuit in Gram-negative bacteria. Transcription of the CueR homologue, PA4778, in *Pseudomonas aeruginosa* (and the copper resistance genes it regulates) is activated directly by LasR, the regulator of the LasR/LasI quorum sensing system (Thaden *et al.*, 2010), which may indicate cell density signalling is involved in copper resistance in some bacteria.

The chromosomal *cus* copper efflux system in *E. coli* is regulated by a two-component regulator system, CusRS – an inner membrane-bound sensor and cyto-plasmic regulator (Munson *et al.*, 2000), and bears similarity to the two-component regulators of plasmid-borne silver and copper resistance regulators (Munson *et al.*, 2000). Another two-component system, CpxRA, is also an important factor in copper stress response in *E. coli* (Fig. 11.2). CpxRA is probably activated by uncharacterized membrane stresses caused by excess copper and regulates the transcription of several membrane and periplasmic proteins which allow the bacterial cell to adapt to copper stress (Yamamoto and Ishihama, 2005).

11.3.4 Copper homeostasis in Gram-positive bacteria

Unlike Gram-negative bacteria, Gram-positive bacteria have reported copper uptake systems as well as copper efflux systems. *Enterococcus hirae* has a system which balances

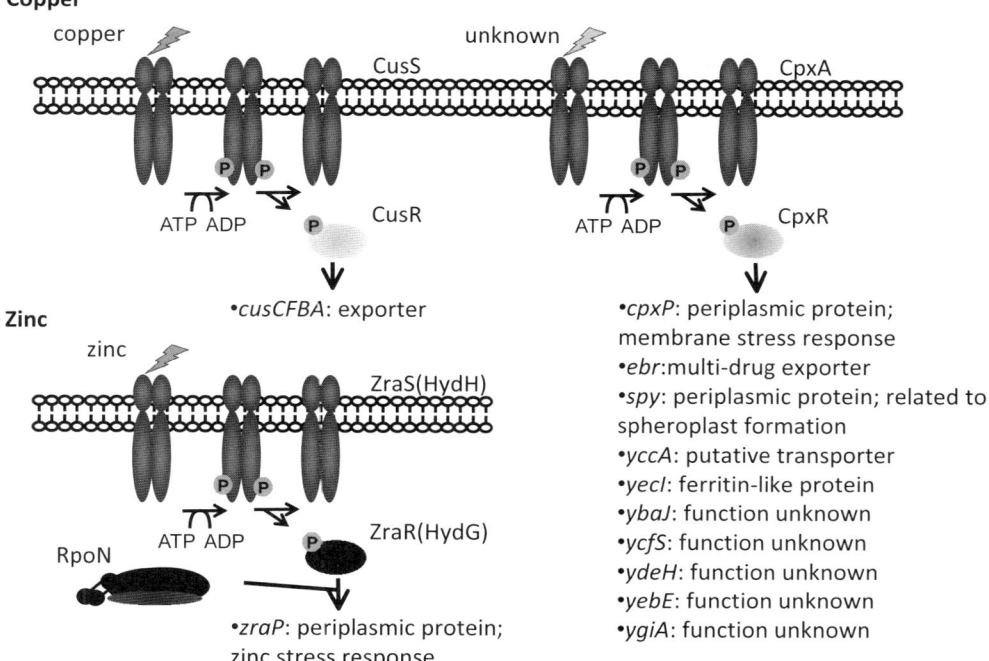

Fig. 11.2. Two-component systems involved in copper and zinc responses in *Escherichia coli*.
CusRS senses copper and induces the *cusCFBA* operon in response to high levels of copper, or under anaerobic conditions. CpxRA responds to membrane stress and also responds to copper stress. Members of the CpxRA regulon include a multi-drug exporter, a ferritin-like protein, putative periplasmic and transport proteins and several proteins of unknown function. The HydH/G (ZraSR) two-component regulator responds to zinc stress by regulating expression of ZraP, a periplasmic protein involved in zinc tolerance.

uptake and export of copper using the CopYZAB system (reviewed in Solioz *et al.*, 2010). CtpA from *Listeria monocytogenes* is a reported copper uptake protein with a role in virulence (Francis and Thomas, 1997a,b), and YcnJ is a putative copper uptake protein in *Bacillus subtilis*; expression of YcnJ was upregulated under copper-limiting conditions and a *ycnJ* deletion mutant grew less well under copper limitation (Chillappagari *et al.*, 2010a). The *copYZAB* operon from *E. hirae* is the best understood of the Gram-positive copper homeostasis systems, using ATPases for both copper import (CopA) and copper export (CopB). CopZ is a copper chaperone and CopY is a copper-responsive repressor, which derepresses expression of the *copYZAB* operon under excess copper (Solioz *et al.*, 2010).

So far, only P-type ATPases have been shown to efflux copper in Gram-positive bacteria. CopA from *B. subtilis* (Banci *et al.*, 2001) and *Lactococcus lactis* (Mangani *et al.*, 2008), as well as CopB from *E. hirae*, are characterized as efflux systems. *L. lactis* CopA is part of the copper-inducible *copRZA* operon, and there is a second unlinked copper ATPase, CopB, which is also regulated by CopR in this organism (Mangani *et al.*, 2008; Solioz *et al.*, 2010). CopZ in *B. subtilis* acts as a chaperone to deliver copper to the CopA ATPase (Gaballa *et al.*, 2003; Smaldone and Helmann, 2007). There is also a plasmid-borne homologue of the *E. hirae copYZAB* system carried on a plasmid from *E. faecium* (Hasman *et al.*, 2006b). Interestingly, recent work has shown that *M. tuberculosis* produces a metallothionein, MymT, which could have the role of binding intracellular copper, similar to metallothioneins in eukaryotes and cyanobacteria; it was demonstrated to bind up to six copper atoms. Expression of MymT was induced over 1000 times by the addition of copper-, cadmium- or nitric oxide-generating compounds (Gold *et al.*, 2008).

11.3.5 Regulation of copper homeostasis in Gram-positive bacteria

Three different copper sensor/regulator systems have been reported for Gram-positive organisms. These are the CsoR and CopY repressors and CueR homologues, though the role of CueR-like regulators in *B. subtilis* remains unclear at present after initial reports that a CueR homologue regulated *copZA* expression have been refuted (Gaballa *et al.*, 2003, Smaldone and Helmann, 2007).

CopY is perhaps the best known of the Gram-positive copper-responsive regulators. CopY is a repressor of transcription of the *copZYAB* in *E. hirae*, as well as in *E. faecium*, *L. lactis* and *Streptococcus* spp. (Odermatt and Solioz, 1995; Solioz *et al.*, 2010). CopY has an N-terminal DNA-binding domain and C-terminal metal-binding domain and acts as a dimer, repressing expression from the *copZYAB* operon by binding to the operator-promoter region of the *cop* operon. CsoR from *M. tuberculosis* is the founding member and archetype of a large family (DUF156) of repressor regulators (Liu *et al.*, 2007; Osman and Cavet, 2010). There is a CsoR homologue in *B. subtilis*, which regulates expression of the *copZA* copper efflux operon by binding the P$_{copZA}$ promoter in the absence of copper ions (Smaldone and Helmann, 2007).

11.3.6 Plasmid-borne copper resistance

Several plasmid-borne copper resistance systems are known. These confer additional resistance to copper above the innate chromosomal resistance or homeostasis mechanisms (Rensing and Grass, 2003). The common feature of the isolation of these plasmids is that they have been found in bacteria isolated from agricultural environ-ments where copper salts have been used as feed supplements or antifungal agents.

The copper resistance from *E. coli* plasmid pRJ1004 (Tetaz and Luke, 1983) contains seven open reading frames, designated *pcoABCDRSE* (Brown *et al.*, 1995; Rouch and Brown, 1997). PcoABCDRS are homologous to the copper-resistance genes from the *cop* system of plasmid pPT23D of *Pseudomonas syringae* pv. *tomato* (Bender and Cooksey, 1986; Mellano and Cooksey, 1988; Cha and Cooksey, 1991). Homologues of the *pco* system have been found in the UK and Australian *E. coli*, *Salmonella* spp. and *Citrobacter freundii* isolates from piggeries (Williams *et al.*, 1993), and homologues of the *cop* system have been found in tomato field isolates of *P. cichorii*, *P. putida*, *P. fluorescens* and *Xanthomonas campestris* pv. *vesicatoria* across the USA (Cooksey *et al.*, 1990; Voloudakis *et al.*, 1993).

Pco and *cop* share four related structural genes: *pco/copABCD*. The *pco* resistance operon possesses an extra gene, *pcoE* downstream of *pcoRS*. The *pcoABCD* genes encode a resistance system that is different from the *cus* and *cue* systems. PcoA is believed to be a multi-copper oxidase family protein and may have a similar function to CueO. PcoB is a predicted outer membrane protein, which may interact with PcoA, and which could either oxidize copper (I) to the less toxic copper (II) or act to sequestrate oxidized catechol siderophores. PcoC and PcoD are required for full resistance, and the model for function of *pco* by Rensing and Grass (2003), which is in part derived from studies on the *cop* system, is that PcoC is a periplasmic copper chaperone and PcoD an inner membrane-spanning protein which import copper into the cytoplasm, possibly for copper loading to PcoA. Expression of *pcoE*, encoding a probable periplasmic copper-binding protein, is controlled by PcoRS, but from a promoter separate to that which controls expression of *pcoABCD*. The *pcoRS* genes encode a two-component regulator system homologous to *copRS*, *cusRS* and the *silRS* regulator from the silver resistance plasmid, pMG101 (Munson *et al.*, 2000). There is also evidence of crosstalk between CusRS and the *pco* system; the *pcoE* promoter can be activated by the CusRS regulator (Munson *et al.*, 2000).

In Gram-positive bacteria, the *tcr* copper-resistance system in *E. faecium* is plasmid-borne and is related to the *copYZAB* operon in *E. hirae*. It has been shown to be selected for in an *in vivo* pig model by increased levels of copper in piglet feed and appears to be co-selected with macrolide and glycopeptide resistance (Hasman and Aarestrup, 2002; Hasman *et al.*, 2006a).

11.4 Copper and Bacterial Infection

Copper-deficient animals are more susceptible to infection from a range of microorganisms, including *Salmonella*, and copper supplementation in cattle increases protection against *E. coli* mastitis (Scaletti *et al.*, 2003). Copper has a role in both adaptive and innate immunology; copper deficiency reduces interleukin secretion and the bactericidal activity of neutrophils and macrophages (White *et al.*, 2009). A *copA* deletion mutant of *E. coli* was found to be hypersensitive to macrophage killing in cell culture models, but this was dependent on an active ATP7A gene in the macrophages (White *et al.*, 2009). *S. enterica* serovar Typhimurium *cueP* and *copA* deletion mutants show reduced survival in macrophages (Osman *et al.*, 2010) and CueO is also required for *S.* Typhimurium infection in murine models (Achard *et al.*, 2010), but deletion of *cueO* in uropathogenic *E. coli* increases their ability to infect in a mouse urinary tract infection model (Tree *et al.*, 2008). In the neonatal meningitis causing *E. coli* K1 strain, a homologue of *cusC*, *ibeB* (97% identical), is involved in invasion of brain microvascular endothelial cells (Huang *et al.*, 1999). Deletion of *cueA*, which encoded a copper efflux ATPase in *P. aeruginosa*, reduced the ability of the mutant to survive *in vitro* on higher levels of copper, but also significantly reduced the mutant's ability to infect mice systemically (Schwan *et al.*, 2005). In *L. monocytogenes*, mutants of *ctpA*, which may encode a copper import ATPase, are linked with reduced virulence (Francis and Thomas, 1997b). In *M. tuberculosis*, the copper-sensitive operon (*cso*) is induced during early infection in a murine model (Talaat *et al.*, 2004) where there are known to be fluctuating levels of copper in the murine alveolar macrophage, the site of primary infection by *M. tuberculosis*. The predicted copper efflux, ATPase CtpV, is induced under high levels of copper, such as can be found in the alveolar macrophage (Ward *et al.*, 2008).

11.4.1 Copper as an antimicrobial

Copper has enjoyed a recent resurgence in interest in its use as an antimicrobial. Concerns about multiple antibiotic resistances in nosocomial infections have prompted the re-examination of some of these 'old' solutions to problems. Copper and copper alloys have been investigated for their contact killing of pathogens, are effective against a wide range of bacteria and viruses (Wilks *et al.*, 2005; Weaver *et al.*, 2008) and have been trialled as replacements for plastic and stainless steel door furniture and other fittings such as bed frames, rails and sanitary fittings in hospital wards (Casey *et al.*, 2010), leading to the launch of a range of antimicrobial products that can be used as direct replacements for standard hospital fittings. Copper is now also being used to impregnate textiles, latex and polymers to endow them with antimicrobial and antifungal properties. Copper and silver are now also used to control *Legionella pneumophila* in hospital water systems (Borkow and Gabbay, 2009).

11.5 Zinc

The importance of zinc to bacterial survival and proliferation is perhaps not as appreciated as that of iron, yet it is drawing an increasing level of interest as a significant factor in bacterial colonization. Zinc, an essential micronutrient for all forms of life, is the only metal to function as a cofactor in all six EC classes of enzyme. Zinc can act as both a catalytic and structural cofactor and a transiently binding regulatory factor. Over 300 enzymes are known to bind zinc, including some necessary for fundamental cell processes, such as RNA polymerase and tRNA synthetases, of which there are thousands of copies per cell (Vallee and Falchuk, 1993).

Several properties of zinc may explain why it plays such an integral role in biological systems. Firstly, its natural abundance in the environment is likely to have influenced its incorporation into enzymatic structures. Additionally, its biochemical properties make it desirable as a cofactor for catalysis; with a small atomic radius of 0.65Å, zinc has a concentrated charge, making it a moderate-strength Lewis acid, second only to copper in the divalent metal ions. Due to its complete d orbital (d^{10}), zinc is technically not a transition metal and will form only 2^+ cations. In contrast, metals such as copper and iron exhibit variable oxidation states, and therefore participate directly in the production of reactive oxygen species. In fact, the role of zinc as an antioxidant is recognized in humans (Powell, 2000), and evidence of this has also been observed in bacteria (Gaballa and Helmann, 2002). Furthermore, zinc has a flexible coordination environment, incorporating four to six protein side chains. The most commonly observed geometry is a slightly distorted tetrahedral of three ligands (McCall *et al.*, 2000) which are rapidly exchangeable, resulting in efficient dissociation and turnover of reaction product. Zinc also contributes to thermodynamically stable protein structures, predominantly via coordination to cysteine and, in some cases, histidine residues.

However, despite the redox-inert properties of zinc, it still remains a source of toxicity if levels within the cell become too high. This toxic effect results from adventitious interactions between zinc and other metal-binding sites. Under normal physiological conditions, levels of 'free' zinc in cells are thought to be extremely low. The total zinc content of *E. coli* or 'zinc quota' has been determined to be 2×10^5 atoms cell^{-1} (Outten and O'Halloran, 2001). However, since *E. coli* possesses zinc-sensing regulatory proteins with femtomolar sensitivity, amounting to less than one atom cell^{-1}, it is believed that no free zinc is available in the cytoplasm. Instead, zinc may be exchanged from an excess of available binding sites. Zinc can be stored in organelles or vesicles in eukaryotic cells. Vesicles containing high concentrations of zinc, termed 'zincosomes' have been identified in mammalian and yeast cells (Haase and Beyersmann, 1999; Devirgiliis *et al.*, 2004). Similarly, plants are able to store zinc in vacuoles, whereas most bacteria rely solely on export and import to control zinc levels.

11.5.1 Human uses of zinc

Zinc is present at 75 mg kg^{-1} of the Earth's crust and concentrations in soil range from <5 to nearly 3000 mg kg^{-1} (Sparks, 2003). Zinc has been used in the manufacture of metal alloys such as brass for at least 2500 years and is still widely used as an anti-corrosive coating, for making alloys and in electrical batteries. Zinc compounds are used as catalysts, pigments, wood preservatives, fungicides and as mild antimicrobial agents. Zinc is a component of many oral hygiene products because of its ability to prevent plaque formation (Phan *et al.*, 2004). Zinc is also commonly added to animal feed to increase growth rates, purportedly by restricting proliferation of pathogenic bacteria in the gut. In the European Union, 250 ppm of zinc, usually delivered as $ZnSO_4$ or ZnO, is permitted as a growth supplement in animal feed. However, no such restrictions apply in the USA and concentrations of zinc from 2500 ppm to 3000 ppm typically are added to piglet feed to prevent post-weaning diarrhoea. Unsurprisingly, decreased susceptibility (minimum inhibitory concentration, MIC, >4 mM) of *E. coli* to treatment with zinc chloride has been recorded in livestock (Aarestrup and Hasman, 2004). Additionally, an association between zinc chloride resistance and methicillin resistance has been observed in *Staphylococcus aureus* in Danish swine (Aarestrup *et al.*, 2010).

11.5.2 Zinc deficiency

The human body contains ~2 g of zinc, making it the second most abundant trace metal after iron. The majority of zinc is stored in the muscle (57%) and bone (29%), with <0.1% present in blood plasma (King *et al.*, 2000). The main site of zinc homeostasis in humans is the gastrointestinal tract, specifically the small intestine, where absorption is controlled (Lee *et al.*, 1989). Homeostasis is also achieved via excretion, primarily in faeces, in which the level of zinc varies depending on the amount of zinc ingested. In contrast, urinary zinc remains relatively constant, with losses of about 0.3 mg day^{-1} (McCance and Widdowson, 1942), representing a highly zinc-deplete environment. In mammals two families of zinc transporters have been identified, ZnT and ZIP. Members of the ZnT family are membrane-bound and possess histidine-rich loops that extend into the cytosol and may be involved in binding zinc (Reilly, 2004). This system allows direct zinc export from the cytosol to the extracellular space. ZnT transporters are part of a larger family of cation diffusion facilitators (CDFs), which are distributed throughout all phylogenetic levels. Intestinal ZnT-1 is expressed constitutively and is located in the basolateral membrane of the villi of the duodenum and jejunum (McMahon and Cousins, 1998), orientated in such a way that exported zinc will enter into the circulation. Znt-1 expression increases in response to higher levels of dietary zinc.

11.5.3 Zinc excess and toxicity

Zinc toxicity is rare, but can occur when an extremely high intake of zinc leads to the suppression of iron and copper uptake. There is also evidence that excess zinc intake impairs the functioning of the immune system (Reilly, 2004). Conversely, zinc deficiency is far more common, particularly in the developing world, and results in reduced immune function and stunted growth in children. Several studies have shown that zinc supplements administered to children in developing countries reduces both the incidence and severity of diarrhoeal episodes (Aggarwal *et al.*, 2007; Lukacik *et al.*, 2008). In mice, lethal milk (*lm*), a recessive mutation on chromosome 2, results in zinc-deficient milk production, which is deadly to nursing pups (Erway and Grider, 1984). Furthermore, an insufficient zinc status is the most prevalent

nutritional deficiency in crops (Broadley *et al.*, 2007).

Despite the abundance of zinc in the body, much of this is unavailable to microorganisms. In blood plasma, 84% of zinc is bound to albumin, 15% to α2-macroglobulin and 1% to amino acids (Tapiero and Tew, 2003). Free zinc in human tissue cells is also limited as it is bound mainly by metalloenzymes. Additionally, all eukaryotes, and some prokaryotes, excluding *E. coli*, produce metallothioneins. Evidently, the human body represents a range of niches with differing zinc availability, and the ability to accumulate zinc and/or tolerate high levels of zinc may explain why some bacterial strains fare better in their specific sites of colonization.

11.6 Bacterial Zinc Homeostasis and Resistance

11.6.1 Zinc uptake systems in *E. coli*

In *E. coli* K-12 zinc enters the periplasm passively via outer membrane porins, where it can be transferred to the cytoplasm via both high-affinity and low-affinity zinc uptake systems. When zinc levels are low, an ABC transporter, ZnuABC, facilitates high-affinity zinc uptake, sourcing energy from ATP hydrolysis (Patzer and Hantke, 1998). ZnuABC is classified into the cluster 9 family of ABC transporters, which are distributed throughout nearly all bacterial species and are in most cases specific to zinc and manganese uptake (Hantke, 2005). The ZnuABC system is composed of three components: a periplasmic-binding protein, ZnuA; a membrane-spanning protein, ZnuB; and an ATPase, ZnuC. ZinT, a zinc-binding protein, expressed under zinc-depleted conditions, has been shown recently to function as an auxiliary component of the ZnuABC system (Petrarca *et al.*, 2010) (Fig. 11.3). While not strictly required for ZnuABC function, ZinT has been shown to provide increased zinc uptake in severely zinc-depleted conditions. Both *znuABC* and *zinT* transcription is under the control of the zinc uptake regulator, Zur.

Zinc uptake

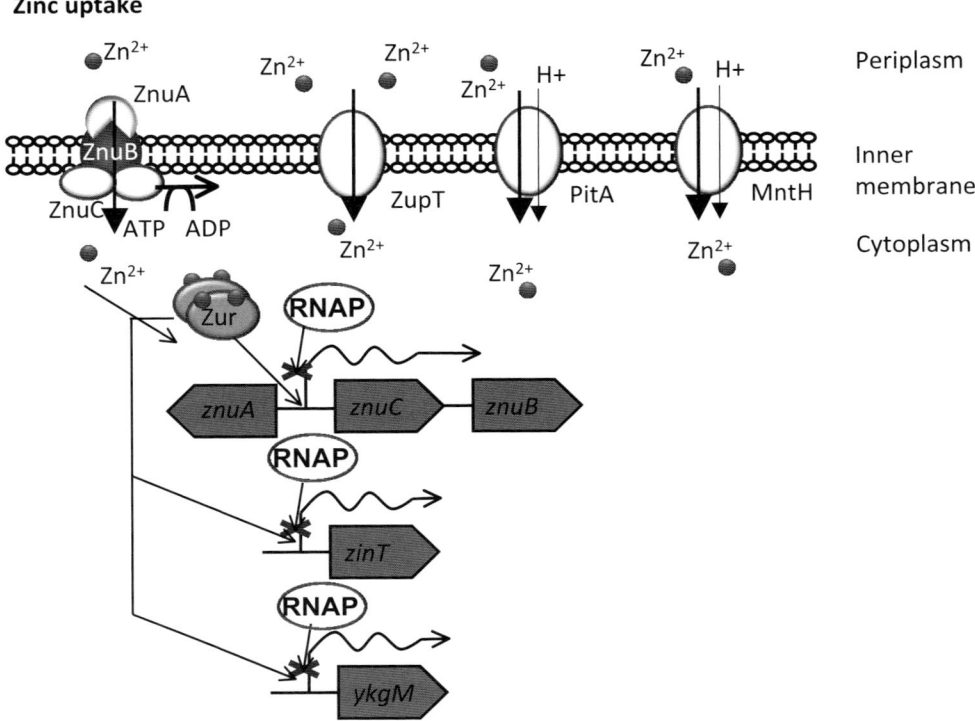

Fig. 11.3. Zinc uptake systems in *Escherichia coli*. ZnuABC is a high-affinity zinc uptake system, composed of a periplasmic zinc-binding protein (ZnuA), an integral membrane protein (ZnuB) and an ATPase (ZnuC). ZupT, a member of the ZIP family of transporters, is a constitutively expressed low-affinity zinc uptake system. The mechanism of uptake is undetermined. Both PitA and MntH are broad-spectrum divalent metal-ion uptake systems, neither of which is specific to zinc uptake. In the presence of zinc, Zur represses the transcription of ZnuABC, the auxiliary component of this system (ZinT) and an alternative ribosomal protein (YkgM). Zur represses transcription by binding to an inverted repeat sequence in the promoter region, occluding RNA polymerase. Each Zur monomer binds two Zn^{2+} ions, one of which may fulfil a structural role and the other regulatory (Outten and O'Halloran, 2001).

In addition to ZnuABC, a low-affinity zinc uptake system, ZupT, has been identified in *E. coli* (Grass *et al.*, 2002). It is a member of the ZIP family of proteins, which typically contain eight transmembrane domains and facilitate the transport of metal ions across membranes into the cytoplasm. ZupT is a broad-range metal transporter, also capable of transporting iron, cobalt and manganese (Grass *et al.*, 2005). Since ZupT is expressed constitutively, it may act as a housekeeping transporter of divalent metal ions. While deletion of *zupT* has been shown to have little effect on the growth of *E. coli*, double mutants of *znuABC* and *zupT* show significantly

reduced growth under zinc-limiting conditions (Grass *et al.*, 2002). This double mutation in the uropathogenic *E. coli* CFT073 was observed to cause a reduction in bladder colonization in a mouse model, to a greater extent than Δ*znuABC* alone (Sabri *et al.*, 2009).

Additionally, the Pit system, which provides low-affinity phosphate uptake via the import of metal-phosphate complexes, may also provide a source of zinc uptake. Typically, the imported metals are magnesium and calcium, but mutations in *pitA* confer increased zinc resistance in *E. coli*, suggesting zinc may also be imported via this system

(Beard *et al.*, 2000). MntH of *E. coli*, which is a homologue of the eukaryotic Nramp superfamily of divalent metal ion transporters, also contributes to zinc uptake, although manganese is the primary substrate (Makui *et al.*, 2000).

11.6.2 Zinc export in *E. coli*

In addition to import systems, *E. coli* possesses a number of export systems to maintain zinc homeostasis. The main route of export for zinc is via the P-type ATPase, ZntA (Beard *et al.*, 1997). ZntA of *E. coli* transports zinc, cadmium and lead ions, and deletion of *zntA* results in strains hypersensitive to these metals. ZntA expression is under the control of the transcriptional regulator, ZntR, which possesses femtomolar sensitivity to zinc (Fig. 11.4) (Outten and O'Halloran, 2001).

Two CDF family proteins, ZitB and YiiP, have also been identified as zinc efflux proteins in *E. coli* (Grass *et al.*, 2001). They possess five transmembrane domains and deliver metal ions to the cytoplasm, driven by the proton motive force (Anton *et al.*, 2004). CzcD, a homologue in the highly metal-resistant *Cupriavidus metallidurans*, is also responsible for the regulation of the expression of the cadmium, zinc and cobalt resistance system, CzcCBA, which is an RND family effluxer that spans the periplasm and transports metals directly from the cytoplasm to the external environment. This efflux system is carried on the pMOL30 plasmid and also carries genes related to the resistance of numerous other metals (Nies *et al.*, 1987).

11.6.3 Regulation of zinc homeostasis in *E. coli*

To maintain zinc homeostasis, the expression of uptake and efflux systems is controlled precisely. Regulation occurs mainly at the level of transcription of import and export systems, although there is evidence of post-

Zinc export

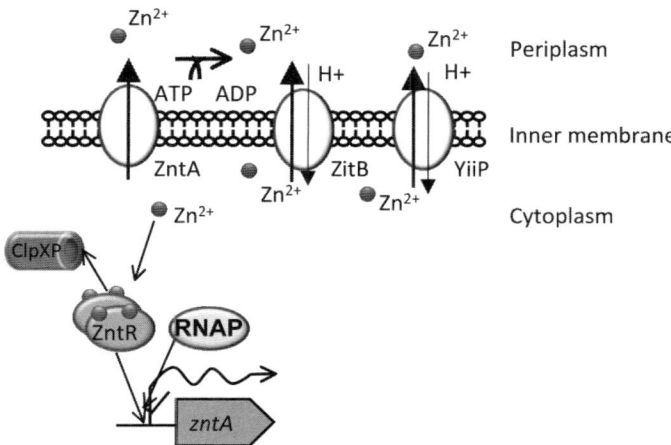

Fig. 11.4. Zinc efflux systems in *Escherichia coli*. ZntA is a high-affinity P-type ATPase which actively transports zinc from the cytoplasm to the periplasm. Both ZitB and YiiP are CDF transporters which expel zinc from the cytoplasm using the proton motive force. ZntR activates the transcription of ZntA in the presence of zinc by changing the DNA conformation at the promoter region to provide optimal spacing for RNA polymerase interaction. Each ZntR monomer can bind two zinc ions. ZntR is known to be a substrate for ClpXP degradation, but in the presence of zinc the half-life of ZntR is increased from 30 min to >60 min (Pruteanu *et al.*, 2007).

translational regulation (Pruteanu *et al.*, 2007). The two main regulators that occupy this role in *E. coli* are Zur (zinc uptake regulator), the regulator of *znuABC*, and ZntR, the regulator of *zntA* expression.

Zur was discovered originally by Patzer and Hantke (1998) due to its ability to complement constitutive mutants of the zinc uptake ABC transporter, *znuABC*. In the presence of Zn^{2+} ions, Zur recognizes and binds to the *znu* operator sequence, a 23 base pair inverted repeat, and prevents transcription by the steric hindrance of RNA polymerase (Outten and O'Halloran, 2001). Zur is a member of the Fur family of transcriptional regulators. It shares 27% sequence identity with Fur, an iron-sensing regulator, which acts both as a repressor and an activator in response to excess Fe^{2+}, and is also involved in protection from oxidative stress and acid tolerance.

In addition to *znuABC* and *zinT*, Zur also regulates the expression of *ykgM*, an alternative ribosomal protein (Graham *et al.*, 2009). Unlike its paralogue, *rpmE*, which is transcribed when zinc levels are replete, *ykgM* does not possess the zinc-binding motif, CXXC. Many other bacterial species possess duplicated ribosomal proteins without these cysteine residues, designated C⁻ (Makarova *et al.*, 2001). It is believed that expression of C⁻ ribosomal proteins instead of C⁺, under zinc-limiting conditions, allows ribosome production to continue and may aid zinc homeostasis by liberating zinc. In *B. subtilis*, expression of the C⁻ ribosomal protein, YtiA, in zinc-deplete conditions has been shown to increase growth significantly when the C⁺ version is the only ribosomal protein present, but not if the C⁻ version is already present (Gabriel and Helmann, 2009).

ZntR is a member of the MerR family of transcriptional regulators (Brocklehurst *et al.*, 1999; Brown *et al.*, 2003). In the presence of Zn^{2+}, ZntR activates *zntA* transcription by changing the conformation of DNA at the *zntA* promoter to provide optimal spacing for RNA polymerase binding, ultimately resulting in increased zinc export. This is the only documented member of the ZntR regulon (Fig. 11.4). ZntR is a target for

degradation by ClpXP and Lon proteases, allowing regulation at the level of protein turnover. Binding to both zinc and DNA have been shown to protect ZntR from degradation, ensuring greater levels of *zntA* transcription (Pruteanu *et al.*, 2007). In *E. coli* there is an additional two-component system, ZraSR, which is activated by zinc and lead stress. ZraSR regulates the expression of *zraP*, which encodes a periplasmic protein and is involved in the tolerance to high zinc concentrations (Leonhartsberger *et al.*, 2001).

Both Zur and ZntR exhibit femtomolar sensitivity to zinc, providing an exceptionally sensitive zinc-binding capacity (Outten and O'Halloran, 2001) and suggesting an absence of free zinc in the cytoplasm under normal physiological conditions. By combining the *in vitro* transcriptional data of Zur and ZntR on *znuABC* and *zntA*, a model for zinc homeostasis was proposed in which the two regulators acted to buffer against either zinc starvation or zinc toxicity, respectively, leaving a narrow range for an optimal free zinc concentration, at which neither regulator was activated. Mathematical modelling of the *in vitro* regulatory response to zinc, incorporating 14 reactions, was found to agree largely with the experimental data (Cui *et al.*, 2008).

11.6.4 Zinc homeostasis in other bacteria

In the Gram-positive bacteria, recent work shows that *Streptococcus pyogenes* encodes a lipoprotein, Lsp, belonging to the cluster 9 family of ligand-binding subunits of ABC transporters, which is important in zinc acquisition and virulence (Weston *et al.*, 2009), while *S. suis* serotype 2 has been reported to carry a Zur homologue (Feng *et al.*, 2008). In *Streptomyces* expression of alternative ribosomal proteins are regulated by Zur (Owen *et al.*, 2007). In *B. subtilis*, the Zur regulon is more extensive and includes an additional member, *yciC*. This gene was found to be regulated by two Zur-binding sites, providing very tight repression (Gabriel *et al.*, 2008). YciC may be a zinc chaperone, but no such homologue exists in *E. coli*. *B. subtilis* possesses an additional zinc uptake system, ZosA, under the regulation of the Fur homologue peroxide stress-sensing regulator, PerR (Gaballa and Helmann, 2002). ZosA is expressed in the presence of H_2O_2, contributing to increased resistance to oxidative stress via zinc uptake. The mechanism of this protection is suggested to occur via the displacement of redox-active metals from adventitious binding sites by redox-inert Zn^{2+} ions. However, in the case of the pathogens *X. oryzae* and *Corynebacterium diptheriae*, deletion of Zur, resulting in increased zinc uptake, has been shown to be associated with increased H_2O_2 sensitivity (Yang *et al.*, 2007; Smith *et al.*, 2008).

11.7 Zinc and Pathogenicity

Zinc availability in host organisms is an important factor in the restriction of bacterial colonization. Systems involved in zinc uptake have been observed as being necessary for infection in several bacterial species, although loss of zinc uptake may have a general negative effect on the fitness of the pathogen, indirectly affecting its ability to colonize or infect. Mutations in *znuA*, a component of the high-affinity zinc uptake system in both *S.* Typhimurium and *Campylobacter jejuni*, resulted in reduced pathogenicity in mice and chicks, respectively (Ammendola *et al.*, 2007; Davis *et al.*, 2008). Additionally, zinc redistribution in response to infection by Gram-negative bacteria has been observed in several species, including humans. Plasma zinc concentrations decrease post-infection in what may be a strategy to restrict bacterial growth. Furthermore, the human immune system has been found to use zinc chelation as a means of defending against invading bacteria. Neutrophils produce a chelating protein called calprotectin, which has been shown to inhibit the growth of *S. aureus* in tissue abscesses via the chelation of zinc and manganese (Corbin *et al.*, 2008). It appears that, much like for iron, there is a significant effort from the host to restrict the availability of this essential micronutrient and, concurrently, several mechanisms employed by the pathogen to obtain it. However, as yet, no

bacterial siderophores for zinc capture and uptake have been identified.

An association between deletion of Zur and virulence is evident in several pathogens. A *zur* mutant of the Gram-negative plant pathogen, *X. oryzae*, displayed a twofold reduction in EPS production and a 65% reduction in leaf lesions compared to the wild type (Yang *et al.*, 2007). A *zur* mutant of another member of the *Xanthomonas* family, *X. campestris*, showed reduced virulence during infection of Chinese radishes (Tang *et al.*, 2005). In *S.* Typhimurium, a reduction in virulence was found to be associated with *zur* mutants when mice were infected intraperitoneally but not orally (Campoy *et al.*, 2002). However, with the Gram-positive *S. aureus*, no loss in virulence was associated with a *zur* mutation when tested in a mouse model (Lindsay and Foster, 2001). In addition to the numerous studies demonstrating reduced virulence on deletion of zinc uptake systems, it is evident that obtaining zinc and regulating its concentration is of significant importance to bacterial fitness and virulence.

11.7.1 Zinc as an antimicrobial

Metals have been known to exhibit antimicrobial properties for centuries, but due to the rise of multiple antibiotic resistances, the use of inorganic antimicrobials to restrict bacterial growth is drawing increasing attention. In particular, research into the use of metal nanoparticles has gathered speed because of the increased antimicrobial efficacy provided by their greater surface area. Nanoparticles of purified zinc oxide powder have shown to be effective at limiting the growth of several pathogens, including *L. monocytogenes*, *S.* Typhimurium, *E. coli* O157:H7 and *S. mutans* (Hernández-Sierra *et al.*, 2008; Jin *et al.*, 2009). The growth of bacterial biofilms on surfaces of medical devices which exhibit greatly reduced susceptibility to antimicrobials is a major public health concern. The coating of silicon-wafers with zinc oxide was shown to reduce *E. coli* ATCC 25922 colonization by 50% when grown in a biofilm reactor (Gittard *et al.*,

2009). However, this was not tested on a pathogenic strain and metal surfaces have been reported as being strongly colonized by biofilm-forming bacteria (Donlan and Costerton, 2002). Furthermore, metal resistance and antibiotic resistance determinants are often found to reside together, particularly on mobile genetic elements, hence co-selection can occur (Baker-Austin *et al.*, 2006); evidence of regulation of antibiotic-resistance determinants by metal ion concentration has been observed. A microarray experiment subjecting *E. coli* K-12 MG1655 to a high level of zinc resulted in increased expression of the multiple drug-resistance system, *mdtABC* (Lee *et al.*, 2005).

Acknowledgements

KH and JLH have been supported by funding from the UK Biotechnology and Biological Sciences Research Council (BBSRC) through project grant BB/E01044X/1, and SRC through a BBSRC studentship. Ongoing collaborative work between JLH and TO has been facilitated by International Joint Project funding from BBSRC, The Royal Society and the Japan Society for the Promotion of Science (JSPS).

References

Aarestrup, F.M. and Hasman, H. (2004) Susceptibility of different bacterial species isolated from food animals to copper sulphate, zinc chloride and antimicrobial substances used for disinfection. *Veterinary Microbiology* 100, 83–89.

Aarestrup, F.M., Cavaco, L. and Hasman, H. (2010) Decreased susceptibility to zinc chloride is associated with methicillin resistant *Staphylococcus aureus* CC398 in Danish swine. *Veterinary Microbiology* 142, 455–457.

Achard, M.E.S., Tree, J.J., Holden, J.A., Simpfendorfer, K.R., Wijburg, O.L.C., Strugnell, R.A., Schembri, M.A., Sweet, M.J., Jennings, M.P. and McEwan, A.G. (2010) The multi-copper-ion oxidase CueO of *Salmonella enterica* serovar Typhimurium is required for systemic virulence. *Infection and Immunity* 78, 2312–2319.

Aggarwal, R., Sentz, J. and Miller, M.A. (2007) Role of zinc administration in prevention of childhood diarrhea and respiratory illnesses: a meta-analysis. *Pediatrics* 119(6), 1120–1130.

Ammendola, S., Pasquali, P., Pistoia, C., Petrucci, P., Petrarca, P., Rotilio, G. and Battistoni, A. (2007) High-affinity Zn^{2+} uptake system ZnuABC is required for bacterial zinc homeostasis in intracellular environments and contributes to the virulence of *Salmonella enterica*. *Infection and Immunity* 75(12), 5867–5876.

Anton, A., Weltrowski, A., Haney, C.J., Franke, S., Grass, G., Rensing, C. and Nies, D.H. (2004) Characteristics of zinc transport by two bacterial cation diffusion facilitators from *Ralstonia metallidurans* CH34 and *Escherichia coli*. *Microbiology* 186(22), 7499–7507.

Bagai, I., Rensing, C., Blackburn, N.J. and McEvoy, M.M. (2008) Direct metal transfer between periplasmic proteins identifies a bacterial copper chaperone. *Biochemistry* 47, 11408–11414.

Baker-Austin, C., Wright, M.S., Stepanauskas, R. and Mcarthur, J.V. (2006) Co-selection of antibiotic and metal resistance. *Trends in Microbiology* 14(4), 176–182.

Banci, L., Bertini, I., Conte, R.D., Markey, J. and Ruiz-Duenas, F.J. (2001) Copper trafficking: the solution structure of *Bacillus subtilis* CopZ. *Biochemistry* 40, 15660–15668.

Beard, S.J., Hashim, R., Membrillo-Hernández, J., Hughes, M.N. and Poole, R.K. (1997) Zinc (II) tolerance in *Escherichia coli* K-12: evidence that the *zntA* gene (*o732*) encodes a cation transport ATPase. *Molecular Microbiology* 25(5), 883–891.

Beard, S.J., Hashim, R., Wu, G., Binet, M.R.B., Hughes, M.N. and Poole, R.K. (2000) Evidence for the transport of zinc(II) ions via the Pit inorganic phosphate transport system in *Escherichia coli*. *FEMS Microbiology Letters* 184, 231–235.

Bender, C.L. and Cooksey, D.A. (1986) Indigenous plasmids in *Pseudomonas syringae* pv. tomato: conjugative transfer and role in copper resistance. *Journal of Bacteriology* 165, 534–541.

Borkow, G. and Gabbay, J. (2009) Copper, an ancient remedy returning to fight microbial, fungal and viral infections. *Current Chemical Biology* 3, 272–278.

Bremner, I. (1998) Manifestations of copper excess. *American Journal of Clinical Nutrition* 67(suppl), 1069S–1073S.

Broadley, M.R., White, P.J., Hammond, J.P., Zelko, I. and Lux, A. (2007) Zinc in plants. *The New Phytologist* 173(4), 677–702.

Brocklehurst, K.R., Hobman, J.L., Lawley, B., Blank, L., Marshall, S.J., Ali, N., Brown, N.L. and Morby, A.P. (1999) ZntR is a Zn(II)-inducible MerR-like transcriptional regulator of *zntA* in *Escherichia coli*. *Molecular Microbiology* 31(3), 893–902.

Brown, N.L., Barrett, S.R., Camakaris, J., Lee, B.T. and Rouch, D.A. (1995) Molecular genetics and transport analysis of the copper-resistance determinant (*pco*) from *Escherichia coli* plasmid pRJ1004. *Molecular Microbiology* 17, 1153–1166.

Brown, N.L., Stoyanov, J.V., Kidd, S.P. and Hobman, J.L. (2003) The MerR family of transcriptional regulators. *FEMS Microbiology Reviews* 27, 145–163.

Campoy, S., Jara, M., Busquets, N., Pérez De Rozas, A.M., Badiola, I. and Barbé, J. (2002) Role of the high-affinity zinc uptake *znuABC* system in *Salmonella enterica* serovar typhimurium virulence. *Infection and Immunity* 70, 4721–4725.

Casey, A.L., Adams, D., Karpanen, T.J., Lambert, P.A., Cookson, B.D., Nightingale, P., Miruszenko, L., Shillam, R., Christian, P. and Elliott, T.S.J. (2010) Role of copper in reducing hospital environment contamination. *Journal of Hospital Infection* 74, 72–77.

Cha, J.S. and Cooksey, D.A. (1991) Copper resistance in *Pseudomonas syringae* mediated by periplasmic and outer membrane proteins. *Proceedings of the National Academy of Sciences of the United States of America* 88, 8915–8919.

Changela, A., Chen, K., Xue, Y., Holschen, J., Outten, C.E., O'Halloran, T.V. and Mondragon, A. (2003) Molecular basis of metal-ion selectivity and zeptomolar sensitivity by CueR. *Science* 301, 1383–1387.

Checa, S.K., Espariz, M., Audero, M.E., Botta, P.E., Spinelli, S.V. and Soncini, F.C. (2007) Bacterial sensing of and resistance to gold salts. *Molecular Microbiology* 63, 1307–1318.

Chillappagari, S., Miethke, M., Trip, H., Kuipers, O.P. and Marahiel, M.A. (2010a) Copper acquisition is mediated by YcnJ and regulated by YcnK and CsoR in *Bacillus subtilis*. *Journal of Bacteriology* 191, 2362–2370.

Chillappagari, S., Seubert, A., Trip, H., Kuipers, O.P., Marahiel, M.A. and Miethke, M. (2010b) Copper stress affects iron homeostasis by destabilizing iron-sulfur cluster formation in *Bacillus subtilis*. *Journal of Bacteriology* 192, 2512–2524.

Cooksey, D.A., Azad, H.R., Cha, J.-S. and Lim, C.-K. (1990) Copper resistance gene homologs in pathogenic and saprophytic bacterial species from tomato. *Applied and Environmental Microbiology* 56, 431–435.

Corbin, B.D., Seeley, E.H., Raab, A., Feldmann, J., Miller, M.R., Torres, V.J., Anderson, K.L., Dattilo, B.M., Dunman, P.M., Gerads, R., Caprioli, R.M., Nacken, W., Chazin, W.J. and Skaar, E.P. (2008) Metal chelation and inhibition of bacterial growth in tissue abscesses. *Science* 319, 962–965.

Cui, J., Kaandorp, J.A. and Lloyd, C.M. (2008) Simulating *in vitro* transcriptional response of zinc homeostasis system in *Escherichia coli*. *BMC Systems Biology* 2, 89.

Davis, L.M., Kakuda, T. and DiRita, V.J. (2008) A *Campylobacter jejuni znuA* orthologue is essential for growth in low-zinc environments and chick colonization. *Journal of Bacteriology* 191(5), 1631–1640.

Devirgiliis, C., Murgia, C., Danscher, G. and Perozzi, G. (2004) Exchangeable zinc ions transiently accumulate in a vesicular compartment in the yeast *Saccharomyces cerevisiae*. *Biochemical and Biophysical Research Communications* 323, 58–64.

Donlan, R.M. and Costerton, J.W. (2002) Biofilms: survival mechanisms of clinically relevant microorganisms. *Clinical Microbiology Reviews* 15(2), 167–193.

Erway, L.C. and Grider, A. (1984) Zinc metabolism in lethal-milk mice. *The Journal of Heredity* 75, 480–484.

Feng, Y., Li, M., Zhang, H., Zheng, B., Han, H., Wang, C., Yan, J., Tang, J. and Gao, G.F. (2008) Functional definition and global regulation of Zur, a zinc uptake regulator in *Streptococcus suis* serotype 2 strain causing streptococcal toxic shock syndrome. *Journal of Bacteriology* 190(22), 7567–7578.

Finney, L.A. and O'Halloran, T.V. (2003) Transition metal speciation in the cell: insights from the chemistry of metal ion receptors. *Science* 300, 931–936.

Foye, C.D., Chaney, R.L. and White, M.C. (1978) The physiology of metal toxicity in plants. *Annual Review of Plant Physiology* 29, 511–566.

Francis, M.S. and Thomas C.J. (1997a) The *Listeria mcnocytogenes* gene *ctpA* encodes a putative P-type ATPase involved in copper transport. *Molecular and General Genetics* 253, 484–491.

Francis, M.S. and Thomas, C.J. (1997b) Mutants of the CtpA copper transporting P-type ATPase reduce virulence of *Listeria monocytogenes*. *Microbial Pathogenesis* 22, 67–72.

Franke, S., Grass, G. and Nies, D.H. (2001) The product of the *ybdE* gene of the *Escherichia coli* chromosome is involved in detoxification of silver ions. *Microbiology* 147, 965–972.

Fraústo Da Silva, J.J.R. and Williams, R.J.P. (2001) Zinc: Lewis acid catalysis and regulation. In: *The Biological Chemistry of the Elements*. Oxford University Press, Oxford, pp. 315–335.

Gaballa, A. and Helmann, J.D. (2002) A peroxide-induced zinc uptake system plays an important role in protection against oxidative stress in *Bacillus subtilis*. *Molecular Microbiology* 45(4), 997–1005.

Gaballa, A., Cao, M. and Helmann, J.D. (2003) Two MerR homologues that affect copper induction of the *Bacillus subtilis copZA* operon. *Microbiology* 149, 3413–3421.

Gabriel, S.E. and Helmann, J.D. (2009) Contributions of Zur-controlled ribosomal proteins to growth under zinc starvation conditions. *Journal of Bacteriology* 191, 6116–6122.

Gabriel, S.E., Miyagi, F., Gaballa, A. and Helmann, J.D. (2008) Regulation of the *Bacillus subtilis yciC* gene and insights into the DNA-binding specificity of the zinc-sensing metalloregulator Zur. *Journal of Bacteriology* 190(10), 3482–3488.

Gaetke, L.M. and Chow, C.K. (2003) Copper toxicity, oxidative stress, and antioxidant nutrients. *Toxicology* 189, 147–163.

Gittard, S.D., Perfect, J.R., Monteiro-Riviere, N.A., Wei, W., Jin, C. and Narayan, R.J. (2009) Assessing the antimicrobial activity of zinc oxide thin films using disk diffusion and biofilm reactor. *Applied Surface Science* 255, 5806–5811.

Gold, B., Deng, H., Bryk, R., Vargas, D., Eliezer, D., Roberts, J., Jiang, X. and Nathan, C. (2008) Identification of a copper binding metallothionein in pathogenic mycobacteria. *Nature Chemical Biology* 4, 609–616.

Graham, A.I., Hunt, S., Stokes, S.L., Bramall, N., Bunch, J., Cox, A.G., McLeod, C.W. and Poole, R.K. (2009) Severe zinc depletion of *Escherichia coli*: roles for high-affinity zinc binding by ZinT, zinc transport and zinc-independent proteins. *The Journal of Biological Chemistry* 284(27), 18377–18389.

Grass, G., Fan, B.I., Rosen, B.P., Franke, S., Nies, D.H. and Rensing, C. (2001) ZitB (YbgR), a member of the cation diffusion facilitator family, is an additional zinc transporter in *Escherichia coli*. *Microbiology* 183(15), 4664–4667.

Grass, G., Wong, M.D., Rosen, B.P., Smith, R.L. and Rensing, C. (2002) ZupT is a Zn(II) uptake system in *Escherichia coli*. *Microbiology* 184(3), 864–866.

Grass, G., Franke, S., Taudte, N., Nies, D.H., Kucharski, L.M., Maguire, M.E. and Rensing, C. (2005) The metal permease ZupT from *Escherichia coli* is a transporter with a broad substrate spectrum. *Journal of Bacteriology* 187, 1604–1611.

Haase, H. and Beyersmann, D. (1999) Uptake and intracellular distribution of labile and total Zn(II) in C6 rat glioma cells investigated with fluorescent probes and atomic absorption. *Biometals* 12, 247–254.

Hantke, K. (2005) Bacterial zinc uptake and regulators. *Current Opinion in Microbiology* 8, 196–202.

Hasman, H. and Aarestrup, F.M. (2002) *tcrB*, a gene conferring transferable copper resistance in *Enterococcus faecium*: occurrence, transferability, and linkage to macrolide and glycopeptide resistance. *Antimicrobial Agents and Chemotherapy* 46, 1410–1416.

Hasman, H., Franke, S. and Rensing, C. (2006a) Resistance to metals used in agricultural production. In: Aarestrup, F.M. (ed.) *Antimicrobial Resistance in Bacteria of Animal Origin*. ASM Press, Washington, DC, pp. 99–114.

Hasman, H., Kempf, I., Chidaine, B., Cariolet, R., Ersboll, A.K., Houe, H., Bruun Hansen, H.C. and Aarestrup, F.M. (2006b) Copper resistance in *Enterococcus faecium*, mediated by the *tcrB* gene, is selected by supplementation of pig feed with copper sulphate. *Applied and Environmental Microbiology* 72, 5784–5789.

Hernández-Sierra, J.F., Ruiz, F., Cruz, D.C., Martínez-Gutiérrez, F., Martínez, A.E., Guillén, A.D., Tapia-Pérez, H. and Martínez Castañón, G. (2008) The antimicrobial sensitivity of *Streptococcus mutans* to nanoparticles of silver, zinc oxide, and gold. *Nanomedicine: Nanotechnology, Biology, and Medicine* 4, 237–240.

Hobman, J.L. (2007) Molecular techniques for the study of toxic metal resistance mechanisms in bacteria. In: Hurst, C.J., Crawford, R.L., Garland, J.L., Lipson, D.A., Mills, A.L. and Stetzenbach, L.D. (eds) *Manual of Environmental Microbiology* (*MEM*), 3rd edn. ASM Press, Washington, DC, pp. 1166–1182.

Hobman, J.L., Wilkie, J. and Brown, N.L. (2005) A design for life: prokaryotic metal-binding MerR family regulators. *Biometals* 18, 429–436.

Hobman, J.L., Yamamoto, K. and Oshima, T. (2007) Transcriptomic responses of bacterial cells to sublethal metal ion stress. In: Nies, D.H. and Silver, S. (eds) *Molecular Microbiology of Heavy Metals*. Springer Verlag Microbial Monographs, Berlin, Heidleberg, pp. 73–116.

Huang, S.-H., Chen, Y.-H., Fu, Q., Stins, M., Wang, Y., Wass, C. and Kim, K.S. (1999) Identification and characterization of an *Escherichia coli* invasion gene locus, *ibeB*, required for penetration of brain microvascular endothelial cells. *Infection and Immunity* 67, 2103–2109.

Jin, T., Sun, D., Su, J., Zhang, H. and Sue, H. (2009) Antimicrobial efficacy of zinc oxide quantum dots against *Listeria monocytogenes*, *Salmonella enteritidis*, and *Escherichia coli* O157:H7. *Journal of Food Science* 74(1), M46–M52.

Kershaw, C.J., Brown, N.L., Constantinidou, C., Patel, M. and Hobman, J.L. (2005) The expression profile of Escherichia coli K-12 in response to minimal, optimal and excess copper concentrations. Microbiology 151, 1187–1198.

Kim, E.-H., Rensing, C. and McEvoy, M.M. (2010) Chaperone-mediated copper handling in the periplasm. Natural Products Reports 27, 711–719.

King, J.C., Shames, D.M. and Woodhouse, L.R. (2000) Zinc homeostasis in humans. Journal of Nutrition 130, 1360–1366.

Lee, H.H., Prasad, A.S., Brewer, G.J. and Owyang, C. (1989) Zinc absorption in human small intestine. The American Journal of Physiology, 256, G87–91.

Lee, L.J., Barrett, J.A. and Poole, R.K. (2005) Genome-wide transcriptional response of chemostat-cultured Escherichia coli to zinc. Journal of Bacteriology 187(3), 1124–1134.

Leonhartsberger, S., Huber, A., Lottspeich, F. and Böck, A. (2001) The hydH/G genes from Escherichia coli code for a zinc and lead responsive two component regulatory system. Journal of Molecular Biology 307, 93–105.

Lindsay, J.A. and Foster, S.J. (2001) Zur: a $Zn^{(2+)}$-responsive regulatory element of Staphylococcus aureus. Microbiology 14, 1259–1266.

Lippard, S.J. and Berg, J.M. (1994) Overview of bioinorganic chemistry. In: Principles of Bioinorganic Chemistry. University Science Books, Mill Valley, California, pp. 1–19.

Liu, T., Ramesh, A., Ma, Z., Ward, S.K., Zhang, L., George, G.N., Talaat, A.M., Sacchettini, J.C. and Giedroc, D.P. (2007) CsoR is a novel Mycobacterium tuberculosis copper-sensing transcriptional regulator. Nature Chemical Biology 3, 60–68.

Lukacik, M., Thomas, R.L. and Aranda, J.V. (2008) A meta-analysis of the effects of oral zinc in the treatment of acute and persistent diarrhea. Pediatrics 121, 326–336.

McCall, K.A., Huang, C. and Fierke, C.A. (2000) Function and mechanism of zinc metalloenzymes. The Journal of Nutrition 130, 1437S–1446S.

McCance, R.A. and Widdowson, E.M. (1942) The absorption and excretion of zinc. Biochemical Journal 36, 692–696.

McMahon, R.J. and Cousins, R.J. (1998) Regulation of the zinc transporter ZnT-1 by dietary zinc. Proceedings of the National Academy of Sciences of the United States of America 95, 4841–4846.

Macomber, L. and Imlay, J.A. (2009) The iron-sulfur clusters of dehydratases are primary targets of intracellular copper toxicity. Proceedings of the National Academy of Sciences of the United States of America 106, 8344–8349.

Macomber, L., Rensing, C. and Imlay, J.A. (2007) Intracellular copper does not catalyze the formation of oxidative DNA damage in Escherichia coli. Journal of Bacteriology 189, 1616–1626.

Magnani, D., Barré O., Gerber, S.D. and Solioz, M. (2008) Characterization of the CopR regulon of Lactococcus lactis IL1403. Journal of Bacteriology 190, 536–545.

Makarova, K.S., Ponomarev, V.A. and Koonin, E.V. (2001) Two C or not two C: recurrent disruption of Zn-ribbons, gene duplication, lineage-specific gene loss, and horizontal gene transfer in evolution of bacterial ribosomal proteins. Genome Biology 2(9), 1–14.

Makui, H., Roig, E., Cole, S.T., Helmann, J.D., Gros, P. and Cellier, M.F. (2000) Identification of the Escherichia coli K-12 Nramp orthologue (MntH) as a selective divalent metal ion transporter. Molecular Microbiology 35(5), 10651078.

Mellano, M.A. and Cooksey, D.A. (1988) Induction of the copper resistance operon from Pseudomonas syringae. Journal of Bacteriology 170, 4399–4401.

Munson, G.P., Lam, D.H., Outten, F.W. and O'Halloran, T.V. (2000) Identification of a copper-responsive two-component system on the chromosome of Escherichia coli K-12. Journal of Bacteriology 182, 5864–5871.

Nies, D., Mergeay, M., Friedrich, B. and Schlegel, H.G. (1987) Cloning of plasmid genes encoding resistance to cadmium, zinc, and cobalt in Alcaligenes eutrophus CH34. Journal of Bacteriology 169(10), 4865–4868.

Nishino K., Yamasaki, S., Hayashi-Nishino, M. and Yamaguchi, A. (2010) Effect of NlpE overproduction on multidrug resistance in Escherichia coli. Antimicrobial Agents and Chemotherapy 54, 2239–2243.

Odermatt, A. and Solioz, M. (1995) Two trans-acting metalloregulatory proteins controlling expression of the ATPases involved in copper homeostasis in Enterococcus hirae. Journal of Biological Chemistry 270, 4349–4354.

Osman, D. and Cavet, J.S. (2008) Copper homeostasis in bacteria. Advances in Applied Microbiology 65, 217–247.

Osman, D. and Cavet, J.S. (2010) Bacterial metal-sensing proteins exemplified by ArsR-SmtB family repressors. Natural Product Reports 27, 668–680.

Osman, D., Waldron, K.J., Denton, H., Taylor, C.M., Grant, A.J., Mastroeni, P., Robinson, N.J. and Cavet, J.S. (2010) Copper homeostasis in Salmonella is atypical and copper-CueP is a major periplasmic metal complex. *Journal of Biological Chemistry* 285, 25259–25268.

Outten, C.E. and O'Halloran, T. (2001) Femtomolar sensitivity of metalloregulatory proteins controlling zinc homeostasis. *Science* 292, 2488–2492.

Outten, F.W., Outten, C.E., Hale, J. and O'Halloran, T.V. (2000) Transcriptional activation of an *Escherichia coli* copper efflux regulon by the chromosomal MerR homologue *cueR*. *Journal of Biological Chemistry* 275, 31024–31029.

Outten, F.W., Huffman, D.L., Hale, J.A. and O'Halloran, T.V. (2001) The independent *cue* and *cus* systems confer copper tolerance during aerobic and anaerobic growth in *Escherichia coli*. *Journal of Biological Chemistry* 276, 30670–30677.

Owen, G.A., Pascoe, B., Kallifidas, D. and Paget, M.S.B. (2007) Zinc-responsive regulation of alternative ribosomal protein genes in *Streptomyces coelicolor* involves *zur* and sigmaR. *Journal of Bacteriology* 189(11), 4078–4086.

Patzer, S. and Hantke, K. (1998) The ZnuABC high-affinity zinc uptake system and its regulator Zur in *Escherichia coli*. *Molecular Microbiology* 28(6), 1199–1210.

Petersen, C. and Moller, L.B. (2000) Control of copper homeostasis in *Escherichia coli* by a P-type ATPase, CopA, and a MerR-like transcriptional activator, CopR. *Gene* 261, 289–298.

Petrarca, P., Ammendola, S., Pasquali, P. and Battistoni, A. (2010) The Zur-regulated ZinT protein is an auxiliary component of the high affinity ZnuABC zinc transporter that facilitates metal recruitment during severe zinc shortage. *Journal of Bacteriology* 192(6), 1553–1564.

Phan, T.-N., Buckner, T., Sheng, J., Baldeck, J.D. and Marquis, R.E. (2004) Physiologic actions of zinc related to inhibition of acid and alkali production by oral *Streptococci* in suspensions and biofilms. *Oral Microbiology and Immunology* 19, 31–38.

Pontel, L.B. and Soncini, F.C. (2009) Alternative periplasmic copper-resistance mechanisms in Gram negative bacteria. *Molecular Microbiology* 73, 212–225.

Powell, S.R. (2000) The antioxidant properties of zinc. *The Journal of Nutrition* 130, 1447S–1454S.

Pruteanu, M., Neher, S.B. and Baker, T.A. (2007) Ligand-controlled proteolysis of the *Escherichia coli* transcriptional regulator ZntR. *Journal of Bacteriology* 189(8), 3017–3025.

Reilly, C. (2004) Zinc. In: *The Nutritional Trace Metals*. Blackwell Publishing, Oxford, UK, pp. 118–134.

Rensing, C. and Grass, G. (2003) *Escherichia coli* mechanisms of copper homeostasis in a changing environment. *FEMS Microbiology Reviews* 27, 197–213.

Rouch, D.A. and Brown, N.L. (1997) Copper-inducible transcriptional regulation at two promoters in the *Escherichia coli* copper resistance determinant *pco*. *Microbiology* 143, 1191–1202.

Sabri, M., Houle, S. and Dozois, C.M. (2009) Roles of the extraintestinal pathogenic *Escherichia coli* ZnuACB and ZupT zinc transporters during urinary tract infection. *Infection and Immunity* 77, 1155–1164.

Sapkota, A.R., Lefferts, L.Y., McKenzie, S. and Walker, P. (2007) What do we feed to food-production animals? A review of animal feed ingredients and their potential impacts on human health. *Environmental Health Perspectives* 115, 663–670.

Scaletti, R.W., Trammell, D.S., Smith, B.A. and Harmon, R.J. (2003) Role of dietary copper in enhancing resistance to *Escherichia coli* mastitis. *Journal of Dairy Science* 86, 1240–1249.

Schwan, W.R., Warrener, P., Keunz, E., Stover, C.K. and Folger, K.R. (2005) Mutations in the *cueA* gene encoding a copper homeostasis P-type ATPase reduce the pathogenicity of *Pseudomonas aeruginosa* in mice. *International Journal of Medical Microbiology* 295, 237–242.

Smaldone, G.T. and Helmann, J.D. (2007) CsoR regulates the copper efflux operon *copZA* in *Bacillus subtilis*. *Microbiology* 153, 4123–4128.

Smith, K.F., Bibb, L.A., Schmitt, M.P. and Oram, D.M. (2008) Regulation and activity of a zinc uptake regulator, Zur, in *Corynebacterium diphtheriae*. *Journal of Bacteriology* 191, 1595–1603.

Solioz, M., Abicht, H.K., Mermod, M. and Mancini, S. (2010) Response of Gram-positive bacteria to copper stress. *Journal of Biological Inorganic Chemistry* 15, 3–14.

Sparks, D.L. (2003) *Environmental Soil Chemistry*. Elsevier, San Diego, California, p. 46.

Stoyanov, J.V., Hobman, J.L. and Brown, N.L. (2001) CueR, (*ybbI*) of *Escherichia coli* is a MerR family regulator controlling expression of the copper exporter CopA. *Molecular Microbiology* 39, 502–511.

Talaat, A.M., Lyons, R., Howard, S.T. and Johnston, S.A. (2004) The temporal expression profile of *Mycobacterium tuberculosis* infection in mice. *Proceedings of the National Academy of Sciences of the United States of America* 101, 4602–4607.

Tang, D., Li, X., He, Y., Feng, J., Chen, B., Tang, J., et al. (2005) The zinc uptake regulator Zur is essential for the full virulence of *Xanthomonas campestris* pv. campestris. *MPMI* 18, 652–658.

Tapiero, H. and Tew, K. (2003) Trace elements in human physiology and pathology: zinc and metallothioneins. *Biomedicine and Pharmacotherapy* 57, 399–411.

Tetaz, T.J. and Luke, R.K. (1983) Plasmid-controlled resistance to copper in *Escherichia coli. Journal of Bacteriology* 154, 1263–1263.

Thaden, J.T., Lory, S. and Gardner, T.S. (2010) Quorum-sensing regulation of a copper toxicity system in *Pseudomonas aeruginosa. Journal of Bacteriology* 192, 2557–2568.

Tree, J.J., Ulett, G.C., Ong, C.L.Y., Trott, D.J., McEwan, A.G. and Schembri, M.A. (2008) Trade-off between iron uptake and protection against oxidative stress: deletion of *cueO* promotes uropathogenic *Escherichia coli* virulence in a mouse model of urinary tract infection. *Journal of Bacteriology* 190, 6909–6912.

Vallee, B.L. and Falchuk, K.H. (1993) The biochemical basis of zinc physiology. *Physiological Reviews* 73(1), 79–118.

Voloudakis, A.E., Bender, C.L. and Cooksey, D.A. (1993) Similarity between copper resistance genes from *Xanthomonas campestris* and *Pseudomonas syringae. Applied and Environmental Microbiology* 59, 1627–1634.

Waldron, K.J. and Robinson, N.J. (2009) How do bacterial cells ensure that metalloproteins get the correct metal? *Nature Reviews Microbiology* 6, 25–35.

Ward, S.K., Hoye, E.A. and Talaat, A.M. (2008) The global response of *Mycobacterium tuberculosis*

to physiological levels of copper. *Journal of Bacteriology* 190, 2939–2946.

Weaver, L., Michels, H.T. and Keevil, C.W. (2008) Survival of *Clostridium difficile* on copper and steel: futuristic options for hospital hygiene. *Journal of Hospital Infection* 68, 145–151.

Weston, B.F., Brenot, A. and Caparon, M.G. (2009) The metal homeostasis protein, Lsp, of *Streptococcus pyogenes* is necessary for acquisition of zinc and virulence. *Infection and Immunity* 77, 2840–2848.

White, C., Lee, J., Kambe, T., Fritsche, K. and Petris, M.J. (2009) A role for the ATP7A copper-transporting ATPase in macrophage bactericidal activity. *Journal of Biological Chemistry* 284, 33949–33956.

Wilks, S.A., Michels, H. and Keevil, C.W. (2005) The survival of *Escherichia coli* O157 on a range of metal surfaces. *International Journal of Food Microbiology* 105, 445–454.

Williams, J.R., Morgan, A.G., Rouch, D.A., Brown, N.L. and Lee, B.T.O. (1993) Copper-resistant enteric bacteria from United Kingdom and Australian piggeries. *Applied and Environmental Microbiology* 59, 2531–2537.

Yamasaki, H., Pilon, M. and Shikanai, T. (2008) How do plants respond to copper deficiency? *Plant Signalling and Behaviour* 3, 231–232.

Yang, W., Liu, Y., Chen, L., Gao, T., Hu, B., Zhang, D. and Liu, F. (2007) Zinc uptake regulator (*zur*) gene involved in zinc homeostasis and virulence of *Xanthomonas oryzae* pv. *oryzae* in rice. *Current Microbiology* 54, 307–314.

Yamamoto, K. and Ishihama, A. (2005) Transcriptional response of *Escherichia coli* to external copper. *Molecular Microbiology* 56, 215–227.

12 Metal Ion Sensing in *Mycobacterium tuberculosis*

Jennifer S. Cavet

Tuberculosis remains a leading killer worldwide, causing 1.8 million deaths and 9 million new cases each year. It is estimated that one-third of the world's population is latently infected with the causative agent *Mycobacterium tuberculosis*. Detecting deficiency and excess of different metal ions is fundamental for *M. tuberculosis* survival in the host. This involves multiple DNA-binding, metal-responsive transcription factors (e.g. IdeR, SirR, FurA, CsoR, Zur, SmtB, NmtR, KmtR and CmtR) which discriminate between elements and trigger expression of genes that mediate appropriate responses. Some of these metal sensors act on a single gene target while others act globally, controlling transcription of regulons in response to fluxes of a particular metal. They not only modulate the expression of genes associated directly with metal homeostasis (such as metal import, export, sequestration and/or storage), but also can control the expression of proteins with other primary roles, such as oxidative stress resistance. This chapter focuses on the different metal-sensing proteins so far identified in *M. tuberculosis* and describes current knowledge of their roles in allowing this important pathogen to adapt to the changing metal levels encountered in the host. These proteins represent potential targets for the development of new anti-tuberculosis drugs.

12.1 Introduction

Transition metals such as iron, zinc, copper, manganese, nickel and cobalt are essential for bacteria due to one-quarter to one-third of all proteins requiring such metals for function (Waldron and Robinson, 2009). However, these elements can also be toxic, even at low concentrations, due to binding to adventitious sites or, in the case of redox active metals such as ferrous and cuprous ions, participation in Fenton chemistry driving the production of highly toxic hydroxyl radicals. All bacteria must therefore somehow sense and adapt rapidly to metal fluxes to survive.

Bacteria encounter metal excess or deficiency due to changes in metal levels in their immediate environment and/or changes in their metal demand. Environmental metal availability can be affected by many factors, including oxygen levels, pH and the presence of organic matter, sulfates and carbonates (Fraústo da Silva and Williams, 2001). Changes in metabolic activity and enzyme production can alter metal demand (Rutherford et al., 1999) and can be influenced by metal availability to reduce consumption of a metal in short supply (McHugh et al., 2003; Gabriel and Helmann, 2009). As a human pathogen, *M. tuberculosis* must adapt to various microenvironmental niches within the host. These include the phagosomes of alveolar macrophages, where it mainly resides and encounters dramatic changes in metal availability (Schnappinger *et al.*, 2003; Olakanmi *et al.*, 2004; Schaible and Kaufmann, 2004; Malik *et al.*, 2005; Wagner *et al.*, 2005a,b; Zhang *et al.*, 2005; Rodriguez and Smith, 2006; Collins, 2008; Legrand *et al.*, 2008). These are in part influenced by the release of metal-binding antimicrobial proteins by the host, such as iron-binding lactoferrin and transferrin (Olakanmi *et al.*, 2004; Legrand *et al.*, 2008), in addition to generalized changes in metal homeostasis during infection, such as the release of hepcidin altering iron homeostasis (Collins, 2008; Wessling-Resnick, 2010). Natural resistance-associated macrophage protein 1 (Nramp1, alias SLC11A1) is a proton divalent cation antiporter that aids *M. tuberculosis* killing by generating metal fluxes across the membrane of the pathogen-containing phagosomal compartment, and hence must be evaded by the more virulent pathogens (Blackwell *et al.*, 2001; Hoal *et al.*, 2004; Malik *et al.*, 2005; Zhang *et al.*, 2005; Techau *et al.*, 2007). In addition, copper levels are reported to increase in macrophages exposed to cytokines, plus other mediators of inflammation, and are accompanied by trafficking of the ATP7A copper transporter from the Golgi apparatus to phagosome-associated vesicles and increased bacterial killing (White *et al.*, 2009).

M. tuberculosis survival within a host therefore requires that it can protect against metal-mediated toxicity while competing effectively for essential metals for its metal-requiring proteins. As a result, this pathogen has evolved elaborate metal-specific regulatory systems that control the intracellular availability of the different metal ions (Fig. 12.1), and which are the focus of this review.

12.2 IdeR and Iron Homeostasis

Iron is required by *M. tuberculosis* as a cofactor for several fundamental enzymes with activities ranging from respiration, DNA replication and oxidative stress resistance. Within host macrophages it is therefore critical that this pathogen maintains an intracellular iron supply. A key survival strategy of *M. tuberculosis* is the ability to disrupt phagosomal development (Warner and Mizrahi, 2007), causing the phagosome to be maintained at an early endosomal stage, where it retains access to the transferrin receptor which delivers iron to the cell (Schaible and Kaufmann, 2004). In addition, *M. tuberculosis* has developed sophisticated iron-acquisition systems. Fundamental to these is the production and release of high-affinity ferric iron-chelating siderophores (mycobactins), which following iron binding are re-internalized via specific cell receptors and *trans*-membrane transporters (Rodriguez and Smith, 2006). The ability to synthesize and transport the siderophores is essential for *M. tuberculosis* virulence (De Voss *et al.*, 2000; Rodriguez and Smith, 2006). A protein belonging to the Nramp transporter family (Mramp or MntH) has also been implicated in iron transport in *M. tuberculosis* (Agranoff *et al.*, 1999; Wagner *et al.*, 2005b), although the Mramp1 metal substrates *in vivo* remain to be established. Furthermore, during iron-replete conditions, *M. tuberculosis* is proposed to deposit intracellular reserves of iron in the iron-storage proteins, bacterioferritin (BfrA) and ferritin (BfrB) (Rodriguez, 2006). The iron stores may then be used to enhance growth when external iron supplies are restricted, as well as protect against iron-mediated oxidative damage that might be caused by excess free iron participating in Fenton chemistry. These responses of *M. tuberculosis* to changes in iron availability are regulated precisely, thereby maintaining sufficient iron to fulfil the needs of iron-requiring proteins while avoiding excessive and potentially toxic levels of iron in the cell.

The iron-dependent regulator, IdeR, is the key transcriptional regulator of iron uptake and homeostasis in *M. tuberculosis*. This protein belongs to the DtxR family of metal-responsive regulators (Osman and Cavet, 2010), where other key representatives of this regulator family include the iron-sensor DtxR (diphtheria toxin regulator), manganese- and cadmium-sensing MntR (manganese transport regulator) and ScaR (*Streptococci* coaggregation-mediating adhesion regulator), and manganese- and zinc-sensing TroR (transport-related operon

Structural family	Regulator	Selectivity	Mode of Action	Function
DtxR	IdeR	Fe(II)	Co-repression/activation +Fe	Iron homeostasis
	SirR	Mn(II)?	Co-repression +Mn	Manganese uptake?
Fur	FurA	Fe(II)	Co-repression +Fe	Peroxide resistance
	Zur	Zn(II)	Co-repression +Zn	Zinc homeostasis
CsoR-RcnR	CsoR	Cu(I)	Derepression +Cu	Copper resistance
SmtB-ArsR	SmtB	Zn(II)	Derepression +Zn	Zinc homeostasis
	NmtR	Ni(II), Co(II)	Derepression +Ni	Nickel and cobalt resistance
	KmtR	Ni(II), Co(II)	Derepression +Ni	Nickel and cobalt resistance
	CmtR	Cd(II), Pb(II)	Derepression +Cd	Cadmium and lead resistance
	ArsR	As(III)	Derepression +As	Arsenic resistance

Fig. 12.1. *Mycobacterium tuberculosis* **cytosolic metalloregulatory proteins.** The founder member(s) of each structural family is/are shown, along with the mechanism of gene regulation and function. The co-repressors include iron-sensing IdeR and the proposed manganese sensor, SirR, belonging to the DtxR family, FurA, which is proposed to use iron as a cofactor to sense peroxide, and the zinc sensor, Zur (alias FurB). Although iron-bound IdeR generally acts as a co-repressor, at some promoters it functions as a DNA-binding activator. The derepressors include copper-sensing CsoR and the SmtB-ArsR family sensors of zinc (SmtB), nickel and cobalt (NmtR and KmtR), cadmium and lead (CmtR) and arsenic (ArsR). Metal-responsive ECF sigma factors and two-component regulatory systems are not included here. Additional metal sensors belonging to these or other, structurally distinct, regulator families may also exist, but remain to be characterized.

regulator) (Posey *et al.*, 1999; Guedon and Helmann, 2003; Hazlett *et al.*, 2003; Stoll *et al.*, 2009). *M. tuberculosis* IdeR is responsible for the regulation of about 40 genes involved in iron uptake and metabolism (Quadri *et al.*, 1998; De Voss *et al.*, 2000; Gold *et al.*, 2001). These include genes involved in the synthesis, export and import of the siderophores, the synthesis of aromatic amino acids, and the bacterioferritin (*bfrA*) and ferritin (*bfrB*) iron-storage genes.

The resolved crystal structures of *M. tuberculosis* IdeR reveal that it exists as a homodimer when in its active metal-bound state (Fig. 12.2a), with each monomer containing three domains (Pohl *et al.*, 1999; Feese *et al.*, 2001; Wisedchaisri *et al.*, 2004). These comprise the N-terminal winged helix-turn-helix DNA-binding domain, followed by a helical dimerization domain connected by a flexible linker region to the C-terminal SH3 (Src homology domain 3)-like domain (Fig. 12.2a). Characterization of the metal (cobalt)-bound form of IdeR has revealed two metal-binding sites per monomer (Fig. 12.2a and b), designated the primary (regulatory) and ancillary sites (Feese *et al.*, 2001).

The ancillary site is pentavalent co-

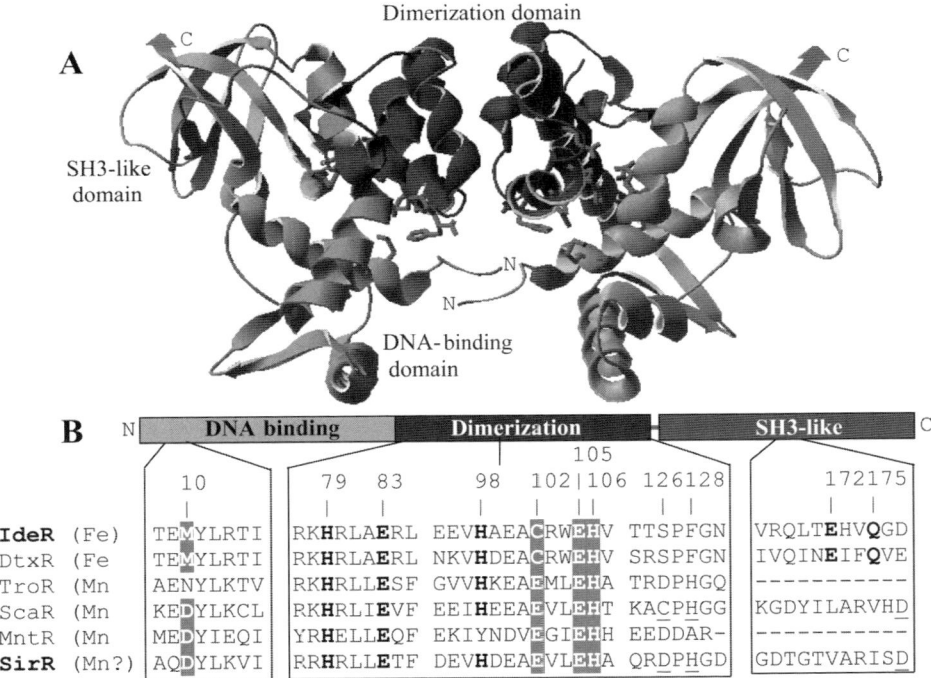

Fig. 12.2. IdeR structure and alignment with other members of the DtxR family of regulators. (a) Structure of the IdeR dimer in its cobalt-bound form (PDB ID FX7). The N-terminal DNA-binding, dimerization and C-terminal SH3-like domains are indicated, with side chains for the regulatory and ancillary metal-binding residues also shown. (b) Sequence alignment of the metal-binding residues of IdeR with other representative members of the DtxR regulator family: DtxR from *Corynebacterium diphtheriae*; TroR from *Treponema pallidum*; ScaR from *Streptococcus gordonii*; MntR from *Bacillus subtilis*; and SirR from *M. tuberculosis* (the *M. tuberculosis* regulators are indicated in bold). The metal specificities of the sensors are indicated, along with residues aligning with the regulatory and ancillary metal-binding sites in IdeR (grey boxes and bold, respectively) and their positions within the DNA-binding, dimerization and C-terminal SH3-like domains of IdeR. The manganese sensors often lack the third SH3-like domain, including Glu[172] and Gln[175], and form metal-independent dimers (Guedon and Helmann, 2003; Kliegman *et al.*, 2006). However, it is suggested that other residues (indicated with an underscore) may contribute to metal binding in these regulators (Stoll *et al.*, 2009).

ordinate with a distorted trigonal bipyramidal geometry involving ligands contributed by His[79], Glu[83], His[98] at the dimerization helix, and Glu[172] and Gln[175] from the C-terminal SH3-like domain. Metal binding at this site is thought to stabilize the dimer and is required for full DNA-binding activity (Wisedchaisri et al., 2004). The regulatory metal-binding site is hexavalent coordinate with an octahedral geometry, involving the side chains of Met[10] from the N-terminal domain with Cys[102], Glu[105], His[106] from the dimerization helix and an ordered water coordinated to the metal ion. Metal binding to the regulatory site is thought to drive a helix-coil transition in the N-terminal region, which is required for metal-dependent allosteric activation of DNA binding, with the suggestion that the unstructured N-terminal region in the apoprotein inhibits DNA binding (White et al., 1998). Although a variety of metals, including nickel, cobalt, cadmium, manganese and zinc, in addition to iron, can bind to IdeR and activate DNA binding in vitro (Schmitt et al., 1995; Semavina et al., 2006), the transcriptional response is highly specific for iron in vivo. The specificity for iron coincides with iron having a high affinity for both of the IdeR metal-binding sites in vitro (Chou et al., 2004), although metal availability in the cytosol is also likely to play an important role in determining iron selectivity in vivo (Guedon and Helmann, 2003; Osman and Cavet, 2010). Metal-bound IdeR binds to a 9-1-9 inverted-repeat consensus sequence (TTAGGTTAG-G-CTAACCTAA) within the operator-promoter regions of its target genes (Gold et al., 2001). IdeR binds as a 'double-dimer', with two dimers bound on opposite sides of the DNA duplex (Pohl et al., 1999; Wisedchaisri et al., 2004). DNA binding by iron-IdeR usually leads to repression of transcription, although iron-bound IdeR activates transcription from the bfrA and bfrB promoters (Gold et al., 2001; Rodriguez, 2006).

IdeR is an essential protein in M. tuberculosis and disruption of ideR in the absence of a second functional copy is lethal (Gold et al., 2001; Rodriguez et al., 2002). Mutation studies have linked this essentiality of IdeR to its regulation of siderophore synthesis and role in oxidative stress resistance (Rodriguez et al., 2002). Furthermore, constitutive expression of IdeR in M. tuberculosis has been shown to attenuate virulence in a murine infection model (Manabe et al., 1999). These studies highlight the requirement for iron sensing and the precise regulation of iron homeostasis in the host environment (Quadri et al., 1998; De Voss et al., 2000).

12.3 SirR and Manganese Homeostasis

M. tuberculosis possesses a second member of the DtxR regulator family, in addition to IdeR, which has been designated SirR (Rodriguez, 2006). M. tuberculosis SirR shares sequence features common to the manganese-sensing members of this regulator family (Guedon and Helmann, 2003; Stoll et al., 2009), including the possession of a deduced regulatory metal-binding site associated with manganese sensing (Fig. 12.2b). This site involves the substitution of Met[10] and Cys[102], involved in metal binding in the iron sensors (numbering is for IdeR), with Asp and Glu, respectively (Fig. 12.2b). Indeed, a DtxR mutant in which Met[10] and Cys[102] are substituted for Asp and Glu, respectively, becomes highly selective for manganese (Guedon and Helmann, 2003). A role for M. tuberculosis SirR in the regulation of genes involved in manganese homeostasis is therefore proposed. No genes associated with manganese transport have been characterized so far in M. tuberculosis. However, in other bacteria, proteins belonging to the Nramp family (MntH) have been associated with the transport of manganese and virulence (Papp-Wallace and Maguire, 2006), and hence it is hypothesized that the M. tuberculosis MntH homologue (Mramp) may also play a role in manganese transport and is a likely target for SirR regulation. In addition, M. tuberculosis carries an ABC-type divalent cation transporter that has yet to be characterized, but based on similarity to the SitABCD manganese transporter in Salmonella Typhimurium, has been suggested to contribute to manganese import (Rodriguez, 2006). Manganese-requiring proteins in

M. tuberculosis include the secreted superoxide dismutase, SodA, which provides protection from reactive oxygen species, thus indicating a requirement for manganese uptake in the host environment.

12.4 FurA and Oxidative Stress Resistance

In addition to members of the DtxR family of regulators, a second family of regulators that control iron homeostasis in bacteria is the Fur family. Proteins of both families are widely distributed in bacteria and although the two families are structurally distinct, the iron-sensing members of these families perform similar functional roles, with Fur proteins also functioning as global regulators of iron homeostasis controlling both the induction of iron uptake functions (under iron limitation) and the expression of iron-storage proteins and iron-utilizing enzymes (under iron sufficiency). Members of the ferric uptake regulator (Fur) family are now known that sense sufficiency of a variety of metals, including iron (Fur), zinc (Zur), manganese (Mur) and nickel (Nur) (Lee and Helmann, 2007). Other family members use metal catalysed oxidation reactions to sense peroxide stress (PerR) or the availability of haem (Irr). Iron-sensing *E. coli* Fur is the prototype for this family and controls the expression of more than 100 genes with roles in iron transport, siderophore biosynthesis, iron storage, the tricarboxylic acid (TCA) cycle and redox stress resistance. Although Fur is commonly thought of as a co-repressor that binds to DNA after binding metal, Fur also activates the expression of many genes by either direct or indirect mechanisms, the latter involving repression of a small regulatory RNA, sRNA, which binds to and decreases the stability of its target mRNAs (Lee and Helmann, 2007).

M. tuberculosis contains two Fur-like proteins, FurA and Zur (alias FurB). FurA appears to be a specialized iron-dependent regulator that controls the *katG* gene, encoding catalase peroxidase, while Zur is a global regulator of zinc homeostasis (see below). The *furA* gene is located immediately upstream of the *katG* gene, and iron-bound FurA binds to a unique sequence upstream of *furA*, repressing expression of both *furA* and *katG* (Sala *et al.*, 2003; Lucarelli *et al.*, 2008). The *furA-katG* genes are induced upon oxidative stress *in vivo* (Milano *et al.*, 2001) and peroxide treatment of FurA abolishes DNA binding *in vitro* (Sala *et al.*, 2003), thus indicating that FurA could represent a metal-dependent peroxide sensor similar to the peroxide regulon repressor, PerR, from *B. subtilis*. PerR can use either iron or manganese as a co-repressor (Chen *et al.*, 1995), with dimeric PerR containing one structural zinc and one regulatory iron/manganese binding site per monomer, although physiological and biochemical studies support a model in which only iron can be used for sensing peroxides *in vivo* (Lee and Helmann, 2007). The exquisite sensitivity of iron-bound PerR to peroxides is proposed to result from oxidation reactions, enabled by the Fe^{2+} co-repressor targeting residues in or near the regulatory metal-binding site.

As well as KatG conferring oxidative stress resistance, it is also the mycobacterial enzyme responsible for the activation of the first-line tuberculosis drug, isoniazid. Isoniazid is able to traverse the complex cell wall of *M. tuberculosis* by passive diffusion, where it becomes activated by KatG, which modifies it into a range of intermediates including NAD^+ and NADP+ adducts (Lucarelli *et al.*, 2008). These then act as potent inhibitors of the NADPH-dependent enoyl acyl carrier protein reductase (InhA), an essential enzyme in cell wall mycolic acid synthesis. A *furA-katG* double mutant of *M. tuberculosis* is highly attenuated in a mouse model of tuberculosis. However, full virulence is retained in a single *furA* mutant in which *katG* is highly expressed, and these cells are also hypersensitive to isoniazid (Pym *et al.*, 2001). These findings have led to FurA attracting interest as a potential drug target, whereby enhancement of *katG* expression could possibly boost isoniazid potency. In addition to *katG*, there is some indication that FurA may also be involved in the regulation of other genes involved in pathogenesis (Pym *et al.*, 2001), although this remains to be confirmed.

12.5 Zinc Sensing and Homeostasis

Zinc is also an essential element for *M. tuberculosis* and is required for a large number of proteins which perform diverse fundamental roles, with zinc being required as a cofactor for enzymatic catalysis, protein structural organization and/or regulation of protein function. Zinc-requiring proteins have roles including DNA replication, transcription, translation, cell division, glycolysis, pH regulation, oxidative stress resistance and the biosynthesis of amino acids, extracellular peptidoglycan and low molecular weight thiols. In fact, zinc is one of the most abundant transition metals in any given bacterium, being estimated at mM concentrations (Outten and O'Halloran, 2001). Despite this, free zinc concentrations in a cell are thought to be maintained at less than one atom cell^{-1}, at least in *E. coli*, and hence all cytosolic zinc is presumed to be bound and buffered (Outten and O'Halloran, 2001). It is likely that this tight regulation of zinc availability relates to zinc being at the upper end of the Irving–Williams stability series for protein–metal binding affinities; zinc therefore having a tendency to bind tightly to metalloproteins, including those that must bind less competitive metals such as iron and manganese (Waldron and Robinson, 2009). *M. tuberculosis* possesses two zinc sensors, Zur (alias FurB) and SmtB (alias Rv2358), which are proposed to control zinc availability in *M. tuberculosis*. Both of the *M. tuberculosis* zinc sensors are transcribed within the *smtB-zur* operon, which is under the control of SmtB and expressed in response to elevated zinc levels (Milano *et al.*, 2004; Canneva *et al.*, 2005; Campbell *et al.*, 2007).

12.5.1 Zur

M. tuberculosis Zur represents a zinc-sensing member of the Fur family of regulators (Maciag *et al.*, 2007). Zur is thought to be responsible for the zinc-dependent repression of 32 genes, organized in 16 operons, which were found to be upregulated in a *zur* mutant (Maciag *et al.*, 2007). Of these, 24 belong to eight putative transcriptional units preceded by a similar 26-bp AT-rich palindrome, which Zur binds to in a zinc-dependent manner. No downregulated genes were identified in the *M. tuberculosis zur*-mutant, suggesting that *M. tuberculosis* Zur acts specifically as a co-repressor (Maciag *et al.*, 2007). A number of the Zur-regulated gene products have been implicated to play roles in zinc homeostasis, including Rv0106 with similarity to the Zur-regulated low-affinity zinc-transporter, YciC, in *B. subtilis*, and two ABC transporter components with some similarity to the ZnuABC zinc-uptake system of *E. coli* (Maciag *et al.*, 2007). In addition, *M. tuberculosis* Zur regulates five ribosomal proteins, three of which have paralogs containing a zinc ribbon in their sequences. Hence, it is tempting to speculate that with analogy to Zur in *B. subtilis* (Gabriel and Helmann, 2009), *M. tuberculosis* Zur may play a role in minimizing zinc consumption during zinc limitation by upregulating genes encoding non-zinc-containing proteins to replace functionally their zinc-requiring counterparts.

The crystal structure of *M. tuberculosis* Zur in complex with zinc shows similarity to other Fur family members, each monomer of the dimer being a two-domain structure with an N-terminal DNA-binding domain composed of a three-helix bundle followed by a short anti-parallel β-sheet and a C-terminal metal-binding and dimerization domain (Lucarelli *et al.*, 2007). Three zinc-binding sites per monomer were identified in the crystal structure of Zur, one representing a putative structural zinc site, one a regulatory zinc site, while the biological role of the third zinc is not clear and might represent a crystallization artefact (Lucarelli *et al.*, 2007). Both *M. tuberculosis* FurA and Zur are proposed to share the same structural zinc site and employ similar mechanisms of action (Lucarelli *et al.*, 2008).

12.5.2 SmtB

M. tuberculosis SmtB is a zinc-sensing member of the SmtB-ArsR family of metal-responsive repressors (Milano *et al.*, 2001; Canneva *et al.*, 2005; Campbell *et al.*, 2007). These proteins act as derepressors, with metal binding impairing DNA binding to alleviate repression (recently reviewed by Osman and Cavet, 2010). Zinc-sensing members of this family in other organisms (e.g. cyanobacterial SmtB, ZiaR and CzrA) regulate genes involved in the response to zinc toxicity including zinc export and/or zinc sequestration. Zinc-sensing cyanobacterial SmtB was the first family member shown to be a winged helix homodimer with helices α3 and α4 predicted to form the DNA-associating helix-turn-helix (Fig. 12.3).

Both *in vitro* and *in vivo* studies revealed cyanobacterial SmtB to possess two pairs of metal-binding sites per dimer (Fig. 12.3); one pair associated with the α3 helices, including two ligands contributed by the N-terminal region of the opposing monomer (α3N sites), and the second pair associated with C-terminal α5 helices (α5 sites) (Turner *et al.*, 1996; VanZile *et al.*, 2002b; Eicken *et al.*, 2003). Both pairs of sites form tetrahedral metal-coordination complexes. However, site-directed mutagenesis studies established that only the α5 sites were required for SmtB inducer responsiveness *in vivo* and *in vitro* (Turner *et al.*, 1996; VanZile *et al.*, 2002a), despite the α3N sites being occupied predominantly in the purified protein (VanZile *et al.*, 2002b). The structures of the apo- and zinc-bound forms of SmtB indicate

that a hydrogen-bond network is established when zinc binds at the α5 site that allosterically couples the zinc- and DNA-binding sites in SmtB to stabilize a low-affinity DNA-binding conformation (Eicken *et al.*, 2003). Recent structural studies involving the closely related sensor CzrA from *Staphylococcus aureus* further revealed that zinc binding drives a quaternary structural switch from a 'closed' DNA-binding state to a more 'open' conformation with reduced DNA affinity, and reduces the mobility of the α5 sites and α4 helices and the conformational heterogeneity (Arunkumar *et al.*, 2009). Zinc binding at the α5 sites therefore appears to 'lock' the structure into a more rigid low DNA affinity conformation. *M. tuberculosis* SmtB possesses residues expected to form a site equivalent to the tetrahedral α5 sites in cyanobacterial SmtB and CzrA (involving Asp[116], His[118], His[129] and His[132]; Fig. 12.4), and the mechanism of *M. tuberculosis* SmtB allosteric regulation by zinc is therefore proposed to be similar to that of the related sensors.

M. tuberculosis SmtB represses transcription from the *smtB-zur* operon by binding to upstream operator-promoter sequences containing a 12-2-12 hyphenated inverted repeat sequence that has similarity to the cyanobacterial SmtB and CzrA binding sites (Canneva *et al.*, 2005; Campbell *et al.*, 2007).

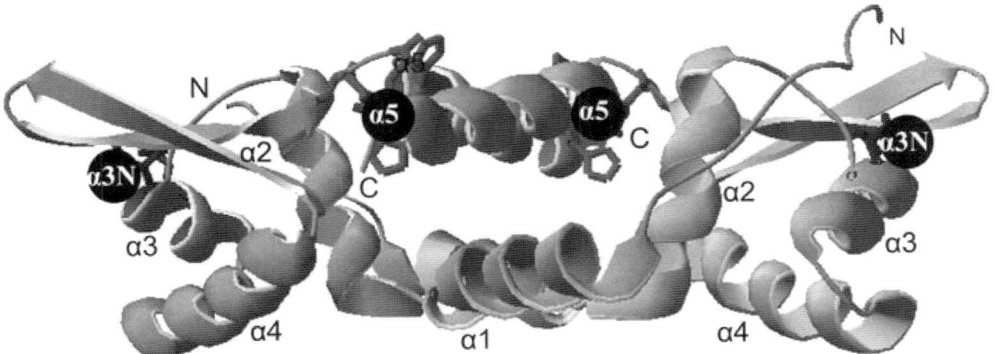

Fig. 12.3. Representation of the winged helix structure of an apo-SmtB homodimer. The resolved structure of cyanobacterial SmtB is shown with helices α3 and α4 forming the helix-turn-helix DNA-binding regions (PDB ID 1R1T). Two pairs of metal-binding sites (α3N and α5, black circles) are located at dimer interfaces, involving: Asp[104], His[106], His[117] and Glu[120] at the regulatory α5 sites; or Cys[61], Asp[64] and two additional residues from the opposing amino-terminal at the α3N sites (not obligatory for metal sensing). It is noted that the amino terminal ~20 amino acids of each monomer were not visible on electron density maps and hence are not included.

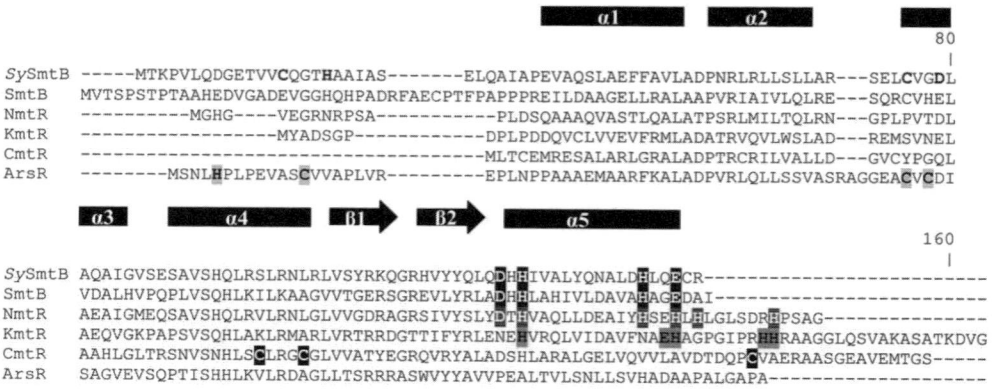

Fig. 12.4. Sensory sites in the SmtB-ArsR family regulators. Alignment of the *M. tuberculosis* metal sensors with cyanobacterial SmtB showing the different metal-sensing sites, involving ligands (boxes) at an α5 site in SmtB, α5C in NmtR, α5-3 in KmtR, α4C in CmtR and predicted α3N in ArsR. The α3N site in cyanobacterial SmtB (residues in bold) is not required for metal responsiveness. The known secondary structure for SmtB (Cook *et al.*, 1998; Eicken *et al.*, 2003) is shown above. Permutations in the effector binding sites of the *M. tuberculosis* sensors allow the detection of different metals within the cytosol, whereas differences in the sensing ligands at α5C and α5-3 in NmtR and KmtR, respectively, are proposed to alter their affinities for nickel and cobalt, allowing these proteins to respond under different surplus nickel and cobalt conditions (Campbell *et al.*, 2007).

Zinc binds to *M. tuberculosis* SmtB and inhibits DNA complex assembly *in vitro* (Campbell *et al.*, 2007), consistent with observed alleviation of SmtB-mediated repression by zinc *in vivo* (Canneva *et al.*, 2005; Campbell *et al.*, 2007). In addition to regulating the *smtB-zur* operon, *M. tuberculosis* SmtB has also been suggested to regulate *Rv2025c* (alias *cdf* or *zitA*) encoding a deduced metal exporter (Riccardi *et al.*, 2008), although previous studies (Campbell *et al.*, 2007) have shown *Rv2025c* to be under the control of a different SmtB-ArsR sensor, KmtR, with KmtR-mediated repression of *Rv2025c* being alleviated specifically by nickel and cobalt (see below). By regulating expression of the low zinc sensor, Zur, in *M. tuberculosis* in response to elevated zinc, it is hypothesized that SmtB plays a role in preventing zinc toxicity and allows tight control of intracellular zinc levels, which may overcome a requirement for a specific zinc effluxer. It will be of interest to calibrate the zinc affinities of both *M. tuberculosis* Zur and SmtB and determine the free zinc concentrations that trigger transcription allosteric regulation by these proteins.

12.6 SmtB-ArsR Metal Sensors (NmtR, KmtR and CmtR) and Metal Export

M. tuberculosis possesses an atypically large number (12) of genes encoding deduced SmtB-ArsR sensors compared to the majority of other sequenced bacteria. In addition to zinc-sensing SmtB described above, the characterization of three more of these sensors (NmtR, CmtR and KmtR) has also been described. Nickel- and cobalt-sensing NmtR regulates the *nmt* operator-promoter, triggering expression of the NmtA metal-transporting P-type ATPase (alias CtpJ) (Cavet *et al.*, 2002). The second nickel and cobalt sensor, KmtR, regulates the *kmtR* and *cdf* operator-promoters, triggering expression of a cation diffusion facilitator family metal exporter (CDF or ZitA) (Campbell *et al.*, 2007). Cadmium lead-sensing CmtR regulates the *cmt* operator-promoter, triggering expression of the CmtA metal-transporting P-type ATPase, CmtA (alias CtpG) and a membrane protein (Rv1993c) of unknown function (Cavet *et al.*, 2003). Studies with the *M. tuberculosis* SmtB-ArsR sensors have given vital insight into determinants of metal specificity among the SmtB-ArsR sensors and correct predictions of metal specificity can now be made based on these findings (reviewed by Osman and Cavet, 2010).

12.6.1 Nickel and cobalt sensing by NmtR and KmtR

NmtR was the first *M. tuberculosis* SmtB-ArsR sensor to be characterized (Cavet *et al.*, 2002). NmtR binds to the *nmt* operator-promoter and represses *nmtA* transcription, with NmtR-mediated repression alleviated by nickel and cobalt, but not zinc or any other metal tested. Site-directed mutagenesis established that NmtR required six ligands for nickel- and cobalt-responsiveness *in vivo* (Cavet *et al.*, 2002). Four of these are provided by α5 sites analogous to those in SmtB, while an extra two ligands are provided by a short carboxyl-terminal extension in NmtR, designated α5C sites (Fig. 12.4). Consistent with these findings, nickel and cobalt absorption spectra indicated octahedral metal coordination by NmtR (Cavet *et al.*, 2002), which was also confirmed by EXAFS (Pennella *et al.*, 2003), and was thus ideal for nickel and cobalt that preferred a higher coordination number. It is notable that while NmtR binds nickel and cobalt, it has a higher affinity for zinc. However, zinc fails to impair NmtR DNA binding *in vitro* or to alleviate repression *in vivo*. This is due to zinc binding to NmtR tetrahedrally and, as such, does not trigger the allostery to inhibit DNA binding, which requires hexadentate coordination (Pennella *et al.*, 2003). The difference in the metals sensed by NmtR and SmtB in *M. tuberculosis* is therefore likely to be due to these proteins demanding different coordination geometries at the α5 helices to mediate allostery rather than discrimination at the level of protein–metal binding.

Metal responsiveness of NmtR has also been studied in a cyanobacterial host (*Synechococcus* PCC 7942). In this heterologous host, NmtR-mediated repression is alleviated by cobalt only and not nickel, despite nickel being the most potent inducer in mycobacterial cells. Quantification of the number of atoms of nickel per cell revealed that the nickel content of mycobacterial cells increased ~40-fold when exposed to maximum permissive nickel concentrations, whereas the nickel content of cyanobacterial cells increased ~threefold. Hence, the effective exclusion of nickel from the cyanobacterial cytosol, compared to the mycobacterial cytosol, was considered the most likely explanation for the lack of nickel sensing by NmtR (Cavet *et al.*, 2002). These studies therefore reveal that different cytosolic metal pools can influence the metals sensed by metal-responsive transcription factors in different cells.

KmtR represents the second, to NmtR, characterized SmtB-ArsR sensor of nickel and cobalt from *M. tuberculosis*. Gene profiling experiments revealed elevated expression of *cdf*, encoding a deduced metal exporter, in *M. tuberculosis* mutants lacking *kmtR*, and KmtR was confirmed to bind to the *cdf* and *kmtR* operator-promoters to repress transcription (Campbell *et al.*, 2007). Comparison of the affinities of NmtR and KmtR for nickel and cobalt revealed that KmtR had tighter affinities for nickel and cobalt than NmtR (Campbell *et al.*, 2007). Site-directed mutagenesis identified five ligands critical for inducer recognition by KmtR, whereas a sixth was found to be required for repressor function and represented a further potential ligand (Fig. 12.4). Present predictions therefore support five- or six-coordinate cobalt and nickel liganding derived from the α5-helix and carboxyl-terminal region of KmtR. These residues are distinct from those in the metal-sensing sites of NmtR (designated α5-3 motif; Fig. 12.4), and hence differences in the sensing ligands at α5C and α5-3 in NmtR and KmtR are proposed to alter their affinities for cobalt and nickel and allow these proteins to regulate gene expression differentially under different surplus cobalt and nickel conditions. Thus, as cytosolic levels of nickel and/or cobalt increase, KmtR detects these metals and confers metal export via a CDF family transporter, whereas only when a higher threshold is reached does NmtR detect these metals and confer export via the P-type ATPase, NmtA. KmtR can also bind zinc *in vitro* and, by analogy to discrimination against zinc by NmtR (Cavet *et al.*, 2002), zinc is expected to bind to a subset of KmtR ligands only and inefficiently trigger allostery.

The existence of two SmtB-ArsR sensors of nickel and cobalt in *M. tuberculosis* implies a particular need to control elevated levels of these ions. Notably, this organism lacks homologues of the low nickel sensor, NikR, found in other bacteria, despite possessing a deduced nickel importer (NicT/NixA) to supply urease. Furthermore, it lacks the other known bacterial sensor of elevated cobalt and nickel, RcnR. Hence, the possession of NmtR and KmtR with differing nickel and cobalt affinities may, at least in part, substitute functionally for the lack of these sensors and act together to control nickel and cobalt availability within tight limits. Within macrophage phagosomes, *M. tuberculosis* survival is dependent on the activity of nickel-dependent urease, with urea hydrolysis contributing to nitrogen availability and environmental pH modulation (Clemens *et al.*, 1995). Production of ammonia by urease can also block phagosome–lysosome fusion, providing further protection from host-killing mechanisms (Gordon *et al.*, 1980). Cobalt is required for vitamin B_{12} biosynthesis, and hence is likely to be required by *M. tuberculosis* in macrophages to protect against vitamin B_{12} binding protein, a component of neutrophil-specific granules that can be acquired by macrophages due to phagocytosis of apoptotic neutrophils to assist in *M. tuberculosis* killing (Tan *et al.*, 2006). It is not yet known whether KmtR and/or NmtR, or indeed any of the other SmtB-ArsR metal sensors in *M. tuberculosis*, contribute to its survival within the human host.

12.6.2 Cadmium and lead sensing by CmtR

CmtR senses cadmium and lead and regulates expression of a metal-exporting P-type ATPase, CmtA. Consistent with a preference for thiophilic cadmium and lead, CmtR forms cysteine thiolate-rich metal coordination complexes. In addition to cadmium and lead, CmtR can also bind zinc *in vitro*, with all three metals being effective allosteric regulators of CmtR-DNA binding *in vitro* (Wang *et al.*,

2005). However, only cadmium and lead can alleviate repression *in vivo* (Cavet *et al.*, 2003). Hence, although zinc is available for detection in mycobacterial cells, at least by SmtB (Campbell *et al.*, 2007), it appears that CmtR is able to acquire zinc from the cytosolic pool far less effectively than SmtB. Site-directed mutagenesis and structural characterization of CmtR established that the effector binding sites of CmtR were unique compared to the other characterized SmtB-ArsR metal-sensing sites (Fig. 12.4), and involved two cysteine residues (Cys^{57} and Cys^{61}) located on the $\alpha 4$ DNA-binding helix of one subunit and a third (Cys^{102}) from the unstructured carboxyl-terminal region of the other subunit, forming a symmetrical pair of trigonal metal-binding sites (Cavet *et al.*, 2003; Wang *et al.*, 2005; Banci *et al.*, 2007). Cys^{102} appears to play an accessory role in stabilizing the coordination complex, as although Cys^{57} and Cys^{61} are required to bind cadmium *in vitro*, Cys^{102} is not and the cadmium affinity is reduced by only one order of magnitude in its absence (Wang *et al.*, 2005). However, substitutions of any of the three cysteines abolishes metal responsiveness *in vivo* (Cavet *et al.*, 2003) and disassembly of CmtR-DNA complexes *in vitro* (Wang *et al.*, 2005), demonstrating that all three are obligatory for metal-mediated allosteric regulation. Mutation of any of the other cysteines in CmtR does not affect metal responsiveness (Cavet *et al.*, 2003). From the solution structure of CmtR, it is apparent that apo-CmtR experiences a large conformational exchange at the dimer interface and within each monomer, whereas cadmium binding by CmtR reduces substantially the conformational heterogeneity of regions within the CmtR subunits, most notably at the dimer interface (Banci *et al.*, 2007). Three apo-CmtR dimers bind one DNA molecule, but cadmium-bound CmtR does not interact with DNA under any ratio. Hence, while DNA can select an optimal conformer from a pool of dynamic apo-homo-dimers, it appears that cadmium binding impairs DNA association by forming a more rigid conformation, thereby excluding the apo-conformers suited to bind DNA (Banci *et al.*, 2007).

It is not clear why an intracellular pathogen such as *M. tuberculosis* has a requirement to sense cadmium and lead. While it has been suggested that *M. tuberculosis* may be exposed to toxic concentrations of cadmium in macrophages due to the presence of cadmium as an air pollutant and in cigarette smoke (Chauhan *et al.*, 2009), it is possible that the possession of CmtR relates to an ancestral extracellular lifestyle. Alternatively, it remains possible that cadmium and lead are not the 'true' effectors of CmtR in the intracellular environment and that CmtR may also respond to other, as yet unidentified, stresses.

12.6.3 Other SmtB-ArsR proteins in *M. tuberculosis*

So far, each characterized sensor from *M. tuberculosis* uses a distinct effector binding site (Fig. 12.4). In addition, we have confirmed recently that the product of open reading frame Rv2642 is responsive to arsenic (hence our designation ArsR, unpublished observations) and regulates the *arsR-arsC* operon with a deduced role in arsenic resistance. ArsR possesses residues corresponding to an α3N metal-sensing site, which presumably it uses to sense arsenic (Fig. 12.4). Seven further *M. tuberculosis* sensors are predicted by the Pfam database HTH-5 family, although three have been excluded in an ensemble of 554 sequences of close relatives (Campbell *et al.*, 2007). The remaining four sensors lack defined metal-sensing sites and provide few clues as to their effectors or target genes. Indeed, it is unclear at this time whether or not all of these proteins do detect metals.

12.7 CsoR and Copper Homeostasis

Copper-requiring proteins in *M. tuberculosis* include an extracytoplasmic superoxide dismutase (SodC protein), which catalyses the dismutation of superoxide into oxygen and hydrogen peroxide (Spagnolo *et al.*, 2004). *M. tuberculosis* must possess systems that ensure a supply of copper to this enzyme, while avoiding copper toxicity. CsoR has been identified as a copper-responsive repressor

that regulates transcription of the *cso* copper efflux operon in *M. tuberculosis*, *csoR-Rv0968-ctpV* (Liu *et al.*, 2007), with *ctpV* and *Rv0968* encoding a deduced copper-exporting P-type ATPase and a membrane protein of unknown function, respectively. Expression of the *cso* operon has been shown to be increased during *M. tuberculosis* infection of mouse lungs (Talaat *et al.*, 2004), consistent with a requirement for copper export in the host environment.

CsoR acts as a derepressor and belongs to a recently identified structural class of metalloregulators (CsoR-RcnR family), along with the nickel and cobalt sensor, RcnR (Iwig and Chivers, 2010). Copper binding to CsoR inhibits DNA binding allosterically, causing alleviation of *cso* repression (Liu *et al.*, 2007; Ma *et al.*, 2009). Structural characterization of CsoR reveals an α-helical homodimer with each monomer composed of three helices. The homodimer contains two symmetry-related subunit bridging copper-binding sites, consisting of two cysteine residues and a histidine, that adopt a trigonal copper-coordination geometry (Liu *et al.*, 2007). A hydrogen-bonding network involving the liganding histidine is proposed to stabilize a copper-bound low-affinity DNA-binding state (Ma *et al.*, 2009).

To identify other genes involved in the response of *M. tuberculosis* to copper stress, microarrays have been used to identify genes expressed in response to short-term copper exposure (Ward *et al.*, 2008). A total of 30 copper-responsive genes were identified which included the *cso* operon. Several of these genes are associated with oxidative stress responses and/or the homeostasis of other metals, including *furA*, *katG*, *mramp*, *cmtR* (*Rv1994c*), *cadI* and *arsR* (*Rv2642*), indicating a generalized stress response in these copper exposed cells (Ward *et al.*, 2008). A metallothionein, MymT, has also been identified in *M. tuberculosis* that binds up to six Cu(I) ions in a solvent shielded core (Gold *et al.*, 2008). MymT (MT0196) is not annotated in the *M. tuberculosis* H37Rv genome database, which likely relates to its small size, 53 amino acids. Copper, cadmium, cobalt, nickel, zinc and compounds that generate nitric oxide or superoxide all induce expression of *mymT*, although copper, cadmium and the nitric oxide liberators are the most potent inducers. *M. tuberculosis* mutants lacking *mymT* were shown to be hypersensitive to copper but not to other divalent metals, reactive oxygen species or reactive nitrogen species, indicating a primary role for MymT in copper detoxification (Gold *et al.*, 2008). Induction of *mymT* expression by stimuli other than copper may therefore be due to copper displacement from MymT and/or other copper-binding proteins. Indeed, copper displacement from MymT and other copper-binding proteins by treatment of cells with nitric oxide has been demonstrated (Gold *et al.*, 2008). The mechanism of regulation of *mymT* by copper is not known. It is possible that this involves CsoR, although sequences corresponding to the CsoR binding site have not been identified upstream of *mymT* (Gold *et al.*, 2008).

12.8 Coordination of the Metal Sensors

The metal-responsive transcription factors commonly regulate genes involved in metal homeostasis and are thereby proposed to influence cellular metal availability. As such, their affinities for their effector metals are presumed to relate to the threshold values for cellular sufficiency or excess of these metals, with metal levels above this value causing them to repress transcription of genes involved in metal uptake and/or trigger the expression of genes involved in metal export, sequestration or storage. It is apparent that the affinities of the metal sensors for their metal effectors increase as they detect metals farther up the Irving–Williams series (Waldron and Robinson, 2009). Thus, the sensors of the more competitive metals (such as zinc and copper) are anticipated to maintain these metals at extremely low concentrations, whereas the sensors of the less competitive metals (such as iron and manganese) will bind and maintain their effectors at higher concentrations (Waldron and Robinson, 2009). As such, the sensors of the more competitive metals may prevent their access to adventitious metal-binding sites, which will include the binding sites of the sensors

for less competitive metals. Differences in access to cytosolic metal pools in different bacteria can alter the *in vivo* metal specificities of the sensor proteins (Cavet *et al.*, 2002; Guedon and Helmann, 2003), and this is consistent with metal access being, at least partly, a function of the relative affinities of the set of metal sensors in a particular organism (Waldron and Robinson, 2009). Other factors that may influence metal access to the sensors include specific interactions with metal-delivery proteins, such as metal importers and/or metallochaperones. Hence, the ability of a sensor to form heterodimers with a specific metal-delivery protein may also determine whether or not they gain access to a particular metal *in vivo* (Cobine *et al.*, 2002). As described above, metal-specific allostery will also contribute to the metal specificities of the sensors and thereby influence metal availability in cells. An additional level of control will also be provided by the metal specificities of the metal homeostatic proteins.

12.9 Conclusions

The survival of *M. tuberculosis* in a host depends on its ability to populate its many metalloproteins with their correct metal cofactors. Multiple metal-responsive transcriptional regulators have been described in this organism that act to control cellular metal availability tightly, thereby ensuring that metalloproteins acquire the right metal cofactors, while also serving other roles such as oxidative stress resistance. These metal sensors must distinguish correctly between the inorganic elements, and studies with the *M. tuberculosis* metal sensors have given vital insight into determinants of metal specificity.

It has been noted that *M. tuberculosis* in particular possesses an atypically large number of genes encoding SmtB-ArsR family sensors and it is plausible that fluctuations in metal concentrations within macrophages have selected for these sensors to detect a range of different metals. Several of the *M. tuberculosis* metal sensors have been associated with the virulence of this organism and a role in virulence now needs to be tested specifically for the others, including the multiple SmtB-ArsR family representatives. In addition, the genome of *M. tuberculosis* possesses a large number of uncharacterized genes encoding further deduced metal sensors and metal homeostatic proteins. Elucidation of the roles of these proteins may provide further clues as to the survival strategies employed by this organism and remains a priority. Understanding the metal stresses encountered by *M. tuberculosis* during infection and the various mechanisms used to adapt to these challenges offers possibilities for the design of novel antimicrobials targeted at these mechanisms, or indeed the use of specific metal chelators in combination with elevated metal levels, to assist host clearance of this important pathogen.

References

Agranoff, D., Monahan, I.M., Mangan, J.A., Butcher, P.D. and Krishna, S. (1999) *Mycobacterium tuberculosis* expresses a novel pH-dependent divalent cation transporter belonging to the Nramp family. *Journal of Experimental Medicine* 190, 717–724.

Arunkumar, A.I., Campanello, G.C. and Giedroc, D.P. (2009) Solution structure of a paradigm ArsR family zinc sensor in the DNA-bound state. *Proceedings of the National Academy of Sciences of the United States of America* 106, 18177–18182.

Banci, L., Bertini, I., Cantini, F., Ciofi-Baffoni, S., Cavet, J.S., Dennison, C., Graham, A.I., Harvie, D.R. and Robinson, N.J. (2007) NMR structural analysis of cadmium sensing by winged helix repressor CmtR. *Journal of Biological Chemistry* 282, 30181–30188.

Blackwell, J.M., Goswami, T., Evans, C.A., Sibthorpe, D., Papo, N., White, J.K., Searle, S., Miller, E.N., Peacock, C.S., Mohammed, H. and Ibrahim, M. (2001) SLC11A1 (formerly NRAMP1) and disease resistance. *Cellular Microbiology* 3, 773–784.

Campbell, D.R., Chapman, K.E., Waldron, K.J., Tottey, S., Kendall, S., Cavallaro, G., Andreini, C., Hinds, J., Stoker, N.G., Robinson, N.J. and Cavet, J.S. (2007) Mycobacterial cells have dual nickel-cobalt sensors: sequence relationships and metal sites of metal-responsive repressors are not congruent. *Journal of Biological Chemistry* 282, 32298–32310.

Canneva, F., Branzoni, M., Riccardi, G., Provvedi,

R. and Milano, A. (2005) Rv2358 and FurB: two transcriptional regulators from *Mycobacterium tuberculosis* which respond to zinc. *Journal of Bacteriology* 187, 5837–5840.

Cavet, J.S., Meng, W., Pennella, M.A., Appelhoff, R.J., Giedroc, D.P. and Robinson, N.J. (2002) A nickel-cobalt-sensing ArsR-SmtB family repressor. Contributions of cytosol and effector binding sites to metal selectivity. *Journal of Biological Chemistry* 277, 38441–38448.

Cavet, J.S., Graham, A.I., Meng, W. and Robinson, N.J. (2003) A cadmium-lead-sensing ArsR-SmtB repressor with novel sensory sites. Complementary metal discrimination by NmtR and CmtR in a common cytosol. *Journal of Biological Chemistry* 278, 44560–44566.

Chauhan, S., Kumar, A., Singhal, A., Tyagi, J.S. and Krishna Prasad, H. (2009) CmtR, a cadmium-sensing ArsR-SmtB repressor, cooperatively interacts with multiple operator sites to autorepress its transcription in *Mycobacterium tuberculosis*. *FEBS Journal* 276, 3428–3439.

Chen, L., Keramati, L. and Helmann, J.D. (1995) Coordinate regulation of *Bacillus subtilis* peroxide stress genes by hydrogen peroxide and metal ions. *Proceedings of the National Academy of Sciences of the United States of America* 92, 8190–8194.

Chou, C.J., Wisedchaisri, G., Monfeli, R.R., Oram, D.M., Holmes, R.K., Hol, W.G. and Beeson, C. (2004) Functional studies of the *Mycobacterium tuberculosis* iron-dependent regulator. *Journal of Biological Chemistry* 279, 53554–53561.

Clemens, D.L., Lee, B.Y. and Horwitz, M.A. (1995) Purification, characterization, and genetic analysis of *Mycobacterium tuberculosis* urease, a potentially critical determinant of host–pathogen interaction. *Journal of Bacteriology* 177, 5644–5652.

Cobine, P.A., George, G.N., Jones, C.E., Wickramasinghe, W.A., Solioz, M. and Dameron, C.T. (2002) Copper transfer from the Cu(I) chaperone, CopZ, to the repressor, Zn(II)CopY: metal coordination environments and protein interactions. *Biochemistry* 41, 5822–5829.

Collins, H.L. (2008) Withholding iron as a cellular defence mechanism – friend or foe? *European Journal of Immunology* 38, 1803–1806.

Cook, W.J., Kar, S.R., Taylor, K.B. and Hall, L.M. (1998) Crystal structure of the cyanobacterial metallothionein repressor SmtB: a model for metalloregulatory proteins. *Journal of Molecular Biology* 275, 337–346.

De Voss, J.J., Rutter, K., Schroeder, B.G., Su, H., Zhu, Y. and Barry, C.E. 3rd (2000) The salicylate-derived mycobactin siderophores of *Mycobacterium tuberculosis* are essential for growth in macrophages. *Proceedings of the National Academy of Sciences of the United States of America* 97, 1252–1257.

Eicken, C., Pennella, M.A., Chen, X., Koshlap, K.M., VanZile, M.L., Sacchettini, J.C. and Giedroc, D.P. (2003) A metal-ligand-mediated intersubunit allosteric switch in related SmtB/ArsR zinc sensor proteins. *Journal of Molecular Biology* 333, 683–695.

Feese, M.D., Ingason, B.P., Goranson-Siekierke, J., Holmes, R.K. and Hol, W.G. (2001) Crystal structure of the iron-dependent regulator from *Mycobacterium tuberculosis* at 2.0-A resolution reveals the Src homology domain 3-like fold and metal binding function of the third domain. *Journal of Biological Chemistry* 276, 5959–5966.

Fraústo da Silva, J.J.R. and Williams, R.J.P. (2001) *The Biological Chemistry of the Elements: The Inorganic Chemistry of Life*, 2nd edn. Oxford University Press, New York.

Gabriel, S.E. and Helmann, J.D. (2009) Contributions of Zur-controlled ribosomal proteins to growth under zinc starvation conditions. *Journal of Bacteriology* 191, 6116–6122.

Gold, B., Rodriguez, G.M., Marras, S.A., Pentecost, M. and Smith, I. (2001) The *Mycobacterium tuberculosis* IdeR is a dual functional regulator that controls transcription of genes involved in iron acquisition, iron storage and survival in macrophages. *Molecular Microbiology* 42, 851–865.

Gold, B., Deng, H., Bryk, R., Vargas, D., Eliezer, D., Roberts, J., Jiang, X. and Nathan, C. (2008) Identification of a copper-binding metallothionein in pathogenic mycobacteria. *Nature Chemical Biology* 4, 609–616.

Gordon, A.H., Hart, P.D. and Young, M.R. (1980) Ammonia inhibits phagosome-lysosome fusion in macrophages. *Nature* 286, 79–80.

Guedon, E. and Helmann, J.D. (2003) Origins of metal ion selectivity in the DtxR/MntR family of metalloregulators. *Molecular Microbiology* 48, 495–506.

Hazlett, K.R., Rusnak, F., Kehres, D.G., Bearden, S.W., La Vake, C.J., La Vake, M.E., Maguire, M.E., Perry, R.D. and Radolf, J.D. (2003) The *Treponema pallidum tro* operon encodes a multiple metal transporter, a zinc-dependent transcriptional repressor, and a semi-autonomously expressed phosphoglycerate mutase. *Journal of Biological Chemistry* 278, 20687–20694.

Hoal, E.G., Lewis, L.A., Jamieson, S.E., Tanzer, F., Rossouw, M., Victor, T., Hillerman, R., Beyers, N., Blackwell, J.M. and Van Helden, P.D. (2004) SLC11A1 (NRAMP1) but not SLC11A2 (NRAMP2) polymorphisms are associated with susceptibility to tuberculosis in a high-incidence community in South Africa. *International Journal of Tubercule Lung Disease* 8, 1464–1471.

Iwig, J.S. and Chivers, P.T. (2010) Coordinating intracellular nickel-metal-site structure–function relationships and the NikR and RcnR repressors. *Natural Products Reproduction* 27, 658–667.

Kliegman, J.I., Griner, S.L., Helmann, J.D., Brennan, R.G. and Glasfeld, A. (2006) Structural basis for the metal-selective activation of the manganese transport regulator of *Bacillus subtilis*. *Biochemistry* 45, 3493–3505.

Lee, J.W. and Helmann, J.D. (2007) Functional specialization within the Fur family of metalloregulators. *Biometals* 20, 485–499.

Legrand, D., Pierce, A., Elass, E., Carpentier, M., Mariller, C. and Mazurier, J. (2008) Lactoferrin structure and functions. *Advances in Experimental Medical Biology* 606, 163–194.

Liu, T., Ramesh, A., Ma, Z., Ward, S.K., Zhang, L., George, G.N., Talaat, A.M., Sacchettini, J.C. and Giedroc, D.P. (2007) CsoR is a novel *Mycobacterium tuberculosis* copper-sensing transcriptional regulator. *Nature Chemical Biology* 3, 60–68.

Lucarelli, D., Russo, S., Garman, E., Milano, A., Meyer-Klaucke, W. and Pohl, E. (2007) Crystal structure and function of the zinc uptake regulator FurB from *Mycobacterium tuberculosis*. *Journal of Biological Chemistry* 282, 9914–9922.

Lucarelli, D., Vasil, M.L., Meyer-Klaucke, W. and Pohl, E. (2008) The metal-dependent regulators FurA and FurB from *Mycobacterium Tuberculosis*. *International Journal of Molecular Science* 9, 1548–1560.

Ma, Z., Cowart, D.M., Ward, B.P., Arnold, R.J., DiMarchi, R.D., Zhang, L., George, G.N., Scott, R.A. and Giedroc, D.P. (2009) Unnatural amino acid substitution as a probe of the allosteric coupling pathway in a mycobacterial Cu(I) sensor. *Journal of the American Chemical Society* 131, 18044–18045.

Maciag, A., Dainese, E., Rodriguez, G.M., Milano, A., Provvedi, R., Pasca, M.R., Smith, I., Palu, G., Riccardi, G. and Manganelli, R. (2007) Global analysis of the *Mycobacterium tuberculosis* Zur (FurB) regulon. *Journal of Bacteriology* 189, 730–740.

McHugh, J.P., Rodriguez-Quinones, F., Abdul-Tehrani, H., Svistunenko, D.A., Poole, R.K., Cooper, C.E. and Andrews, S.C. (2003) Global iron-dependent gene regulation in *Escherichia coli*. A new mechanism for iron homeostasis. *Journal of Biological Chemistry* 278, 29478–29486.

Malik, S., Abel, L., Tooker, H., Poon, A., Simkin, L., Girard, M., Adams, G.J., Starke, J.R., Smith, K.C., Graviss, E.A., Musser, J.M. and Schurr, E. (2005) Alleles of the *NRAMP1* gene are risk factors for pediatric tuberculosis disease. *Proceedings of the National Academy of Sciences of the United States of America* 102, 12183–12188.

Manabe, Y.C., Saviola, B.J., Sun, L., Murphy, J.R. and Bishai, W.R. (1999) Attenuation of virulence in *Mycobacterium tuberculosis* expressing a constitutively active iron repressor. *Proceedings of the National Academy of Sciences of the United States of America* 96, 12844–12848.

Milano, A., Forti, F., Sala, C., Riccardi, G. and Ghisotti, D. (2001) Transcriptional regulation of *furA* and *katG* upon oxidative stress in *Mycobacterium smegmatis*. *Journal of Bacteriology* 183, 6801–6806.

Milano, A., Branzoni, M., Canneva, F., Profumo, A. and Riccardi, G. (2004) The *Mycobacterium tuberculosis* Rv2358-*furB* operon is induced by zinc. *Research in Microbiology* 155, 192–200.

Olakanmi, O., Schlesinger, L.S., Ahmed, A. and Britigan, B.E. (2004) The nature of extracellular iron influences iron acquisition by *Mycobacterium tuberculosis* residing within human macrophages. *Infection and Immunity* 72, 2022–2028.

Osman, D. and Cavet, J.S. (2010) Bacterial metal-sensing proteins exemplified by ArsR-SmtB family repressors. *Natural Products Reproduction* 27, 668–680.

Outten, C.E. and O'Halloran, T.V. (2001) Femtomolar sensitivity of metalloregulatory proteins controlling zinc homeostasis. *Science* 292, 2488–2492.

Papp-Wallace, K.M. and Maguire, M.E. (2006) Manganese transport and the role of manganese in virulence. *Annual Reviews in Microbiology* 60, 187–209.

Pennella, M.A., Shokes, J.E., Cosper, N.J., Scott, R.A. and Giedroc, D.P. (2003) Structural elements of metal selectivity in metal sensor proteins. *Proceedings of the National Academy of Sciences of the United States of America* 100, 3713–3718.

Pohl, E., Holmes, R.K. and Hol, W.G. (1999) Crystal structure of the iron-dependent regulator (IdeR) from *Mycobacterium tuberculosis* shows both metal binding sites fully occupied. *Journal of Molecular Biology* 285, 1145–1156.

Posey, J.E., Hardham, J.M., Norris, S.J. and Gherardini, F.C. (1999) Characterization of a manganese-dependent regulatory protein, TroR, from *Treponema pallidum*. *Proceedings of the National Academy of Sciences of the United States of America* 96, 10887–10892.

Pym, A.S., Domenech, P., Honore, N., Song, J., Deretic, V. and Cole, S.T. (2001) Regulation of catalase-peroxidase (KatG) expression, isoniazid sensitivity and virulence by *furA* of *Mycobacterium tuberculosis*. *Molecular Microbiology* 40, 879–889.

Quaeri, L.E., Sello, J., Keating, T.A., Weinreb, P.H. and Walsh, C.T. (1998) Identification of a *Mycobacterium tuberculosis* gene cluster encoding the biosynthetic enzymes for assembly of the virulence-conferring siderophore mycobactin. *Chemical Biology* 5, 631–645.

Riccardi, G., Milano, A., Pasca, M.R. and Nies, D.H. (2008) Genomic analysis of zinc homeostasis in *Mycobacterium tuberculosis*. *FEMS Microbiology Letters* 287, 1–7.

Rodriguez, G.M. (2006) Control of iron metabolism in *Mycobacterium tuberculosis*. *Trends in Microbiology* 14, 320–327.

Rodriguez, G.M. and Smith, I. (2006) Identification of an ABC transporter required for iron acquisition and virulence in *Mycobacterium tuberculosis*. *Journal of Bacteriology* 188, 424–430.

Rodriguez, G.M., Voskuil, M.I., Gold, B., Schoolnik, G.K. and Smith, I. (2002) *ideR*, An essential gene in *Mycobacterium tuberculosis*: role of IdeR in iron-dependent gene expression, iron metabolism, and oxidative stress response. *Infection and Immunity* 70, 3371–3381.

Rutherford, J.C., Cavet, J.S. and Robinson, N.J. (1999) Cobalt-dependent transcriptional switching by a dual-effector MerR-like protein regulates a cobalt-exporting variant CPx-type ATPase. *Journal of Biological Chemistry* 274, 25827–25832.

Sala, C., Forti, F., Di Florio, E., Canneva, F., Milano, A., Riccardi, G. and Ghisotti, D. (2003) *Mycobacterium tuberculosis* FurA autoregulates its own expression. *Journal of Bacteriology* 185, 5357–5362.

Schaible, U.E. and Kaufmann, S.H. (2004) Iron and microbial infection. *Nature Reviews Microbiology* 2, 946–953.

Schmitt, M.P., Predich, M., Doukhan, L., Smith, I. and Holmes, R.K. (1995) Characterization of an iron-dependent regulatory protein (IdeR) of *Mycobacterium tuberculosis* as a functional homolog of the diphtheria toxin repressor (DtxR) from *Corynebacterium diphtheriae*. *Infection and Immunity* 63, 4284–4289.

Schnappinger, D., Ehrt, S., Voskuil, M.I., Liu, Y., Mangan, J.A., Monahan, I.M., Dolganov, G., Efron, B., Butcher, P.D., Nathan, C. and Schoolnik, G.K. (2003) Transcriptional adaptation of *Mycobacterium tuberculosis* within macrophages: insights into the phagosomal environment. *Journal of Experimental Medicine* 198, 693–704.

Semavina, M., Beckett, D. and Logan, T.M. (2006) Metal-linked dimerization in the iron-dependent regulator from *Mycobacterium tuberculosis*. *Biochemistry* 45, 12480–12490.

Spagnolo, L., Toro, I., D'Orazio, M., O'Neill, P., Pedersen, J.Z., Carugo, O., Rotilio, G., Battistoni, A. and Djinovic-Carugo, K. (2004) Unique features of the *sodC*-encoded superoxide dismutase from *Mycobacterium tuberculosis*, a fully functional copper-containing enzyme lacking zinc in the active site. *Journal of Biological Chemistry* 279, 33447–33455.

Stoll, K.E., Draper, W.E., Kliegman, J.I., Golynskiy, M.V., Brew-Appiah, R.A., Phillips, R.K., Brown, H.K., Breyer, W.A., Jakubovics, N.S., Jenkinson, H.F., Brennan, R.G., Cohen, S.M. and Glasfeld, A. (2009) Characterization and structure of the manganese-responsive transcriptional regulator ScaR. *Biochemistry* 48, 10308–10320.

Talaat, A.M., Lyons, R., Howard, S.T. and Johnston, S.A. (2004) The temporal expression profile of *Mycobacterium tuberculosis* infection in mice. *Proceedings of the National Academy of Sciences of the United States of America* 101, 4602–4607.

Tan, B.H., Meinken, C., Bastian, M., Bruns, H., Legaspi, A., Ochoa, M.T., Krutzik, S.R., Bloom, B.R., Ganz, T., Modlin, R.L. and Stenger, S. (2006) Macrophages acquire neutrophil granules for antimicrobial activity against intracellular pathogens. *Journal of Immunology* 177, 1864–1871.

Techau, M.E., Valdez-Taubas, J., Popoff, J.F., Francis, R., Seaman, M. and Blackwell, J.M. (2007) Evolution of differences in transport function in Slc11a family members. *Journal of Biological Chemistry* 282, 35646–35656.

Turner, J.S., Glands, P.D., Samson, A.C. and Robinson, N.J. (1996) Zn^{2+}-sensing by the cyanobacterial metallothionein repressor SmtB: different motifs mediate metal-induced protein-DNA dissociation. *Nucleic Acids Research* 24, 3714–3721.

VanZile, M.L., Chen, X. and Giedroc, D.P. (2002a) Allosteric negative regulation of *smt* O/P binding of the zinc sensor, SmtB, by metal ions: a coupled equilibrium analysis. *Biochemistry* 41, 9776–9786.

VanZile, M.L., Chen, X. and Giedroc, D.P. (2002b) Structural characterization of distinct alpha3N and alpha5 metal sites in the cyanobacterial zinc sensor SmtB. *Biochemistry* 41, 9765–9775.

Wagner, D., Maser, J., Lai, B., Cai, Z., Barry, C.E. 3rd, Honer Zu Bentrup, K., Russell, D.G. and Bermudez, L.E. (2005a) Elemental analysis of *Mycobacterium avium-*, *Mycobacterium tuberculosis-*, and *Mycobacterium smegmatis*-containing phagosomes indicates pathogen-induced microenvironments within the host cell's endosomal system. *Journal of Immunology* 174, 1491–1500.

Wagner, D., Maser, J., Moric, I., Boechat, N., Vogt, S., Gicquel, B., Lai, B., Reyrat, J.M. and Bermudez, L. (2005b) Changes of the phagosomal elemental concentrations by *Mycobacterium tuberculosis* Mramp. *Microbiology* 151, 323–332.

Waldron, K.J. and Robinson, N.J. (2009) How do bacterial cells ensure that metalloproteins get the correct metal? *Nature Reviews Microbiology* 7, 25–35.

Wang, Y., Hemmingsen, L. and Giedroc, D.P. (2005) Structural and functional characterization of *Mycobacterium tuberculosis* CmtR, a PbII/CdII-sensing SmtB/ArsR metalloregulatory repressor. *Biochemistry* 44, 8976–8988.

Ward, S.K., Hoye, E.A. and Talaat, A.M. (2008) The global responses of *Mycobacterium tuberculosis* to physiological levels of copper. *Journal of Bacteriology* 190, 2939–2946.

Warner, D.F. and Mizrahi, V. (2007) The survival kit of *Mycobacterium tuberculosis*. *Nature Medicine* 13, 282–284.

Wessling-Resnick, M. (2010) Iron homeostasis and the inflammatory response. *Annual Reviews Nutrition* 30, 105–122.

White, A., Ding, X., vanderSpek, J.C., Murphy, J.R. and Ringe, D. (1998) Structure of the metal-ion-activated diphtheria toxin repressor/*tox* operator complex. *Nature* 394, 502–506.

White, C., Lee, J., Kambe, T., Fritsche, K. and Petris, M.J. (2009) A role for the ATP7A copper-transporting ATPase in macrophage bactericidal activity. *Journal of Biological Chemistry* 284, 33949–33956.

Wisedchaisri, G., Holmes, R.K. and Hol, W.G. (2004) Crystal structure of an IdeR-DNA complex reveals a conformational change in activated IdeR for base-specific interactions. *Journal of Molecular Biology* 342, 1155–1169.

Zhang, W., Shao, L., Weng, X., Hu, Z., Jin, A., Chen, S., Pang, M. and Chen, Z.W. (2005) Variants of the natural resistance-associated macrophage protein 1 gene (*NRAMP1*) are associated with severe forms of pulmonary tuberculosis. *Clinical Infectious Diseases* 40, 1232–1236.

13 *Salmonella* and the Host in the Battle for Iron

Elisa Deriu, Janet Z. Liu and Manuela Raffatellu*

13.1 Introduction

Eukaryotic and most prokaryotic cells require iron for survival and proliferation as a constituent of many prosthetic groups, including haem and iron-sulfur (Fe-S) proteins. In addition, several proteins use iron in other functional groups to carry out essential housekeeping roles for cellular metabolism. Plasma concentrations and adequate cellular iron levels are preserved through a strict regulation of absorption, transport storage and recycling (Andrews and Schmidt, 2007; De Domenico *et al.*, 2008). Both iron deficiency and iron overload cause severe disorders in humans (De Domenico *et al.*, 2008). Iron deficiency may result in microcytic anaemia, characterized by a reduction in haemoglobin concentration, red blood cell number and size. At the other end of the spectrum, severe iron overload as observed in haemochromatosis patients may result in damage of parenchymal organs, including the liver, heart and pancreas. Thus, iron homeostasis requires a fine modulation to ensure iron availability for essential functions, while preventing the toxicity associated with excess.

Iron is considered a key regulator of host–pathogen interactions. Hosts and pathogens have evolved sophisticated systems to compete for the acquisition and utilization of this essential micronutrient. Consequently, disturbances in iron availability and homeostasis have dramatic effects on pathogen virulence and on the host immune response to infection. In this chapter, we will describe the mechanisms by which the pathogen *Salmonella* acquires iron during infection.

13.2 Mammalian Iron Metabolism and Its Regulation

The majority of plasma iron is derived from the destruction of senescent erythrocytes through reticulo-endothelial (RE) macrophages, providing for 90% of erythropoietic need. Dietary iron (Fe^{3+}), which contributes around 10% of the total body iron content, is transported across the luminal surface of duodenal absorptive cells (enterocytes) by the transporter Nramp2 (also known as DMT1 and Slc11a2) after reduction to the ferrous state (Fe^{2+}) by an apical membrane ferrireductase (McKie *et al.*, 2001). The absorbed iron is released across the basolateral membrane of the enterocyte through the export protein ferroportin FPN (Slc40a1 or

* Corresponding author.

MTP1). Ceruloplasmin, and its intestinal homologue hephaestin, oxidize Fe^{2+} after cellular iron export for loading on to transferrin (Tf). Tf-bound iron is captured by the ubiquitously expressed Tf receptor (TfR1) and internalized by receptor-mediated endocytosis (Richardson, 2005). The low pH in the endosomal compartment allows iron released from transferrin to be incorporated into the iron-sequestration protein, ferritin. The relative levels of TfR1 and ferritin are regulated post-transcriptionally by iron-regulatory proteins 1 and 2 (IRP1 and IRP2), which sense the cytosolic iron concentration (Rouault, 2006). Iron overloading results in increased expression of the hepatocyte-derived peptide hormone, hepcidin, which binds FPN and mediates its internalization and degradation, resulting in the decreased export of cellular iron. Thus, regulation of FPN expression by hepcidin has a central role in iron homeostasis (Ganz, 2003).

13.2.1 Iron metabolism during infection and inflammation

As iron is frequently important for microbial growth and pathogenicity, mammals have developed mechanisms to withhold iron from microorganisms. The concentration of free iron is limited in the host as a mechanism of defence (Weinberg, 1984). In the absence of inflammation, iron is largely bound to transferrin. In the presence of inflammation, other iron-binding proteins play a central role in regulating iron availability. One of these proteins, lactoferrin, is a transferrin homologue that is secreted by glandular epithelial cells and is found in many mucosal secretions such as tears and saliva (Jenssen and Hancock, 2009). It is also present in secondary granules of neutrophils and is released at sites of inflammation (Masson *et al.*, 1969). Lactoferrin has a higher affinity for ferric iron at lower pH, as found in an inflammatory state. Thus, both transferrin and lactoferrin play a role in reducing the availability of free iron for microorganisms.

The host can also reduce iron availability through the induction of hepcidin (Nemeth *et al.*, 2004b). Hepcidin mRNA is upregulated strongly in the liver after exposure of hepatocytes to either LPS or IL-6 (Nemeth *et al.*, 2003, 2004a). Hepcidin inhibits iron uptake in the duodenum and iron release from macrophages in the spleen, resulting in anaemia of inflammation (previously called anaemia of chronic disease) (Ganz, 2003). Thus, anaemia of inflammation is likely a mechanism of host defence, with the goal of reducing iron availability to invading pathogens.

Another protein that recently has been found to bind iron is the neutrophil gelatinase-associated lipocalin (NGAL, also called lipocalin 2 or siderocalin in mammals) (Goetz *et al.*, 2002; Yang *et al.*, 2002). Lipocalins are a family of small secreted proteins that bind to low molecular weight ligands (Flower, 1996; Akerstrom *et al.*, 2000) and play diverse molecular roles in many biological processes including kidney development, olfaction, pheromone transport, retinol transport, prostaglandin synthesis, modulation of cell growth and metabolism, regulation of the immune response and animal behaviour (Flower, 1996). Human NGAL was discovered initially as a component of neutrophil granules, but was found subsequently to be expressed in epithelial cells in response to inflammatory signals (Kjeldsen *et al.*, 1993). The murine NGAL ortholog 24p3 (uterocalin) has been identified as a protein induced in response to various proliferative signals and is highly expressed in uterine luminal fluids and epithelial cells. 24p3 does not interact with iron directly, but instead iron is bound through 24p3-associated small molecular weight siderophores (iron-binding mol-ecules), as described in detail later (Goetz *et al.*, 2002; Yang *et al.*, 2002).

24p3 has also been implicated in intracellular trafficking of iron and apoptosis (Devireddy *et al.*, 2001). Internalization of iron-loaded 24p3 (holo-24p3) and its receptor, 24p3R, leads to the uptake of iron from the siderophore-iron complex in a process that requires acidification of an endosomal-like compartment and the transport of the released iron via DMT1. The increase in intracellular iron regulates TfR1 expression negatively, while leading to an increase in ferritin levels without promoting apoptosis.

In contrast, the endocytosis of Apo-24p3 leads to a depletion of iron from the cell and upregulation of the proapoptotic molecule, Bim, resulting in apoptosis (Devireddy *et al.*, 2005).

13.3 Iron Acquisition by *Salmonella*

Commensal bacteria and pathogens like *Salmonella* have evolved multiple mechanisms directly to acquire iron selectively in its dicationic form or when bound to exogenous or endogenous siderophores. Siderophores are low molecular weight, high-affinity iron ligands with a variety of structures, although most may be classified as hydroxamates or catechols (Leong and Neilands, 1976). *Salmonella* can uptake iron from several siderophores (or siderochromes) (Luckey *et al.*, 1972). Enterochelin (also known as enterobactin) is a highly conserved enterobacterial siderophore made from a cyclic trimer of 2,3-dihydroxybenzoyl-serine (DHB-serine). Enterochelin is synthesized via the expression of the *entCEBA*, *entD* and *entF* genes in the enterochelin locus. Synthesis begins when chorismic acid is converted to 2,3-dihydroxybenzoate (DHB) via the activities of EntC, an isochorismate synthatase, EntB, a synthetase, and EntA, a dehydrogenase (Raymond *et al.*, 2003). DHB is then linked to L-serine and through the activities of EntD, EntE and EntF eventually made into the completed siderophore. Secretion of enterochelin is achieved through EntS, an efflux pump belonging to the major facilitator superfamily (Furrer *et al.*, 2002).

Enterochelin is a strong chelator of ferric iron. Once bound to iron, siderophores are transported into the cell by outer membrane proteins such as FhuA for ferrichrome and FepA for enterochelin. The FepA receptor consists of a 22-strand transmembrane β-barrel domain and an N-terminal plug domain made up of 148 residues where enterochelin is bound in its iron-laden state (Usher *et al.*, 2001). The N-terminal plug blocks iron-laden enterochelin from entering the cell until a conformational change in the plug structure allows the siderophore to pass through. The energy necessary to transport

siderophores into the periplasm is provided by a proton motive force generated by ABC (ATP-binding cassette) transporters on the inner membrane. The force generated by the ABC transporter is conducted to the TonB complex, consisting of TonB, ExbB and ExbD (Larsen *et al.*, 1999). The N-terminal domain of enterochelin possesses a TonB box that binds to the β-barrel domain of TonB (Buchanan *et al.*, 1999). This interaction causes a conformational change in the FepA plug, allowing the iron-laden siderophore to enter the periplasmic space. There, FepB binds and transfers the enterochelin-iron complex to the transmembrane pore made up of FepD and FepG (Shea and McIntosh, 1991). Energy for transport through the pore is provided by the FepC cytoplasmic ATPase. Upon internalization, iron is released from the iron-laden enterochelin complex via Fes-mediated hydrolysis of enterochelin's trilactone ring backbone, with Fe^{3+} subsequently reduced to Fe^{2+}.

Mutants in the enterochelin synthetic pathway have a growth defect in low iron media; however, they can be rescued by the addition of ferrichrome, a cyclic hexapeptide containing ferric iron complexed to three hydroxamate groups produced by smut fungi (Pollack *et al.*, 1970; Luckey *et al.*, 1972). To this end, it is apparent that *Salmonella* has developed a scavenger ability to utilize siderophores that are secreted into the environment by other microorganisms.

While all *Salmonella* isolates produce enterochelin, some isolates also produce the siderophore, aerobactin (Rabsch and Reissbrodt, 1988). Aerobactin production has been linked to cases of nosocomial bacteraemia (Kingsley *et al.*, 1995). Host-adapted serotypes like *Salmonella enterica* serovar Typhi and strains isolated from patients with inflammatory diarrhoea do not produce aerobactin. Aerobactin, unlike enterochelin, is a hydroxamate consisting of 6-(*N*-acetyl-*N*-hydroxyamino)-2-aminohexanoic acid conjugated to citric acid (Crosa, 1989). As in *Escherichia coli* and *Enterobacter aerogenes*, the aerobactin uptake system in *Salmonella* is plasmid encoded (Colonna *et al.*, 1985). Aerobactin production and uptake is mediated by an operon encoding for five proteins: firstly, IucC (a synthetase), IucB (an acetylase), IucC (a synthetase) and IucD (an oxygenase) are involved in aerobactin synthesis, while IutA is the outer membrane receptor that binds iron-laden aerobactin. Once taken up into the periplasm, aerobactin depends on the FhuBCD system and the TonB complex to cross into the cytoplasm. Both enterochelin and aerobactin have the ability to chelate iron from transferrin. While enterochelin has a higher stability constant (10^{52}) than aerobactin (10^{23}), aerobactin exhibits a faster rate of iron removal from transferrin in serum albumin (Konopka *et al.*, 1982). This characteristic may account for the increased virulence of strains expressing aerobactin during septicaemia, and thus for the acquisition of aerobactin by pathogens causing systemic infections (Kingsley *et al.*, 1995).

Salmochelin was discovered initially in the 1970s as a substance which, when included in diet of mice in the form of dried hen egg white, protected mice from infection compared to mice fed a synthetic diet (Schneider, 1967). It was later recognized that salmochelin was a C-glucosylated derivative of enterochelin and that the synthesis, transport and utilization of salmochelin was orchestrated by five genes (*iroBCDE* and *iroN*) in the *iroA* locus (Muller *et al.*, 2009). IroB, a C-glucosyltransferase, adds glucosyl groups to enterochelin to generate salmochelin (Fischbach *et al.*, 2005). Once synthesized, salmochelin is transported out of the bacterium through the ATP-driven IroC ABC transporter (Crouch *et al.*, 2008). IroN, which possesses 52% homology to FepA, is the receptor that recognizes and imports iron-laden salmochelin into the periplasm (Bäumler *et al.*, 1998; Hantke *et al.*, 2003). With such a high degree of similarity, it comes as no surprise that the uptake process of salmochelin is very similar to that of enterochelin. After IroN-mediated transport into the periplasm, salmochelin is linearized by the IroE esterase, then transported into the cytoplasm via the FepBCDG transport system. Once in the cytoplasm, iron-bound salmochelin is broken down further by another esterase, IroD, to release the iron cargo (Lin *et al.*, 2005). In contrast to the promiscuity of the FepBCDG transport system, IroN and FepA exhibit greater substrate specificity (Rabsch *et al.*, 1999).

13.4 Regulation of Iron Uptake Systems

Bacterial synthesis of siderophores is regulated in response to intracellular iron levels. The master siderophore regulator was first discovered in a mutant strain of *S. enterica* serovar Typhimurium, which expressed high levels of enterochelin and exhibited constitutive uptake of iron-bound enterochelin and ferrichrome (Ernst *et al.*, 1978). Known as Fur, for Fe uptake regulation, this protein has been shown to be highly expressed in a variety of bacterial species, including *E. coli*, *Bacillus subtilis*, *Helicobacter pylori*, *Pseudomonas aeruginosa* and *Yersinia pestis*, among many others (Carpenter *et al.*, 2009). Fur is a 15 kDa protein that dimerizes when iron is readily available in the bacterial cell. First described in *E. coli* (de Lorenzo *et al.*, 1987), dimeric Fur binds to a consensus DNA sequence and prevents RNA polymerase from accessing the promoter of the Fur-regulated gene, thereby preventing transcription.

Fur regulates a number of iron-uptake genes, either directly or indirectly (Stojiljkovic *et al.*, 1994; Tsolis *et al.*, 1995). The IroA locus originally was identified in *S.* Typhi by screening for Fur-regulated genes absent from *E. coli* (Bäumler *et al.*, 1996). In *E. coli* the expression of several iron-binding enzymes and iron-storage molecules is downregulated by RyhB, a small non-coding RNA whose transcription is regulated negatively by Fur and thus derepressed when iron is limiting (Masse and Gottesman, 2002). In *Salmonella* it has been shown that regulation of the *sitABCD* transport system in response to iron is regulated directly by iron-bound Fur (Ikeda *et al.*, 2005). While the major metal transported by *sitABCD* is manganese, mutations in the putative binding sites for Fur in *Salmonella* result in an increased iron concentration inside the bacteria. In addition to regulating iron acquisition, Fur regulates expression of the *Salmonella* pathogenicity island I invasion genes through activation of HilD (Ellermeier and Slauch, 2008).

While Fur traditionally has been known for its role in suppressing gene transcription in environments where iron is abundant, in *H. pylori* it has also been found to act as an apo-protein (suppressing gene expression in the absence of iron), as well as functioning as a gene activator in both its iron-bound and apo-protein forms (Ernst *et al.*, 2005). In *Salmonella*, Foster and Hall discovered that six gene products were downregulated in a *fur* mutant in a low iron environment, indicating they were activated by apo-Fur (Foster and Hall, 1992). Thus, while Fur is known primarily to be a negative regulator, in some instances it may also function as a positive regulator, either directly or indirectly.

13.5 Role of Iron Uptake During *Salmonella* Infection

The concentration of iron in the host changes susceptibility to bacterial infections, including *Salmonella* infections. Several studies have investigated the role of iron uptake in *Salmonella* pathogenesis by using several mouse models with different susceptibility to infection. One of the most important genetic factors that determine susceptibility of mice to intracellular pathogens is Nramp1 (Slc11a1). Nramp1 is a phagosomal divalent metal ion transporter that represents a critical host defence mechanism against *Salmonella* by limiting its replication, though the mechanism(s) by which this is accomplished remain unclear (Wyllie *et al.*, 2002). Mice lacking a functional Nramp1 transporter are highly susceptible to *Salmonella* bacteraemia (Bellamy, 1999). It has been suggested that this protein pumps iron and manganese out of the phagosomal compartment, reducing the growth of intraphagosomal pathogens (Fortier *et al.*, 2005; Cellier *et al.*, 2007; Huynh and Andrews, 2008). It has also been proposed that Nramp1 limits iron availability to *Salmonella* while enhancing phagocytic function, thereby reducing growth of this pathogen by two different mechanisms (Nairz *et al.*, 2009a). Below, we discuss the outcome

of *Salmonella* infection in several mouse models, noting the Nramp1 allele status of the mice used in each study.

Clinical observations prompted investigations into the role of iron acquisition in the pathogenesis of *Salmonella*. Increased susceptibility to *Salmonella* infection is observed in patients with iron overload (beta-thalassaemia) or iron deficiency (sickle cell anaemia). Desferrioxamine (DF) is a bacterial siderophore widely used as an iron chelator in patients with iron overload. However, normal and iron-overloaded Swiss Webster (Nramp1+) mice treated with DF are more susceptible to S. Typhimurium infections because DF can be used by S. Typhimurium as an iron source, thereby promoting growth of this pathogen (Jones *et al.*, 1977). In addition, infected and iron-injected C3D2F1 (Nramp1+) mice succumb more rapidly to S. Typhimurium infection than infected and saline-injected mice, with acute iron overload resulting in increased S. Typhimurium growth in all tissue compared to mock treated mice (Sawatzki *et al.*, 1983). Taken together, these studies demonstrate that excess iron abolishes the normal clearance of *Salmonella* and enhances its growth. This is in agreement with clinical studies indicating that the host responds to infection by limiting iron availability to pathogens (Jurado, 1997), including *Salmonella*.

When human volunteers were infected experimentally with S. Typhi, the aetiologic agent of typhoid fever, serum iron and zinc concentrations became depressed just before the onset of fever, subsequently returning to normal after several days of therapy with chloramphenicol (Pekarek *et al.*, 1975). These data were also reproduced in rhesus macaques, likely indicating that the availability of trace metabolites was altered in response to infections as a mechanism of host defence. In spite of the iron-limiting conditions, *Salmonella* must acquire iron to grow in the host. Because siderophores are secreted during iron starvation *in vitro*, several studies have looked at siderophore-mediated iron uptake during *Salmonella* infection, with somewhat inconclusive results until recently.

Several studies have demonstrated that iron-uptake mutants are attenuated *in vivo*.

Growth of *S.* Typhi iron-uptake mutants is inhibited by human serum, and these mutants are also attenuated in an *in vivo* mucin mouse model (CFW mice, which are outbred Swiss Webster and Nramp1+) in which *S.* Typhi is re-suspended in mucin to allow colonization of mouse peritoneal macrophages (Furman *et al.*, 1994). An *S.* Typhimurium mutant in enterochelin synthesis exhibited an increase in the LD_{50} when injected intraperitoneally in CFW mice, a defect which was remedied in a dose-dependent fashion by injecting enterochelin (Yancey *et al.*, 1979). While these studies point towards a role of enterochelin (and salmochelin) in the pathogenesis of *Salmonella* infection, Benjamin *et al.* found that several inbred and outbred mice (likely with mixed Nramp1 allele status) infected intravenously with enterochelin mutants still developed mouse typhoid. This led to the hypothesis that enterochelin might not be necessary once the infection became intracellular and might be more important for the extracellular phase of the infection (Benjamin *et al.*, 1985). In agreement with this, deficiency in iron acquisition through sidero-phores resulting from a *tonB* deletion had no effect on virulence when *S.* Typhimurium was injected intraperitoneally in Nramp1-Balb/c mice. However, dissemination of the *tonB* mutant to the mesenteric lymph nodes was reduced when this strain was administered by oral gavage (Tsolis *et al.*, 1996).

Recently, Crouch *et al.* have looked at the relative role of enterochelin and salmochelin in pathogenesis of *S.* Typhimurium infection in a typhoid model by using C3H/HeN Nramp1+ mice. In this study, mutants in *entC* and *entB* (which do not produce either enterochelin or salmochelin) or in *iroB* (which produces enterochelin but does not produce salmochelin) are attenuated during systemic *S.* Typhimurium infection (Crouch *et al.*, 2008). In addition, it was shown in a mouse model of inflammatory diarrhoea that an *S.* Typhimurium mutant in the salmochelin receptor *iroN* was attenuated compared to wild type during either single or competitive infection in Nramp1-C57BL/6 mice (Raffatellu *et al.*, 2009). Although the interplay between Nramp1 and iron uptake is not understood

completely, these studies indicate that salmochelin-mediated iron uptake promotes *S.* Typhimurium growth *in vivo* in both a typhoid model and a gastroenteritis model of infection.

13.6 Lipocalin-2 and Iron Withholding Responses

As described earlier, in mammals iron is bound largely by transferrin and, if neutrophils are recruited at the site of infection, lactoferrin. However, all Entero-bacteriaceae, including commensal *E. coli*, secrete the siderophore enterochelin in response to iron starvation (Rutz *et al.*, 1991). While this has been known since the 1970s, it was only in 2000 that a protein produced by mammals was found to bind to enterochelin.

Known as lipocalin-2 in humans, Goetz *et al.* have shown that this host protein co-purifies with a bacterial chromophore (Goetz *et al.*, 2002). The original lipocalin-2 structure was obtained in baculovirus cells and lacked the 'third molecule' present in the lipocalin-2 purified from *E. coli* (Goetz *et al.*, 2000). Atomic absorption confirmed that the red chromophore bound to lipocalin-2 was iron, and the bound molecule was identified as enterochelin. As a result of binding to enterochelin, lipocalin-2 was shown to be bacteriostatic for *E. coli in vitro* (Goetz *et al.*, 2002). A few years later, a seminal study from Flo and colleagues demonstrated that lipocalin-2 represented a new component of the innate immune system and that its expression was activated *in vivo* on stimulation of Toll-like receptors (Flo *et al.*, 2004). Lipocalin-2 expression is protective against infection with commensal *E. coli* and binds to certain catecholate-type (but not hydroxamate-type) siderophores (Flo *et al.*, 2004; Berger *et al.*, 2006).

In spite of the similarity between salmochelin and enterochelin, lipocalin-2 does not bind salmochelin (Fischbach *et al.*, 2006). As a consequence, many pathogenic Gram-negative bacteria harbouring the *iroBCDE iroN* gene cluster (including pathogenic *E. coli*, *Salmonella* and *Klebsiella*) are resistant to lipocalin-2 and evade this component of the innate immune response (Caza *et al.*, 2008; Crouch *et al.*, 2008; Bachman *et al.*, 2009). An *S.* Typhimurium *iroN* mutant is outcompeted by wild type in the inflamed gut, but not in the absence of inflammation or lipocalin-2 (Raffatellu *et al.*, 2009). Thus, the siderophore salmochelin promotes *Salmonella* growth in iron-limiting condition and evasion of lipocalin-2-mediated iron withholding. During murine nasal colonization, *K. pneumoniae* is able to circumvent the effect of lipocalin-2 with both yersiniabactin and salmochelin (Bachman *et al.*, 2009). Long fatty acid chain carboxymycobactins produced by *Mycobacteria* (Holmes *et al.*, 2005) and petrobactin made by *B. anthracis* (Abergel *et al.*, 2006) also cannot be bound by lipocalin-2. Thus, while lipocalin-2 limits growth of a subset of bacteria, several pathogens have evolved to evade lipocalin-2 activity by acquiring iron with additional siderophores.

Of interesting note, it has been shown recently that lipocalin-2 binds to a mammalian siderophore, 2,5-dihydroxybenzoic acid (2,5-DHBA), which is similar to the iron-binding component of enterochelin (2,3-DHBA) (Bao *et al.*, 2010; Devireddy *et al.*, 2010). Knockdown of BDH2, an enzyme involved in 2,5-DHBA synthesis, results in reduced iron import to mitochondria (Devireddy *et al.*, 2010). Mitochondria have thus retained a similar mechanism as bacteria to acquire iron, thereby providing a link between iron transport in mammalian cells and host defence mechanisms.

13.6.1 Regulation of lipocalin-2 during infections at mucosal surfaces

During the course of infections initiated by a number of different microorganisms, Toll-like receptors on immune cells stimulate the transcription, translation and secretion of lipocalin-2; lipocalin-2 represents one of the most highly induced antimicrobial responses against *S.* Typhimurium infection. When *S.* Typhimurium is injected into ligated ileal loops of monkeys, lipocalin-2 expression is induced as early as 5 h post-infection, together with other pro-inflammatory genes.

One of the major characteristics of *Salmonella* and other closely related pathogens that cause inflammatory diarrhoea is their capability to invade host cells and survive in an intracellular environment. With regards to *Salmonella*, the invasion process is mediated by the type III secretion system (T3SS) encoded within *Salmonella* pathogenicity island 1 (SPI-1), while intracellular survival is dependent on the T3SS encoded on *Salmonella* pathogenicity island 2 (SPI-2) (reviewed in Zhou and Galan, 2001; Waterman and Holden, 2003). After invasion, *S.* Typhimurium pathogen-associated molecular patterns (PAMPs) such as lipopolysaccharide (LPS) and flagellin are detected by pattern recognition receptors of the host innate immune system (Takeuchi *et al.*, 1999; Tapping *et al.*, 2000; Gewirtz *et al.*, 2001; Hayashi *et al.*, 2001; Reed *et al.*, 2002). In response to these signals, epithelial cells and macrophages secrete pro-inflammatory cytokines and chemokines. Although the direct interaction of *S.* Typhimurium with host cells is important for initiating the host response, a large subset of pro-inflammatory responses is triggered in cells with no direct contact with bacteria.

Paracrine amplification loops involving interactions between antigen-presenting cells (APCs) and T cells have been described recently. Infected APCs secrete the cytokine interleukin (IL)-18, which induces interferon-g (IFN-γ) production by T cells (Mastroeni *et al.*, 1999; Iwai *et al.*, 2008). IL-18 secretion is dependent on activation of the inflammasome, which is itself dependent on the recognition of flagellin by the host receptor Nlrc4 (Ipaf) during *S.* Typhimurium infection (Franchi *et al.*, 2006; Miao *et al.*, 2006). APCs also secrete the cytokine IL-23, which in turn stimulates several subsets of T cells, including $\alpha\beta$ and $\gamma\delta$ T cells, to produce the cytokines IL-17 and IL-22 (Ye *et al.*, 2001a,b; Happel *et al.*, 2003, 2005; Aujla *et al.*, 2008; Godinez *et al.*, 2008, 2009). Receptors for IL-17, IL-22 and IFN-γ are found in epithelial cells, which in turn respond to stimulation with these cytokines by secreting chemokines, mucins and molecules with antimicrobial activity, including lipocalin-2 (Bevins, 2004; Muller *et al.*, 2005).

Lipocalin-2 is highly induced in the intestinal lumen during *S.* Typhimurium infection, and IL-17 and IL-22 have a synergetic effect in controlling its production in the inflamed intestine (Raffatellu *et al.*, 2009). Intestinal epithelial cells are a major source of lipocalin-2 *in vivo* during *S.* Typhimurium infection and *in vitro* upon stimulation with IL-17 and IL-22 (Raffatellu *et al.*, 2008, 2009). Similarly, IL-17 and/or IL-22 induce lipocalin-2 expression in lung epithelial cells and osteoblasts (Shen *et al.*, 2006; Aujla *et al.*, 2008). Lung and nasal infection with *K. pneumoniae* (Aujla *et al.*, 2008; Bachman *et al.*, 2009; Chan *et al.*, 2009), oral cavity infection with *Candida albicans* (Conti *et al.*, 2009), nasal infection with *Streptococcus pneumoniae* and *Haemophilus influenzae* (Nelson *et al.*, 2005) and gastric infection with *H. pylori* (Hornsby *et al.*, 2008) also result in the upregulation of lipocalin-2. Macrophages are also a source of lipocalin-2 during bacterial infections. It was shown recently that both lung macrophages and epithelial cells were a source of lipocalin-2 in response to

Mycobacterium tuberculosis infection (Saiga *et al.*, 2008). In addition, both IFN-γ stimulation and *S.* Typhimurium invasion induce lipocalin-2 expression in RAW264.7 macrophages (Nairz *et al.*, 2007, 2008). Mice lacking Hfe (a non-classical MHC-I molecule mutated in hereditary haemochromatosis) are protected from *S.* Typhimurium septicaemia because of enhanced production of lipocalin-2, which reduced the availability of iron to *S.* Typhimurium in macrophages (Nairz *et al.*, 2009b). In addition to epithelial cells and macrophages, neutrophils accumulating at the site of an infection in the gut and intestinal lumen can release lipocalin-2 (Carlson *et al.*, 2002). Various aspects of iron metabolism and infection are shown in Fig. 13.1.

13.7 Conclusions

It has been known for over 30 years that iron uptake is fundamental for growth of most bacteria and that the host responds to bacterial infection by restricting availability of this

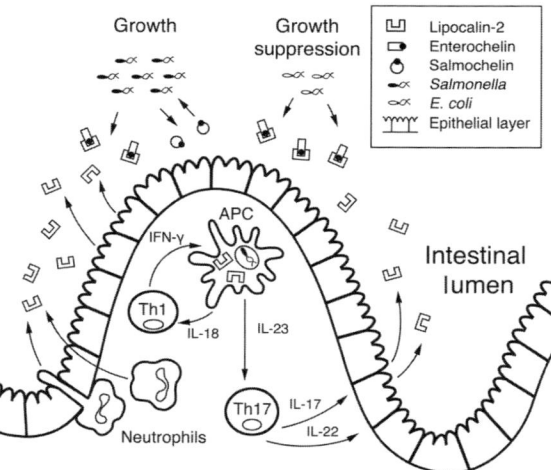

Fig. 13.1. Current model of *S.* Typhimurium iron uptake in the inflamed gut. *S.* Typhimurium invades the intestinal mucosa, where it resides in macrophages. Infected macrophages activated by IFN-γ produce lipocalin-2. Activated Th17 cells secrete IL-17 and IL-22, which stimulate intestinal epithelial cells to release lipocalin-2. Neutrophils recruited to the mucosa are an additional source of lipocalin-2. Lipocalin-2 binds to enterochelin, thereby reducing growth of commensal *E. coli*. In contrast, lipocalin-2 does not bind to salmochelin, thereby favouring growth of *S.* Typhimurium and other bacteria that can utilize this siderophore to acquire iron.

essential nutrient. What has become clearer in recent years is that pathogens have exploited inflammation and host-induced iron starvation to gain an edge in competing with susceptible species including the commensal microbiota (Lupp *et al.*, 2007; Stecher *et al.*, 2007); pathogens like *S.* Typhimurium thrive in the inflamed gut while the microbiota is suppressed by the host response. High levels of colonization ultimately result in greater host-to-host transmission via the faecal–oral route (Lawley *et al.*, 2008). Lipocalin-2 mediated iron withholding promotes the growth of *S.* Typhimurium because this pathogen can acquire iron through additional siderophores like salmochelin (Raffatellu *et al.*, 2009). Thus, not only do pathogens like *S.* Typhimurium respond to iron withholding by acquiring iron via different mechanisms, but they also take advantage of iron withholding for colonizing a niche while suppressing growth of competing microbes. To this end, future studies are necessary to identify how pathogens utilize iron starvation to promote their own growth and win the battle for iron, especially against their competitors.

Acknowledgements

We would like to thank Sean-Paul Nuccio, Andreas J. Bäumler and Renee M. Tsolis for helpful discussions. Work in M. Raffatellu's laboratory is supported by NIH grants AI083619 and AI083663 and by an IDSA ERF/ NIFID Astellas Young Investigator Award. JZL is supported by NIH Immunology Research Training Program grant NIH T32 AI60573.

References

Abergel, R.J., Wilson, M.K., Arceneaux, J.E., Hoette, T.M., Strong, R.K., Byers, B.R. and Raymond, K.N. (2006) Anthrax pathogen evades the mammalian immune system through stealth siderophore production. *Proceedings of the National Academy of Sciences of the United States of America* 103, 18499–18503.

Akerstrom, B., Flower, D.R. and Salier, J.P. (2000)

Lipocalins: unity in diversity. *Biochimical et Biophysica ACTA* 1482, 1–8.

Andrews, N.C. and Schmidt, P.J. (2007) Iron homeostasis. *Annual Review of Physiology* 69, 69–85.

Aujla, S.J., Chan, Y.R., Zheng, M., Fei, M., Askew, D.J., Pociask, D.A., Reinhart, T.A., Mcallister, F., Edeal, J., Gaus, K., Husain, S., Kreindler, J.L., Dubin, P.J., Pilewski, J.M., Myerburg, M.M., Mason, C.A., Iwakura, Y. and Kolls, J.K. (2008) IL-22 mediates mucosal host defense against Gram-negative bacterial pneumonia. *Nature Medicine* 14, 275–281.

Bachman, M.A., Miller, V.L. and Weiser, J.N. (2009) Mucosal lipocalin-2 has pro-inflammatory and iron-sequestering effects in response to bacterial enterobactin. *PLoS Pathogens* 5, e1000622.

Bao, G., Clifton, M., Hoette, T.M., Mori, K., Deng, S.X., Qiu, A., Viltard, M., Williams, D., Paragas, N., Leete, T., Kulkarni, R., Li, X., Lee, B., Kalandadze, A., Ratner, A.J., Pizarro, J.C., Schmidt-Ott, K.M., Landry, D.W., Raymond, K.N., Strong, R.K. and Barasch, J. (2010) Iron traffics in circulation bound to a siderocalin (Ngal)-catechol complex. *Nature Chemical Biology* 6, 602–609.

Bäumler, A.J., Tsolis, R.M., Van Der Velden, A.W., Stojiljkovic, I., Anic, S. and Heffron, F. (1996) Identification of a new iron regulated locus of *Salmonella typhi. Gene* 183, 207–213.

Bäumler, A.J., Norris, T.L., Lasco, T., Voight, W., Reissbrodt, R., Rabsch, W. and Heffron, F. (1998) IroN, a novel outer membrane siderophore receptor characteristic of *Salmonella enterica. Journal of Bacteriology* 180, 1446–1453.

Bellamy, R. (1999) The natural resistance-associated macrophage protein and susceptibility to intracellular pathogens. *Microbes and Infection* 1, 23–27.

Benjamin, W.H. Jr, Turnbough, C.L. Jr, Posey, B.S. and Briles, D.E. (1985) The ability of *Salmonella typhimurium* to produce the siderophore enterobactin is not a virulence factor in mouse typhoid. *Infection and Immunity* 50, 392–397.

Berger, T., Togawa, A., Duncan, G.S., Elia, A.J., You-Ten, A., Wakeham, A., Fong, H.E., Cheung, C.C. and Mak, T.W. (2006) Lipocalin 2-deficient mice exhibit increased sensitivity to *Escherichia coli* infection but not to ischemia-reperfusion injury. *Proceedings of the National Academy of Sciences of the United States of America* 103, 1834–1839.

Bevins, C.L. (2004) The Paneth cell and the innate immune response. *Current Opinion in Gastro-enterology* 20, 572–580.

Buchanan, S.K., Smith, B.S., Venkatramani, L., Xia, D., Esser, L., Palnitkar, M., Chakraborty, R., Van Der Helm, D. and Deisenhofer, J. (1999) Crystal structure of the outer membrane active transporter FepA from *Escherichia coli*. *Nature Structural Biology* 6, 56–63.

Carlson, M., Raab, Y., Seveus, L., Xu, S., Hallgren, R. and Venge, P. (2002) Human neutrophil lipocalin is a unique marker of neutrophil inflammation in ulcerative colitis and proctitis. *Gut* 50, 501–506.

Carpenter, B.M., Whitmire, J.M. and Merrell, D.S. (2009) This is not your mother's repressor: the complex role of fur in pathogenesis. *Infection and Immunity* 77, 2590–2601.

Caza, M., Lepine, F., Milot, S. and Dozois, C.M. (2008) Specific roles of the *irobCDEN* genes in virulence of an avian pathogenic *Escherichia coli* O78 strain and in production of salmochelins. *Infection and Immunity* 76, 3539–3549.

Cellier, M.F., Courville, P. and Campion, C. (2007) Nramp1 phagocyte intracellular metal withdrawal defense. *Microbes and Infection* 9, 1662–1670.

Chan, Y.R., Liu, J.S., Pociask, D.A., Zheng, M., Mietzner, T.A., Berger, T., Mak, T.W., Clifton, M.C., Strong, R.K., Ray, P. and Kolls, J.K. (2009) Lipocalin 2 is required for pulmonary host defense against *Klebsiella* infection. *Journal of Immunology* 182, 4947–4956.

Colonna, B., Nicoletti, M., Visca, P., Casalino, M., Valenti, P. and Maimone, F. (1985) Composite IS1 elements encoding hydroxamate-mediated iron uptake in FIme plasmids from epidemic *Salmonella* spp. *Journal of Bacteriology* 162, 307–316.

Conti, H.R., Shen, F., Nayyar, N., Stocum, E., Sun, J.N., Lindemann, M.J., Ho, A.W., Hai, J.H., Yu, J.J., Jung, J.W., Filler, S.G., Masso-Welch, P., Edgerton, M. and Gaffen, S.L. (2009) Th17 cells and IL-17 receptor signaling are essential for mucosal host defense against oral candidiasis. *Journal of Experimental Medicine* 206, 299–311.

Crosa, J.H. (1989) Genetics and molecular biology of siderophore-mediated iron transport in bacteria. *Microbiology Reviews* 53, 517–530.

Crouch, M.L., Castor, M., Karlinsey, J.E., Kalhorn, T. and Fang, F.C. (2008) Biosynthesis and IroC-dependent export of the siderophore salmochelin are essential for virulence of *Salmonella enterica* serovar Typhimurium. *Molecular Microbiology* 67, 971–983.

De Domenico, I., Mcvey Ward, D. and Kaplan, J. (2008) Regulation of iron acquisition and storage: consequences for iron-linked disorders. *Nature Reviews Molecular and Cellular Biology* 9, 72–81.

De Lorenzo, V., Wee, S., Herrero, M. and Neilands, J.B. (1987) Operator sequences of the aerobactin operon of plasmid ColV-K30 binding the ferric uptake regulation (Fur) repressor. *Journal of Bacteriology* 169, 2624–2630.

Devireddy, L.R., Teodoro, J.G., Richard, F.A. and Green, M.R. (2001) Induction of apoptosis by a secreted lipocalin that is transcriptionally regulated by IL-3 deprivation. *Science* 293, 829–834.

Devireddy, L.R., Gazin, C., Zhu, X. and Green, M.R. (2005) A cell-surface receptor for lipocalin 24p3 selectively mediates apoptosis and iron uptake. *Cell* 123, 1293–1305.

Devireddy, L.R., Hart, D.O., Goetz, D.H. and Green, M.R. (2010) A mammalian siderophore synthesized by an enzyme with a bacterial homolog involved in enterobactin production. *Cell* 141, 1006–1017.

Ellermeier, J.R. and Slauch, J.M. (2008) Fur regulates expression of the *Salmonella* pathogenicity island 1 type III secretion system through HilD. *Journal of Bacteriology* 190, 476–486.

Ernst, F.D., Bereswill, S., Waidner, B., Stoof, J., Mader, U., Kusters, J.G., Kuipers, E.J., Kist, M., Van Vliet, A.H. and Homuth, G. (2005) Transcriptional profiling of *Helicobacter pylori* Fur- and iron-regulated gene expression. *Microbiology* 151, 533–546.

Ernst, J.F., Bennett, R.L. and Rothfield, L.I. (1978) Constitutive expression of the iron-enterochelin and ferrichrome uptake systems in a mutant strain of *Salmonella typhimurium*. *Journal of Bacteriology* 135, 928–934.

Fischbach, M.A., Lin, H., Liu, D.R. and Walsh, C.T. (2005) In vitro characterization of IroB, a pathogen-associated C-glycosyltransferase. *Proceedings of the National Academy of Sciences of the United States of America* 102, 571–576.

Fischbach, M.A., Lin, H., Zhou, L., Yu, Y., Abergel, R.J., Liu, D.R., Raymond, K.N., Wanner, B.L., Strong, R.K., Walsh, C.T., Aderem, A. and Smith, K.D. (2006) The pathogen-associated *iroA* gene cluster mediates bacterial evasion of lipocalin 2. *Proceedings of the National Academy of Sciences of the United States of America* 103, 16502–16507.

Flo, T.H., Smith, K.D., Sato, S., Rodriguez, D.J., Holmes, M.A., Strong, R.K., Akira, S. and Aderem, A. (2004) Lipocalin 2 mediates an innate immune response to bacterial infection by sequestrating iron. *Nature* 432, 917–921.

Flower, D.R. (1996) The lipocalin protein family: structure and function. *Biochemical Journal* 318(Pt 1), 1–14.

Fortier, A., Min-Oo, G., Forbes, J., Lam-Yuk-Tseung, S. and Gros, P. (2005) Single gene effects in mouse models of host: pathogen interactions. *Journal of Leukocyte Biology* 77, 868–877.

Foster, J.W. and Hall, H.K. (1992) Effect of *Salmonella typhimurium* ferric uptake regulator (Fur) mutations on iron- and pH-regulated protein synthesis. *Journal of Bacteriology* 174, 4317–4323.

Franchi, L., Amer, A., Body-Malapel, M., Kanneganti, T.D., Ozoren, N., Jagirdar, R., Inohara, N., Vandenabeele, P., Bertin, J., Coyle, A., Grant, E.P. and Nunez, G. (2006) Cytosolic flagellin requires Ipaf for activation of caspase-1 and interleukin 1beta in *Salmonella*-infected macrophages. *Nature Immunology* 7, 576–582.

Furman, M., Fica, A., Saxena, M., Di Fabio, J.L. and Cabello, F.C. (1994) *Salmonella typhi* iron uptake mutants are attenuated in mice. *Infection and Immunity* 62, 4091–4094.

Furrer, J.L., Sanders, D.N., Hook-Barnard, I.G. and Mcintosh, M.A. (2002) Export of the siderophore enterobactin in *Escherichia coli*: involvement of a 43 kDa membrane exporter. *Molecular Microbiology* 44, 1225–1234.

Ganz, T. (2003) Hepcidin, a key regulator of iron metabolism and mediator of anemia of inflammation. *Blood* 102, 783–788.

Gewirtz, A.T., Navas, T.A., Lyons, S., Godowski, P.J. and Madara, J.L. (2001) Cutting edge: bacterial flagellin activates basolaterally expressed TLR5 to induce epithelial proinflammatory gene expression. *Journal of Immunology* 167, 1882–1885.

Godinez, I., Haneda, T., Raffatellu, M., George, M.D., Paixao, T.A., Rolan, H.G., Santos, R.L., Dandekar, S., Tsolis, R.M. and Bäumler, A.J. (2008) T cells help to amplify inflammatory responses induced by *Salmonella enterica* serotype Typhimurium in the intestinal mucosa. *Infection and Immunity* 76, 2008–2017.

Godinez, I., Raffatellu, M., Chu, H., Paixao, T.A., Haneda, T., Santos, R.L., Bevins, C.L., Tsolis, R.M. and Bäumler, A.J. (2009) Interleukin-23 orchestrates mucosal responses to *Salmonella enterica* serotype Typhimurium in the intestine. *Infection and Immunity* 77, 387–398.

Goetz, D.H., Willie, S.T., Armen, R.S., Bratt, T., Borregaard, N. and Strong, R.K. (2000) Ligand preference inferred from the structure of neutrophil gelatinase associated lipocalin. *Biochemistry* 39, 1935–1941.

Goetz, D.H., Holmes, M.A., Borregaard, N., Bluhm, M.E., Raymond, K.N. and Strong, R.K. (2002) The neutrophil lipocalin NGAL is a bacteriostatic agent that interferes with siderophore-mediated iron acquisition. *Molecules and Cells* 10, 1033–1043.

Hantke, K., Nicholson, G., Rabsch, W. and Winkelmann, G. (2003) Salmochelins, siderophores of *Salmonella enterica* and uropathogenic *Escherichia coli* strains, are recognized by the outer membrane receptor IroN. *Proceedings of the National Academy of Sciences of the United States of America* 100, 3677–3682.

Happel, K.I., Zheng, M., Young, E., Quinton, L.J., Lockhart, E., Ramsay, A.J., Shellito, J.E., Schurr, J.R., Bagby, G.J., Nelson, S. and Kolls, J.K. (2003) Cutting edge: roles of Toll-like receptor 4 and IL-23 in IL-17 expression in response to *Klebsiella pneumoniae* infection. *Journal of Immunology* 170, 4432–4436.

Happel, K.I., Dubin, P.J., Zheng, M., Ghilardi, N., Lockhart, C., Quinton, L.J., Odden, A.R., Shellito, J.E., Bagby, G.J., Nelson, S. and Kolls, J.K. (2005) Divergent roles of IL-23 and IL-12 in host defense against *Klebsiella pneumoniae*. *Journal of Experimental Medicine* 202, 761–769.

Hayashi, F., Smith, K.D., Ozinsky, A., Hawn, T.R., Yi, E.C., Goodlett, D.R., Eng, J.K., Akira, S., Underhill, D.M. and Aderem, A. (2001) The innate immune response to bacterial flagellin is mediated by Toll-like receptor 5. *Nature* 410, 1099–1103.

Holmes, M.A., Paulsene, W., Jide, X., Ratledge, C. and Strong, R.K. (2005) Siderocalin (Lcn 2) also binds carboxymycobactins, potentially defending against mycobacterial infections through iron sequestration. *Structure* 13, 29–41.

Hornsby, M.J., Huff, J.L., Kays, R.J., Canfield, D.R., Bevins, C.L. and Solnick, J.V. (2008) *Helicobacter pylori* induces an antimicrobial response in rhesus macaques in a cag pathogenicity island-dependent manner. *Gastroenterology* 134, 1049–1057.

Huynh, C. and Andrews, N.W. (2008) Iron acquisition within host cells and the pathogenicity of Leishmania. *Cellular Microbiology* 10, 293–300.

Ikeda, J.S., Janakiraman, A., Kehres, D.G., Maguire, M.E. and Slauch, J.M. (2005) Transcriptional regulation of *sitABCD* of *Salmonella enterica* serovar Typhimurium by MntR and Fur. *Journal of Bacteriology* 187, 912–922.

Iwai, Y., Hemmi, H., Mizenina, O., Kuroda, S., Suda, K. and Steinman, R.M. (2008) An IFN-gamma-IL-18 signaling loop accelerates memory CD8+ T cell proliferation. *PLoS ONE* 3, e2404.

Jenssen, H. and Hancock, R.E. (2009) Antimicrobial properties of lactoferrin. *Biochimie* 91, 19–29.

Jones, R.L., Peterson, C.M., Grady, R.W.,

Kumbaraci, T., Cerami, A. and Graziano, J.H. (1977) Effects of iron chelators and iron overload on *Salmonella* infection. *Nature* 267, 63–65.

Jurado, R.L. (1997) Iron, infections, and anemia of inflammation. *Clinical Infectious Diseases* 25, 888–895.

Kingsley, R., Rabsch, W., Stephens, P., Roberts, M., Reissbrodt, R. and Williams, P.H. (1995) Iron supplying systems of *Salmonella* in diagnostics, epidemiology and infection. *FEMS Immunology and Medical Microbiology* 11, 257–264.

Kjeldsen, L., Johnsen, A.H., Sengelov, H. and Borregaard, N. (1993) Isolation and primary structure of NGAL, a novel protein associated with human neutrophil gelatinase. *Journal of Biological Chemistry* 268, 10425–10432.

Konopka, K., Bindereif, A. and Neilands, J.B. (1982) Aerobactin-mediated utilization of transferrin iron. *Biochemistry* 21, 6503–6508.

Larsen, R.A., Thomas, M.G. and Postle, K. (1999) Protonmotive force, ExbB and ligand-bound FepA drive conformational changes in TonB. *Molecular Microbiology* 31, 1809–1824.

Lawley, T.D., Bouley, D.M., Hoy, Y.E., Gerke, C., Relman, D.A. and Monack, D.M. (2008) Host transmission of *Salmonella enterica* serovar Typhimurium is controlled by virulence factors and indigenous intestinal microbiota. *Infection and Immunity* 76, 403–416.

Leong, J. and Neilands, J.B. (1976) Mechanisms of siderophore iron transport in enteric bacteria. *Journal of Bacteriology* 126, 823–830.

Lin, H., Fischbach, M.A., Liu, D.R. and Walsh, C.T. (2005) *In vitro* characterization of salmochelin and enterobactin trilactone hydrolases IroD, IroE, and Fes. *Journal of the American Chemical Society* 127, 11075–11084.

Luckey, M., Pollack, J.R., Wayne, R., Ames, B.N. and Neilands, J.B. (1972) Iron uptake in *Salmonella typhimurium*: utilization of exogenous siderochromes as iron carriers. *Journal of Bacteriology* 111, 731–738.

Lupp, C., Robertson, M.L., Wickham, M.E., Sekirov, I., Champion, O.L., Gaynor, E.C. and Finlay, B.B. (2007) Host-mediated inflammation disrupts the intestinal microbiota and promotes the overgrowth of Enterobacteriaceae. *Cell Host and Microbe* 2, 119–129.

McKie, A.T., Barrow, D., Latunde-Dada, G.O., Rolfs, A., Sager, G., Mudaly, E., Mudaly, M., Richardson, C., Barlow, D., Bomford, A., Peters, T.J., Raja, K.B., Shirali, S., Hediger, M.A., Farzaneh, F. and Simpson, R.J. (2001) An iron-regulated ferric reductase associated with the absorption of dietary iron. *Science* 291, 1755–1759.

Masse, E. and Gottesman, S. (2002) A small RNA regulates the expression of genes involved in iron metabolism in *Escherichia coli*. *Proceedings of the National Academy of Sciences of the United States of America* 99, 4620–4625.

Masson, P.L., Heremans, J.F. and Schonne, E. (1969) Lactoferrin, an iron-binding protein in neutrophilic leukocytes. *Journal of Experimental Medicine* 130, 643–658.

Mastroeni, P., Clare, S., Khan, S., Harrison, J.A., Hormaeche, C.E., Okamura, H., Kurimoto, M. and Dougan, G. (1999) Interleukin 18 contributes to host resistance and gamma interferon production in mice infected with virulent *Salmonella typhimurium*. *Infection and Immunity* 67, 478–483.

Miao, E.A., Alpuche-Aranda, C.M., Dors, M., Clark, A.E., Bader, M.W., Miller, S.I. and Aderem, A. (2006) Cytoplasmic flagellin activates caspase-1 and secretion of interleukin 1beta via Ipaf. *Nature Immunology* 7, 569–575.

Muller, C.A., Autenrieth, I.B. and Peschel, A. (2005) Innate defenses of the intestinal epithelial barrier. *Cellular and Molecualr Life Sciences* 62, 1297–1307.

Muller, S.I., Valdebenito, M. and Hantke, K. (2009) Salmochelin, the long-overlooked catecholate siderophore of *Salmonella*. *Biometals* 22, 691–695.

Nairz, M., Theurl, I., Ludwiczek, S., Theurl, M., Mair, S.M., Fritsche, G. and Weiss, G. (2007) The co-ordinated regulation of iron homeostasis in murine macrophages limits the availability of iron for intracellular *Salmonella typhimurium*. *Cellular Microbiology* 9, 2126–2140.

Nairz, M., Fritsche, G., Brunner, P., Talasz, H., Hantke, K. and Weiss, G. (2008) Interferon-gamma limits the availability of iron for intra-macrophage *Salmonella typhimurium*. *European Journal of Immunology* 38, 1923–1936.

Nairz, M., Fritsche, G., Crouch, M.L., Barton, H.C., Fang, F.C. and Weiss, G. (2009a) Slc11a1 limits intracellular growth of *Salmonella enterica* sv. Typhimurium by promoting macrophage immune effector functions and impairing bacterial iron acquisition. *Cellular Microbiology* 11, 1365–1381.

Nairz, M., Theurl, I., Schroll, A., Theurl, M., Fritsche, G., Lindner, E., Seifert, M., Crouch, M.L., Hantke, K., Akira, S., Fang, F.C. and Weiss, G. (2009b) Absence of functional Hfe protects mice from invasive *Salmonella enterica* serovar Typhimurium infection via induction of lipocalin-2. *Blood* 114, 3642–3651.

Nelson, A.L., Barasch, J.M., Bunte, R.M. and Weiser, J.N. (2005) Bacterial colonization of nasal mucosa induces expression of siderocalin, an iron-sequestering component of innate immunity. *Cellular Microbiology* 7, 1404–1417.

Nemeth, E., Valore, E.V., Territo, M., Schiller, G., Lichtenstein, A. and Ganz, T. (2003) Hepcidin, a putative mediator of anemia of inflammation, is a type II acute-phase protein. *Blood* 101, 2461–2463.

Nemeth, E., Rivera, S., Gabayan, V., Keller, C., Taudorf, S., Pedersen, B.K. and Ganz, T. (2004a) IL-6 mediates hypoferremia of inflammation by inducing the synthesis of the iron regulatory hormone hepcidin. *Journal of Clinical Investigations* 113, 1271–1276.

Nemeth, E., Tuttle, M.S., Powelson, J., Vaughn, M.B., Donovan, A., Ward, D.M., Ganz, T. and Kaplan, J. (2004b) Hepcidin regulates cellular iron efflux by binding to ferroportin and inducing its internalization. *Science* 306, 2090–2093.

Pekarek, R.S., Kluge, R.M., Dupont, H.L., Wannemacher, R.W. Jr, Hornick, R.B., Bostian, K.A. and Beisel, W.R. (1975) Serum zinc, iron, and copper concentrations during typhoid fever in man: effect of chloramphenicol therapy. *Clinical Chemistry* 21, 528–532.

Pollack, J.R., Ames, B.N. and Neilands, J.B. (1970) Iron transport in *Salmonella typhimurium*: mutants blocked in the biosynthesis of enterobactin. *Journal of Bacteriology* 104, 635–639.

Rabsch, W. and Reissbrodt, R. (1988) Investigations of *Salmonella* strains from different clinical-epidemiological origin with phenolate and hydroxamate (aerobactin) – siderophore bioassays. *Journal Hygiene, Epidemiology, Microbiology and Immunology* 32, 353–360.

Rabsch, W., Voigt, W., Reissbrodt, R., Tsolis, R.M. and Bäumler, A.J. (1999) *Salmonella typhimurium* IroN and FepA proteins mediate uptake of enterobactin but differ in their specificity for other siderophores. *Journal of Bacteriology* 181, 3610–3612.

Raffatellu, M., Santos, R.L., Verhoeven, D.E., George, M.D., Wilson, R.P., Winter, S.E., Godinez, I., Sankaran, S., Paixao, T.A., Gordon, M.A., Kolls, J.K., Dandekar, S. and Bäumler, A.J. (2008) Simian immunodeficiency virus-induced mucosal interleukin-17 deficiency promotes *Salmonella* dissemination from the gut. *Nature Medicine* 14, 421–428.

Raffatellu, M., George, M.D., Akiyama, Y., Hornsby, M.J., Nuccio, S.P., Paixao, T.A., Butler, B.P., Chu, H., Santos, R.L., Berger, T., Mak, T.W., Tsolis, R.M., Bevins, C.L., Solnick, J.V., Dandekar, S. and Bäumler, A.J. (2009) Lipocalin-2 resistance confers an advantage to *Salmonella enterica* serotype Typhimurium for growth and survival in the inflamed intestine. *Cell Host and Microbe* 5, 476–486.

Raymond, K.N., Dertz, E.A. and Kim, S.S. (2003) Enterobactin: an archetype for microbial iron transport. *Proceedings of the National Academy of Sciences of the United States of America* 100, 3584–3588.

Reed, K.A., Hobert, M.E., Kolenda, C.E., Sands, K.A., Rathman, M., O'Connor, M., Lyons, S., Gewirtz, A.T., Sansonetti, P.J. and Madara, J.L. (2002) The *Salmonella typhimurium* flagellar basal body protein FliE is required for flagellin production and to induce a proinflammatory response in epithelial cells. *Journal of Biological Chemistry* 277, 13346–13353.

Richardson, D.R. (2005) Molecular mechanisms of iron uptake by cells and the use of iron chelators for the treatment of cancer. *Current Medical Chemistry* 12, 2711–2729.

Rouault, T.A. (2006) The role of iron regulatory proteins in mammalian iron homeostasis and disease. *Nature Chemical Biology* 2, 406–414.

Rutz, J.M., Abdullah, T., Singh, S.P., Kalve, V.I. and Klebba, P.E. (1991) Evolution of the ferric enterobactin receptor in Gram-negative bacteria. *Journal of Bacteriology* 173, 5964–5974.

Saiga, H., Nishimura, J., Kuwata, H., Okuyama, M., Matsumoto, S., Sato, S., Matsumoto, M., Akira, S. Yoshikai, Y., Honda, K., Yamamoto, M. and Takeda, K. (2008) Lipocalin 2-dependent inhibition of Mycobacterial growth in alveolar epithelium. *Journal of Immunology* 181, 8521–8527.

Sawatzki, G., Hoffmann, F.A. and Kubanek, B. (1983) Acute iron overload in mice: pathogenesis of *Salmonella typhimurium* infection. *Infection and Immunity* 39, 659–665.

Schneider, H.A. (1967) Ecological ectocrines in experimental epidemiology. A new class, the 'pacifarins', is delineated in the nutritional ecology of mouse salmonellosis. *Science* 158, 597–603.

Shea, C.M. and McIntosh, M.A. (1991) Nucleotide sequence and genetic organization of the ferric enterobactin transport system: homology to other periplasmic binding protein-dependent systems in *Escherichia coli*. *Molecular Microbiology* 5, 1415–1428.

Shen, F., Hu, Z.H., Goswami, J. and Gaffen, S.L. (2006) Identification of common transcriptional regulatory elements in Interleukin-17 target genes. *Journal of Biological Chemistry* 281, 24138–24148.

Stecher, B., Robbiani, R., Walker, A.W., Westendorf, A.M., Barthel, M., Kremer, M., Chaffron, S., Macpherson, A.J., Buer, J., Parkhill, J., Dougan, G., Von Mering, C. and Hardt, W.D. (2007) *Salmonella enterica* serovar Typhimurium exploits inflammation to compete with the intestinal microbiota. *PLoS Biology* 5, 2177–2189.

Stojiljkovic, I., Bäumler, A.J. and Hantke, K. (1994) Fur regulon in Gram-negative bacteria. Identification and characterization of new iron-regulated *Escherichia coli* genes by a Fur titration assay. *Journal of Molecular Biology* 236, 531–545.

Takeuchi, O., Hoshino, K., Kawai, T., Sanjo, H., Takada, H., Ogawa, T., Takeda, K. and Akira, S. (1999) Differential roles of TLR2 and TLR4 in recognition of Gram-negative and Gram-positive bacterial cell wall components. *Immunity* 11, 443–451.

Tapping, R.I., Akashi, S., Miyake, K., Godowski, P.J. and Tobias, P.S. (2000) Toll-like receptor 4, but not Toll-like receptor 2, is a signaling receptor for *Escherichia* and *Salmonella* lipopolysaccharides. *Journal of Immunology* 165, 5780–5787.

Tsolis, R.M., Bäumler, A.J., Stojiljkovic, I. and Heffron, F. (1995) Fur regulon of *Salmonella typhimurium*: identification of new iron-regulated genes. *Journal of Bacteriology* 177, 4628–4637.

Tsolis, R.M., Bäumler, A.J., Heffron, F. and Stojiljkovic, I. (1996) Contribution of TonB- and Feo-mediated iron uptake to growth of *Salmonella typhimurium* in the mouse. *Infection and Immunity* 64, 4549–4556.

Usher, K.C., Ozkan, E., Gardner, K.H. and Deisenhofer, J. (2001) The plug domain of FepA, a TonB-dependent transport protein from *Escherichia coli*, binds its siderophore in the absence of the transmembrane barrel domain. *Proceedings of the National Academy of Sciences of the United States of America* 98, 10676–10681.

Waterman, S.R. and Holden, D.W. (2003) Functions and effectors of the *Salmonella* pathogenicity island 2 type III secretion system. *Cellular Microbiology* 5, 501–511.

Weinberg, E.D. (1984) Iron withholding: a defence against infection and neoplasia. *Physiology Reviews* 64, 65–102.

Wyllie, S., Seu, P. and Goss, J.A. (2002) The natural resistance-associated macrophage protein 1 Slc11a1 (formerly Nramp1) and iron metabolism in macrophages. *Microbes and Infection* 4, 351–359.

Yancey, R.J., Breeding, S.A. and Lankford, C.E. (1979) Enterochelin (enterobactin): virulence factor for *Salmonella typhimurium. Infection and Immunity* 24, 174–180.

Yang, J., Goetz, D., Li, J.Y., Wang, W., Mori, K., Setlik, D., Du, T., Erdjument-Bromage, H., Tempst, P., Strong, R. and Barasch, J. (2002) An iron delivery pathway mediated by a lipocalin. *Molecules and Cells* 10, 1045–1056.

Ye, P., Garvey, P.B., Zhang, P., Nelson, S., Bagby, G., Summer, W.R., Schwarzenberger, P., Shellito, J.E. and Kolls, J.K. (2001a) Interleukin-17 and lung host defence against *Klebsiella pneumoniae* infection. *American Journal of Respiratory and Cellular Molecular Biology* 25, 335–340.

Ye, P., Rodriguez, F.H., Kanaly, S., Stocking, K.L., Schurr, J., Schwarzenberger, P., Oliver, P., Huang, W., Zhang, P., Zhang, J., Shellito, J.E., Bagby, G.J., Nelson, S., Charrier, K., Peschon, J.J. and Kolls, J.K. (2001b) Requirement of Interleukin 17 receptor signaling for lung CXC chemokine and granulocyte colony-stimulating factor expression, neutrophil recruitment, and host defense. *Journal of Experimental Medicine* 194, 519–527.

Zhou, D. and Galan, J. (2001) *Salmonella* entry into host cells: the work in concert of type III secreted effector proteins. *Microbes and Infection* 3, 1293–1298.

Index

Note: page numbers in **bold** refer to figures and tables